AF479227

Geophysical Monograph Volumes

1 **Antarctica in the International Geophysical Year** *A. P. Crary, L. M. Gould, E. O. Hulburt, Hugh Odishaw, and Waldo E. Smith (Eds.)*

2 **Geophysics and the IGY** *Hugh Odishaw and Stanley Ruttenberg (Eds.)*

3 **Atmospheric Chemistry of Chlorine and Sulfur Compounds** *James P. Lodge, Jr. (Ed.)*

4 **Contemporary Geodesy** *Charles A. Whitten and Kenneth H. Drummond (Eds.)*

5 **Physics of Precipitation** *Helmut Weickmann (Ed.)*

6 **The Crust of the Pacific Basin** *Gordon A. Macdonald and Hisashi Kuno (Eds.)*

7 **Antarctica Research: The Matthew Fontaine Maury Memorial Symposium** *H. Wexler, M. J. Rubin, and J. E. Caskey, Jr. (Eds.)*

8 **Terrestrial Heat Flow** *William H. K. Lee (Ed.)*

9 **Gravity Anomalies: Unsurveyed Areas** *Hyman Orlin (Ed.)*

10 **The Earth Beneath the Continents: A Volume of Geophysical Studies in Honor of Merle A. Tuve** *John S. Steinhart and T. Jefferson Smith (Eds.)*

11 **Isotope Techniques in the Hydrologic Cycle** *Glenn E. Stout (Ed.)*

12 **The Crust and Upper Mantle of the Pacific Area** *Leon Knopoff, Charles L. Drake, and Pembroke J. Hart (Eds.)*

13 **The Earth's Crust and Upper Mantle** *Pembroke J. Hart (Ed.)*

14 **The Structure and Physical Properties of the Earth's Crust** *John G. Heacock (Ed.)*

15 **The Use of Artificial Satellites for Geodesy** *Soren W. Henricksen, Armando Mancini, and Bernard H. Chovitz (Eds.)*

16 **Flow and Fracture of Rocks** *H. C. Heard, I. Y. Borg, N. L. Carter, and C. B. Raleigh (Eds.)*

17 **Man-Made Lakes: Their Problems and Environmental Effects** *William C. Ackermann, Gilbert F. White, and E. B. Worthington (Eds.)*

18 **The Upper Atmosphere in Motion: A Selection of Papers With Annotation** *C. O. Hines and Colleagues*

19 **The Geophysics of the Pacific Ocean Basin and Its Margin: A Volume in Honor of George P. Woollard** *George H. Sutton, Murli H. Manghnani, and Ralph Moberly (Eds.)*

20 **The Earth's Crust: Its Nature and Physical Properties** *John C. Heacock (Ed.)*

21 **Quantitative Modeling of Magnetospheric Processes** *W. P. Olson (Ed.)*

22 **Derivation, Meaning, and Use of Geomagnetic Indices** *P. N. Mayaud*

23 **The Tectonic and Geologic Evolution of Southeast Asian Seas and Islands** *Dennis E. Hayes (Ed.)*

24 **Mechanical Behavior of Crustal Rocks: The Handin Volume** *N. L. Carter, M. Friedman, J. M. Logan, and D. W. Stearns (Eds.)*

25 **Physics of Auroral Arc Formation** *S.-I. Akasofu and J. R. Kan (Eds.)*

26 **Heterogeneous Atmospheric Chemistry** *David R. Schryer (Ed.)*

27 **The Tectonic and Geologic Evolution of Southeast Asian Seas and Islands: Part 2** *Dennis E. Hayes (Ed.)*

28 **Magnetospheric Currents** *Thomas A. Potemra (Ed.)*

29 **Climate Processes and Climate Sensitivity (Maurice Ewing Volume 5)** *James E. Hansen and Taro Takahashi (Eds.)*

30 **Magnetic Reconnection in Space and Laboratory Plasmas** *Edward W. Hones, Jr. (Ed.)*

31 **Point Defects in Minerals (Mineral Physics Volume 1)** *Robert N. Schock (Ed.)*

32 **The Carbon Cycle and Atmospheric CO_2: Natural Variations Archean to Present** *E. T. Sundquist and W. S. Broecker (Eds.)*

33 **Greenland Ice Core: Geophysics, Geochemistry, and the Environment** *C. C. Langway, Jr., H. Oeschger, and W. Dansgaard (Eds.)*

34 **Collisionless Shocks in the Heliosphere: A Tutorial Review** *Robert G. Stone and Bruce T. Tsurutani (Eds.)*

35 **Collisionless Shocks in the Heliosphere: Reviews of Current Research** *Bruce T. Tsurutani and Robert G. Stone (Eds.)*

36 **Mineral and Rock Deformation: Laboratory Studies—The Paterson Volume** *B. E. Hobbs and H. C. Heard (Eds.)*

37 **Earthquake Source Mechanics (Maurice Ewing Volume 6)** *Shamita Das, John Boatwright, and Christopher H. Scholz (Eds.)*

38 **Ion Acceleration in the Magnetosphere and Ionosphere** *Tom Chang (Ed.)*

39 **High Pressure Research in Mineral Physics (Mineral Physics Volume 2)** *Murli H. Manghnani and Yasuhiko Syono (Eds.)*

40 **Gondwana Six: Structure Tectonics, and Geophysics** *Gary D. McKenzie (Ed.)*

41 **Gondwana Six: Stratigraphy, Sedimentology, and Paleontology** *Garry D. McKenzie (Ed.)*

42 **Flow and Transport Through Unsaturated Fractured Rock** *Daniel D. Evans and Thomas J. Nicholson (Eds.)*

43 **Seamounts, Islands, and Atolls** *Barbara H. Keating, Patricia Fryer, Rodey Batiza, and George W. Boehlert (Eds.)*

44 **Modeling Magnetospheric Plasma** *T. E. Moore and J. H. Waite, Jr. (Eds.)*

45 **Perovskite: A Structure of Great Interest to Geophysics and Materials Science** *Alexandra Navrotsky and Donald J. Weidner (Eds.)*

46 **Structure and Dynamics of Earth's Deep Interior (IUGG Volume 1)** *D. E. Smylie and Raymond Hide (Eds.)*

47 **Hydrological Regimes and Their Subsurface Thermal Effects (IUGG Volume 2)** *Alan E. Beck, Grant Garven, and Lajos Stegena (Eds.)*

48 **Origin and Evolution of Sedimentary Basins and Their Energy and Mineral Resources (IUGG Volume 3)** *Raymond A. Price (Ed.)*

49 **Slow Deformation and Transmission of Stress in the Earth (IUGG Volume 4)** *Steven C. Cohen and Petr Vaníček (Eds.)*

50 **Deep Structure and Past Kinematics of Accreted Terranes (IUGG Volume 5)** *John W. Hillhouse (Ed.)*

51 **Properties and Processes of Earth's Lower Crust (IUGG Volume 6)** *Robert F. Mereu, Stephan Mueller, and David M. Fountain (Eds.)*

52 **Understanding Climate Change (IUGG Volume 7)** *Andre L. Berger, Robert E. Dickinson, and J. Kidson (Eds.)*

53 **Plasma Waves and Istabilities at Comets and in Magnetospheres** *Bruce T. Tsurutani and Hiroshi Oya (Eds.)*

54 **Solar System Plasma Physics** *J. H. Waite, Jr., J. L. Burch, and R. L. Moore (Eds.)*

55 **Aspects of Climate Variability in the Pacific and Western Americas** *David H. Peterson (Ed.)*

56 **The Brittle-Ductile Transition in Rocks** *A. G. Duba, W. B. Durham, J. W. Handin, and H. F. Wang (Eds.)*

57 **Evolution of Mid Ocean Ridges (IUGG Volume 8)** *John M. Sinton (Ed.)*

58 **Physics of Magnetic Flux Ropes** *C. T. Russell, E. R. Priest, and L. C. Lee (Eds.)*

59 **Variations in Earth Rotation (IUGG Volume 9)** *Dennis D. McCarthy and Williams E. Carter (Eds.)*

60 **Quo Vadimus Geophysics for the Next Generation (IUGG Volume 10)** *George D. Garland and John R. Apel (Eds.)*

61 **Cometary Plasma Processes** *Alan D. Johnstone (Ed.)*

62 **Modeling Magnetospheric Plasma Processes** *Gordon K. Wilson (Ed.)*

63 **Marine Particles Analysis and Characterization** *David C. Hurd and Derek W. Spencer (Eds.)*

64 **Magnetospheric Substorms** *Joseph R. Kan, Thomas A. Potemra, Susumu Kokubun, and Takesi Iijima (Eds.)*

65 **Explosion Source Phenomenology** *Steven R. Taylor, Howard J. Patton, and Paul G. Richards (Eds.)*

66 **Venus and Mars: Atmospheres, Ionospheres, and Solar Wind Interactions** *Janet G. Luhmann, Mariella Tatrallyay, and Robert O. Pepin (Eds.)*

67 **High-Pressure Research: Application to Earth and Planetary Sciences (Mineral Physics Volume 3)** *Yasuhiko Syono and Murli H. Manghnani (Eds.)*

68 **Microwave Remote Sensing of Sea Ice** *Frank Carsey, Roger Barry, Josefino Comiso, D. Andrew Rothrock, Robert Shuchman, W. Terry Tucker, Wilford Weeks, and Dale Winebrenner*

69 **Sea Level Changes: Determination and Effects (IUGG Volume 11)** *P. L. Woodworth, D. T. Pugh, J. G. DeRonde, R. G. Warrick, and J. Hannah*

70 **Synthesis of Results from Scientific Drilling in the Indian Ocean** *Robert A. Duncan, David K. Rea, Robert B. Kidd, Ulrich von Rad, and Jeffrey K. Weissel (Eds.)*

71 **Mantle Flow and Melt Generation at Mid-Ocean Ridges** *Jason Phipps Morgan, Donna K. Blackman, and John M. Sinton (Eds.)*

72 **Dynamics of Earth's Deep Interior and Earth Rotation (IUGG Volume 12)** *Jean-Louis Le Mouël, D.E. Smylie, and Thomas Herring (Eds.)*

73 **Environmental Effects on Spacecraft Positioning and Trajectories (IUGG Volume 13)** *A. Vallance Jones (Ed.)*

74 **Evolution of the Earth and Planets (IUGG Volume 14)** *E. Takahashi, Raymond Jeanloz, and David Rubie (Eds.)*

75 **Interactions Between Global Climate Subsystems: The Legacy of Hann (IUGG Volume 15)** *G. A. McBean and M. Hantel (Eds.)*

76 **Relating Geophysical Structures and Processes: The Jeffreys Volume (IUGG Volume 16)** *K. Aki and R. Dmowska (Eds.)*

77 **The Mesozoic Pacific: Geology, Tectonics, and Volcanism—A Volume in Memory of Sy Schlanger** *Malcolm S. Pringle, William W. Sager, William V. Sliter, and Seth Stein (Eds.)*

78 **Climate Change in Continental Isotopic Records** *P. K. Swart, K. C. Lohmann, J. McKenzie, and S. Savin (Eds.)*

79 **The Tornado: Its Structure, Dynamics, Prediction, and Hazards** *C. Church, R. Davies-Jones, C. Doswell, D. Burgess (Eds.)*

Maurice Ewing Volumes

1 **Island Arcs, Deep Sea Trenches, and Back-Arc Basins** *Manik Talwani and Walter C. Pitman III (Eds.)*

2 **Deep Drilling Results in the Atlantic Ocean: Ocean Crust** *Manik Talwani, Christopher G. Harrison, and Dennis E. Hayes (Eds.)*

3 **Deep Drilling Results in the Atlantic Ocean: Continental Margins and Paleoenvironment** *Manik Talwani, William Hay, and William B. F. Ryan (Eds.)*

4 **Earthquake Prediction—An International Review** *David W. Simpson and Paul G. Richards (Eds.)*

5 **Climate Processes and Climate Sensitivity** *James E. Hansen and Taro Takahashi (Eds.)*

6 **Earthquake Source Mechanics** *Shamita Das, John Boatwright, and Christopher H. Scholz (Eds.)*

IUGG Volumes

1 **Structure and Dynamics of Earth's Deep Interior** *D. E. Smylie and Raymond Hide (Eds.)*

2 **Hydrological Regimes and Their Subsurface Thermal Effects** *Alan E. Beck, Grant Garven, and Lajos Stegena (Eds.)*

3 **Origin and Evolution of Sedimentary Basins and Their Energy and Mineral Resources** *Raymond A. Price (Ed.)*

4 **Slow Deformation and Transmission of Stress in the Earth** *Steven C. Cohen and Petr Vaníček (Eds.)*

5 **Deep Structure and Past Kinematics of Accreted Terrances** *John W. Hillhouse (Ed.)*

6 **Properties and Processes of Earth's Lower Crust** *Robert F. Mereu, Stephan Mueller, and David M. Fountain (Eds.)*

7 **Understanding Climate Change** *Andre L. Berger, Robert E. Dickinson, and J. Kidson (Eds.)*

8 **Evolution of Mid Ocean Ridges** *John M. Sinton (Ed.)*

9 **Variations in Earth Rotation** *Dennis D. McCarthy and William E. Carter (Eds.)*

10 **Quo Vadimus Geophysics for the Next Generation** *George D. Garland and John R. Apel (Eds.)*

11 **Sea Level Changes: Determinations and Effects** *P. L. Woodworth, D. T. Pugh, J. G. DeRonde, R. G. Warrick, and J. Hannah (Eds.)*

12 **Dynamics of Earth's Deep Interior and Earth Rotation** *J.-L. Le Mouël, D. E. Smylie, and T. Herring (Eds.)*

13 **Environmental Effects on Spacecraft Positioning and Trajectories** *A. Vallance Jones (Ed.)*

14 **Evolution of the Earth and Planets** *E. Takahashi, Raymond Jeanloz, and David Rubie (Eds.)*

15 **Interactions Between Global Climate Subsystems: The Legacy of Hann** *G. A. McBean and M. Hantel (Eds.)*

16 **Relating Geophysical Structures and Processes: The Jeffreys Volume** *K. Aki and R. Dmowska (Eds.)*

Mineral Physics Volumes

1 **Point Defects in Minerals** *Robert N. Schock (Ed.)*

2 **High Pressure Research in Mineral Physics** *Murli H. Manghnani and Yasuhiko Syona (Eds.)*

3 **High Pressure Research: Application to Earth and Planetary Sciences** *Yasuhiko Syono and Murli H. Manghnani (Eds.)*

Geophysical Monograph 80

Auroral Plasma Dynamics

Robert L. Lysak

Editor

American Geophysical Union

Published under the aegis of the AGU Books Board.

Library of Congress Cataloging-in-Publication Data

Auroral plasma dynamics / Robert L. Lysak, editor.
     p.   cm. — (Geophysical monograph ; 80)
    Includes bibliographical references.
    ISBN 0-87590-039-9
    1. Auroras. 2. Space plasmas. Plasma dynamics.   I. Lysak, Robert L.
   II. Series.
    QC971.A865 1993
    538'.768—dc20
                                                      93-43462
                                                        CIP

ISSN: 0065-8448

ISBN: 0-87590-039-9

This book is printed on acid-free paper. ∞

CONTENTS

Preface
R. L. Lysak ix

Optical, Ultra-violet and X-Ray Observations of the Aurora

1 Observations of the Pulsation Phase of Auroras Observed at Minneapolis During the Peak of Solar Cycle 22
J. R. Winckler and R. J. Nemzek 1

2 The Discovery of Auroral X-Rays by Balloon-Borne Detectors and Their Contributions to Magnetospheric Research
G. K. Parks, T. J. Freeman, M. P. McCarthy, and S. H. Werden 17

3 Optical Measurements of the Fine Structure of Auroral Arcs
J. E. Borovsky and D. M. Suszcynsky 25

4 Some UV Dayside Auroral Morphologies
R. D. Elphinstone, D. J. Hearn, J. S. Murphree, L. L. Cogger, M. L. Johnson, and H. B. Vo 31

5 Auroral Expansion Into the Dayside Polar Cap: Ground and Satellite Observations in the Prenoon Sector
P. E. Sandholt, J. Moen, D. Opsvik, W. F. Denig, and W. J. Burke 47

Auroral Currents, Convection, and Fields

6 High-Latitude Electrodynamics and Aurorae During Northward IMF
L. G. Blomberg and G. T. Marklund 55

7 Convection and Electrodynamic Signatures in the Vicinity of a Sun-Aligned Arc: Results From the Polar Acceleration Regions and Convection Study (Polar ARCS)
L A. Weiss, E. J. Weber, P. H. Reiff, J. R. Sharber, J. D. Winningham, F. Primdahl, I. S. Mikkelsen, C. Seifring, and E. M. Wescott 69

8 Plasma Flows Associated With an Auroral Arc at the Polar Cap Boundary
H. A. Gallagher Jr., R. L. Carovillano, E. J. Weber, and J. F. Vickrey 81

9 Plasma Convection and Currents in the Auroral Zone
L. Zhang and R. L. Carovillano 89

10 Auroral Weak Double Layers: A Critical Assessment
H. E. J. Koskinen and A. M. Mälkki 97

11 Are Weak Double Layers Important for Auroral Particle Acceleration?
A. I. Eriksson and R. Boström 105

12 The Strong-Double-Layer Model of Auroral Arcs: An Assessment
J. E. Borovsky 113

13 Generalized Model of the Ionospheric Alfvén Resonator
R. L. Lysak 121

14 Laboratory Work on Transient Currents and Its Application to Auroral Arc Currents
J. M. Urrutia and R. L. Stenzel 129

Electron Acceleration

15 From Balloons to Chemical Releases—What Do Charged Particles Tell Us About the Auroral Potential Region?
R. A. Hoffman 133

CONTENTS

16 On the High- and Low-Altitude Limits of the Auroral Electric Field Region
P. H. Reiff, G. Lu, J. L. Burch, J. D. Winningham, L. A. Frank, J. D. Craven,
W. K. Peterson, and R. A. Heelis   143

17 The Acceleration of Electrons by Electromagnetic Ion Cyclotron Waves
M. Temerin, C. Carlson, and J. P. McFadden   155

18 Statistical Distributions of the Auroral Electron Albedo in the Magnetosphere
D. M. Klumpar   163

19 Diffusion of Echo 7 Electron Beams During Bounce Motion
R. J. Nemzek   173

Ion Acceleration

20 Ion Acceleration in the Low- and Mid-Altitude Auroral Ionosphere
A. W. Yau and B. A. Whalen   183

21 Transverse Ion Acceleration by Active Experiments
R. L. Arnoldy   195

22 Ion Heating by Low Frequency Waves
T. Chang and M. André   207

23 Interaction of Ion Beams in the Auroral Acceleration Region
R. Bergmann, P. C. Gray, M. K. Hudson, and I. Roth   213

24 Effects of Solar Cycle on Auroral Particle Acceleration
C. A. Cattell, T. Nguyen, M. Temerin, W. Lennartsson, and W. Peterson   219

25 Sounding Rocket Observations of Ion Injections in the Morning Convection Reversal Region
J. H. Clemmons and C. W. Carlson   227

26 Centrifugal Flow Reversal in the Equatorial Magnetosphere
D. C. Delcourt, J. A. Sauvaud, and T. E. Moore   233

Waves in the Auroral Zone

27 DE 1 Particle and Wave Observations in an AKR Source Region
J. D. Menietti and J. L. Burch   239

28 Acceleration and Radiation From Auroral Cavitons
R. Pottelette, R. A. Treumann, G. Holmgren, N. Dubouloz, and M. Malingre   253

29 Elusive Upper Hybrid Waves in the Auroral Topside Ionosphere
R. F. Benson   267

30 SCEX 3 Observations of HF Z-mode Emissions From the Aurora
R. T. Goerke, P. J. Kellogg, S. D. Bale, S. J. Monson, H. R. Anderson, D. W. Potter,
E. P. Szuszczewicz, and G. D. Earle   275

31 Cherenkov Whistler Emissions From Tethers and Magnetic Antennas in Relative Motion to Plasmas
R. L. Stenzel and J. M. Urrutia   283

32 Effects of Transverse, Localized, DC Electric Fields on Current-Driven Ion-Cyclotron Waves
M. E. Koepke, W. E. Amatucci, J. J. Carroll III, M. J. Alport, and T. E. Sheridan   287

PREFACE

The auroral zone is a region in which complex plasma dynamics take place. This region has been studied in great detail by sounding rocket and satellite experiments, as well as through ground-based observations. The primary focus of this volume is advances made in auroral research in the last 10 years. Main areas of concern are the global morphology of the auroral zone, auroral acceleration processes, parallel electric fields, magnetosphere-ionosphere coupling and waves and turbulence in the auroral zone, as well as the role played by active auroral experiments. The recent results from the Viking and Akebono satellites are considered, and the current state of the field is assessed in preparation for new missions such as Fast, Freja, and the ISTP program. This volume will serve to disseminate these results to the wider space physics community.

The aurora is the most visible manifestation of the complex plasma physics processes which occur in space. In many ways, all of magnetospheric physics is focussed on phenomena which ultimately give rise to the aurora. To the astrophysicist, the auroral zone provides an easily accessible region in which the cosmically important mechanisms of particle acceleration and the generation of radiation may be studied. Within the last two decades, it has become commonly accepted that the aurora is caused by the presence of an electric field parallel to the geomagnetic field in the altitude regime up to roughly 15,000 km, the so-called auroral acceleration region. This volume concentrates on recent developments in describing the plasma processes which take place in this acceleration region.

A previous AGU Monograph, *Physics of Auroral Arc Formation*, was published about 12 years ago. At that time, the existence of the parallel potential drop along auroral field lines was just gaining acceptance, thanks in part to observations from the S3-3 satellite. Since then, new understanding of the aurora has been gained from observations from the pair of Dynamics Explorer satellites, from the Japanese Akebono satellite, and from the Swedish Viking satellite. In addition, advances in instrumentation on these satellites as well as on sub-orbital sounding rockets have made new and higher-quality observations possible.

These advances have led to a detailed picture of many of the particle acceleration mechanisms. In addition to establishing the parallel potential drop model for the primary auroral acceleration, recent results have shown many detailed features of the particle distributions which involve time-dependent currents and fields and interactions between waves and particles. It has also become recognized that in addition to creating the visible aurora, auroral processes also play a major role in populating the magnetosphere with ionospheric ions, including singly ionized oxygen ions.

Most of the papers presented here are based on contributions to a Chapman Conference on Auroral Plasma Dynamics, which was held in Minneapolis in October 1991. This conference, convened by R. L. Lysak, K. A. Anderson, and G. K. Parks, brought 120 auroral researchers from eight countries together to discuss these advances in auroral research. The conference contained sessions on Global Auroral Morphology, Active Experiments, Particle Acceleration, Parallel Electric Fields, Magnetosphere-Ionosphere Coupling, Substorms and Aurora, and Waves and Turbulence. In addition, there was a half-day scientific symposium in honor of John R. Winckler, a pioneer in auroral research as well as many other aspects of space and atmospheric physics. Winckler received a Distinguished Service Award from NASA during the conference banquet, as well as beginning the conference on an exciting note by showing stereoscopic images of the aurora taken from near Minneapolis.

The production of this volume was facilitated by the assistance of many members of the auroral community who agreed to referee the articles contained within. These referees include K. Anderson, M. André, R. Arnoldy, K. Baker, E. Basinska, R. Benson, J. Birn, L. Blomberg, J. Borovsky, W. Burke, R. Carovillano, T. Chang, J. Craven, P. Dusenbery, R. Ergun, D. Fairfield, D. Gallagher, G. Ganguli, D. Gorney, J. Green, T. Hallinan, N. Hershkowitz, R. Hoffman, M. Hudson, P. Kellogg, D. Klumpar, M. Lockwood, W. Lotko, J. Maggs, G. Marklund, S. Mende, C. Meng, D. Menietti, F. Mozer, P. Newell, W. Peterson, P. Reiff, J. Retterer, I. Roth, N. Rynn, P. Sandholt, C.

Seyler, N. Singh, G. Siscoe, R. Smith, T. Speiser, D. Stern, M. Temerin, L. Weiss, J. Winckler, K. Wong, and A. Yau. Although many of the articles contain elements of a review or tutorial nature, the efforts of these referees assured that the quality of the articles was consistent with the high standards of AGU publications. In addition, the efforts of Associate Editor Thomas E. Moore are gratefully acknowledged. His contributions were crucial in relieving some of the burden of producing this volume.

As a final note, the Chapman Conference on which this volume is based was the last chance that most of us in the auroral community had to interact with Christoph Goertz, who was murdered in Iowa City less than one week after the conclusion of the conference. Chris had been an active participant in auroral research as well as other areas of space plasma physics. This monograph is dedicated to his memory.

Robert L. Lysak
Editor

Observations of the Pulsation Phase of Auroras Observed at Minneapolis During the Peak of Solar Cycle 22

JOHN R. WINCKLER

School of Physics and Astronomy, University of Minnesota
Minneapolis, Minnesota

ROBERT J. NEMZEK[1]

Los Alamos National Laboratory,
Los Alamos, New Mexico

Ground-based observations have been made of about 30 auroras from an observatory near Minneapolis (geomag. lat 55°, L=3.5) over the maximum of Solar Cycle 22 particularly to study the auroral pulsations using TV and conventional cameras and photometers, including stereoscopic real-time TV over a 16 km baseline. These sub-auroral zone auroras always appeared in three phases which we have called the Diffuse, the Breakup-Rayed and the Pulsating. The pulsating forms developed only during and after the rayed phase, equally in evening and morning hours, and showed a wide variety of forms and pulsating frequencies from 0.05 Hz to 3 Hz and scale sizes of 10 to 100 km. Pulsating auroras appeared stereoscopically as short field aligned forms pulsating independently whose bottoms formed a "ceiling" near 100 km altitude extending over a range of latitudes. Their origin is consistent with free energy stored during the rayed phase being released over many hours during the pulsating phase, during which the pulsation activity increased and subsided in 10-15 min intervals. Both rayed and pulsating forms were observed to ExB drift, which may explain this behavior by bringing active local regions into view quasi-periodically. Time-intensity profiles, power spectra and images of typical events including sub-visual zenith pulsations, long-coherence 0.15 Hz and 3 Hz pulsations and "frequency bursts" are presented. The pulsating forms executed upward field-aligned motion during each pulse. This could be produced by energy or pitch-angle dispersion of the electrons forming the pulsation, during motion from their distant source. Current theoretical ideas related to our observations are briefly discussed.

INTRODUCTION

We have made ground-based observations of auroras visible over Minnesota during about 30 magnetic storms which occurred near the maximum years of the present Solar Cycle 22. The observations were made from the University of Minnesota O'Brien Observatory at Marine-on-the-St. Croix, a site near the Twin Cities (Lat 45° N, Long 93° W, Geomagnetic Lat 55°, L=3.5) using photometers, color photography and low-light-level TV cameras. For some of the events stereoscopic images using a 16-km, E-W baseline have been obtained using photography and real-time TV. Ground-based and satellite magnetic and Solar data were used in event forecasting and magnetic storm analysis. Our

investigation began seriously towards the end of 1989. During that year the three large SKYFLASH program telescopic photometers [Nemzek and Winckler, 1989; Winckler et al., 1993] located at the observing site had from time to time picked up auroral pulsations at the zenith, often below visual threshold, at times when active auroral forms were visible to the north. The pulsating auroral luminosity produced such intense patterns on the SKYFLASH chart recorders in the frequency range 0.05 Hz to 5 Hz and occurred so frequently that it was decided to investigate this phenomenon in its own right. It was realized that although pulsating auroras in the polar regions had been observed and reported since Størmer's time (e.g. the pioneering work of Cresswell and Davis [1966], see also Ohmolt [1971] page 156), the sub-auroral zone events had been little studied. Yet it is our impression, and that of other observers [e.g. Ohmolt 1971, page 165] that these sub-auroral zone storms are much more potent sources of pulsations than auroral zone events. The paper of Hall [1974] when published was apparently the only study of mid-latitude pulsating auroras. The recent review of pulsating auroras by

---

[1]Formerly at University of Minnesota, Minneapolis, MN.

Auroral Plasma Dynamics
Geophysical Monograph 80

Davidson [1990] appeared in a timely manner, and documented a rather extensive literature on the subject of auroral zone pulsations. A previous review has been published by Johnstone [1978]. The low-latitude events were of special interest to us because the high energy range of the precipitating electrons (10-100 keV), the low L- value range of our observations (3-5.4), and the long time duration (many hours) of this kind of precipitation showed that the processes involved must play a major role in the physics of the outer Van Allen radiation belt, a role heretofore not widely appreciated.

GENERAL OBSERVATIONS

The observations in 1989 and into Fall 1990 used the SKYFLASH photometers and high-speed color film, but when the first low-light level TV camera was put into use during the 9-10 October 1990 event, its advantages became immediately obvious. The cameras (XYBION intensified CCD Model ISS-255) had a sensitivity of 10^{-5} Lux and made bright images of auroral features not well seen even by the dark-adapted eye. The second camera was acquired in time for the 24-25 March 1991 event, when many hours of stereoscopic TV images of the pulsating phase of this great event were obtained. As part of the TV image analysis we have played back the video tapes at 30X normal speed, which condensed an entire night's data, often including several substorms, into 10 or 15 minutes. On viewing the tapes in this manner, it became clear that most of the substorms, but particularly those in the pre-midnight sector, went through three distinct phases characterized by type of aurora. We have labeled these the Diffuse Phase, the Rayed Phase and the Pulsating Phase. The transition from the diffuse to the rayed phase is appropriately called "Breakup", a term which historically originated with observers in the auroral zone to describe the sudden, dramatic change from a quiet arc to a rayed form. Our description of the three phases corresponds almost exactly to that given by Størmer [1955, p. 336]:

> The succession of events in a common Aurora Borealis display: that it starts as a homogeneous arc, which may persist unchanged for hours except that it may move slowly southward. Then its lower border grows more sharp and suddenly it changes to a rayed arc, followed by curtains, rays, and corona. After this outburst of rayed forms, the aurora fades and changes to diffuse surfaces, pulsating forms, and flames.

These three stages are described in detail below in the description of Figures 1 and 3, and are included because the diffuse and rayed phases "set the stage" for the extensive production of pulsations. Our breakup-rayed phase is analogous to the expansion phase of auroral-zone events as described by Akasofu [1977].

The fast video tape play-back also clearly showed the large-scale Magnetospheric ExB drift of the entire aurora, usually westward before midnight and eastward after. Both rayed and pulsating forms drifted with the same apparent speed, which suggests a common region of origin in the distant Magnetosphere. This ExB drift has been observed to the south of the station on large events, showing that the Plasmapause had contracted down to L=3 or less. Auroral zone pulsating auroras drifting with the

ExB velocity have been previously reported by Oguti et al. [1981] and by Scourfield et al. [1983].

The fast play-back was particularly effective in identifying the pulsating forms. The first pulsating patch was seen to start 5 or 10 minutes after the rayed phase began, and was located south of the equatorward boundary of the rayed forms. There was then a transition period in which more and more pulsations appeared, and the rayed structures faded, until the aurora consisted almost entirely of low-altitude, short, field-aligned pulsating structures and pulsating luminous blobs or patches. The fact that pulsating forms begin soon after breakup has been pointed out by Davidson [1990].

In order to give the general reader a graphic idea of the appearance of the three phases as seen from our sub-auroral- zone point of observation we show in Figure 1 color photos of a moderate aurora (Kp max = 6, Dst max = -70 gammas) which occurred 13-14 May 1991. The views were selected to be about midway through each phase. The 35 mm camera, which used a 28 mm lens, had a field of view extending vertically from 5° elevation to 50° (just above Polaris, 45° elevation), and horizontally for 60°. The film was 800 ASA Ektachrome exposed from 20 sec to several minutes at f:2.8. The diffuse phase is characterized by a featureless band of luminosity extending across the northern sky as shown in Figure 1A, usually with an apparent sharp lower (or northern) border and a much more diffuse upper (or southern) limit. The color is a whitish green, so that the emission must contain the 557.7 nm oxygen air-glow line, but is generally devoid of the 630 nm red oxygen emission. The brightness was less than 10 kRayleighs. This type of luminosity was often already apparent on a magnetically active night when evening twilight had faded sufficiently, but it also appeared following a dark sky at later times, and could be seen to re-form after an active period of breakup and pulsations, even into morning twilight. Some long-period (10-30 sec) pulsations were usually observed from the diffuse arc, and one or more pseudo-breakups with an occasional weak tall ray which dissolved before the final bright breakup, were also characteristic of this phase. Our observations suggest that the elevation angle of the diffuse phase viewed from our site (i.e. the southern extension of its apparent upper edge as shown diagrammatically in Figure 2) is a good indicator of the intensity of the ensuing rayed phase and pulsating phase aurora and its low-latitude extension. The low-latitude extent of diffuse arcs has been related to the magnitude of the Dst ring current parameter by Akasofu and Chapman [1963], and our qualitative observations are in general agreement with theirs. The duration of the diffuse phase of Figure 1A was 30 minutes, a typical value.

The breakup and rayed phase is very dramatic in these low-latitude events as shown in Figure 1B. Breakup would often first appear as a "clumping" of the diffuse arc, with each "clump" forming the base for a group of very tall rays in which the intensity was at first very moderate, not much brighter than the diffuse arc. Very rapidly, however, the luminosity increased, and in a few minutes the rayed structures stood out with a 10x or more brightness increase and rapid, irregular transverse motions. A marked characteristic of low-latitude breakups, which seems to

Figure 1. Color photographic views of the three phases of a moderate aurora on 13-14 May, 1991.

us to distinguish them from auroral-zone events, is the great spreading at breakup of the aurora over a range of latitudes, notably equatorward, and the forming of a literal "forest" of very tall disjoint ray groups not necessarily along any well-defined band or arc. An aurora which as breakup began was confined to the northern sky at 20-30 degree elevation angles would suddenly form a rayed corona at the magnetic zenith! Statistically, from an analysis of our 31 events, a corona at $L = 3.5$ was probable for

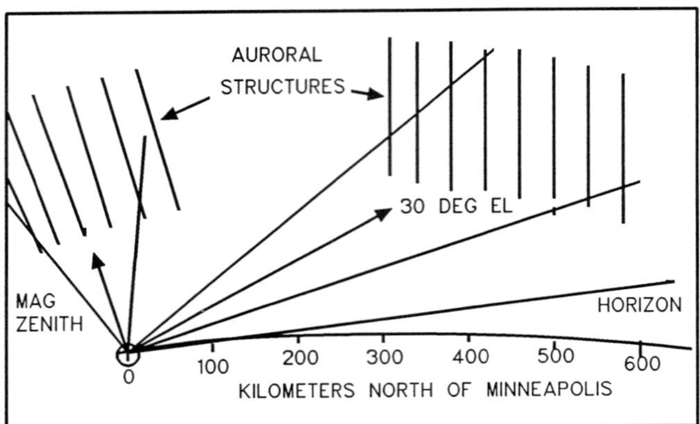

Figure 2. Schematic of sky view directions from the observation site.

a maximum Kp for the event of 6 or higher and a maximum Dst of -110 gammas. Events with Kp = 4 or less were confined to the northern sky. Large events frequently spread to the south of our station, such as the very low-latitude events of 23-24 March 1991. The duration of the rayed phase of Figure 1B, defined as the interval between breakup and the time when all quasi-steady rays were replaced by pulsating forms, was 75 minutes. The bright part of this phase was however much shorter, about 15 minutes.

The rayed phase of even moderate auroras usually developed the red (630 nm) Oxygen emission, forming the upper part of tall rays, but sometimes in an amorphous region above the rays. This red color was often at or below visual level for the dark-adapted eye, but was readily apparent photographically as shown by Figure 1B. Both the authors observed this event, from the two stereoscopic observing points, and the red color clearly shown in the photo was very difficult to see.

Figure 1B is one of a stereo pair, which when viewed stereoscopically showed that the tall rayed structures had their bottoms near the northern horizon (L=5), while some fainter rays and the still-visible diffuse layer stretched far to the south, thereby appearing at higher elevation angles than the ray bottoms.

In contrast with this moderate event of 13-14 May 1991 shown here, large events produced red-rayed coronas, like the storm of 19-21 October 1989 [and the more famous event of February 1958, e.g. Winckler et al., 1959]. In such events the red emission was very intense and easily seen.

The pulsating phase was much less luminous than the preceding rayed phase, but the fact that a given pulsating feature would reappear repeatedly in the same location within a time span of up to one minute produced a pattern of luminosity which could be photographed with time exposures of up to one minute duration, as shown in Figure 1C. The view in this figure appears like patches, green in color (557.7 nm Oxygen emission), as if projected against a flat screen. However, this photo is one of a stereo pair which when examined in a stereo viewer appeared like a "vast festooned ceiling" extending over a wide range of latitude, almost reaching the zenith at the observing site. The "festoons" consisted of localized forms with mostly a short

vertical extent, showing the magnetic field direction, and with their bottoms all at about the same height creating the appearance of the "ceiling" (Note: Brown et al. [1976] have estimated from triangulation measurements that the bottoms of auroral zone pulsations lie near 100 km, and sometimes lower). We have viewed pulsating forms like those shown in Figure 1C in stereo real-time video. The individual features could then be seen to pulse independently. Such behavior has been noted by other observers [see Davidson, 1990].

Our observations show that the pulsating phase was not at all restricted to the morning hours, but began during the rayed phase whenever (or wherever) it occurred. Of the 31 auroras studied here, 15 pulsating phases began before local midnight, including the example in Figure 3 and the last example discussed in Figure 11. This contradicts the widely held idea that pulsating auroras occur <u>only</u> in the morning hours (as implied by Davidson [1990] p. 55, based on the work of Cresswell and Davis [1966] and Omholt and Berger [1967] and as stated by Johnstone [1983] p. 402). The opposite conclusion is supported not only by our observations but those of Oguti et al. [1981], which showed that pulsations can occur as early as 6 hours prior to midnight during periods of high Kp. Furthermore, our video images clearly show that in a large fraction (about half) of the cases studied, the pulsations from an early rayed phase persisted and overlapped in time a new event. For example, a pulsating corona from substorm A was still in progress when a new breakup appeared in the north from substorm B during the 13-14 July 1991 event discussed below. This accumulative tendency, which is due to the long duration of many pulsating phases, would tend to load later times with pulsating phase auroras, i.e. into the morning hours as observed. Examples of typical pulsating forms are analyzed in detail below.

We include here Figure 2, which shows roughly to scale the sky viewing angles and distances from the observation point. Actually, our northern view horizon (about 5° elevation) corresponded to 100-km-altitude auroral bottoms about 700 km north of the station, at L=5.4, while the magnetic zenith is at L=3.5. Figure 2 is intended to emphasize the difference between viewing auroral forms in the northern sky and at the zenith, if stereo viewing is not used. In the northern view the auroral rays or pulsations over a wide latitude range are superposed or projected into a single view plane, making analysis difficult. At the zenith this problem is absent, but then only the bottoms of forms may be seen, and the observed luminosity is an integral along the view direction up the magnetic field, and vertical motion can not be detected. It was because of these problems, which are particularly baffling for the pulsating phase, that we initiated the stereo coverage both photographically and with TV. The effort was successful during about seven cases beginning on 23 March 1991, including several very good pulsating phases. Unfortunately we have not yet found a generally suitable journal format for stereo viewing, and no stereo view pairs are included in this publication.

Views of the three phases of a much larger event (Kp max =9, Dst max = -185 γ) which occurred on 13-14 July 1991 are shown in Figure 3 in the same format as Figure 1, but were reproduced from the TV images after some minor processing

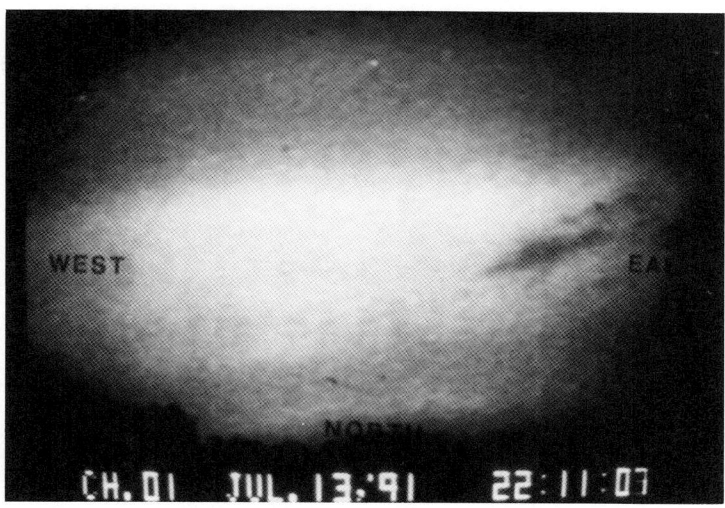

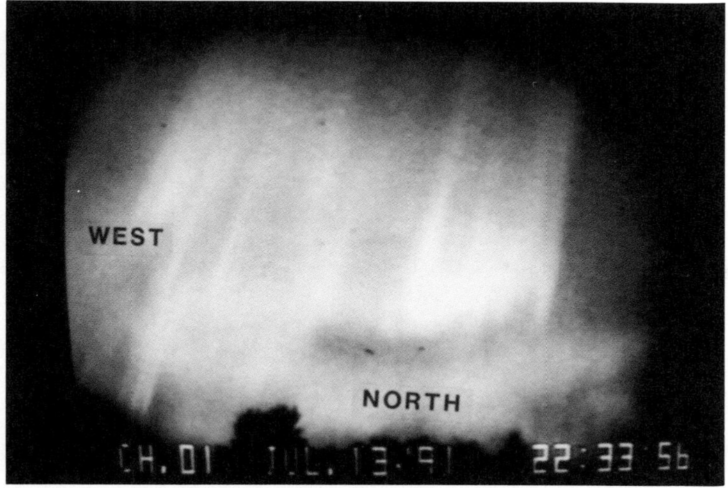

Figure 3. The three phases of the large aurora of 14-15 July 1991. Shown are reproductions of the TV images.

instead of direct color photography. In this case the diffuse phase showed a multiple-arc structure (there is a faint arc at the top of Figure 1A, at 50° elevation angle). The rayed phase (Figure 3B) was extremely bright and active, and produced a rayed corona over O'Brien Observatory almost immediately. A vigorous pulsating phase (Figure 3C) began during the rayed phase, and by 2250 CST 13 July, well before local midnight, was fully developed. This pulsating phase continued for many hours, overlapping the next breakup which occurred at about 0130 CST 14 July. A sample of the pulsating corona from this event is analyzed in the next section.

ANALYSIS OF INDIVIDUAL PULSATION EVENTS

The following is an attempt to translate the live video, photographic and photometric coverage of the auroral pulsating phase into a more quantitative journal format. We show first an example of a nearly subvisual pulsating corona (Figures 4 and 5) observed on 13 July 1991, a very common type of event for larger storms as described in the introduction, but which to our knowledge is not previously mentioned in the literature. This event had a brightness below 1 kilo-Rayleigh. Most of the events described here have been analyzed by converting small regions on the TV monitor image (indicated by the circles in Figure 4B and 5) to intensity-time plots in a simple way using CdS photo-receptors and a computer data system. Because of the ability to select at will interesting pulsation features in small regions of the video image, this technique has proved superior to the on-line photometers, whose field of view was too broad and not well identified. Similar schemes have been previously used by several investigators [e.g. Hansen et al., 1988]. The interior of a single circle in Figure 4B covers a region approximately 5 km in dimension at the 100 km zenith, and proportionally larger for events in the northern sky, typically 10 km near 30° elevation angle. This size is smaller than most individual pulsating features.

Figure 4A shows that the three regions selected as defined in Figure 4B pulsate incoherently, but with similar pulse types. Region 2 is located as close as possible to the corona center (i.e. the magnetic zenith vanishing point), and gives the frequency-power spectrum shown in Figure 4C, which exhibits a peak near 0.3 Hz. The part labeled Section A (Figure 4A) where this frequency dominates has been enlarged in the center panel of Figure 5, which represents the time variation of intensity within the circle marked 2 for a 10 sec interval. A series of images like Figure 4B has been extracted by camera from the TV screen and selected frames are included in Figure 5 at times of intensity minima (bottom set) and maxima (top set) as defined by region 2. An irregular-shaped pulsating feature of 10-20 km transverse size overlaps the vanishing point at the Region 2 circle and generally follows in brightness the intensity-time profile in the center panel (see particularly the middle upper image). There is no transverse motion at the center of the corona during these pulsations (the large-scale ExB drift does not appear on this time scale) but most pulsations viewed transversely away from the magnetic zenith exhibit vertical motion, as will be shown in later examples. The time-intensity plot of Figure 5 shows an exponential decay for each pulsation, a result which seems unique to

viewing parallel to the magnetic field (see further remarks in the Discussion section).

We show in Figure 6 a moderately strong event (Kp = 7) which was observed through clouds in the northern sky. Thus no distinct auroral features are shown by the TV image reproduced photographically in Figure 6B, but on the live TV screen, despite the clouds, rather dramatic pulsations were evident which moved upward on each pulse, giving the appearance of upward waves. The time-intensity plots shown in Figure 6A for the three regions defined in Figure 6B show a complex structure, with peaks at 0.2, 0.6 and 0.9 Hz in the power spectrum (Figure 6C), and as usual large power at very low frequencies near 0 Hz. Figure 6A shows, however, a certain coherence in the 4-second-period features between regions 1 and 2. A magnification of the 10-s sample of Figure 6A labeled Section A is shown in Figure 7, and clearly identifies the upward moving feature as a phase displacement in which region 1 peaks out always before region 2 by about 0.3 s. At this time the pulsations had a rather well-defined frequency peak at 1.3 Hz. In the case of slow pulsations, this kind of brief upward motion of the luminosity during each pulse has generally been termed "Flaming" [Ohmolt, 1971, p. 165]. Our results show that it is a feature of every pulsating form encountered, for example the 3 Hz event described next. It contains an essential clue to the origin of the pulsations (see Discussion section).

The data just described were taken during the interval 0130-0131 CST, but about one minute later in the interval 0132-0133 the frequency character of the event changed abruptly. As shown in Figure 8A and the power spectrum of Figure 8D a strong, relatively coherent oscillation appeared centered at 3.0 Hz. The enlargement of the 10-s sample marked Section A in Figure 8A, shown in Figure 8B, reveals the same sense of the phase delay from region 1 to region 2 shown in the preceding minute, but now occurring for the higher frequency. The frequency peak for Section A (Figure 8C) is also at 3.0 Hz. This pulsation retained its coherence for the order of one minute, which is unusually long for a frequency lying in the upper range of pulsation frequencies. The observed 3.0 Hz frequency implies a rather fast optical response of the atmosphere to the precipitating electrons, certainly much faster than the 0.7 s lifetime of the O 557.7 nm transition. This fact, and the high quenching rate for the O emission at the low altitude of pulsating forms makes it improbable that this line can contribute strongly, and our observations are probably tied to N_2 391.4 nm and other molecular emissions (note however that color photos of pulsations such as Figure 1C do show some green color, but probably for lower frequency pulsations). The sudden onset of a high frequency pulsating regime has been frequently noted during major events by simply watching the pen motion of an on-line photometer chart recorder. In fact, the long-coherence 3 Hz event described above was first identified in this manner. However, high frequency regimes as displayed by a sky photometer were usually more complex in frequency structure than this case.

Another type of pulsation with remarkable features which we have called a "frequency burst" comprised a patch of luminosity which appeared in a region previously dark, pulsed a few times

FAINT ZENITH PULSATIONS

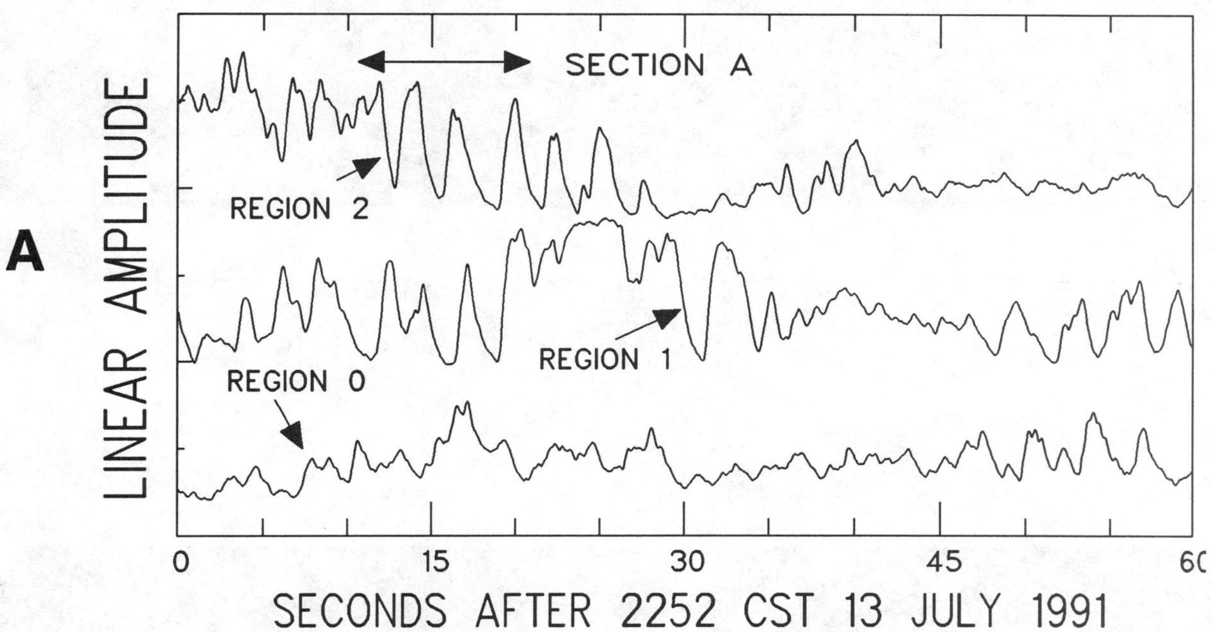

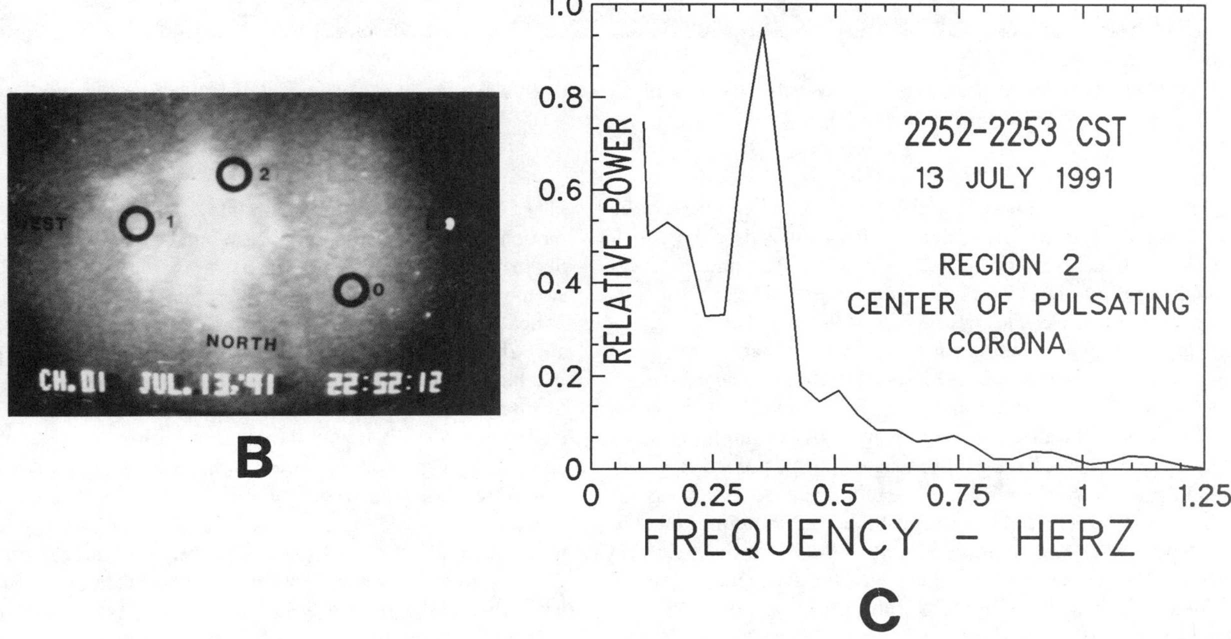

Figure 4. Time-intensity and spectral analysis plots for the faint zenith pulsations of 13 July 1991. The regions are defined in the photo of Figure 4B. One minute data sample.

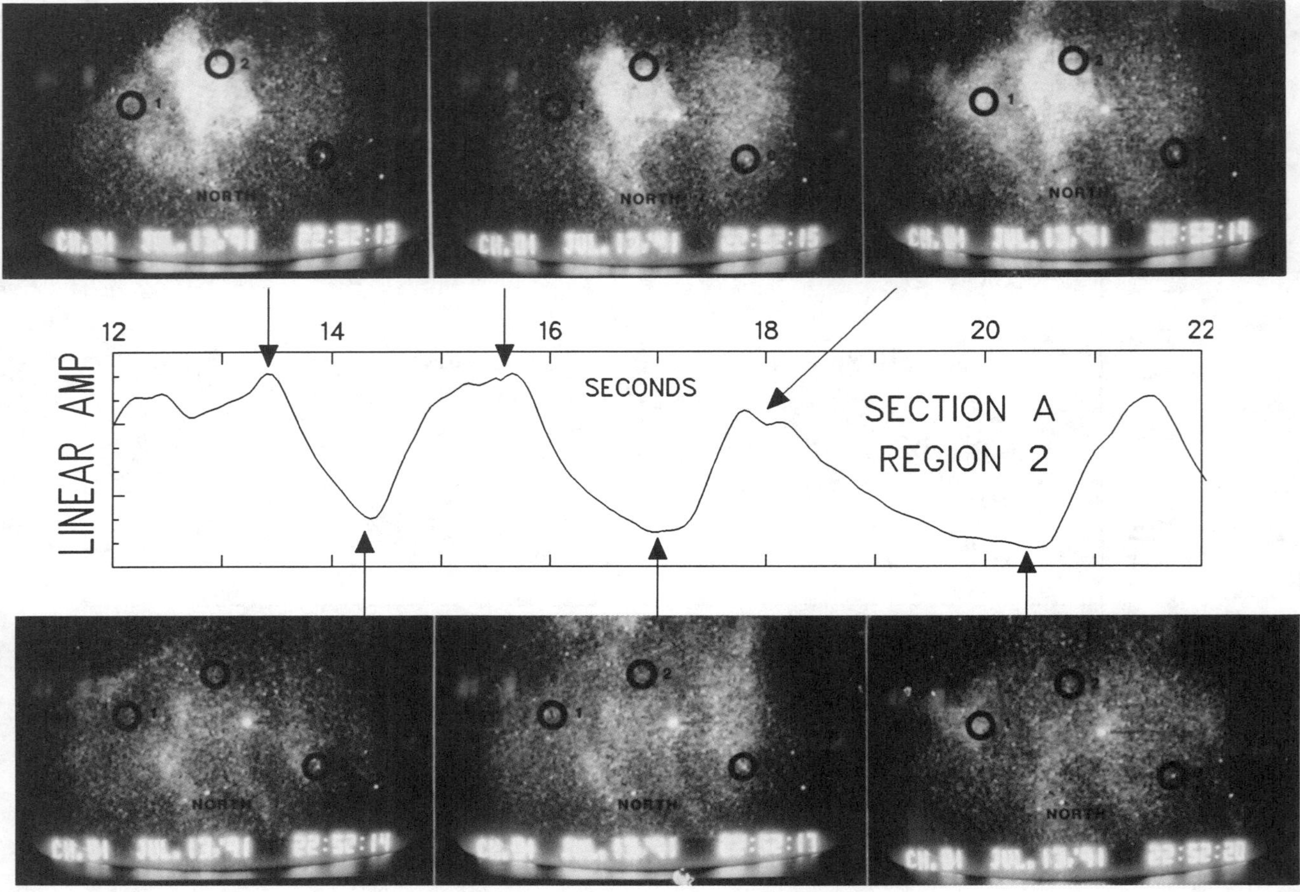

Figure 5. A detail of the time-intensity variation over a 10 s interval of the faint zenith pulsations with indexed views from the TV images. The corona center is at region 2.

rhythmically, and disappeared, only to reappear and repeat the performance. A good example is shown analyzed in Figures 9 and 10, from the event of 9-10 October 1990. The pulsation was located about 200 km distant, and appears behind the horizontal group of three circle markers in Figure 9C at about 25° elevation. It is a single pulsating form, and we estimate that its size transverse to the magnetic field was 30-40 km, and its height along the field about 10 km like other pulsations. The time series of intensities in the three regions (Figure 9A) exhibit the typical "frequency burst" behavior of groups of pulses recurring quasi-periodically about every 25 s. The subsequent behavior was for the groups to spread and overlap in time, and the pulsation then disappeared after two or three minutes. A 20 s sample, Section A Figure 9A, is enlarged in Figure 9B, and frequency analyzed in Figure 9D, where the principal pulsing frequency within the burst is seen to be about 0.9 Hz. The same 20 s interval, Section A, was also transcribed with the three regions arranged vertically across the pulsation as in Figure 6B, to sample the vertical intensity profile, with the result shown in Figure 10. Now the

phase displacement from region 0 to 1 to 2 indicating an upward motion of each pulsation within the burst can clearly be seen. The photos in the top row of Figure 10 were during the pulsation and each photo was made at the point on the intensity-time plot indicated by the arrows. The photos in the bottom row are before and after the event when the intensity was near background. Although this type of event has been noted mixed with other forms, it is likely to be spatially rather isolated, as in the present example, where it lay to the south of other, more continuous-appearing forms shown in the lower parts of the photos. Large, slowly repeating frequency bursts as shown in the present case seem characteristic of the late stages of the pulsating phase, and we have often observed these events against morning twilight. They seem to be a part of what is called "flaming aurora". We have observed some cases against morning twilight as much as 100 km in east-west extent. A careful review of our real-time stereo sequenced video tapes, for example the big event of 23-24 March 1991, shows that "frequency bursts" are a very common type of pulsation, and may occur with burst repetition intervals

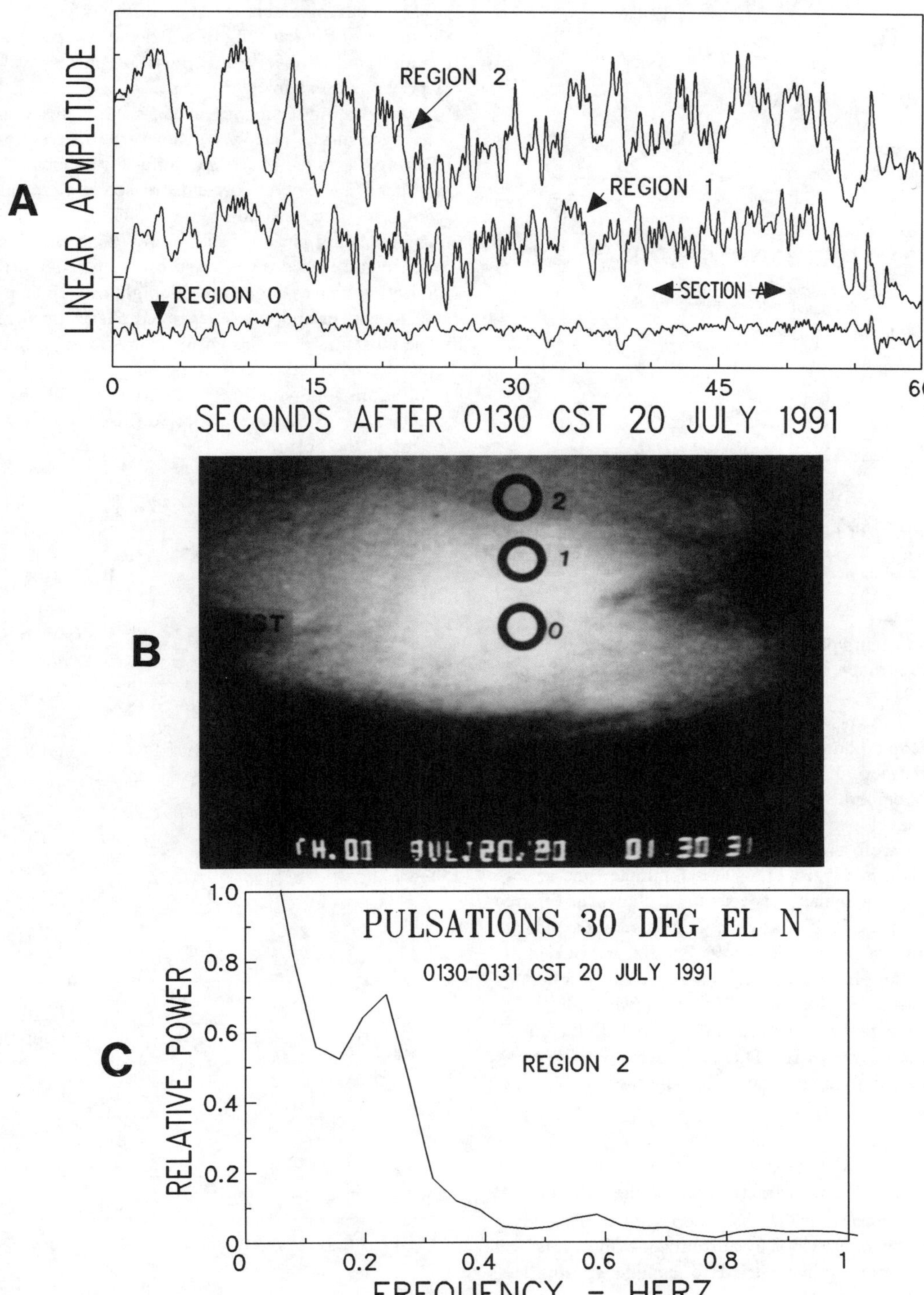

Figure 6. Time-intensity and spectral analysis plots for a vigorous pulsating phase aurora seen through clouds on 19-20 July 1991. The regions are defined by the photo, derived from a TV image. One-minute data sample.

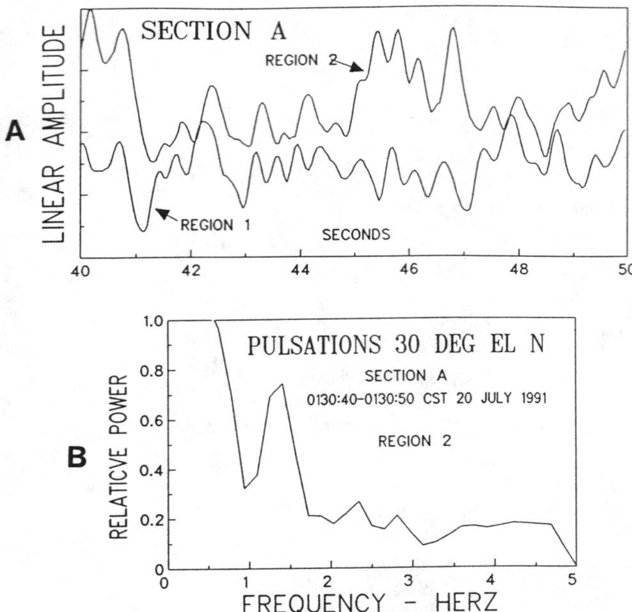

Figure 7. A 10 s sample of Figure 6, showing the phase displacement from region 1 to region 2 associated with the upward motion of the pulses. The spectral peak is near 1.3 Hz.

of one or two seconds, and pulse frequencies within a single burst of 3-4 Hz (and often much smaller scale sizes than the example in Figures 9 and 10).

The final event shown was recorded by one of the large SKYFLASH telescopic photometers directed at 30° elevation to the north during the period 2220-2228 CST on 25 March 1990, before the first TV camera was acquired. This was another case of a completely developed pulsating phase well before local midnight. The data were analyzed by digitizing a chart record with the result shown in Figure 11A. This 8-minute time series has been converted to a dynamic spectrum as shown in Figure 11B. There is a complete lack of any single persistent frequency in the resolved range of frequencies above 0.5 Hz, but buried in the large-amplitude low-frequency end is a remarkably coherent oscillation lasting for about 1 minute. This is shown in detail in Figure 11C, with the power spectrum in Figure 11D having a strong, isolated peak near 0.15 Hz. Due to the lack of video coverage, the exact feature giving this rather low-frequency coherent pulsation has not been identified.

DISCUSSION

We first list the important characteristics of the sub-auroral zone pulsating phase auroras as they have appeared in this study. Some of these have been described by auroral zone observers for decades; in these cases we have tried to include appropriate literature references in the detailed descriptions above. Others are new, or have been observed in greater detail than previously. All are helpful or necessary in forming a physical picture of the underlying phenomena:

1) Sub-auroral zone auroras go through three distinct phases appropriately called diffuse, rayed and pulsating.
2) The southern extent of the diffuse phase is a qualitative indicator of the intensity of the rayed and pulsating phases.
3) In all 31 cases observed no pulsation phase developed without a previous breakup-rayed phase.
4) New pulsating forms appeared 5-10 minutes after the start of the rayed phase, usually equatorward of rayed forms.
5) During the rayed phase, pulsating forms increased in number and intensity as rayed forms decreased, until the aurora entered completely the pulsating phase.
6) The pulsating phase of sub-auroral zone events appears more intense and varied than the auroral-zone counterpart.
7) Pulsating phases began pre-midnight as well as post-midnight, i.e. whenever and wherever rayed phases occurred.
8) The pulsation phase lasted for many hours, often with distinct episodes of 10-15 minute duration.
9) Pulsations following one breakup were still observable at the zenith (L = 3.5) when a new diffuse-phase-rayed phase had begun in the north.

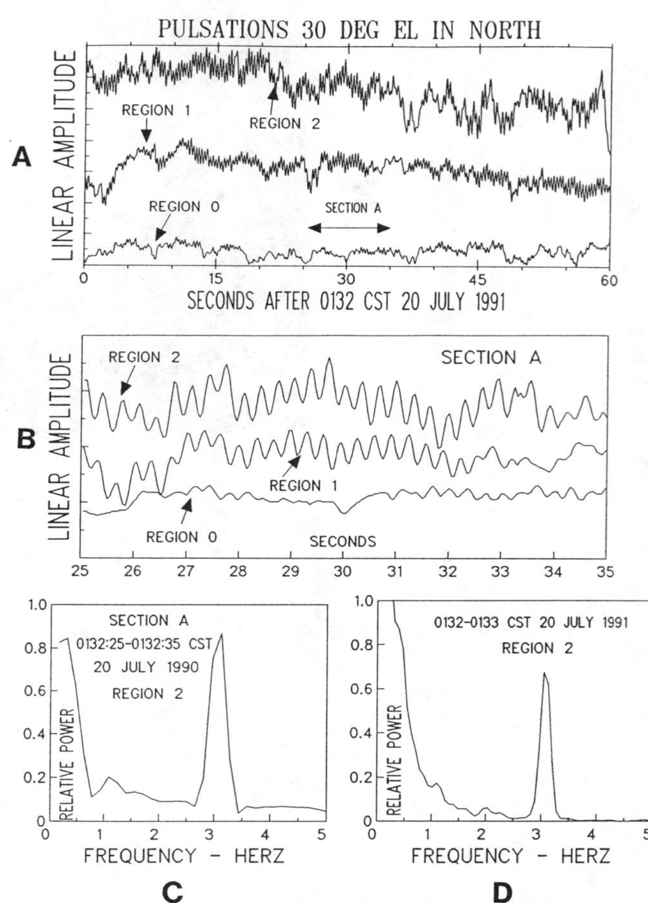

Figure 8. A one-minute sample two minutes after Figure 6 showing the sudden appearance of a very coherent 3.0 Hz frequency (panels C and D). The detail in panel B shows the upward motion of the pulsation by the phase shift from regions 1 to 2 to 3.

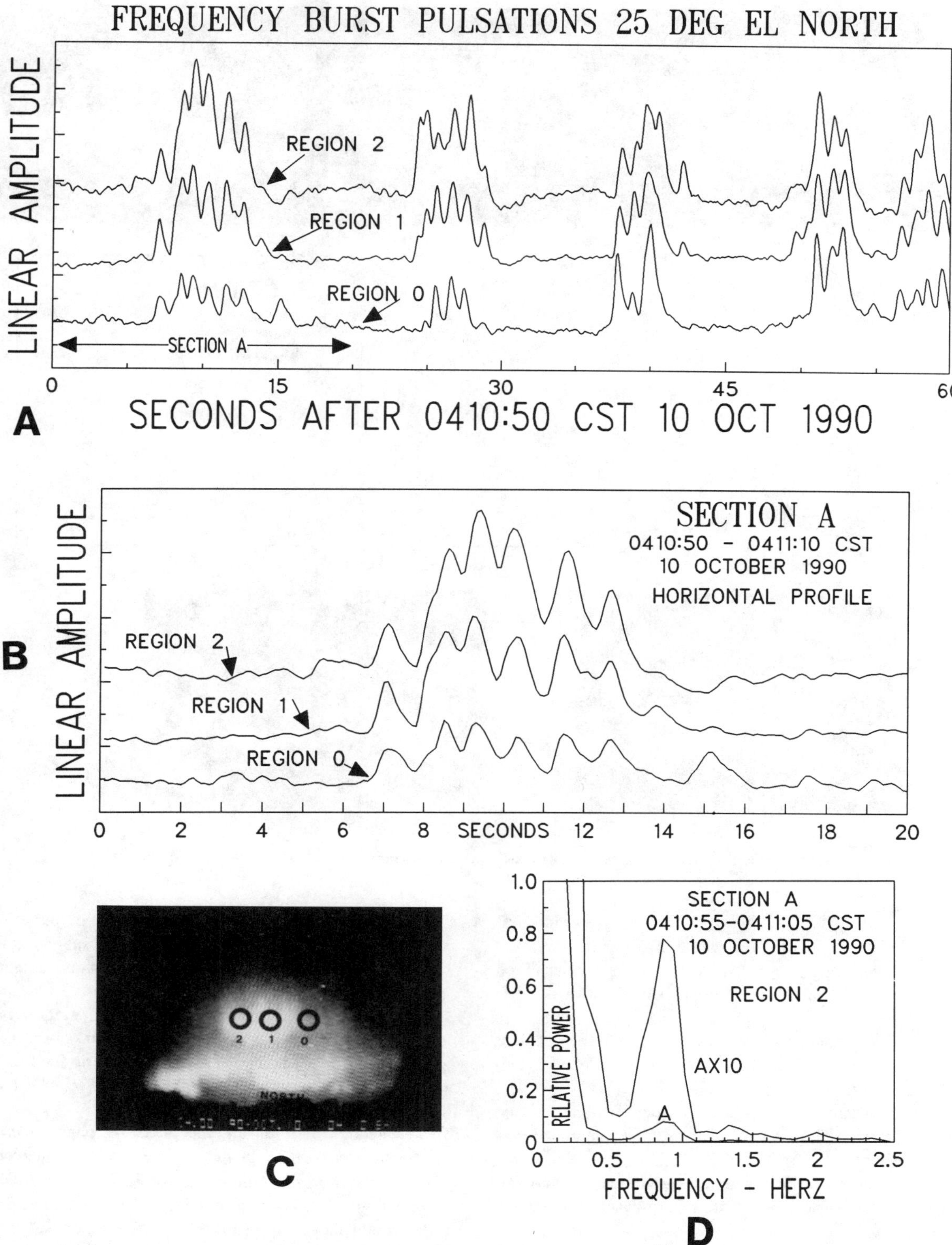

Figure 9. A) A one minute sample showing frequency burst pulsations. B) 20 s detail of a single burst showing the phase shifts associated with upward motion. The regions are defined in C). The principal component of frequency is near 1 Hz (D).

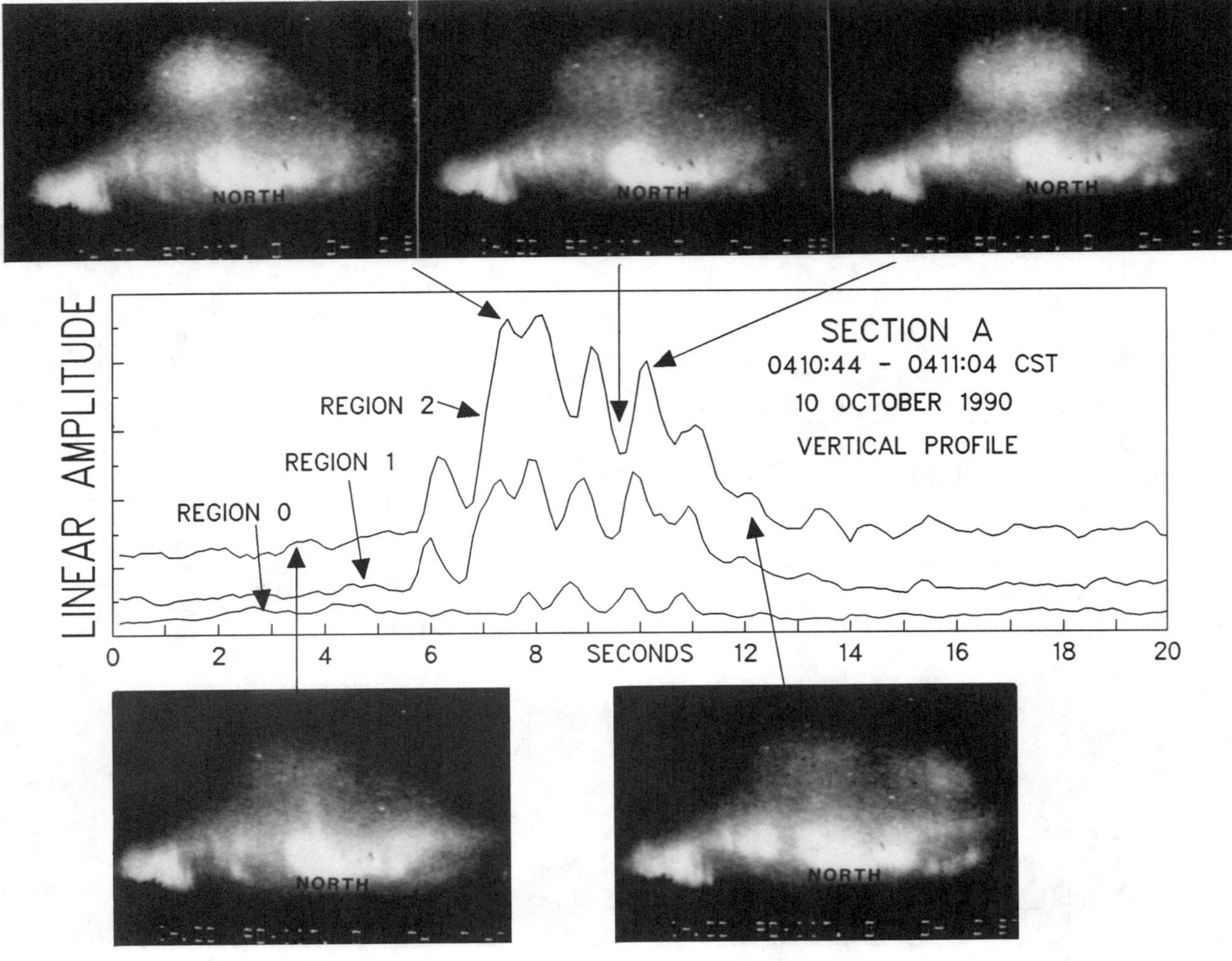

Figure 10. Photos of the frequency burst (from TV) which has a dimension of about 40 km, indexed to the burst profile.

10) The large-scale, typical two-cell Magnetospheric ExB convection drift was readily observable in these auroras over the entire sky. Pulsating forms drifted at the same rate as rayed forms, usually westward pre-midnight and eastward post-midnight.

11) Rayed phase or pulsating phase auroras reached the zenith at L = 3.5 for storms in which the maximum Kp reached 6, and maximum Dst -110 γ.

12) Faint or sub-visual pulsations have frequently been detected at the zenith at L = 3.5. A typical transverse scale size was 10-20 km.

13) The intensity-time profile for zenith pulsations exhibited an exponential rise and decay not seen in pulsations observed transversely.

14) In the pulsation phase, individual forms pulsated independently.

15) Pulsation frequencies of 0.05-5 Hz were measured.

16) The "frequency burst" is a common type of pulsation, with a scale size of 10-100 km transverse to the magnetic field.

17) Most, or perhaps all, pulsating forms show a momentary upward field-aligned motion during the pulse.

18) In our data set, no transverse pulsating motion has been discovered, either by stereo methods or observations of pulsating coronas (except for Magnetospheric ExB drift).

19) Strong pulsations coherent for the order of minutes have been observed at 0.15 Hz as well as at 3 Hz.

20) The sudden shift during the pulsating phase from a low-frequency to a high-frequency regime is a common occurrence.

Considered in broad outline, the above facts seem to show that the pulsating phase is a consequence of a major modification of the Magnetosphere during the breakup-rayed phase, in which a store of free energy is generated, which can supply the instabilities responsible for the pulsations for many hours. In the absence of further substorms this energy source is exhausted after a period

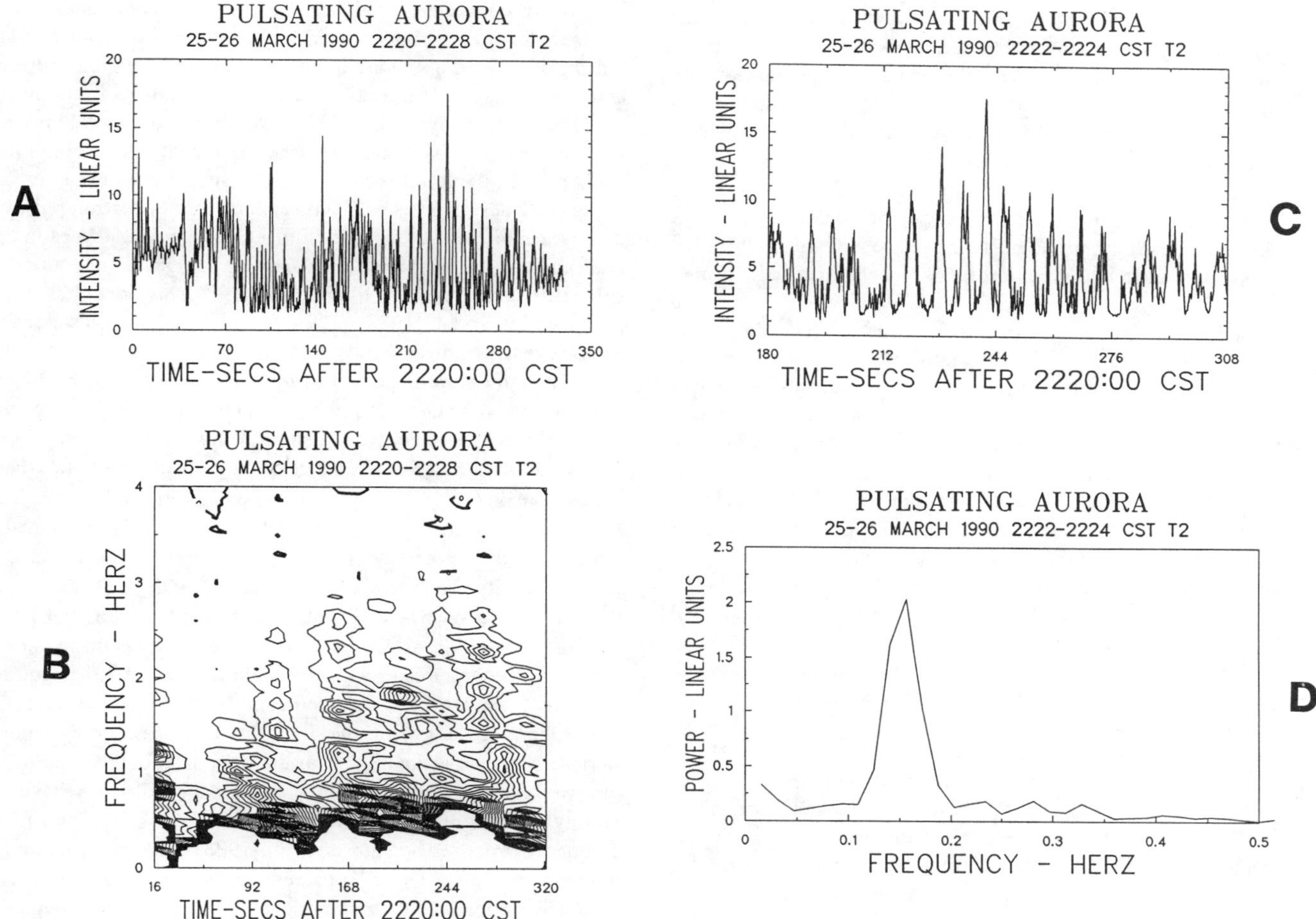

Figure 11. A) 7-min time-intensity sample and B) dynamic spectrum showing lack of continuous frequencies. However, the 2-min sub-sample in C) shows the very coherent low-frequency (0.15 Hz) pulsation. Power spectrum is in D).

which may be 12 hours or more, and the Magnetosphere returns to its quiescent level. This free energy has been attributed by several investigators (e.g., Johnstone, [1983], and also as discussed in Davidson [1990] and Demekhov and Trakhtengerts [1992]) to hot electrons of substorm origin interacting with a cooler (but not necessarily cold) plasma. Observations at geostationary orbit have shown that both hot plasma [DeForest and McIlwain, 1971] and energetic electrons [Parks and Winckler, 1968, Lezniak and Winckler, 1970] are injected into the trapping region during the dipolarization of the magnetic field in the nightside Magnetosphere as a part of substorms. The ExB drift of the plasma driven by the solar wind and the much higher speed eastward gradient-curvature drift of the electrons of energy above 50 keV, carries these two components through one another so that they may generate pulsating type precipitation by interacting. Although the pre-midnight westward convection can move plasma towards the afternoon sector, the newly-injected hot electrons would have to circle the Earth to reach this region, and may not share a common drift shell. No observations of pulsa-

tions in the afternoon sector have been reported, to our knowledge.

It became clear to us by scanning through our early chart records of pulsations that the activity went through periods of maxima and minima every 10 to 15 minutes. Since the video tape play-back clearly identifies the large-scale ExB drift in the entire aurora including pulsations, these quasi-periodic recurrences of activity may be attributed to regions of higher instability in the distant Magnetosphere, whose footprint drifted past the point of observation. This gives observational support to the frequently-made assumption of the role of unstable drifting plasma in the pulsation phase.

Nevertheless, a continuing energy input from the Solar wind which could energize the pulsating phase cannot be ruled out. Such a continuing transfer from the Solar wind is clearly shown by the frequent development of new substorms during the long-continuing pulsation phase of a previous substorm. Continued Solar wind coupling is also shown by the ExB large-scale Magnetospheric convection drift observed in essentially all the auroras in this study. The sub-auroral zone auroras studied here are particularly well connected to definite Solar events and to the ensuing Solar wind enhancements at Earth. Although the rayed or expansion phase of substorms when the energy dissipation is high seems traceable to a global Magnetospheric instability, the diffuse and pulsating phases which represent a more steady, lower-level input may be tied more directly to the Solar wind. It would thus seem to be important to know if the region of Magnetospheric modification resulting in the pulsating phase is restricted to areas previously invaded by rayed auroral forms characteristic of high energy dissipation.

Although the observations discussed here were very limited in being made only from a single location, it is our impression that the region of the Earth's surface covered by the pulsating phase precipitation is similar to the region into which rayed forms spread during the rayed phase. However, we have often observed pulsations at the zenith simultaneously with more steady rayed forms in the northern sky, and generally when both pulsations and rayed forms were present together, the pulsations were located equatorward of the groups of rays. The origin and exact nature of the "unstable" plasma and energetic electrons and their subsequent distribution to form the pulsating regions and individual forms is an important item for further study.

Various schemes for the interactions which produce individual pulsations have been discussed by Davidson [1986, 1990], Davidson and Chiu [1991], Demekhov and Trakhtengerts [1992], Hansen et al. [1988] and previously e.g. by Bryant [1983] and Johnstone [1983]. The diffusion of electron pitch angles into the loss cone in a regular series of pulses is a necessary consequence of all the approaches. Some current theories treat a type of relaxation oscillation involving the build-up of plasma waves due to an instability, assumed near the equatorial plane, which scatters electrons into the loss cone, saturates and decays during each cycle of the pulsing. With these theories it seems possible to fit some of the observations, for example the familiar 3 Hz pulsation. To fit the "frequency burst" type of pulsation described in this paper, which we find to be quite common, either a double instability operating simultaneously at two frequencies would be required, or a more or less uniformly spaced group of hot electron filaments drifting past a single localized plasma region to produce the burst repetition rate, and a relaxation oscillation at each passage to give the frequency within the burst. The long coherence period of two cases reported here at opposite ends of the frequency scale at 0.15 Hz and 3 Hz may also pose a problem for these theories as they would require very constant plasma conditions for long periods. On the other hand, the theories require a filamentary type structure in the plasma to obtain the observed scale size of pulsations from 5 to 100 km in the ionosphere [Davidson, 1990]. It may not be possible to reconcile these requirements [see Johnstone 1978].

When viewed with a component of the view direction transverse to the magnetic field, our real-time stereo video images of pulsations always show during each pulse an upward, field-aligned motion of the lower border of the pulsating form. Intensity samples spaced vertically along a pulsating form on our video images show a progressive time delay of the peak intensity recorded at consecutively higher points in the image, as shown in Figure 8 for 3 Hz pulsations and Figure 9 for "frequency bursts" pulsing at 1 Hz. This behavior has been called "Flaming" aurora. Our video images of pulsating aurora at low angles in the north resemble distant forest fires. We have observed upward moving features reaching 100 km in east-west extent; the example in Figure 9 is about 40 km in extent. Such behavior is rather completely described as a time dispersion of the electron pulse as it moves into the atmosphere from its source near the geomagnetic equator, with high energies arriving before low energies and penetrating deeper into the atmosphere. This idea has been proposed by Bryant [1983] on the basis of electron pulsations at various energies measured at rocket altitudes in one of the very few direct measurements yet made. In his book "The Optical Aurora", p. 165, Omholt [1971] describes measurements of the apparent upward velocity of "Flaming" aurora by Carleton, who found 1000 km/s apparent speed. Omholt clearly describes these forms, fairly rare in the auroral zone, in terms of a bunch of electrons with finite energy spread liberated near the equator which time disperse while traveling to the polar atmosphere. Our present results are in complete agreement with these ideas, but they show that probably <u>all</u> pulsating forms have this behavior. A purely relaxation-oscillation type source may contain an energy spread of the precipitating electrons limited only by the requirement that the travel-time dispersion from the equator to the atmosphere is less than the half the measured optical pulse period. Thus we estimate that in the case of a 3 Hz pulsation an electron energy spectrum centered at 10 keV could have an energy width of $\pm$ 4 keV, and at 40 keV, $\pm$ 25 keV. The bounce resonance proposed by Hansen et al. [1988] could not account for our observations, as their precipitating electrons would be monoenergetic. The upward motions we have observed require a range of precipitated energies to provide dispersion. The theory of Demekhov and Trakhtengerts [1992] actually predicts energy dispersion at the source, in addition to any occurring along the path to the ionosphere.

It is also true however that an upward movement of pulsations might be produced by a progressive migration of pitch angles out of the low-altitude loss cone, as it empties to produce the pulsation.

Since the frequencies predicted by relaxation oscillation type theories depend on the plasma parameters at large distance in the Magnetosphere, and since these are not known for specific events, comparisons of theory and experiment are difficult. The assumed source region near the equatorial plane is accessible only to orbiting vehicles, whose speed through the plasma is too great to detect locally pulsating regions. An exception to this is observations at geostationary orbit, where under the proper conditions of ExB drift the measuring instruments might be temporarily at rest in the pulsation source region [see Johnstone, 1983, Fig. 9].

Finally, the pulsating phase electron precipitation represents a major loss mechanism for the trapped electrons in the outer Van Allen region, and may well destroy most of the flux added during the breakup-rayed phase by precipitating it into the atmosphere!

Final Note

It is difficult to appreciate the nature of these low-latitude pulsations without direct viewing of a video tape. Accordingly a real-time video tape of the events described herein, including photos and video in stereo format, is being prepared. Please contact J. R. Winckler, School of Physics and Astronomy, University of Minnesota, 116 Church St. SE, Minneapolis, MN 55455, 612-624-5086 for further details.

Acknowledgments. This work has been supported by the Space Physics Division of NASA Headquarters under Grant NSG 5088, and by the Atmospheric Sciences Division of NSF under Grant NSF/ATM-9019839. We are indebted to Prof. Laurence Cahill and to the University of Minnesota Space Science Center for the use of the Magnetic Calibration Facility at O'Brien Observatory for these studies. R. J. Nemzek is supported at Los Alamos by the Department of Energy.

References

Akasofu, S.-I. and S. Chapman, The lower limit of latitude (US sector) of northern quiet auroral arcs, and its relation to Dst(H), *J. Atmos. Terr. Phys., 25,* 9-12, 1963.

Akasofu, S.-I., *Physics of Magnetospheric Substorms,* 599p., D. Reidel Publ. Co., Dordrecht, Holland, 1977.

Brown, N. B., T. N. Davis, T. J. Hallinan, and H. C. Stenbaek-Nielsen, Atitude of pulsating aurora determined by a new instrumental technique, *Geophys. Res. Lett., 3,* 403-406, 1976.

Bryant, D. A., Magnetospheric processes reflected in the pulsating aurora, presented at the Chapman Conference on Waves in Magnetospheric Plasmas, Kona Coast, Hawaii, Feb. 1983.

Cresswell, G. R. and T. N. Davis, Observations on pulsating auroras, *J. Geophys. Res., 71,* 3155-3163, 1966.

Davidson, G. T., Pitch angle diffusion in morningside aurorae 2. The formation of repetitive auroral pulsations, *J. Geophys. Res., 91,* 4429-4436, 1986.

Davidson, G. T., Pitch-angle diffusion and the origin of temporal and spatial structures in morningside aurora, *Space Sci. Rev., 53,* 45-82, 1990.

Davidson, G. T., and Y. T. Chiu, An unusual nonlinear system in the magnetosphere: a possible driver for auroral pulsations, *J. Geophys. Res., 96,* 19353-19362, 1991.

DeForest, S. E. and C. E. McIlwain, Plasma clouds in the magnetosphere, *J. Geophys. Res., 76,* 3587-3610, 1971.

Demekhov, A. G., and V. Yu. Trakhtengerts, A mechanism of formation of pulsating aurorae, submitted to *J. Geophys. Res.,* Dec., 1992.

Hall, W. N., Mid-Latitude Pulsating Auroras, *Planet. Space Sci., 22,* 1315-1321 (1974).

Hansen, H. J., E. Mravlag and M. W. J. Scourfield, Coupled 3- and 1.3-Hz components in auroral pulsations, *J. Geophys. Res. 93,* 10029-10034, 1988.

Johnstone, A. D., Pulsating aurora, *Nature, 274,* 119-126, 1978.

Johnstone, A. D., The mechanism of pulsating aurora, *Ann. Geophys., 1,* 397-410, 1983.

Lezniak, T. W. and J. R. Winckler, Experimental study of magnetospheric motions and the acceleration of energetic electrons during substorms, *J. Geophys. Res., 75,* 7075-7098, 1970.

Nemzek, R. J. and J. R. Winckler, Observation and interpretation of fast sub-visual light pulses from the night sky, *Geophys. Res. Lett., 16,* 1015-1018, 1989.

Oguti, T., S. Kokubun, K. Hayashi, K. Tsuruda, S. Machida, T. Kitamura, O. Saka, and T. Watanabe, Statistics of pulsating auroras on the basis of all-sky TV data from five stations I. Occurrence frequency, *Can. J. Phys., 59,* 1150-1157, 1981.

Ohmolt, A. *The Optical Aurora,* 198p., Springer-Verlag, Berlin, Heidelburg, New York, 1971.

Ohmolt, A. and Berger, S., The occurrence of auroral pulsations in the frequency range 0.01-0.1 c/s over Tromsø, *Planet. Space Sci., 15,* 1075-1080, 1967.

Parks, G. K., and J. R. Winckler, Acceleration of energetic electrons observed at the synchronous altitude during magnetospheric substorms, *J. Geophys. Res., 73,* 5786-5791, 1968.

Scourfield, M. W. J., J. G. Keys, E. Nielson, C. K. Goertz, and H. Collin, Evidence for the ExB drift of pulsating auroras, *J. Geophys. Res., 88,* 7983-7988, 1983.

Størmer, C., *The Polar Aurora,* 403 p., The Clarendon Press, Oxford, 1955.

Winckler, J. R., L. Peterson, R. Hoffman, and R. Arnoldy, Auroral X-rays, cosmic rays and related phenomena during the storm of February 10-11, 1958, *J. Geophys. Res., 64,* 597-610, 1959.

Winckler, J. R., R. C. Franz and R. J. Nemzek, Fast low-level light pulses from the night sky observed with the SKYFLASH program, to be published, *J. Geophys Res. (Atmospheres),* 1993.

J. R. Winckler, School of Physics and Astronomy, University of Minnesota, 116 Church St SE, MInneapolis, MN 55455.

R. J. Nemzek, Group SST-9, MS-D436, Los Alamos National Laboratory, Los Alamos, NM 87545.

The Discovery of Auroral X-Rays by Balloon-Borne Detectors and Their Contributions to Magnetospheric Research

GEORGE K. PARKS[1], THEODORE J. FREEMAN[2], MICHAEL P. MCCARTHY, AND SCOTT H. WERDEN

Geophysics Program, University of Washington, Seattle, Washington

Balloon-borne auroral X-rays were discovered over Minneapolis during a magnetic storm on July 1, 1957. These X-rays were atmospheric bremsstrahlung X-rays produced by precipitated Van Allen energetic electrons interacting with atmospheric constituents. Now, more than 35 years later, X-ray observations are still providing useful results. How do X-ray observations contribute to the study of precipitation and substorm mechanisms? What is the significance of the diverse spatial and temporal characteristics of X-rays? X-rays can be observed by detectors carried on balloons, rockets, and satellites. This article will emphasize the results of balloon-borne observations. A short discussion of auroral X-rays, their role in magnetospheric research, including new information that has been obtained on electron precipitation and substorms by a balloon-borne X-ray pinhole camera, will be presented.

1. DISCOVERY OF AURORAL X-RAYS

Finding that a "wiggly" line in the data was due to a previously undiscovered natural phenomenon was a serendipitous event. This was the case when a balloon-borne particle experiment flown from Minneapolis on July 1, 1957 detected an anomalous increase of radiation [Winckler and Peterson, 1957]. A careful analysis of the data, obtained by an experiment originally designed to study solar effects on cosmic rays, indicated that the particle instruments worked properly and the quality of data was good. Winckler and his co-workers studied the responses of the detectors to various types of radiation and concluded that the anomalous radiation was most probably due to penetrating X-rays. This interpretation was later proven correct by experiments flown into other magnetic storms---for example, the great storm of February 10-11, 1958 [Winckler, Peterson, Arnoldy, and Hoffman, 1958]. The Minnesota group further deduced that the X-rays were atmospheric bremsstrahlung X-rays produced by 50-200 keV energetic electrons. The X-rays were closely correlated with the appearance of visual auroras. Moreover, recognizing that these X-rays could be directly related to the soft radiation observed by rocket-borne detectors in the auroral zone [Meredith, Gottlieb, and Van Allen, 1955; Van Allen, 1957], further observations began in the auroral zone

[Anderson, 1958]. These investigations gave birth to auroral X-ray research.

The discovery of X-rays preceded the discovery of the Van Allen radiation belts, and the connection of energetic electrons to the radiation belts could not be made at that time (the source was thought to be the Sun). Nevertheless, it was recognized early on that X-ray observations provided a powerful means to study the behavior of primary electrons. X-ray observations then and now have emphasized obtaining accurate information on spatial, temporal and energy spectral characteristics. X-ray observations provide a unique means for studying the dynamics of the primary electrons since X-ray energy spectra can be inverted by means of the well-established bremsstrahlung theory to obtain energy spectra of the source electrons [Anderson and Enemark, 1960].

X-ray observations in the early 1960s led to several important discoveries which are central to the understanding of magnetospheric phenomena. For example in 1962, Winckler and his colleagues calculated from X-rays that the energy flux in precipitation was higher than the trapped radiation, thereby concluding that "...during periods of magnetic disturbance a large energy flux passes through the radiation belt. This flux of energy must originate in the solar plasma stream and be transferred by the magnetic field to the trapped particles, which are then accelerated and precipitated into the atmosphere, where the original energy is finally deposited. The radiation belt may be considered the residue of accelerated and deflected particles remaining in trapped trajectories after the perturbed period." [from Winckler, Bhavsar, and Anderson, 1962]. This conclusion was the basis for the "splasher catcher" [O'Brien, 1964] and pitch-angle diffusion models [Kennel, 1969]. That the outer Van Allen radiation belt particles and precipitation particles had a

---

[1] Also at Atmospheric Sciences and Physics Departments, University of Washington
[2] Also at Physics Department, University of Washington

Auroral Plasma Dynamics
Geophysical Monograph 80
Copyright 1993 by the American Geophysical Union.

common particle source was later shown by an experiment which observed that energetic particle injection in the outer Van Allen radiation belts occurs simultaneously with electron precipitation on auroral field lines [Parks and Winckler, 1968]. Other important discoveries included observations of precipitation over extended regions of latitude and longitude and the nearly periodic electron precipitation phenomena [Winckler, Bhavsar, and Anderson, 1962].

These early X-ray observations initiated new experiments which led to many other important observations. A concise summary of some of these balloon-borne observations is given below. We conclude with a short discussion of the current balloon-borne X-ray measurements and how future X-ray observations can contribute to magnetospheric research.

2. Periodic and Impulsive Precipitation Phenomena

The X-rays detected by Winckler et al. [1958] showed rapid intensity variations. Although the cause of these variations was not known, their importance was recognized because they contained information on the dynamics of electron precipitation. It was also recognized that measurements made by balloon-borne detectors could provide unambiguous information on temporal variations since balloons are virtually stationary platforms. Information on rapid time variations associated with energetic electron precipitation comes primarily from X-ray observations.

Figure 1 shows an example of the X-ray time profile and a power spectrum of a portion of the data obtained over Minneapolis during a magnetic storm on October 1, 1961. Note the impulsive nature of the precipitation. Each burst is typically a few tenths of a second. Significant peaks were found at frequencies corresponding to periods of ≈ 0.8 and 1.6 seconds. That electrons can precipitate impulsively and "periodically" was an important discovery. Extensive auroral electron precipitation and X-ray observational programs have since identified other precipitation forms and time scales, including microbursts [Anderson and Milton, 1964], 1-15 second pulsations [Anger, Barcus, Brown, and Evans, 1963; Brown, Barcus, and Parsons, 1965], 75- and 100-second periodic pulsations on the dayside [Evans, 1963; Barcus and Christensen, 1965], multiple component energy spectra [Barcus and Rosenberg, 1966], substructures in microbursts [Lampton, 1967; Parks, 1974], millisecond structures [Parks, Milton, and Anderson, 1967] and relativistic electron precipitation on the dayside [West and Parks, 1984] and the nightside [Imhof, Voss, Mobilia, Datlowe, and Gaines, 1991]. The nightside relativistic precipitation is bursty and resembles microbursts on the morning side [Imhof, Voss, Mobilia, Datlowe, Gaines, and McGlennon, 1992].

Winckler and his colleagues interpreted the 0.8 second period in terms of particle bounce times. They also recognized that impulsive precipitation required bunching of electrons which would involve plasma instabilities and wave particle interactions. These interpretations still hold today for microburst-like events. Other temporal forms are interpreted as modulations of precipitation by plasma and hydromagnetic waves. Our understanding of these observations are however still qualitative and further

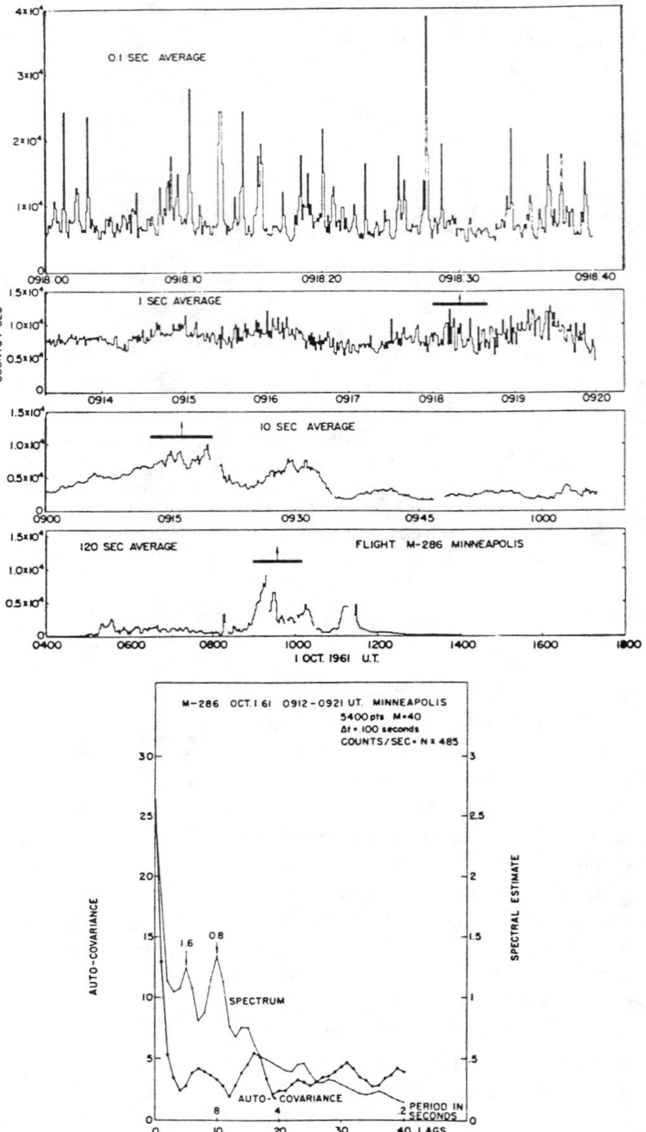

Fig. 1. The top four panels show examples of the time structure of the electron precipitation observed at Minneapolis on October 1, 1961, for the 20 keV X-ray channel. The intensity scale is linear, and beginning with the fourth panel from the top, the region covered by the black bar is expanded into the section just above. Note that the maximum variance seems to occur in the very short intervals where extreme intensity variations occur in 0.1 sec. The bottom panel shows the results of a power spectrum analysis of a section of the high-intensity data. The data used were 0.1 sec averages [from Winckler et al., 1962].

research is needed to identify the physical processes that are responsible for these dynamic features.

3. X-ray Observations and Magnetospheric Substorms

Auroral X-ray research during the early 1960s was exploratory and nearly every balloon flight yielded new results. One of the earliest studies was an attempt to establish relationships of X-rays to other auroral phenomena. These studies yielded results that were not always understood as there were many contrasting observations. For example, while Winckler et al. [1958] showed that X-rays were correlated with visual auroral activities, Anderson [1960] observed that X-rays were present during most of the high latitude auroral zone flights even in the absence of magnetic storms. Although the intense auroral X-ray fluxes were correlated with geomagnetic storms, thus supporting Winckler et al. [1958], "weak" X-ray fluxes did not always appear to be associated with the geomagnetic activity. The same conclusion was reached by Brown [1961]. Another study that related X-rays to auroral electrojet activity showed that some X-ray events were more intense in the vicinity of an auroral electrojet, but X-rays were often observed in the absence of the electrojet activity [Barcus and Brown, 1962; Brown and Campbell, 1962].

The observations of Winckler et al. [1958] showed that X-rays appeared in close correlation with visual auroras. Motivated by these observations, and further noting that visual auroras were organized by the auroral substorm model [Akasofu, 1964], the Berkeley group under the leadership of Kinsey Anderson began to explore an idea that microburst precipitation on the morning side occurred simultaneously with auroral breakup near local midnight. This idea was shown to be correct and led to the concept that X-ray and other auroral data obtained from a wide region covering many hours of local time could be systematically organized by a global model. Figure 2 taken from Parks, Coroniti, McPherron, and Anderson [1968] shows the temporal behavior of particle precipitation around the auroral oval during substorms. This diagram was derived using many hundreds of hours of X-ray observations in the auroral zone at various local times. It shows that the various temporal forms of X-rays are local features of a global disturbance phenomenon. The Berkeley group also showed the relationship of X-rays and micropulsation activity [McPherron, Parks, Coroniti, and Ward, 1968]. These studies ultimately led to the concept of magnetospheric substorms [Coroniti, McPherron, and Parks, 1968].

At about the same time the Berkeley group was developing the magnetospheric substorm model, Jelly and Brice [1967] were attempting to explain the causes of two precipitation peaks that were observed: one near local midnight and the other on the dayside [McDiarmid and Burrows, 1964; Hartz and Brice, 1967]. Using Alouette 1 satellite and ground-based magnetogram and riometer data, Jelly and Brice [1967] found that increases in precipitation in these two regions frequently show a one-to-one correspondence. Moreover, the maximum flux of semitrapped electrons (E>40 keV) measured by the satellite in the morning sector was larger on the average by a factor of $\approx$ 5 shortly after the onset of a negative magnetic bay compared with shortly before. Thus, Jelly and Brice [1967] concluded that "auroral precipitation is not a spatially isolated phenomenon but is intimately connected with large-scale processes that occupy a substantial part of the magnetosphere." The term "elementary magnetos-

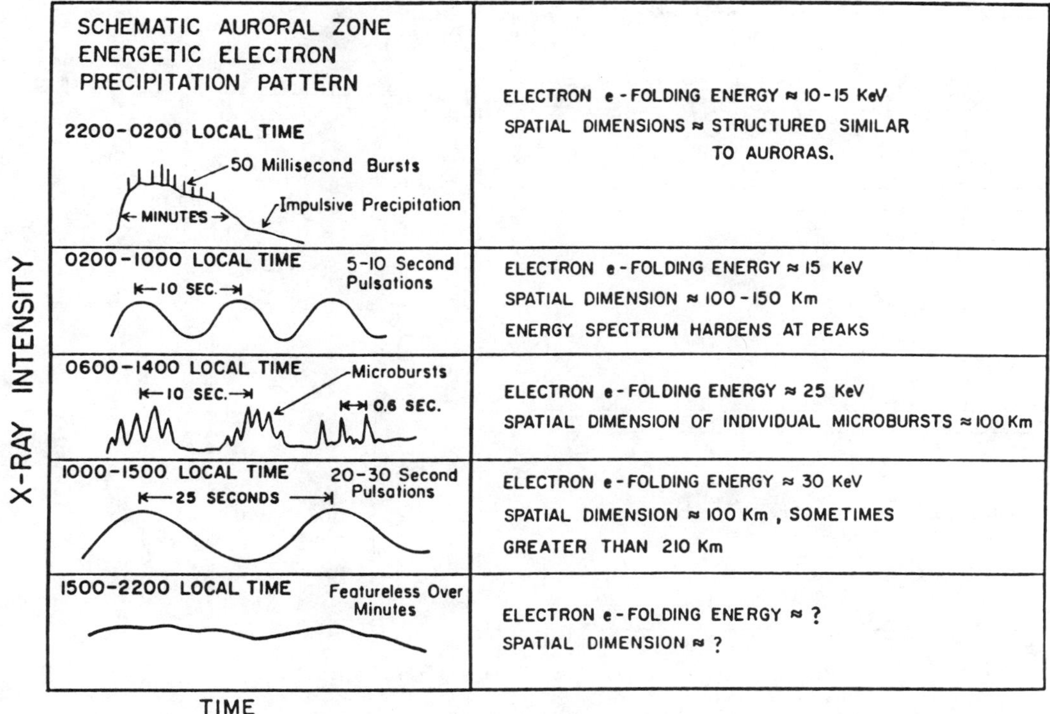

Fig. 2. A schematic diagram showing local time characteristics of the various forms of energetic electron precipitation around the auroral oval. Each form occurs at its respective local time sector during an auroral breakup near local midnight [from Parks et al., 1968].

pheric substorm" was suggested by Jelly and Brice [1967] to describe this global disturbance.

These investigations have clearly indicated a magnetospheric substorm involves the entire magnetosphere and includes particles and fields over wide energy and frequency ranges. Visual auroras, X-rays, micropulsations, electrojets and all other activities in the auroral zone and the magnetosphere are seen as different manifestations of a single large-scale magnetospheric disturbance phenomenon. Precipitation is a part of both acceleration and injection processes which affect a broad range of pitch-angles, including the loss cone from which energetic electrons precipitate.

4. CURRENT RESEARCH IN AURORAL X-RAYS

The 1970s and 1980s saw marked improvement in X-ray observational techniques. Instruments were flown on polar orbiting spacecraft to obtain large-scale images of X-rays [Imhof, Nakano, Johnson, and Reagan, 1974; Mizera, Luhmann, Kolasinski, and Blake, 1978]. These images considerably improved our knowledge of large-scale structures, first observed by detectors flown on several balloons separated by several hundred kilometers [Winckler, Bhavsar, and Anderson, 1962; Brown, Barcus, and Parsons, 1965]. We will not discuss X-ray observations from spacecraft-borne detectors in detail other than to mention that arc-like forms and patches have been identified [Datlowe, Imhof, and Voss, 1988]. These X-ray images complement the visual and ultraviolet images. X-ray energy spectra also yield the primary energy spectra of precipitated electrons which can include several components (see below). Note that X-rays have also been detected by rocket-borne detectors [Goldberg, Barcus, Treinish, and Vondrak, 1982]. The remainder of this section will focus on small-scale X-ray structures that have been observed by the most recent development in balloon-borne X-ray observations.

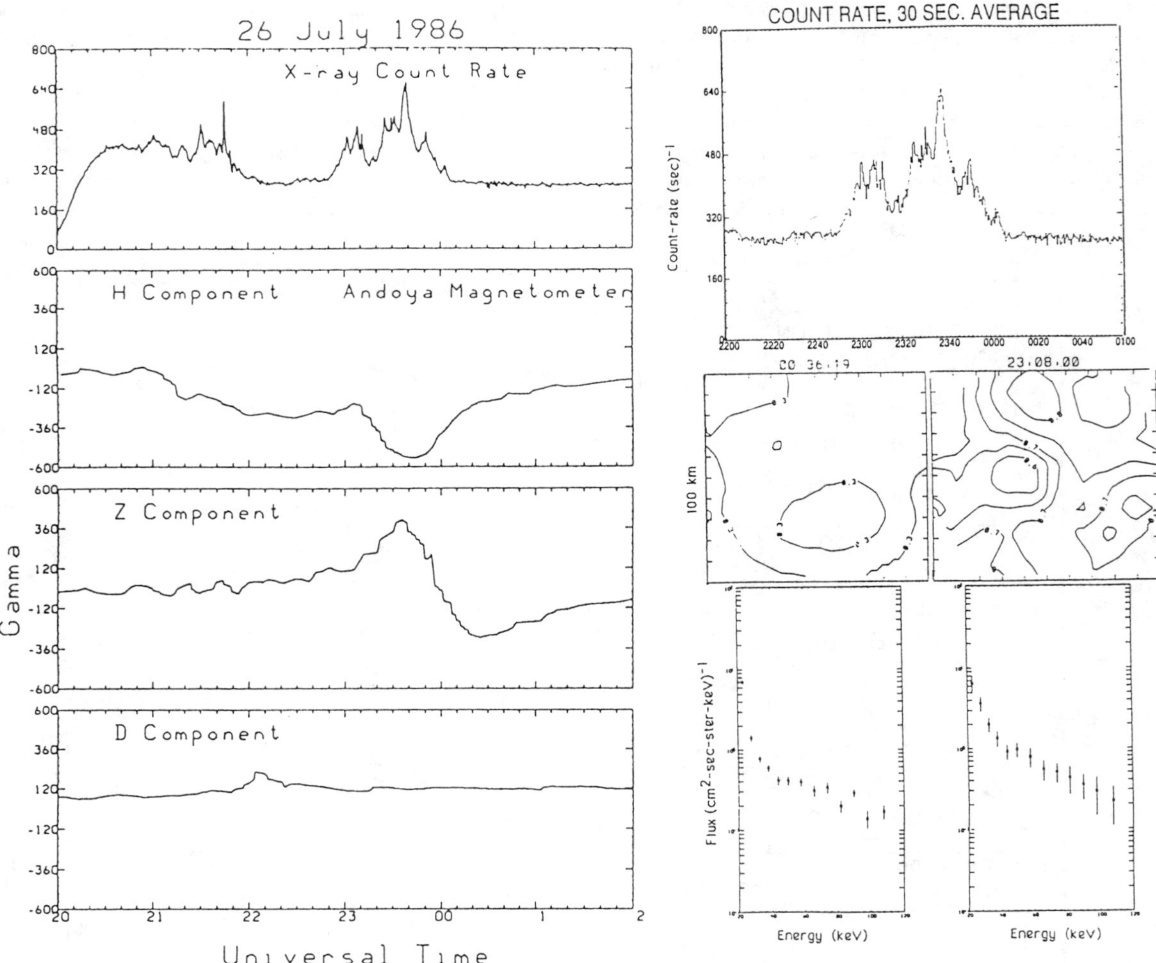

Fig. 3. The left panels show a summary of auroral X-rays encountered by a balloon-borne X-ray pinhole camera (top). The bottom three panels show magnetic records from Andoya, Norway. The right-hand panels show the X-ray event from which images are constructed (top) and two examples of images, one from a quiet period (left) and the other from the initial X-ray burst (right). Energy spectra constructed from these images are shown below.

A balloon-borne X-ray pinhole camera was developed in the late 1970s to detect energetic electron precipitation sources [Mauk, Chin, and Parks, 1981; Werden and Parks, 1987; Werden, 1988; see also Yamagami, Miyaoka, Nakamoto, Hirasima, Ohta, Namiki, Murakami, Sato, Fujii, Okudaira, Nishimura, and Kodama, 1990]. This new tool permits more detailed studies of X-rays than those done in the 1960s. For example, the X-ray pinhole camera has the capability to remotely sense distant magnetospheric regions with spatial resolutions of 10-20 km at ionospheric heights. Current balloon-borne X-ray research objectives continue to stress obtaining precise spatial and temporal information and complementing satellite and other geophysical observations.

The X-ray pinhole camera detects X-rays in the energy range $\approx$20–120 keV. To illustrate the kinds of information we can now obtain, we show in Figure 3 (left panel) an X-ray event that was intercepted by our camera during a substorm that occurred on July 26, 1986. The event between 2100-2200 UT was detected while the balloon was still ascending and will not be discussed any further. The event we will discuss in detail below occurred just before local midnight (LT=UT+1). This event began with an initial burst which was then followed by a series of more intense bursts that coincided with the northward expansion of the electrojet that began around 2250 UT. As indicted by the reversal of the Z-component magnetic field at Andoya, Norway, the electrojet passed over Andoya at 2330 UT.

Two examples of X-ray images from this event are shown in the right panels of Figure 3. The images (middle panels) come from the time when the fluxes were near the background level (left) and during the initial burst (right). The energy spectra constructed from these images shown in the bottom panels have been corrected for atmospheric absorption. Each image consists of contours of constant 20-120 keV X-ray flux at an assumed source height of 100 km. The field of view (FOV) of the camera is 90 degrees, and projects to a 100 km by 100 km area at 100 km altitude. The image view is looking downward on the auroral zone with north at the top and east to the right. The exposure time of the image and the corresponding energy spectrum (bottom panels) is 30 seconds.

The quiet time image taken at 0036 UT shows uniform precipitation over the FOV of the camera. The energy spectrum from the quiet period includes two components with a break at around 40 keV. Assuming an exponential form for each component yields X-ray e-folding energies of $\approx$16 and 88 keV. The quiet image and spectrum are typical of "quiet" time precipitation and they establish a standard of reference for comparisons to data taken during active periods. The X-rays during the quiet interval are interpreted to come from the "drizzle" form of precipitation.

The X-ray image from 2308 UT during the initial burst shows fairly smooth contours with no significant flux gradient, indicating that precipitation during this interval was also uniform over the FOV of our camera. The contribution of X-ray fluxes during the initial burst came mainly from >40 keV X-rays. Note that only the "hard" spectral component was active during this interval. X-rays with energies >100 keV were detected. These high energy X-rays could involve electrons with relativistic energies and may be related to precipitation of relativistic electrons from the trapping boundary [Imhof, Voss, Mobilia, Datlowe, and Gaines, 1991].

Figure 4 shows a sequence of six X-ray images and their energy spectra taken during the northward expansion (2320-2400 UT). Precipitation here was very dynamic and X-ray features were observed to vary in both space and time. This contrasts with the uniform features observed during the initial burst (Figure 3). Large flux gradients occurred over the 100 km precipitation region, and the source included many sharp structures within the 100 km region viewed by our camera. Some of the structures had dimensions as small as $\approx$15-20 km, which was the limit of the instrument resolution. Note the rapid growth and decay patterns

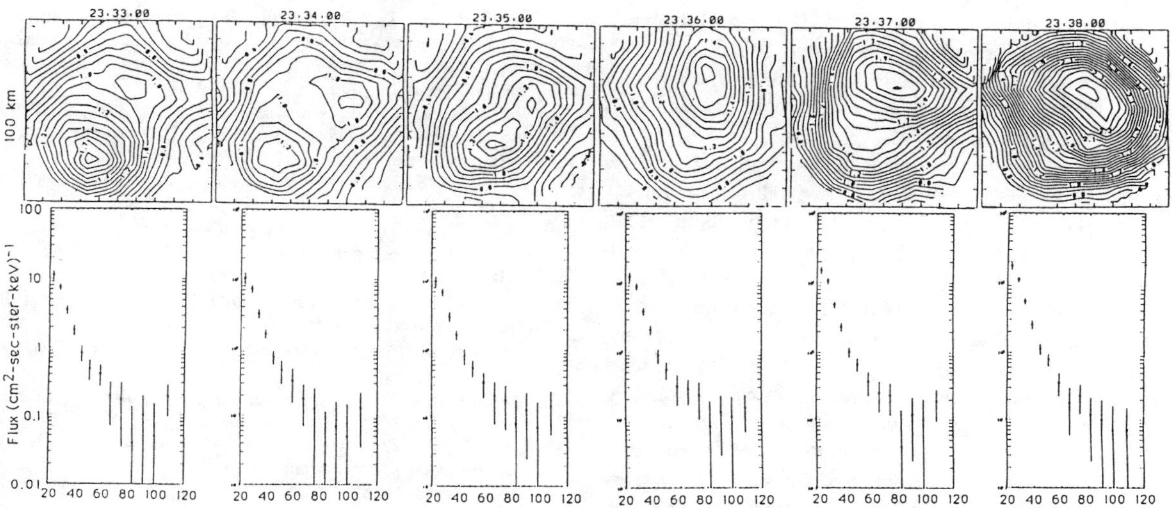

Fig. 4. A sequence of six X-ray images and their corresponding energy spectra obtained during the main X-ray burst. See text for detailed discussions.

as well as the apparent motion of the source from southwest to northeast.

The energy spectra shown below each image were constructed from X-ray counts accumulated from the most intense regions of precipitation in each image. This unique capability of imaging detectors enables us to study different precipitation features as a function of precipitated energy. For the images under consideration, the small scale features were mainly associated with X-rays <40 keV. The fluxes >40 keV remained essentially constant during these times.

The results of the X-ray camera data briefly summarized above show that the substorm on July 26, 1986 was initiated with a burst of X-rays with energies >40 keV, and precipitation during this interval was uniform. Following this burst, larger fluxes of X-rays had energies <40 keV, and electron precipitation included sharp small-scale structures. It is interesting to note that the two electron sources were active at different phases of the substorm. This behavior suggests that substorms possibly involve a two-step process. Note that a two population energy source similar to the event discussed above has been observed previously by Mauk, Chin and Parks [1981] and Sauvaud, Treilhou, Saint-Marc, Dandouras, Reme, Korth, Kremser, Parks, Zaitzev, Petrov, Lazutine, and Pellinen [1987].

5. CONCLUSION

X-ray observations which began in the 1960s have revealed many important features about magnetospheric plasma dynamics and paved the way for other experiments. The motivation to continue research of energetic electron precipitation phenomena by means of X-rays comes from observations that precipitated electrons are accelerated from both the ionosphere and distant magnetosphere. The current understanding is that electrons responsible for visual auroras have energies typically <10 keV [Hoffman and Evans, 1968; Arnoldy, 1981; Lin and Hoffman, 1979] and their source is in the ionosphere [Hoffman and Evans, 1968; Frank and Ackerson, 1971]. On the other hand, electrons responsible for X-rays observed on balloons have energies >20 keV and their primary source is the magnetosphere [Parks and Winckler, 1968]. Since the two components appear to be active possibly at different phases of a substorm, it is important to study both low and high energy electron precipitation sources.

Visual and UV auroral features have shown small spatial scales of several tens of meters [Creswell and Davis, 1966] to features global in size [Frank, Craven, Burch, and Winningham, 1982; Meng and Huffman, 1984; Anger, Murphree, Jones, King, Broadfoot, Cogger, Creutzberg, Gattinger, Gustafsson, Harris, Haslett, Llewellyn, McConnell, McEwen, Richardson, Rostoker, Sandel, Shepherd, Venkatesan, Wallis, and Witt, 1987]. The visible and UV features are produced by both high and low energy particles and the emission data alone cannot separate which precipitating component is contributing. X-ray observations can help resolve this problem. Although X-ray observations of small-scale features are few, they have been observed on several occasions [Mauk and Parks, 1981; Mauk, Chin and Parks, 1981; Werden and Parks, 1987]. It is important to identify these small-scale structures and to learn about the relationship of high energy (>20

keV) to low energy (<10 keV) precipitation observed near local midnight during substorms.

In conclusion, auroral X-ray observations which began in the early 1960s have contributed in a significant way toward our understanding of magnetospheric phenomena. X-ray observations yield a rich variety of data, are unique, complement other types of auroral and magnetospheric observations, and should continue to make important contributions to magnetospheric research.

Acknowledgments. We would like to thank Professor Kinsey Anderson for providing us with accurate information on dates of early X-ray observations. The pinhole camera was developed by grants from the Atmospheric Sciences Section of the National Science Foundation to the University of Washington, ATM-8420220 and ATM-8710314.

REFERENCES

Akasofu, S.-I., The development of the auroral substorm, *Plan. Space Sci., 12,* 273-282, 1964.

Anderson, K. A., Soft radiation events at high altitude during the magnetic storm of August 29-30, 1957, *Phys. Rev., 111,* 1397-1405, 1958.

Anderson, K. A., Balloon observations of X-rays in the auroral zone I, *J. Geophys. Res., 65,* 551-564, 1960.

Anderson, K. A., and D. C. Enemark, Balloon observations of X-rays in the auroral zone II, *J. Geophys. Res., 65,* 3521-3538, 1960.

Anderson, K. A., and D. W. Milton, Balloon observations of X-rays in the auroral zone, 3, High time resolution studies, *J. Geophys. Res., 69,* 4457-4480, 1964.

Anger, C. D., J. R. Barcus, R. R. Brown, and D. S. Evans, Auroral zone X-ray pulsations in the 1- to-15-second period range, *J. Geophys. Res., 68,* 1023-1030, 1963.

Anger, C. D., J. S. Murphree, A. V. Jones, R. A. King, A. L. Broadfoot, L. L. Cogger, F. Creutzberg, R. L. Gattinger, G. Gustafsson, F. R. Harris, J. W. Haslett, E. J. Llewellyn, J. C. McConnell, D. J. McEwen, E. H. Richardson, G. Rostoker, B. R. Sandel, G. G. Shepherd, D. Venkatesan, D. D. Wallis, and G. Witt, Scientific results from the Viking ultraviolet imager: An introduction, *Geophys. Res. Lett., 14,* 383-386, 1987.

Arnoldy, R., Review of auroral particle precipitation, in *Physics of Auroral Arc Formation,* edited by S.-I. Akasofu and J. R. Kan, pp. 56-66, American Geophys. Union, Washington, DC, 1981.

Barcus, J. R., and R. R. Brown, Electron precipitation accompanying ionospheric current systems in the auroral zone, *J. Geophys. Res., 67,* 2673-2680, 1962.

Barcus, J. R., and T. J. Rosenberg, Energy spectrum for auroral zone X-rays, 1, Diurnal and type effects, *J. Geophys. Res., 71,* 803-824, 1966.

Barcus, J. R., and A. Christensen, A 75-second periodicity in auroral-zone X rays, *J. Geophys. Res., 70,* 5455-5459, 1965.

Brown, R. R., J. R. Barcus, and N. R. Parsons, Balloon observations of auroral zone X-rays in conjugate regions, 2, microbursts and pulsations, *J. Geophys. Res., 70,* 2599-2612, 1965.

Brown, R. R., Balloon observations of auroral zone X-rays, *J. Geophys. Res., 66,* 1379-1388, 1961.

Brown, R. R., and W. H. Campbell, An auroral zone electron precipitation event and its relationship to a magnetic bay, *J. Geophys. Res., 67,* 1357-1366, 1962.

Coroniti, F. V., R. L. McPherron, and G. K. Parks, Studies of the magnetospheric substorm, 3, Concept of the magnetospheric substorm and its relation to electron precipitation and micropulsations, *J. Geophys. Res., 73,* 1715-1722, 1968.

Cresswell, G., and N. Davis, Observations on pulsating auroras, *J. Geophys. Res., 71,* 3155-3164, 1966.

Datlowe, D. W., W. L. Imhof, and H. D. Voss, X-ray spectral images of energetic electrons precipitating in the auroral zone, *J. Geophys. Res., 93,* 8662-8680, 1988.

Evans, D. S., A pulsating auroral-zone X-ray event in the 100-second period range, *J. Geophys. Res., 68,* 395-400, 1963.

Frank, L. A., J. D. Craven, J. L. Burch, and J. D. Winningham, Polar views of the Earth's aurora with Dynamics Explorer, *Geophys. Res. Lett., 9,* 1001-1004, 1982.

Frank, L. A., and K. L. Ackerson, Observations of charged particle precipitation in the auroral zone, *J. Geophys. Res., 76,* 3612-3643, 1971.

Goldberg, R. A., J. R. Barcus, L. A. Treinish, and R. R. Vondrak, Mapping of auroral X-rays from rocket overflights, *J. Geophys. Res., 87,* 2509-2524, 1982.

Hartz, T. and N. Brice, The general pattern of auroral particle precipitation, *Planet. Space Sci., 15,* 301-329, 1967.

Hoffman, R. A., and D. S. Evans, Field-aligned electron bursts at high latitudes observed by OGO 4, *J. Geophys. Res., 73,* 6201-6214, 1968.

Imhof, W. L., G. H. Nakano, R. G. Johnson, and J. B. Reagan, Satellite observations of bremsstrahlung from widespread energetic electron precipitation events, *J. Geophys. Res., 79,* 565-574, 1974.

Imhof, W. L., H. D. Voss, J. Mobilia, D. W. Datlowe, and E. E. Gaines, The precipitation of relativistic electrons near the trapping boundary, *J. Geophys. Res., 96,* 5619-5630, 1991.

Imhof, W. L., H. D. Voss, J. Mobilia, D. W. Datlowe, E. E. Gaines, and J. P. McGlennon, Relativistic electron microbursts, *J. Geophys. Res., 97,* 13829-13837, 1992.

Jelly, D., and N. Brice, Changes in Van Allen radiation associated with polar substorms, *J. Geophys. Res., 72,* 5919-5931, 1967.

Kennel, C. P., Consequences of a magnetospheric plasma, *Rev. Geophys., 7,* 379-420, 1969.

Lampton, M., Daytime observations of energetic auroral-zone electrons, *J. Geophys. Res., 72,* 5817-5823, 1967.

Lin, C. S., and R. A. Hoffman, Characteristics of the inverted-V event, *J. Geophys. Res., 84,* 1514-1524, 1979.

Mauk, B. H., and G. K. Parks, X-ray images of an auroral breakup, in *Physics of Auroral Arc Formation,* edited by S.-I. Akasofu and J. R. Kan, pp. 129-135, American Geophys. Union, Washington, DC, 1981.

Mauk, B. H., J. Chin, and G. K. Parks, Auroral X-ray images, *J. Geophys. Res., 86,* 6827-6835, 1981.

McDiarmid, I. and J. Burrows, Diurnal intensity variations in the outer radiation zone at 1000 km., *Can. J. Phys. 42,* 1135-1148, 1964.

McPherron, R. L., G. K. Parks, F. V. Coroniti, and S. H. Ward, Studies of the magnetospheric substorm, 2, Correlated magnetic micropulsations and electron precipitation occurring during auroral substorms, *J. Geophys. Res., 73,* 1697-1714, 1968.

Meng, C.-I. and R. E. Huffman, Ultraviolet imaging from space of the aurora under full sunlight, *Geophys. Res. Lett., 11,* 315-318, 1984.

Meredith, L. H., M. B. Gottlieb, and J. A. Van Allen, Direct detection of soft radiation above 50 kilometers in the auroral zone, *Phys. Rev., 97,* 201-205, 1955.

Mizera, P. F., J. G. Luhmann, W. A. Kolasinski, and J. B. Blake, Correlated observations of auroral arcs, electrons, and X-rays from a DMSP satellite, *J. Geophys. Res., 83,* 5573-5578, 1978.

O'Brien, B. J., High-latitude geophysical studies with satellite Injun 3. 3.

Precipitation of electrons into the atmosphere, *J. Geophys. Res., 69,* 13-44, 1964.

Parks, G. K., D. W. Milton, and K. A. Anderson, Auroral zone X-rays bursts of 5- to 25-millisecond duration, *J. Geophys. Res., 72,* 4587-4589, 1967.

Parks, G. K., F. V. Coroniti, R. L. McPherron, and K. A. Anderson, Studies of the magnetospheric substorm, 1, Characteristics of modulated energetic electron precipitation occurring during auroral substorms, *J. Geophys. Res., 73,* 1685-1696, 1968.

Parks, G. K. and J. R. Winckler, Simultaneous observations of 5- to 15-second period modulated energetic electron fluxes at the synchronous altitude and the auroral zone, *J. Geophys. Res., 74,* 4003-4017, 1968.

Parks, G. K., Auroral zone microbursts, substructures, and a model for microburst precipitation, in *Proceedings of X-rays in Space,* edited by D. Venkatesan, pp. 849-874, The University of Calgary, 1974.

Sauvaud, J. A., J. P. Treilhou, A. Saint-Marc, J. Dandouras, H. Reme, A. Korth, G. Kremser, G. K. Parks, A. N. Zaitzev, V. Petrov, L. Lazutine, and R. Pellinen, Large scale response of the magnetosphere to a southward turning of the interplanetary magnetic field, *J. Geophys. Res., 92,* 2365-2376, 1987.

Van Allen, J. A., Direct detection of auroral radiation with rocket equipment, *Proc. Natl. Acad. Sci., U. S., 43,* 57-62, 1957.

Werden, S. H., and G. K. Parks, Auroral X-ray images during a pulsation event, in *Proceedings of 8th ESA symposium on European Rocket and Balloon Programmes and Related Research, ESA SP-270,* pp. 71-78, 1987.

Werden, S. H., *Energetic Electron Precipitation in the Aurora as Determined by X-ray Imaging,* Ph.D. thesis, University of Washington, December, 1988.

West, R. H., and G. K. Parks, ELF emissions and relativistic electron precipitation, *J. Geophys. Res., 89,* 159-167, 1984.

Winckler, J. R., and L. Peterson, Large auroral effect on cosmic ray detectors observed at 8 g/cm^2 atmospheric depth, *Phys. Rev., 108,* 903-904, 1957.

Winckler, J. R., L. Peterson, R. Arnoldy, and R. A. Hoffman, X-rays from visible aurorae at Minneapolis, *Phys. Rev., 110,* 1221-1231, 1958.

Winckler, J. R., P. D. Bhavsar, and K. A. Anderson, A study of the precipitation of energetic electrons from the geomagnetic field during magnetic storms, *J. Geophys. Res., 67,* 3717-3736, 1962.

Yamagami, T., H. Miyaoka, A. Nakamoto, Y. Hirasima, S. Ohta, M. Namiki, H. Murakami, N. Sato, R. Fujii, K. Okudaira, J. Nishimura, and M. Kodama, Two-dimensional auroral X-ray image observation at a balloon altitude in the northern auroral zone, *J. Geomag. Geoelectr., 42,* 1175-1191, 1990.

G. Parks, T. Freeman, M. McCarthy, and S. Werden, Geophysics Program AK-50, University of Washington, Seattle, WA 98195.

Optical Measurements of the Fine Structure of Auroral Arcs

Joseph E. Borovsky and David M. Suszcynsky

Space and Atmospheric Sciences Group, Los Alamos National Laboratory

An optical system was designed, constructed, and fielded and auroral arcs were imaged. A hierarchy of spatial scales was seen in auroral arcs, with the smallest scale being the sheet-like fine structures. Auroral-arc fine structure was found to be a real and robust phenomena, with brightnesses that are significantly higher than the background brightnesses within arcs. Typical thicknesses for the fine structures are 100 m. The observations support the findings of *Maggs and Davis* [1968], which have been largely unappreciated.

1. Motivation

One of the interesting properties of an auroral arc is its latitudinal extent (thickness). Measurements of arc thicknesses can provide information about the mechanisms that accelerate auroral electrons, about generator mechanisms that supply energy to auroral-arc field lines, and about the mechanisms that transfer energy electrodynamically between the magnetosphere and the ionosphere. The measurements might also provide information about gradient scale sizes in the nightside magnetosphere.

For some time it has been known that auroral arcs are thin, but the degree of their thinness has been in dispute. Early ground-based optical measurements obtained a variety of thickness values: 7.4 km [*Stormer*, 1955], 250 m [*Elvey*, 1957], 114 m and 336 m [*Akasofu*, 1961], and 3.5 - 18.2 km [*Kim and Volkman*, 1963]. The variety of values may reflect the fact that there is a hierarchy of thicknesses associated with a single auroral arc and the various observers may have measured various aspects of the arcs. The larger of these values could also be attributed to poor spatial resolution of the cameras or to the auroral arcs not being in the magnetic zenith when the measurements were made, *i.e.* they were not observed edge on.

Maggs and Davis [1968] systematically measured the thicknesses of auroral arcs with the use of an image-orthicon television system that was pointed into the magnetic zenith. Their system had a resolution of 70 m (one line of a video image being equivalent to 70 m at ionospheric altitudes) and they adjusted their imaging system to enhance the finest-scale structures in each arc that passed through their field of view [*J. Maggs, private communication*, 1991]. The distribution of auroral-arc thicknesses that they obtained is reproduced in Figure 1. In this distribution, the number of arcs observed increases with decreasing thickness until the resolution limit of 70 m is encountered. If the system resolution did not effect this distribution, then a typical arc thickness is somewhat greater than 70 m. However, it appears that the system resolution did cut off the distribution, indicating that some auroral arcs may be even thinner than 70 m.

To a large extent, these results of *Maggs and Davis* [1968] have been ignored by the auroral research community. The observation of auroral-arc fine structures with 70-m thicknesses is at odds with every theory of auroral arcs [*Borovsky*, 1993a], but little criticism of the various theories is made concerning this discrepancy. At times the results of *Maggs and Davis* have been dismissed, owing to the possibility that the fine structures that they observed were not dynamically robust, *i.e.* that they constituted optical-brightness variations of only a few per cent in much-thicker auroral arcs.

Because the *Maggs and Davis* [1968] measurements disagree so strongly with theoretical expectations of the thicknesses of auroral arcs, it is important to confirm or disprove the measurements. For that purpose, a new optical imaging system was designed, constructed, and fielded and the thicknesses of auroral arcs were measured. As the reader will find, the preliminary measurements taken with the new system support the results of Maggs and Davis.

2. Measuring Auroral-Arc Thicknesses

The thickness of an auroral arc is measured by imaging the arc edge on. Since the region of airglow of an arc is in the form of a magnetic-field-aligned sheet, the arc must be imaged in the direction of the observer's magnetic zenith.

Auroral Plasma Dynamics
Geophysical Monograph 80

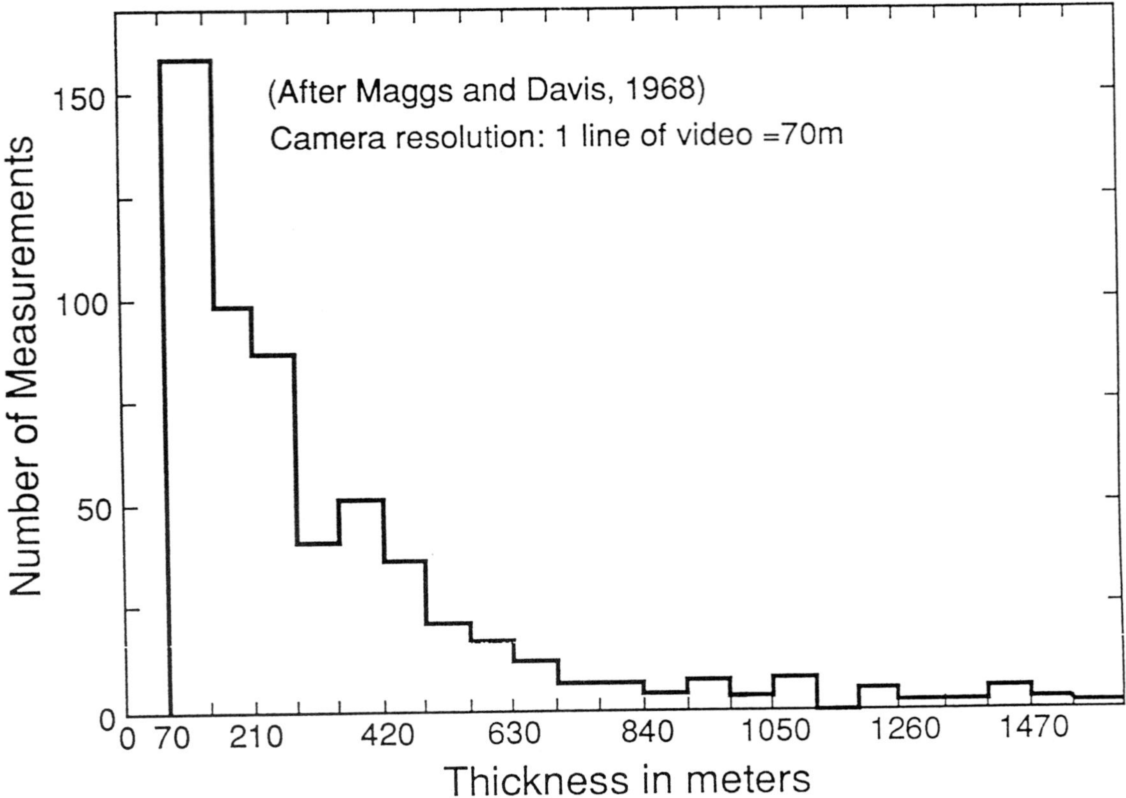

Fig. 1. The distribution of auroral-arc thicknesses measured by *Maggs and Davis* [1968].

To do this, one points an imaging system into the magnetic zenith and records auroral arcs as they drift through the field of view of the system.

To be successful, an imaging system must (1) be able to spatially resolve auroral arcs, (2) be able to temporally capture auroral-arc images in "snapshot" fashion to prevent blurring caused by arc motion, and (3) be sensitive enough to obtain auroral-arc images during the restricted capture time.

In resolving auroral arcs, an improvement over the 70-m resolution of *Maggs and Davis* [1968] is warranted, since arc thicknesses of 70 m were observed. An estimate of the maximum resolution that will be needed is obtained by calculating the minimum-possible thickness for an airglow structure caused by the precipitation of energetic electrons into the upper atmosphere. An airglow structure has such a minimum thickness owing to the Coulomb scattering of the energetic electrons by the molecules of the upper atmosphere; even if an auroral arc is caused by electrons in a sheet beam that has zero thickness, the airglow region of the arc will have a nonzero thickness owing to this electron scattering. To calculate the broadening of sheet beams of precipitating electrons, a computer code was developed and run. The code follows the path of a single electron as it moves downward into the atmosphere: the electron spirals in the terrestrial magnetic field owing to the $\vec{v} \times \vec{B}$ force, it slows down owing to the energy it transfers to atmospheric gas molecules, and it abruptly and randomly changes its direction owing to large-angle Coulomb scattering off the gas molecules. (Scattering owed to collective effects [*e.g. Hallinan et al.*, 1978] is ignored in this minimum-scattering calculation.) In the code, the atmospheric composition and density as functions of height are estimated from the MSIS-86 atmospheric model [*Hedin*, 1987] and the terrestial magnetic field is taken to be uniform in strength and tilted $11°$ away from normal into the atmosphere. The slowing down of the electron is estimated from a nonrelativistic Bethe stopping power formula [*Marmier and Sheldon*, 1969]. The random large-angle Coulomb scattering of the electrons is accounted for by developing scattering-probability coefficients from theoretical elastic-scattering cross sections for electron impact on nitrogen atoms [*Riley et al.*, 1975; *Fink and Ingram*, 1972]. The details of the code are contained in the Appendix of *Borovsky et al.* [1991]. By following the paths of many electrons and calculating where the electron energy is lost into the atmospheric gas, the width of the region of airglow caused by an infinitely thin sheet of precipitating electrons is obtained. The results are displayed in Figure 2 for 5-keV electrons and 25-keV electrons. For the lower-energy electrons, a full width of the energy-deposition region of 10 m is

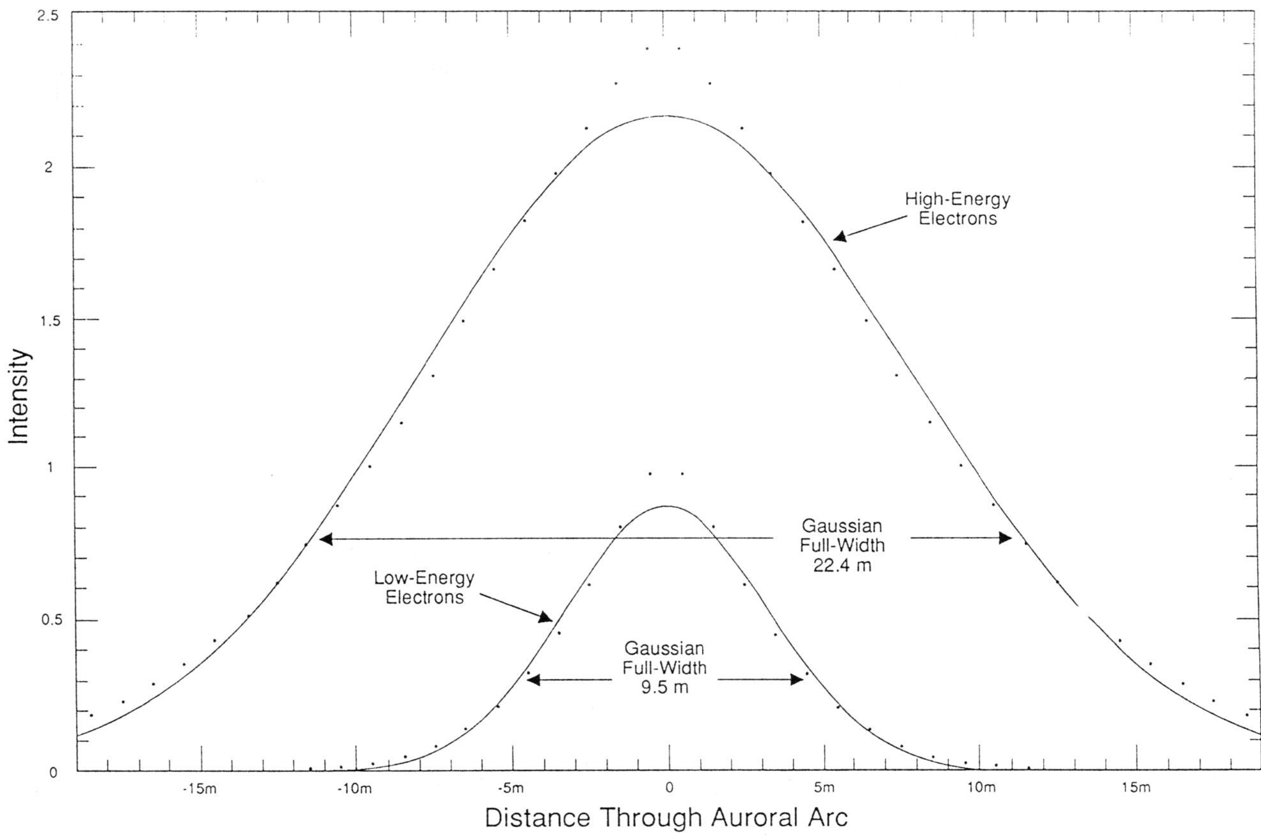

Fig. 2. The spatial distribution of airglow intensity produced by two infinitely thin sheet beams of electrons precipitating into the upper atmosphere. To obtain each curve, 1000 electrons were followed.

obtained. Hence the minimum-possible thickness of an auroral arc is about 10 m, and so a successful optical system should be able to resolve a 10-m structure, but need not be designed to resolve structures much smaller. The new imaging system that was fielded had a resolution of 2.4 m at an altitude of 105 km.

Auroral arcs drift, and so to obtain an unblurred image of an arc, a short exposure time must be used. Arcs typically drift latitudinally with velocities of 100 - 200 m/sec [*Omholt*, 1971]. If an arc is 10-m thick and moves at 200 m/sec, an exposure time of 1/20 sec will result in an image of the arc that is blurred to twice its actual thickness. Therefore, for a successful imaging system, an exposure time that is less than 1/20 sec is needed. The imaging system that was fielded had an exposure time of 1/60 sec.

To be able to image an auroral arc with an exposure time of only 1/60 sec, the imaging system must be sensitive to low light levels. Two factors are involved: the photon sensitivity of the camera and the light-gathering ability of the optics. Because the brightness of auroral-arc fine structure was unknown, the most-sensitive practical imaging system was chosen.   The system that was fielded consisted of a

microchannel-plate-intensified video camera mounted on a 10" Schmidt-Cassegrain telescope configured to $f/6.3$.

To point the telescope and to monitor the passage of arcs through the field of view of the telescope, a second imaging system was utilized. This wider-angle spotting system consisted of a microchannel-plate-intensified television camera and a 50-mm $f/1.4$ lens mounted on the side telescope.

Data from both imaging systems were recorded continuously on VCR tape and arc-thickness measurements were made from the recorded images.

An issue to consider when making edge-on images of moving auroral arcs is the blurring caused by the nonzero lifetimes of the emission lines. Of particular concern is the bright 5577-Å line of atomic oxygen, which has a 0.7-sec lifetime. If emission from longer-lifetime lines must be filtered out, then the imaging system will have less light to work with and the sensitivity issue becomes more critical. To determine the severity of the blurring caused by nonzero lifetimes, a computer code was developed and run. In the code energy was deposited into air by an electron sheet beam of a specified width drifting with a specified velocity and the deposited energy from each portion of the air was released at the following relative rates: 248 units of intensity (kilo-Rayleighs) of emission was released with a lifetime of 10^{-4} sec to represent band emission with lifetimes of 10^{-4} sec and faster, 100 units with a lifetime of 0.7 sec to represent the 5577-Å emission of atomic oxygen, 6 units with a lifetime of 44 sec to represent the 7060-Å line of molecular oxygen, and 400 units are released with a lifetime of 110 sec to represent the 6300-Å and 6364-Å lines of atomic oxygen. For three different arcs with widths of 20 m moving with three different velocities v_{arc}, the instantaneous brightness as a function of distance normal to the arcs is plotted in the three panels of Figure 3. In each case the arc moves toward the left and Figure 3 shows the light intensity emitted from the arc fine structure and its wake. The morphology of the airglow is a bright region 20-m wide caused by fast-line emissions and an approximately-exponential tail $\sim e^{-x/(v_{\mathrm{arc}}0.7\mathrm{sec})}$. As can be seen, the 20-m thick arc is not obscured by the blurring (exponential emission) arising from the slow emission lines. Therefore, arc-thickness measurements will not be affected by the long-lifetime line emission and these lines need not be blocked from the imaging system. Consistent with this finding, "white-light" (unfiltered) was used to image auroral arcs.

A decision of when to use the imaging system must be made. In the auroral zone of the Northern Hemisphere, summer is unacceptable because of the absence of dark skies. The high-resolution imaging system cannot be placed in an enclosed dome to protect it because of the distortion intro-duced to images by the dome. This makes winter observing very difficult for the equipment and for the operators. Hence spring and fall are the only practical observing seasons.

A final consideration is the selection of a site to field the imaging system. The imager must be located under the auroral arcs, so to study quiet arcs the system must be fielded within the quiet-time auroral zone. In North America this zone cuts across northern Alaska, sweeps down to the south-west corner of the Northwest Territories, across the northern portion of Manitoba, then across Hudson bay over the northern portion of Quebec. Of the larger towns located within the auroral zone, Fort Smith, N.W.T. has the optimal observing weather as determined by the number of cloudless days during the spring and fall observing seasons [Hare and Hay, 1974]. In terms of the number of clear days, spring is superior to fall for the entire North-American auroral zone. The imaging system was fielded in Fort Smith and data were obtained for 4 out of 4 nights in April, 1990. An upgrade of the system (cameras with higher sensitivity, a 10" $f/3.1$ telescope, and improved VCR recorders) was fielded in Fort Smith in October, 1990 and obtained 0 out of 10 nights of data owing to clouds.

3. Observations

As seen by the telescope and by the wider-angle spotting imager, auroral arcs are characterized by at least three latitudinal spatial scales. The largest-scale feature of an arc

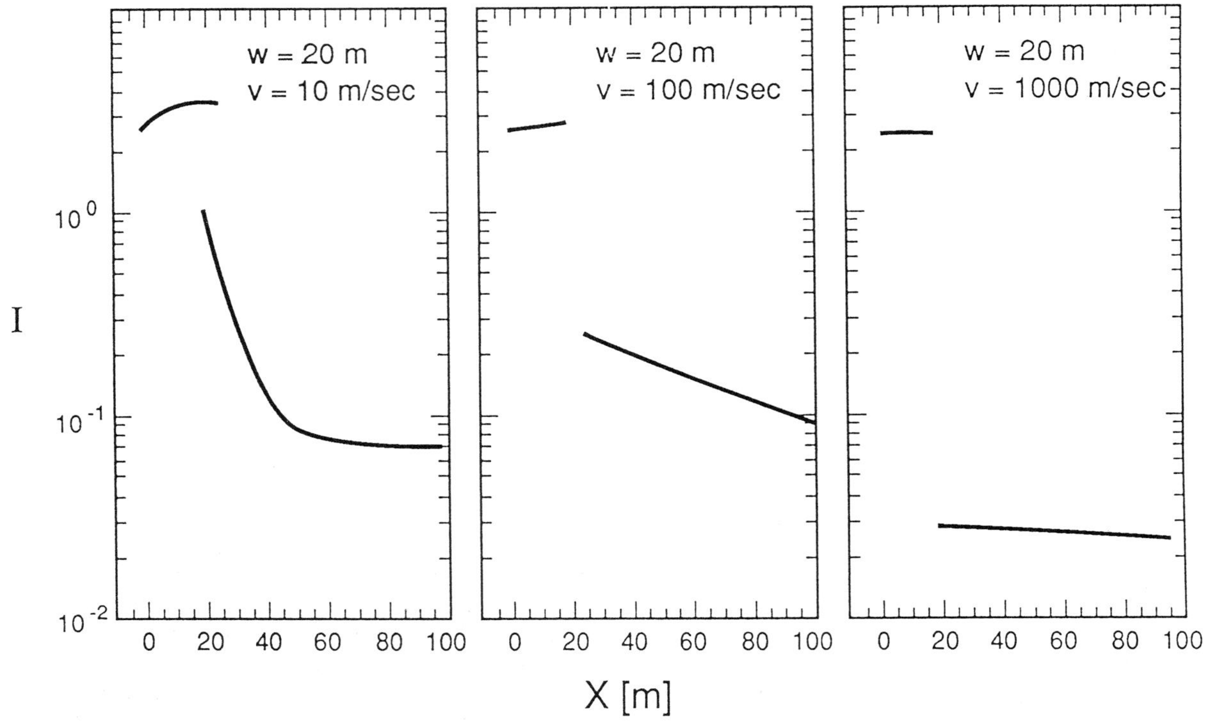

Fig. 3. The theoretical intensity profile of three moving auroral arcs is plotted.

consists of a broad channel of diffuse emission that is on the order of 10 km in width. The broad channel is seen in both white light and in the 10^{-7} - sec-lifetime 3914-Å line of N_2^+. The shape of this channel varies slowly with time (many seconds), but the intensity of the channel may vary rapidly as brightenings sweep along the arc. The diffuse channel can also have a finer-scale quasi-stationary pattern of emission to it. The middle-spatial-scale feature is a bright dynamic sheet of emission that resides within the broad channel. This bright feature is about 1 km wide. The bright feature is seen in both white light and 3914-Å. Within the bright 1-km-wide dynamic feature, narrow structures of emission with thicknesses of on the order of 100 m are observed. These narrow structures are the auroral-arc fine structures.

The auroral-arc fine structures are oriented more-or-less in an east-west fashion and they move in the north-south direction. Typical widths are on the order of 100 m and the velocities are of the order of 100 - 200 m/sec. The lengths of the fine structures are at least 500 m, which is the east-west extent of the telescope's field of view. The fine structures at times are multiple. At times they appear to be temporally modulated in intensity, however they pass through the field-of-view of the telescope in a few seconds so they are not sampled for very long. They are seen in white light, but they have not been observed in the 3914-Å line; the fine structures were barely imaged in the telescope when it was operated with white light and they may have simply been too faint in 3914 to be seen.

Defining the x-axis to be normal to the fine structure, the relative intensity of one fine structure versus x is plotted in Figure 4. (A photograph of this structure appears in Fig. 8 of *Borovsky et al.* [1991].) The structure was observed to drift at 220 m/sec, assuming that its height was 100 km. The structure moves toward the left in Figure 4. The intensity profile of this auroral-arc fine structure can be fit by a Gaussian with a full width of 110 m and an exponential $e^{-x/240\text{ m}}$ on the trailing side. Comparing the emission profile of this structure to the expected auroral-arc profiles in Figure 3, the Gaussian portion of the emission probably originates from very-fast-line emission and the exponential portion probably is 5577-Å emission from atomic oxygen. The Gaussian portion is the nearly instantaneous emission from air that is being bombarded by energetic electrons, so it is a measure of the width of the electron sheet beam in the upper atmosphere. Since the sheet beam is expected to broaden only by 10 - 20 m, the true width of the energetic-electron beam is about 100 m. This is very consistent with the findings of *Maggs and Davis* [1968] (see Figure 1). The afterglow of the arc in Figure 4 was fit by $e^{-x/240\text{ m}}$, which corresponds to a temporal exponential decay in intensity of 0.9 sec for a structure moving at 220 m/sec. This 0.9-sec value is close to the 0.7-sec value expected for the 5577-Å line of atomic oxygen.

Note that the intensity within the arc structure in Figure 4 is approximately an order of magnitude higher than the intensity ahead (to the left) of the arc. This demonstrates that the arc is very robust: it is not a few-percent increase

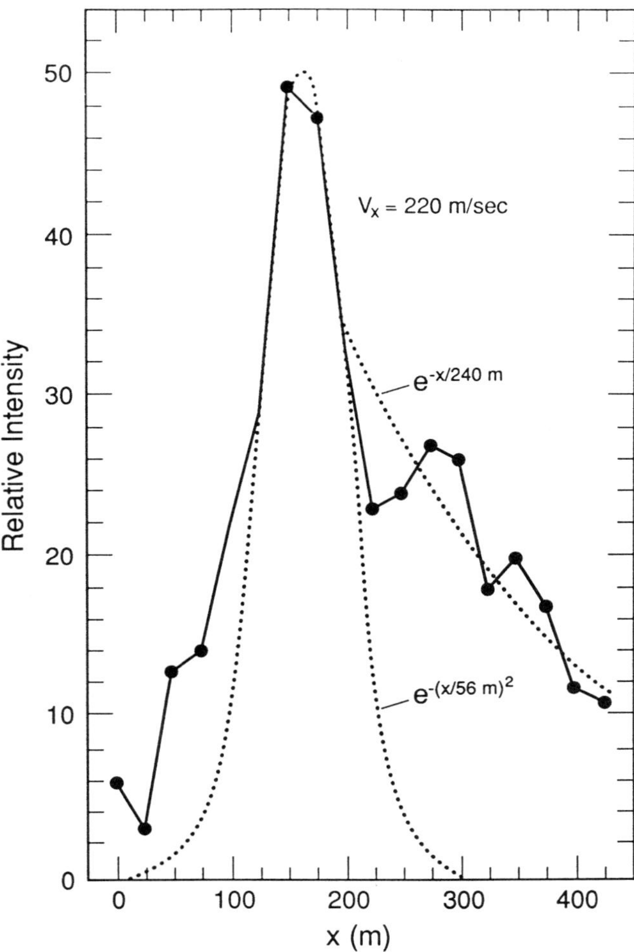

Fig. 4. The relative intensity of the sky is plotted for a cut through an auroral-arc fine structure imaged above Fort Smith, N.W.T.

in intensity within a broad arc — it is a significant increase in intensity. This demonstration dilutes the criticism that the *Maggs and Davis* [1968] small-scale structures were instrumental artifacts associated with very-small brightenings of larger arcs. Maggs and Davis were reporting a new and important class of auroral phenomena.

4. SUMMARY AND IMPLICATIONS

An optical imaging system with high spatial resolution was constructed and fielded and small-scale auroral arcs were observed. The data obtained during four nights of observations support the observations of *Maggs and Davis* [1968]. The conclusions are that small-scale structures in auroral arcs (1) are real, (2) they are robust in that they are much brighter than the background brightness of the arcs, (3) they are very thin (typically ~ 100 m), (4) they move with velocities of 100 - 200 m/sec in the upper atmosphere,

(5) they are at times multiple, and (6) they may have temporally varying brightnesses.

Auroral-arc structures with latitudinal thicknesses of 100 m or less is contrary to the predictions of every theory of auroral arcs. As catalogued in *Borovsky* [1993a], auroral-arc accelerator and generator theories predict auroral-arc thicknesses in the upper atmosphere of 1 km or larger. This is at least an order of magnitude larger than the observed arc-structure thickness. Theories cannot be disqualified by these optical-thickness measurements because there are other spatial scales associated with arcs that the theories may be correctly describing. A number of possible origins of the fine-scale structures were systematically explored in *Borovsky* [1993b], but none of the possibilites considered seemed likely. Since no theories describe the fine structures, it can be concluded that the fine structures are new phenomena that are not understood. Thicknesses of 100 m in the upper atmosphere map to a few Debye lengths or a few electron gyroradii in the magnetosphere. It is difficult to construct theoretical mechanisms to accelerate electrons to significant energies that do not span many Debye lengths and it is difficult to imagine gradients in plasma properties that do not span many electron gyroradii. In *Borovsky* [1993b] it was argued that the source of the fine-scale structures resides in the ionosphere or the upper atmosphere. Clearly, much theoretical work is called for to explain this phenomena first described by *Maggs and Davis* [1968].

Acknowledgments. The authors wish to thank Hal DeHaven and Mel Buchwald for their assistance with the imaging system. This work was supported by the U.S. Department of Energy and by the NASA Research and Analysis Program and CRRES Program.

The authors wish to dedicate this paper to Chris Goertz, who gave us final memories in Minneapolis.

REFERENCES

Akasofu, S.-I., "Thickness of an Active Auroral Curtain"", *J. Atmosph. Terr. Phys., 21*, 287, 1961

Borovsky, J. E., "Auroral-Arc Thicknesses as Predicted by Various Theories", *J. Geophys. Res., 98*, 6101, 1993a.

Borovsky, J. E., "Fine-Scale Structures in Auroral Arcs: An Unexplained Phenomenon", submitted to *AGU Monograph on Micro and Meso Scale Phenomena in Space Plasmas*, American Geophysical Union, Washington, 1993b.

Borovsky, J. E., D. M. Suszcynsky, M. I. Buchwald, and H. V. DeHaven, "Measuring the Thicknesses of Auroral Curtains", *Arctic, 44*, 231, 1991.

Elvey, C. T.. "Problems of Auroral Morphology", *Proc. Nat. Acad. Sci., 43*, 63, 1957

Fink, M., and Ingram, J., "Theoretical Electron Scattering Amplitudes and Spin Polarizations", *Atomic Data 4*, 129, 1972.

Hallinan, T. J., H. C. Stenbaek-Nielsen, and J. R. Winkler, "The Echo 4 Electron Beam Experiment: Television Observation of Artificial Auroral Streaks Indicating Strong Beam Interactions in the High-Latitude Magnetosphere", *J. Geophys. Res., 83*, 3263, 1978.

Hedin, A. E., "MSIS-86 Thermospheric Model" , *J. Geophys. Res., 92*, 4649, 1987.

Kim, J. S., and Volkman, R. A., "Thickness of Zenithal Auroral Arc over Fort Churchill, Canada", *J. Geophys. Res., 68*, 3187, 1963.

Maggs, J. E., and Davis, T. N., "Measurements of the Thicknesses of Auroral Structures", *Plan. Space Sci., 16*, 205, 1968.

Marmier, P. and Sheldon, E., *Physics of Nuclei and Particles. Volume I.*, Academic, New York, 1969, Sec. 4.9.2.

Omholt, A., *The Optical Aurora*, Springer-Verlag, New York, 1971, Sec. 2.6.

Stormer, C., *The Polar Aurora*, Oxford, London, 1955, Chap. VIII.

J. E. Borovsky and D. M. Suszcynsky, Space and Atmospheric Group, Mail Stop D466, Los Alamos National Laboratory, Los Alamos, NM 87545.

Some UV Dayside Auroral Morphologies

R. D. ELPHINSTONE, D. J. HEARN, J. S. MURPHREE, L. L. COGGER
M. L. JOHNSON AND H. B. VO

Department of Physics and Astronomy, University of Calgary, Calgary, Alberta, Canada

A variety of dayside auroral morphologies as recorded by the Viking ultraviolet imager are described. Each of these basic auroral morphologies may be linked to distinctive processes in the magnetosphere. Large scale asymmetries in the dayside distribution are found to be consistent with IMF B_y convection asymmetries. Candidate "FTE-like" auroral forms lasting about 5 minutes in the post noon sector for IMF $B_y < 0$ are reported. A specific example is shown where the event is accompanied by a brightening and an equatorward drift of the main auroral distribution. There is a bipolar ground magnetic signature associated with the auroral signature. Auroral forms observed to occur northward of the cusp near the poleward boundary of the DMSP plasma mantle indicate that current systems separate from the regular region 1 and 2 system do indeed exist. In the afternoon sector during IMF $B_y < 0$ and magnetically active conditions a splitting between the day and night sectors occurs which may indicate a division between dayside boundary layer processes, driven by the solar wind, and those driven from the nightside magnetotail. Also during this time omega-like bands are observed in the early morning hours along with periodically spaced features closer to noon. Their association with the most equatorward portion of the double oval links them to a near Earth source. The associated instability has a wavelength of about .6 MLT. Observations are presented showing the unwinding of a dayside spiral. These support the view that spirals are related to the winding up of magnetic field lines due to an upward field aligned current perturbation. Impulsive low latitude auroral forms are described for the first time and associated with magnetic pulsations. The accompanying ground based pulsations often occur before the arrival of the discrete auroral forms. This indicates much of the initial current is carried by low energy electrons to which the UV emissions are not as sensitive. It is proposed that the motion and spacing of morning fan arcs can be accounted for by a simple model whereby antisunward disturbances propagating perpendicular to the magnetic field couple to Alfvén waves which modulate the ground electric and magnetic fields at some multiple of the Alfvén transit time to the ionosphere. The morphology which results consists of periodically spaced arc systems which shift antisunward and poleward.

INTRODUCTION

Ground-based optical observations of the dayside auroral distribution are limited to winter conditions when the sky is not sunlit. Further there are few auroral observatories (e.g., Svalbard/Ny Alesund) which are at high enough latitude to observe the noon sector auroral zone under quiet conditions. Nevertheless over the years general statistical patterns of dayside auroral precipitation have been inferred (Shepherd, 1979) and terminologies such as cleft and cusp developed (Heikkila and Winningham, 1971; Heikkila, 1985). Large scale static

auroral morphologies such as the "mid-day gap" (Cogger et al., 1977; Dandekar and Pike, 1978) and fan arcs (Akasofu, 1981) have been developed and attempts at their interpretation have been made in terms of magnetospheric processes (Crooker, 1990; Reiff, 1990). Motions of the dayside auroral oval have been described and related to magnetic activity, tilt angle and the conditions in the solar wind (Burch, 1972; Vorobjev et al., 1976; Sandholt et al., 1988). Fewer details however are available concerning the dynamics of specific auroral forms. Rezhenov et al. (1979) first showed, using all sky camera data, that the auroral region splits in the afternoon sector. Elphinstone et al. (1991a) recorded the dynamics of such a splitting and related it to the substorm process. More recently Egeland et al. (1992) have shown the existence of arc systems in the

Auroral Plasma Dynamics
Geophysical Monograph 80

dawn sector which are spaced at equal intervals (1.1 degrees) and which propagate poleward. Viking observations have provided some information regarding the dynamics and characteristics of auroral intensifications in the afternoon sector (Lui et al., 1987; Potemra et al., 1990; Vo, 1992) as well as reporting on a new auroral morphology called "high latitude dayside forms" (Murphree et al., 1990). Perhaps the greatest wealth of information concerning the dynamics of the dayside aurora and its relation to other geophysical data sets can be found about the "dayside auroral breakup" events recorded at Nyalesund (Sandholt et al., 1990a, Jacobsen et al., 1991). The criteria for identifying such a breakup can be found in Sandholt et al. (1990b). These auroral forms move poleward out of the pre-existing auroral distribution and have dependencies on the interplanetary magnetic field (IMF) which make them candidates for the ionospheric signatures of flux transfer events (Lockwood, 1990).

Two dimensional nearly instantaneous images can be useful in resolving controversies concerning the ionospheric signatures of magnetospheric processes. In this paper we use Viking UV auroral image data (Anger et al., 1987) for describing the variety of auroral morphologies seen on the dayside. The paper will use Figure 1 as a summary of the various auroral morphologies recorded by the Viking imager. Each morphology shown in the Figure will be briefly described and relevant references to previous work given. Limited statistics for each morphology relating it to the interplanetary magnetic field (IMF) conditions will be presented (see Table 1). A specific event associated with the candidate fte's (upper left panel of Figure 1) will be described using ground based magnetometer field data and Viking image data. The relationship between this and the previously observed dayside auroral breakup will be discussed. An example of the high latitude auroral form (middle panel of Figure 1) will be presented illustrating its

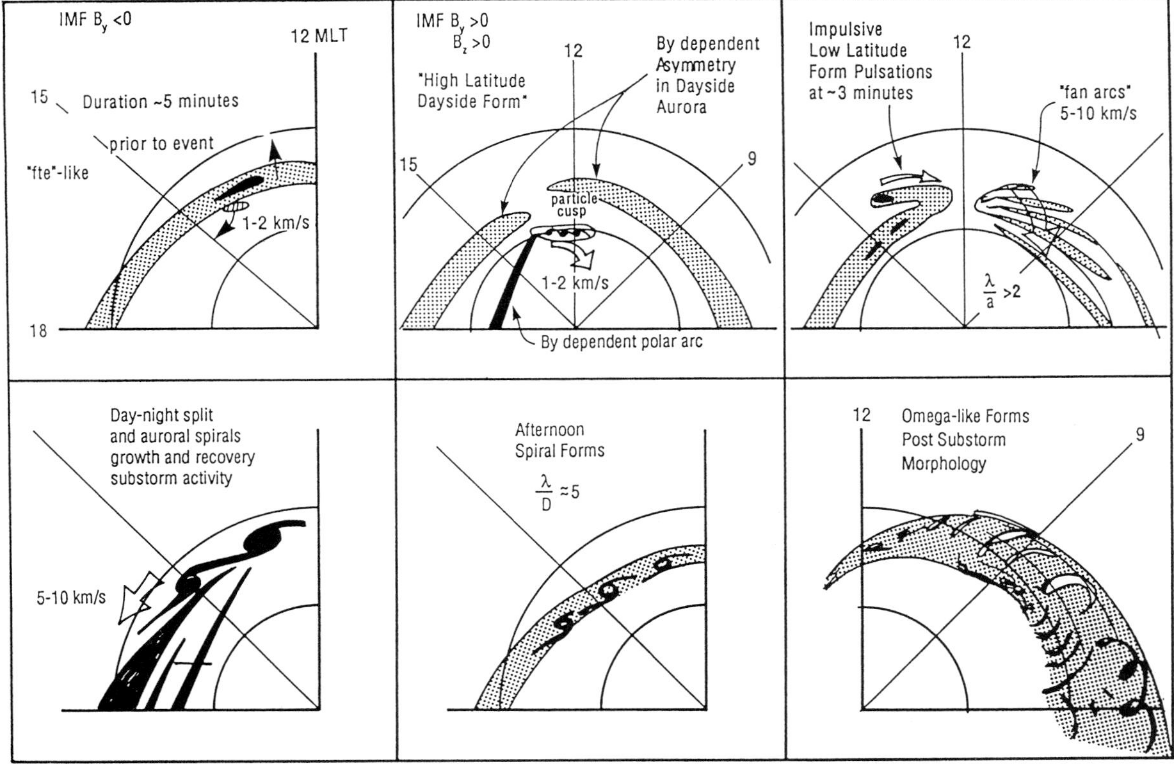

Fig. 1. A variety of auroral morphologies each potentially related to a different magnetospheric phenomenon. Top left panel illustrates a typical candidate FTE event as observed from Viking UV auroral data. Top middle panel: A high latitude dayside feature indicative of processes occurring poleward of the cusp. Top right panel: An impulsive low latitude auroral form and fan arcs which may be related to propagating compressional waves coupled to Alfvén waves modulating the ionospheric field aligned current system. Bottom left panel: A separation between the day and night auroral systems in the afternoon sector. This probably indicates separate dynamo systems in action. The ionospheric meeting point of the systems often creates a spiral form. Bottom middle panel: A vortex street of spirals wound up counter-clockwise when viewed in the direction of B. These spirals often connect two parallel arcs separated in latitude. Bottom right panel: Post substorm auroral forms. Near noon these features are small in scale (≈2-300 km) but they grow in the antisunward direction to scale sizes >500 km. These forms exist on the most equatorward of the two ovals during the recovery phase of an auroral substorm.

relationship to DMSP particle boundaries. A new form of auroral morphology termed a low latitude impulsive form will be briefly discussed and associated with pulsations in the solar wind near the bow shock and in ground based magnetic field data. Fan arcs observed in the morning sector are given a new interpretation as being due to coupling of an anti-sunward propagating disturbance to Alfvén waves which modulate the field aligned currents in the ionosphere. Finally auroral forms in the prenoon sector are associated with the more familiar omega-like bands in the morning sector. The double oval morphology during this time allows one to make inferences concerning the mapping of these features and the omega bands to the near Earth region.

BASIC DAYSIDE AURORAL MORPHOLOGIES

Using auroral image data acquired by the Viking spacecraft a number of auroral morphologies on the dayside have been identified. These have been represented in a schematic form in Figure 1. While other interesting morphologies do exist, the types presented in Figure 1 are ones which occur regularly. A proper understanding of the associated magnetospheric processes for each one of these is likely to greatly enhance our understanding of boundary layer processes as well their mappings to the ionosphere. The general statistical occurrences of these morphologies with the interplanetary magnetic field is given in Table 1. This table was constructed by identifying orbits during which a morphology was observed and then calculating the average IMF conditions for the 40 minutes prior to and including the time of the orbit. The number of events found for IMF B_y and B_z positive and negative are listed. The following is a qualitative description of each morphology.

THE "CANDIDATE FTE" EVENT
(TOP LEFT PANEL, FIGURE 1)

Ground-based observers can study these types of events in some detail (e.g., Sandholt et al., 1992a). They term events of a similar type midday auroral breakup events. From a ground-based viewpoint these forms appear between 9 and 15 MLT and, when IMF B_z is negative, they propagate poleward. Ground based observers have dominantly recorded such forms in the pre-noon sector and associate them with IMF $B_y > 0$ convection. Such prenoon forms have also been found in the Viking image data for IMF $B_y > 0$. In general however because of the more broken up character of the morning sector aurora the events as seen in the Viking images are most obvious in the post noon sector near 14 MLT (preferentially for IMF $B_y < 0$). Events have also been seen on the ground for IMF $B_y < 0$ (eg. Sandholt et al., 1992a, 1992b). It is unclear whether the events seen from the ground are the same as the ones reported here since the ground based observations seem to occur closer to local noon.

The upper left panel of Figure 1 is a schematic of the type of events observed in the Viking data in the afternoon sector. Based on 19 events when solar wind data was available, 16 of the events occurred when IMF B_y was negative consistent with convection asymmetries. If one includes magnetosheath data for the determination of the B_z component, then the number of events increased to 26 and it was found that these were approximately equally distributed for $B_z > 0$ and $B_z < 0$ or changing (14 for $B_z < 0$ and 12 for $B_z > 0$: see Table 1). A future study is necessary to determine if the poleward motion is a function of IMF B_z as has been reported by Sandholt et al. (1992b).

An example of this type of event when the IMF B_z was northward can be found in Murphree and Elphinstone (1988). Another such event is given in Figure 2. For this example, IMF B_z turned southward at 1743 UT at the GSE location of the IMP-8 satellite (34 R_E, 1.4 R_E, -8.4 R_E). IMF B_y was positive until 1800 UT at which time it became negative briefly and then steadily by 1805 UT. The average solar wind speed during the time period of interest was 600 km/s and the solar wind density was 3.6 cm^{-3}. The change in IMF B_z would first appear at the subsolar point at about 0747 UT and in the ionosphere by about 0749 or 0750 UT. Shown in Figure 3 are the magnetometer traces from the four ground stations displayed in Figure 2. Both Godhaven and Narssarssuaq show an enhancement of an eastward electrojet (equatorward of Godhaven and poleward of Narssarssuaq) beginning at about 0757 UT. This eastward electrojet seems therefore to be developing in association with the change in the IMF B_z component. At 1805 UT an impulse was seen at Iqualuit. It seems that the bipolar signature seen at this time may be related to the IMF B_y change which occurred at 1800 UT. At Iqualuit a small decrease in the H component (about 20 nT) occurred followed by a positive deflection of about 100 nT reaching a peak by 1809 to 1810 UT. This was accompanied by a 300 nT positive spike in the D component beginning at 1805 and maximizing by

TABLE 1. IMF Dependencies of the Dayside Morphologies

Auroral Type	# of Events for Average IMF Condition			
	IMF $B_y > 0$	IMF $B_y < 0$	IMF $B_z < 0$ or changing to negative	IMF $B_z > 0$
High latitude dayside form	7	0	1	6
Omega-like forms	0	7	7	0
Candidate fte's (afternoon sector)	3	16	14	12
Day-night split	0	13	4	9
Fan arcs	2	11	4	9
Low latitude impulsive form	2	6	5	3

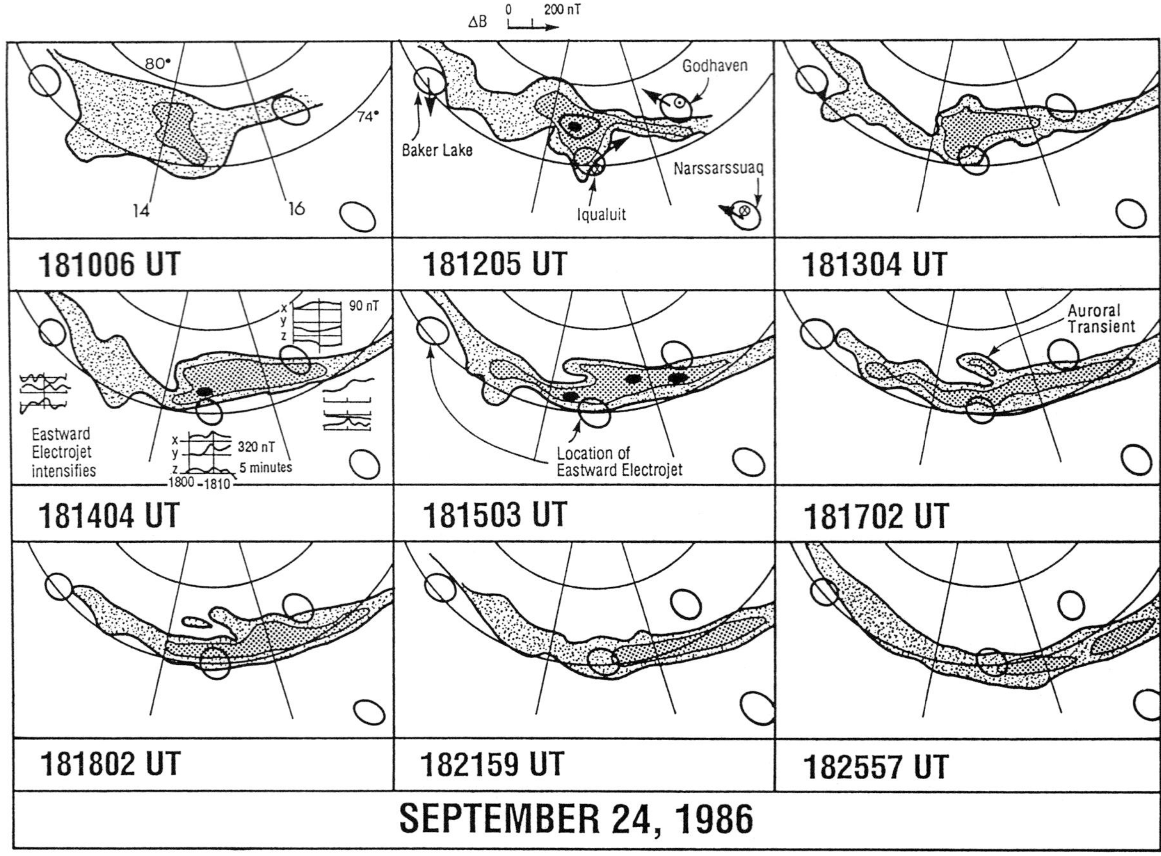

Fig. 2. A schematic of the afternoon sector aurora on September 24, 1986 between 1810:06 and 1825:57 UT. 15 MLT (CGM 1980) is at the centre of each image, 74° and 80° Mlat are shown for reference. 45° zenith angle field of views for Baker Lake (far left circle), Iqaluit (middle), Godhaven (top right) and Narssarssuaq (bottom right) are also shown. The approximate magnetic perturbations for each station are shown in the top middle panel at 1812:05 UT. The magnetic field traces for each station are shown in the middle left panel (labelled 1814:04 UT) and also in Figure 3. The auroral transient of interest is the small form rotating poleward out of the auroral "oval" between 1815 and 1818 UT (near 78° Mlat and 15 MLT). This occurs after a bipolar signature is seen at Iqaluit.

1809 UT. This impulse at Iqaluit lasted until about 1812 UT. At the time of the impulse a broad region of diffuse auroral luminosity appeared near 14 MLT.

The other ground stations nearby (Baker Lake, Godhaven and Narssarssuaq) indicate ΔB perturbations which are consistent with an eastward electrojet near the equatorward edge of the UV aurora and an upward field aligned current presumably associated with the 14 MLT enhancement. The top middle panel of Figure 2 shows the approximate ΔB perturbations at each station. The counterclockwise sense of the magnetic perturbations implies an upward current system.

The first major enhancement of UV emission occurred between 1810:06 UT and 1812:05 UT near 14 MLT and 77 Mlat (CGM 1980 coordinates) coincident with an equatorward drift of the auroras of 450 m/s ± 51 m/s (Elphinstone et al., 1991a). A dayside auroral transient formed after the first auroral brightening at 1815 UT (middle panel of Figure 2). This form can be seen to rotate out from the main UV oval between 1815:03 and 1818:02 UT and has disappeared by 1821:59 UT. The location of and sense of rotation of the auroral transient is consistent with the convective flow which one might expect from merging during a time of IMF B_y negative. This interpretation is also consistent with the IMF B_y becoming steadily negative after 1805 UT (corresponding to a change in the ionosphere at sometime after 1811 UT). Although such an event may not be associated with a "flux transfer event" we feel that it is appropriate to label it an as a "candidate fte". The entire sequence of events lasts about 15 minutes although each separate part only lasts for about 5 minutes (i.e., the magnetometer impulse, the UV dayside enhancement, and the dayside transient). As noted by Elphinstone et al. (1991a) this enhancement of dayside auroral luminosity and equatorward motion was followed by enhancements at later local times and substorm onset at 1828 UT.

One striking aspect of this event is that the main auroral brightening and the subsequent transient occur a few minutes

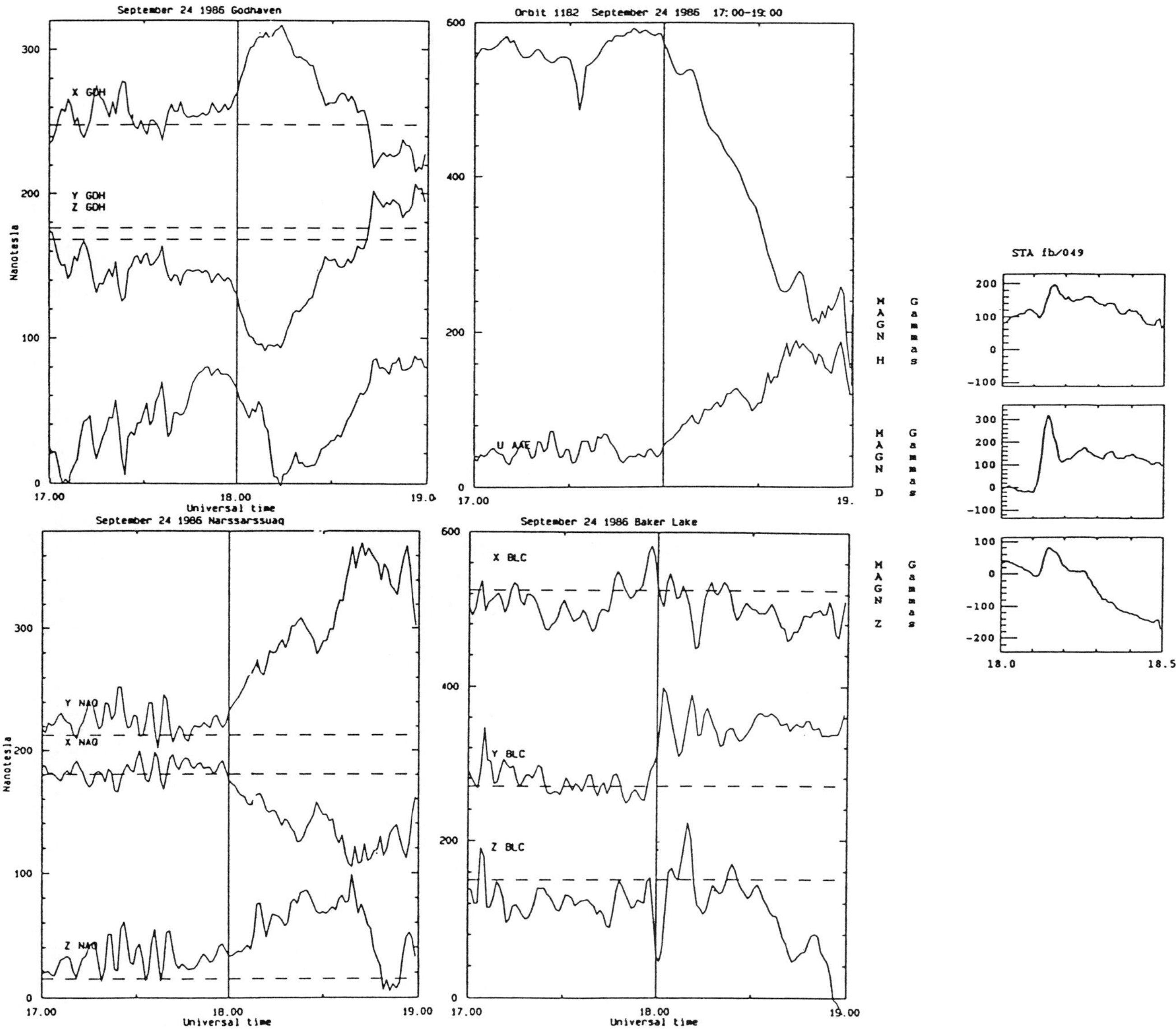

Fig. 3. Ground magnetic field data (x, y, z components) for the event on September 24, 1986 between 17 and 19 UT. The provisional AL (top curve) and AU (bottom curve) indices are displayed in the upper right panel. The zero line for the AL index is the top of the graph while the zero line for the AU index is the bottom of the graph. This example illustrates the directly driven component of the AU index very clearly. An impulse seen at Iqaluit is shown on the far right of the figure (H, D, and Z components are shown). The other panels show the X, Y and Z components from Godhaven (upper left), Narssarssuaq (bottom left) and Baker Lake (bottom right). Zero lines of the traces are given by dashed lines. The locations of these ground stations are shown in Figure 2.

after the magnetometer impulse. This appears to be a relatively consistent feature of dayside auroral events and could be explained as follows: The particles carrying the initial current arriving at the ionosphere are relatively low energy electrons (less than .5 keV). After the Alfvén wave carrying the current has been reflected a few times the precipitating electrons are accelerated to high enough energy for the UV intensities to be enhanced (Goertz and Boswell, 1979). Elphinstone et al. (1992) have shown that the LBH camera on board Viking is relatively insensitive to low fluxes of precipitating electrons whose energies are less than about .5 keV. The data concerning the "low latitude impulsive auroral form" shown later in this paper provide further evidence for such an interpretation.

DAYSIDE AURORAL ASYMMETRIES (IMF B_y CONTROL)

The middle panel of Figure 1 illustrates that the dayside aurora has an asymmetry (Elphinstone and Hearn, 1992) consistent with the IMF B_y control of the northern ionosphere convection throat (e.g., Moses et al., 1987). For IMF B_y positive (negative) the dusk (dawn) sector auroral oval near noon tends to occur poleward of the dawn (dusk) side auroral oval. This is consistent with mapping considerations for IMF B_y positive whereby a merging region exists which forms a throat oriented from low latitude dusk to high latitude dawn (Elphinstone et al., 1992). There is also an enhanced level of emission on dawn versus dusk for IMF B_y positive. This is what one might expect if there existed a narrow strong dawn cell and a broad dusk cell. A less pronounced effect exists for IMF

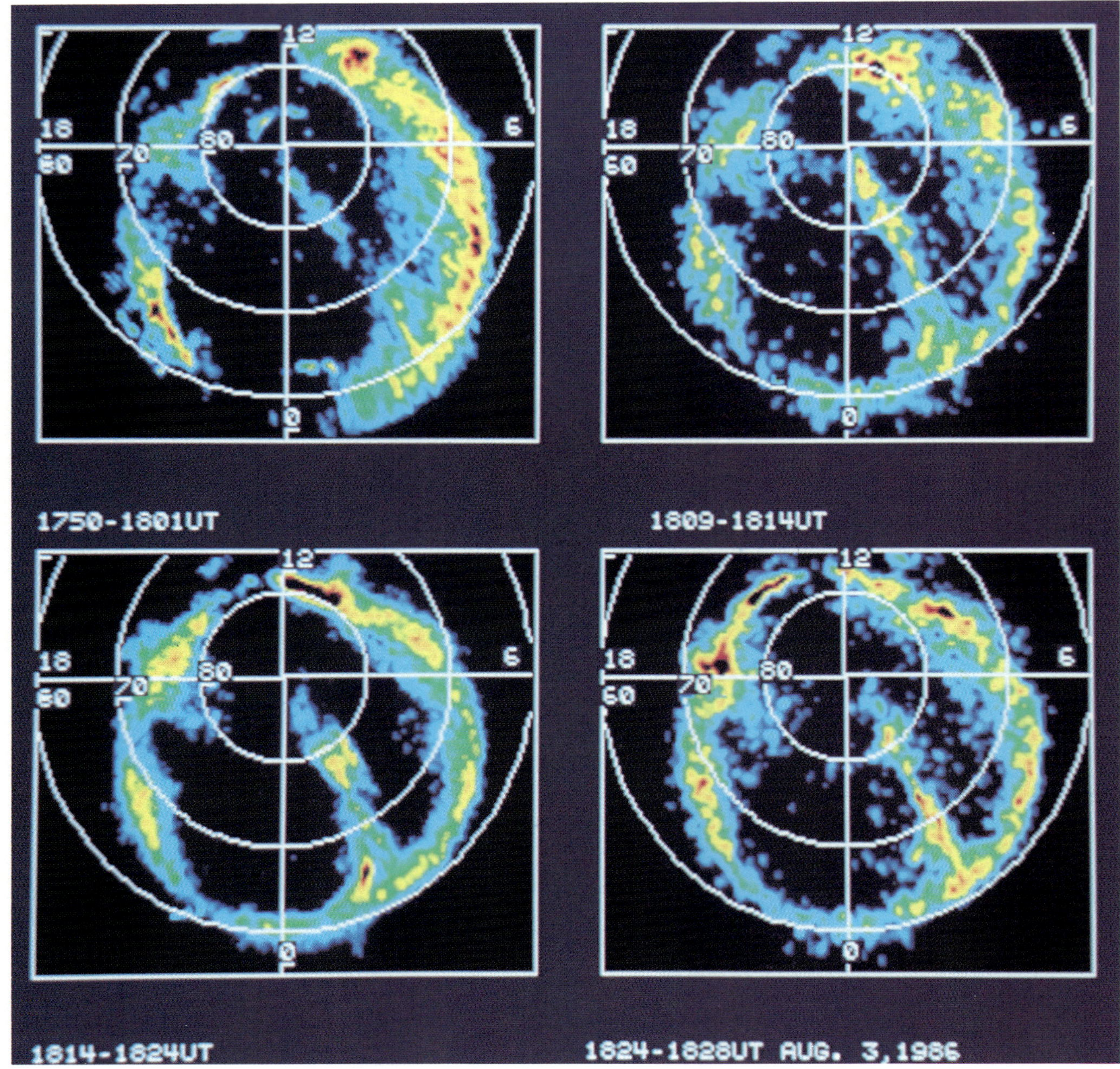

Plate 1. UV auroral images have been averaged in CGM (1980) coordinates for the time intervals shown below each panel. The IMF B_y component has changed from positive to negative just prior to this sequence of images. The dayside asymmetry (top of images) between dusk (left) and dawn (right) is very evident in the upper left panel consistent with IMF $B_y > 0$. A transition to a $B_y < 0$ auroral distribution occurs over the next 30 minutes.

$B_y < 0$ where the dusk sector aurora lies equatorward of the dawn sector aurora near noon. The effect is less evident probably because the aurora in the dawn and dusk sectors are linked to the region 2 and region 1 field aligned current systems respectively. Thus an asymmetry also exists independently of IMF B_y which puts the dawn aurora at lower latitudes than at dusk.

While static configurations exist for steady IMF conditions it is also interesting to consider the dynamics of such an asymmetry. Plate 1 illustrates this for an event on August 3, 1986 between 1750 and 1828 UT. The IMP-8 satellite was located at GSE position (32 R_E, -14 R_E, -1.5 R_E). IMF B_y and B_z changed from positive to negative at 1742 UT, at the same time as the number density changed from 20 to 40 cm^{-3}. The solar wind speed was about 475 km/s. The IMF B_z and B_y remained negative after this time. Allowing for propagation to the ionosphere, effects associated with this change should begin at about 1750 UT. Plate 1 illustrates how the UV auroral distribution changes over time. The dayside auroral distribution at 1750 UT is asymmetric, consistent with an IMF B_y positive. The dawn sector oval is broad and lies equatorward of the dusk sector oval. The dawn sector is greatly enhanced in intensity. Over the next 30 minutes the following changes occur:

1)   The equatorward edge of the dawn sector oval near noon moves poleward by about 3 degrees Mlat. This may be in part due to the large pressure change which occurs in the solar wind. The change in this low latitude auroral precipitation should have a counterpart in convection.

2)   A minimum in the auroral intensities which existed near noon (probably indicating the throat region) disappears during the interval 1809 to 1814 UT. The location of this ''throat'' is in the post noon region at the end of the interval while at the beginning it was nearly symmetric about noon.

3)   The dusk sector oval near noon shifts equatorward by about the same amount as the dawn sector moved poleward. This motion is most evident after about 1814 UT.

4)   The dawn sector emissions weaken by 1828 UT and become of comparable intensity to those at dusk.

5)   The sun-aligned auroral arc system in the northern hemisphere (which is occurring during a period of IMF B_z negative) moves towards the dawn sector (Craven et al., 1991).

These changes appear to be relatively consistent with what one might expect due a change from an IMF B_y positive condition, to a B_y negative condition. The equatorward motion of the dusk sector aurora is consistent with the addition of open flux to that sector. During an IMF $B_y < 0$ condition one would expect to observe a narrow cell in the dusk sector and a broad cell at dawn (the reverse is true for IMF $B_y > 0$). Indeed a comparison of the upper left panel to the lower right panel of Plate 1 shows that a transition from a $B_y > 0$ system

to that of an IMF $B_y < 0$ appears to be evident directly in the auroral data. The dusk and dawn sectors have reversed their relative importance with the dusk sector aurora now being the more intense (consistent with a narrower dusk cell). Thus these observations provide evidence of the dynamical changes the throat region undergoes during a change in the IMF B_y component. It seems that the transition takes place over about a half hour period with the dawn sector (for IMF B_y changing to a negative value) aurora being modified first and the dusk sector changes beginning some 5 minutes later.

HIGH LATITUDE DAYSIDE AURORAL FORMS
(IMF $B_y > 0$, $B_z \geq 0$)

An illustration of this auroral form is given in the top middle panel of Figure 1. The new aspects of this auroral morphology are the following:

1)   It exists poleward of but parallel to the main UV auroral oval near noon (Murphree et al., 1990).

2)   It occurs at the interface region between the plasma mantle (defined by DMSP particle data) and the polar rain (see Figure 4 and Elphinstone et al., 1992). At the location where the ion fluxes have decreased to a low level a burst of relatively high energy electrons (≈ 1 keV) is observed coincident with these forms. While the main aurora occurs equatorward of the particle cusp, this form exists poleward of it.

3)   It is often coincident with the noon sector connection point of polar arcs.

4)   Westward and antisunward motions of this form are consistent with antiparallel merging occurring poleward of the cusp on the front surface of the magnetotail.

Figure 4 illustrates an intensity vs. latitude profile of UV emissions for one such event. DMSP particle boundaries

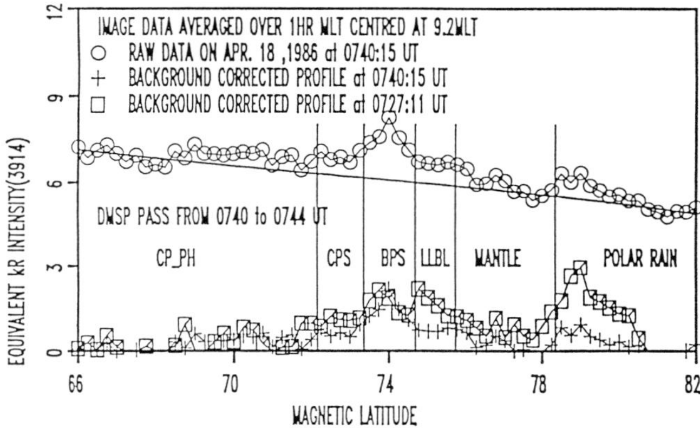

Fig. 4. Viking intensity profiles taken at 0727:11 and 0740:15 UT on April 18, 1986 averaged over one hour local time centred at 9.2 MLT (eccentric dipole 1985 coordinates). Boundaries defined from DMSP-F7 particle data (between 0740 and 0744 UT) using a neural network have been superimposed on the profile. The high latitude dayside feature is seen to occur at the poleward edge of the plasma mantle. A distinct arc system can be seen evolving in the LLBL region.

derived from a neural network (Newell et al., 1990) are superimposed to show the relation of the auroral forms to particle precipitation regions. The main UV auroral oval in this case corresponds to the particle BPS with fainter emissions corresponding to the CPS precipitation just equatorward. Poleward of this lies the LLBL. A separate auroral form can be associated with this LLBL region at 0727 UT. By the time of the DMSP overflight however the discrete arc form has faded and only a diffuse form remained (corresponding to the low energy LLBL signature seen by the DMSP F-7 spacecraft). Poleward of this auroral activity lies the plasma mantle which has little if any auroral luminosity (in the LBH bands) associated with it. At the poleward edge of this mantle precipitation is an intense region of auroral luminosity (at about 78 to 80° Mlat). This corresponds to the high latitude dayside auroral form. This example serves to illustrate that each DMSP particle region has its own UV auroral signature.

These "high latitude dayside auroral forms" strongly support the existence of the additional mantle current system first described by Bythrow et al. (1988). The dayside auroral asymmetry can be linked to asymmetries in convection and to asymmetries in the Region 1 and 2 field aligned current systems. Independent of whether some portion of these currents are related to the cusp currents it appears clear that an additional independent current system exists poleward of these and must be linked to the high latitude dayside auroral feature.

MORNING FAN ARCS AND AFTERNOON IMPULSIVE LOW LATITUDE AURORAL FORMS (TOP RIGHT PANEL OF FIGURE 1)

Fan arcs were first described schematically by Akasofu (1981) and later by Meng and Lundin (1986). The motion of these forms was reported by Elphinstone et al. (1991a) to be poleward and antisunward. These arc systems occur predominantly in the morning sector at periodically spaced intervals and are accompanied by magnetic pulsations. They point towards magnetic noon and on the basis of magnetic field mappings, the point from which they emanate is probably a good indicator of the location of the magnetic cusp (Elphinstone et al., 1992). A simple scenario can be given for the generation of these arc systems which explains the magnetic pulsations, the arc spacing and motions. Suppose first that a disturbance moves along the morning sector magnetopause in the antisunward direction. As a new voltage source is established at each point (i) in the equatorial plane an Alfvén wave will be launched to the ionosphere after a delay time, t_i. This is the time from when the disturbance reaches the subsolar point to when it reaches the point i, in the equatorial plane. If the ionosphere is not matched to the magnetosphere the currents and electric fields in the ionosphere associated with the field line, f_i, will be modulated with a period $2\tau_i$ (where τ_i is the Alfvén transit time from the equatorial plane to the ionosphere (Goertz and Boswell, 1979)). We assume also that the auroral intensity will be modulated in a similar manner. Elphinstone and Hearn (1993) used this to show that arc sys-

tems would result which would propagate poleward and were periodically spaced in latitude. These systems would have greater separations at higher latitudes. Considering adjacent arc systems, they also showed that if the higher latitude arc system had a longer Alfvén transit time than the more equatorward one, then over time, the distance between the arc systems would decrease. Associated with these systems would be magnetic pulsations whose period would increase with increasing magnetic latitude. The time over which these forms lasted should determine the relationship between the apparent ionospheric wavelengths, period and velocities involved. It is interesting to speculate that this scenario could allow arc systems to exist down to scale sizes limited only by the ability to be able to differentiate transit times on "adjacent" field lines. Many of the predictions made by this model still need to be observationally verified.

The development of fan arcs seems to be limited primarily to the morning sector hours. In the afternoon sector a different morphology exists that may however be related (see upper right panel of Figure 1). In this morphology an impulsive auroral form first appears approximately three to five degrees equatorward of the main UV auroral oval. Associated with it are magnetic pulsations in the Y component with about a 3 minute period. A surprising characteristic of these forms is that they begin to intensify first at their equatorward edge and then disappear there, with the intensification moving poleward towards the main auroral distribution. This is illustrated in Plate 2 for an event on September 27, 1986 between 0240 UT and 0252 UT. The IMP-8 satellite was located at about (6 R_E, 26.9 R_E, 9.8 R_E) GSE, near the location of the nominal bow shock. IMF B_z was northward, IMF B_x positive and IMF B_y was variable. The solar wind density was 3.5 cm^{-3} and the solar wind speed was on average 670 km/s. The solar wind density increased at 0229 UT to 4.9 cm^{-3} at which time the solar wind speed was 663 km/s indicating a pressure enhancement of about 25% from the previous minute. By 0233 UT the density was down to 3.3 cm^{-3} and the solar wind speed was still about 660 km/s. The associated pressure decrease from 0229 UT was about 33%. During this general time interval (between 0140 and 0300 UT) the solar wind density was modulated at about 4 mHz (corresponding to a period of about 4.1 minutes) and 6.9 mHz (corresponding to a period of 2.4 minutes). (Samples of this data were acquired about every minute). Beginning immediately after the pressure decrease the solar wind density showed particularly clear oscillations. These modulations lasted until about 0251 UT which coincided with the end of the ionospheric event. Accompanying the density modulations were periodic variations in the solar wind IMF B_z component. Similar to the density perturbations, spectral peaks were seen at 3.4 to 4 mHz and at 6.9 mHz (an equivalent period of 2.4 minutes). (The magnetic field values were sampled every 15-16 s). In addition to these peaks, the dominant oscillation in the IMF B_z occurred at 5.2 mHz (corresponding to a period of about 3.2 minutes). The location of IMP-8 near the bow shock precludes being able to say definitively if these solar wind changes were responsible for the auroral event or were generated in association with it.

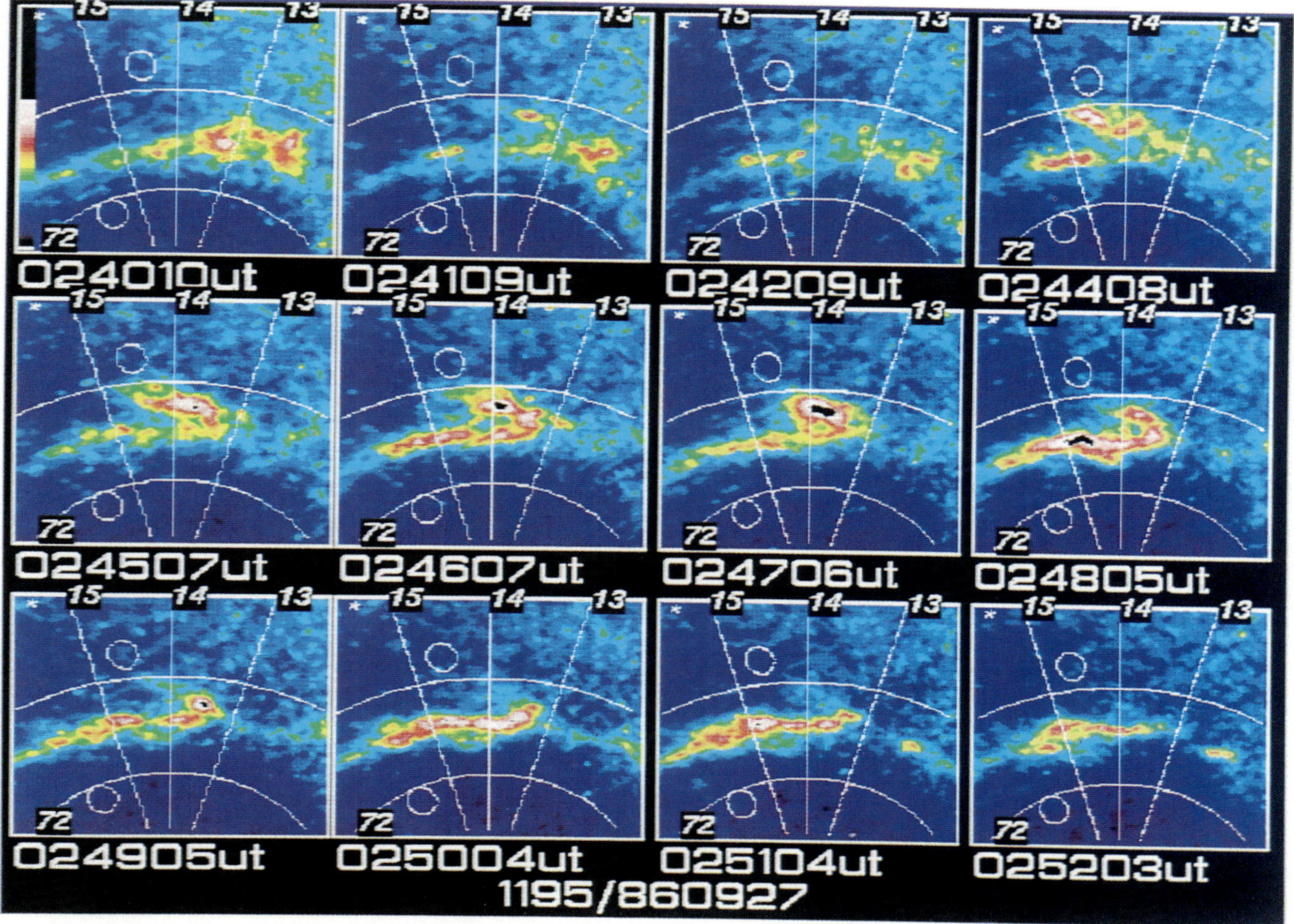

Plate 2. Observations of a low latitude impulsive auroral transient on September 27, 1986. The data is displayed in CGM 1980 coordinates with 12 MLT to the right, and the afternoon sector to the left. 72° and 80° Mlat are shown as well as 45° zenith angle field of views (at 120 km altitude) for the ground stations Mould Bay (the poleward station) and Barrow (the more equatorward one at about 15 MLT and 70 Mlat). The event begins equatorward of the oval and later develops on the oval itself. Magnetic pulsations are observed for about 3 cycles (beginning at about 0231 UT) before the discrete auroral form appears between 0241 and 0243 UT. The dynamics of such forms may be related to differing Alfvén transit times between different regions of the dayside boundary layer. IMF B_z is northward.

The more equatorward ground station (Barrow) in Plate 2 showed the onset of magnetic pulsations in the Y-component with a period of about 3 minutes beginning at 0231 UT and ending about 20 minutes later. Between 0200 and 0300 UT a spectral peak in the Y component occurred with a frequency between 5.2 and 5.6 mHz (corresponding to a period between 3.0 and 3.2 minutes). This frequency is about the same as the frequency associated with IMF B_z at IMP-8 but is between the frequencies found in the density oscillations. A very faint UV auroral signature of this onset was visible at 0231 UT (not shown in the Plate). The ground magnetometer pulsation continued for 3 cycles before the auroral signature (resembling a sideways hook shape) became pronounced at about 0241:09 UT (second panel of Plate 2). The aurora further intensified

after another cycle (0244:08 UT). Over the next six minutes the auroral form retreated from low latitudes until, by 0251 UT, only the previously existing auroral distribution was left. It is also apparent that the main auroral distribution did not brighten until about 0246 UT and that intensities on this oval also appeared to be modulated (compare the images in the middle row of Plate 2).

This event is further evidence that the ground magnetic pulsations are initiated before the discrete UV auroral signature begins, supporting the contention that the pulsations are due to propagating Alfvén waves whose current is initially carried predominantly by low energy electrons (< .5 keV). The trigger for this event appears to be a density impulse in the solar wind. It is unclear whether the modulations in the ionospheric currents are related to the oscillations in the solar wind density or to those in the IMF or to both. In any case the solar wind fluctuations appear to be driving compressional waves in the magnetosphere which in turn are coupling to the ionospheric current system via Alfvén waves along the field lines. The wave frequencies inside (ω_i) and outside (ω_o) the magnetosphere are related by the expression $\omega_i = |\omega_o + k_o \cdot v_o|$ (Wolfe and Kaufmann, 1975), where v_o is the plasma flow velocity and k_o is the propagation vector of the wave. Since ω_i is between 5.2 and 5.6 mHz and ω_o is 5.2 mHz, $k_o \cdot v_o$ must be relatively small.

It appears that 3 cycles of the wave were required to generate sufficiently high energies to allow the UV imager to detect the auroral signature. The appearance of the intensified form first at low latitudes may be explained as follows. A compressional or rarefaction wave with a narrow frequency range moves inward from the magnetopause region launching Alfvén waves towards the ionosphere. Since the Alfvén transit time well inside the magnetopause is shorter than that near the magnetopause, the effect in the ionosphere may be to first see a low latitude intensification. The short lived aspect of this impulsive low latitude auroral form (i.e., only about 2 cycles of the driving frequency) may be due to the lack of the source region being able to support the implied current.

We propose that this auroral form is initiated by a surface wave with a perpendicular wave number k_y acting at the magnetopause driven by external conditions. The appearance of the low latitude form is a result of the wave coupling to lower L shells at some distance from the boundary. The initial appearance at lower L shells is linked to the faster Alfvén transit time there than at the boundary. When the external driver is turned off the auroral feature also disappears, first from the most equatorward point. The difference between this morphology and the fan arcs according to this scenario would be that the former is due to a surface wave driving pulsations at a constant frequency while the latter is due to a continuum of waves driving L shells at their own local resonant frequency.

THE DAY-NIGHT SPLIT AND AURORAL SPIRALS (BOTTOM LEFT PANEL OF FIGURE 1)

The split between the dayside and nightside auroral systems has been described by Vorobjev et al. (1976) and later by Meng and Lundin (1986). Table 1 shows that these events occur predominantly for IMF $B_y < 0$ conditions. Elphinstone et al. (1991a) found that this splitting, at least on one occasion, preceded an auroral substorm by just a few minutes and was associated with magnetic pulsations with a period of about 3-4 minutes. These in turn were linked to an enhancement of the AU index. It was also found that the splitting began from the dayside sector and propagated toward the nightside at speeds of 5-10 km/s. After reaching about 17 MLT the dayside system merged with the nightside one and substorm onset occurred about two minutes later. Elphinstone et al. (1993) have noted this auroral morphology may be an ideal way to

differentiate between dayside and nightside boundary layer mappings. If this were correct the dayside boundary layers would have a limited ionospheric local time extent (to about 17 MLT). It also appears that there is an interesting connection between these splittings and auroral spirals. Four separate examples of this morphology can be found in Elphinstone et al. (1993). As illustrated in Figure 1 (bottom left panel) the nightside system lies poleward of the dayside one and points to auroral spirals which are located on the dayside system. Spirals such as illustrated in the bottom left and middle panels of Figure 1, have frequently been attributed to a flow shear in the boundary layers. While this mechanism is a viable one to explain the source for a perturbation on an upward field aligned current sheet, it fails in a fundamental way to explain the auroral observations.

As described by Murphree et al. (1989) auroral spirals observed by Viking with scale sizes greater than about 200 km consistently have a counter clockwise sense when viewed in the direction of the magnetic field (i.e., downward when studied in the northern ionosphere). A velocity shear however generates a vortex which is "wound" in exactly the wrong sense when viewed in the ionosphere. This is illustrated in the left panel of Figure 5. One should note however that flow in the ionosphere associated with such spirals may be in the sense implied by this velocity shear and so the Kelvin Helmholtz instability theory may be a partial explanation for the source mechanism. Thus while the spiral is oriented in an anticlockwise sense when viewed in the direction of the magnetic field, the convective flow and auroral intensifications within the spiral may in fact move in a clockwise manner. Hallinan (1976) has given the simplest and most satisfactory explanation for the shape of these spiral forms. This is illustrated on the right side of Figure 5. In this scheme any mechanism which allows a perturbation on an upward field aligned current sheet is sufficient to generate the spiral shape. With this perturbation established it is a magnetic vorticity which causes field lines to "wrap" up in the correct sense when viewed in the ionosphere. A constraint on the mechanism for generating this disturbance is implied by the observation that dayside ionospheric spirals have a ratio of their diameter to the spacing between them (i.e., the wavelength) of about 1/5.

One prediction of the above simple model is that if a steady current is applied the spiral will not change, but if the current is increased the spiral will wind up, and if it is decreased, the spiral will unwind. An example of such a case is given in Plate 3 for an event on July 30, 1986. In this case a well developed spiral form near 13 MLT can be seen in the first images between 1408 and 1410 UT. The anticlockwise sense of the spiral can be seen in the upper left panel where the most poleward arc system is clearly on the east side of the intensification while the most equatorward arc system is to the west. Over the next five minutes the spiral form appears to unwind and disappear. The poleward edge of the spiral moves equatorward and the equatorward position fades and shifts poleward. When these images are seen in a motion picture format the sense of unwinding is clearly seen. The sense of motion

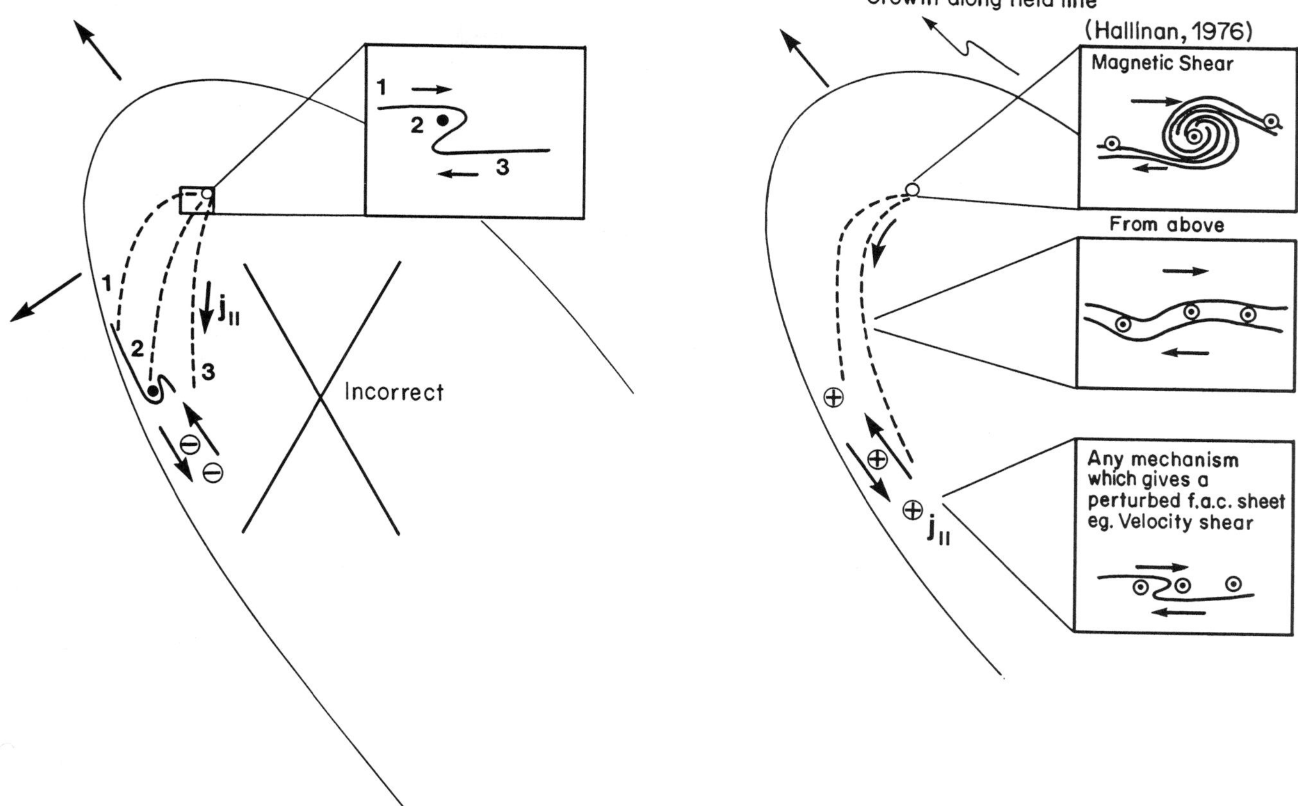

Fig. 5. The left side of the figure illustrates that a shear in velocity will create a vortex which if projected directly in the iono-sphere has exactly the wrong sense to the spirals observed by Viking. The right panel illustrates that even though the velocity shear (lower inset panel) gives the spiral the wrong sense, the twisting of fieldlines due to the associated upward current could give rise to a spiral wound up correctly. It is important to note that the view for all the inset panels is toward the direction of the magnetic field.

of auroral forms within the spiral tends to be clockwise (when viewed in the direction of the magnetic field) even though the spiral is wrapped in an anticlockwise sense. Researchers doing simulations of the Kelvin Helmholtz instability involving ionospheric coupling should, in the future, attempt to include this effect.

THE OMEGA-LIKE FORMS AND THE DOUBLE OVAL (BOTTOM RIGHT PANEL OF FIGURE 1)

During substorm recovery phase when IMF B_z is negative and IMF B_y is also negative the morning sector auroral distri-bution is filled with auroral forms. Two basic types of auroral forms occur in this morphology. The first occurs at high lati-tudes in the poleward edge of the "double oval". Multiple arc systems aligned more or less in the north-south direction are found spaced at regular intervals just equatorward of the most poleward arc system. The spacing between the arcs increases toward the nightside.

The other form occurs in the poleward portion of the most equatorward oval. These latter forms resemble omega bands, with the symbol Ω being repeated in the early morning hours. At later local times the morphology undergoes a smooth transi-tion to a set of arcs periodically spaced and aligned from northeast to southwest (see bottom right panel of Figure 1). The north-south extent of these forms decreases closer to noon. It is probable that these omega like forms have a source in the near Earth region since they are found in the poleward portion of the most equatorward oval (Elphinstone et al., 1993).

Plate 4 shows four examples of this auroral morphology. The top row shows the development of the forms over a fif-teen minute interval and illustrates how dynamic these features can be. In this example large scale north-south aligned forms can be seen developing in the most poleward portion of the double oval near 6 MLT. On the most equatorward oval mul-tiple auroral forms can be seen from about 3 MLT to 12 MLT. These forms tend to develop in an eastward direction. The bot-tom panels of Plate 4 illustrate three other examples of this active dayside morphology. In all the cases the double oval is

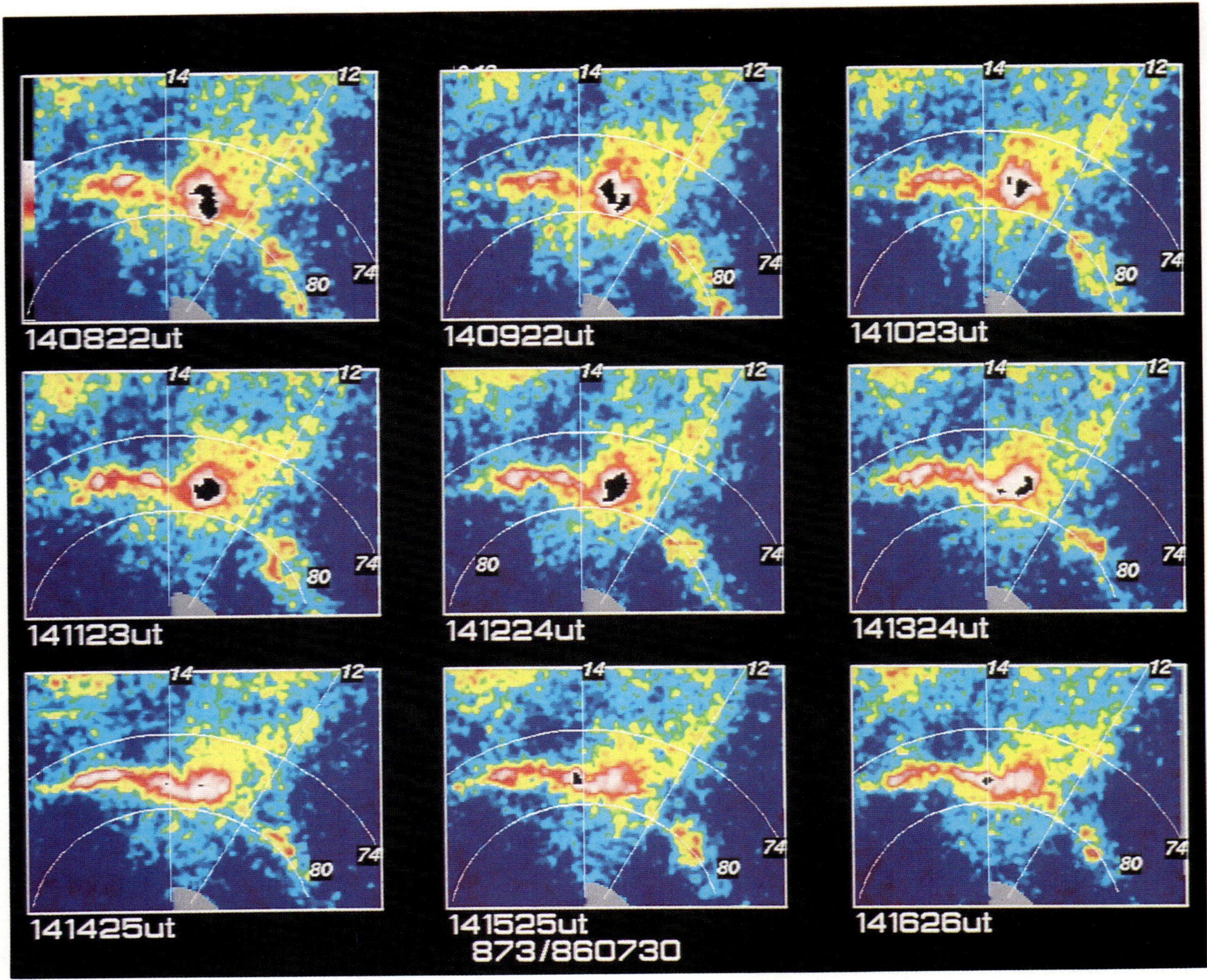

Plate 3. Close-up images of a dayside auroral spiral as it unwinds over an 8 minute period. Noon and 14 MLT (CGM 1980) are shown. Lower latitudes are towards the top of the images. The spiral form is "wrapped" up anti-clockwise when viewed in the direction of the magnetic field (down in the northern hemisphere).

present and, in the most equatorward oval, wave-like structures are evident with wavelengths on the order of 0.6 MLT. The most equatorward oval in the morning sector can probably be associated with with a region 2 field aligned current there, and so the wave instability can presumably be linked to a near Earth ring current process during substorm recovery. One should also note that these events seem to occur when IMF B_y and B_z are negative and that they occur at very low magnetic latitudes (eg. the example in the lower right panel of Plate 4 occurs at about 70 to 72 CGM 1980 Mlat).

SUMMARY

A variety of dayside auroral morphologies have been described (see Figure 1). Each of these morphologies probably has a distinct magnetospheric mechanism responsible for its generation. Some of the basic results of this paper are given below:

1) The candidate FTE events presented here occur during periods of IMF $B_y < 0$ in the postnoon sector. They do not appear to have a dependence on IMF B_z. These

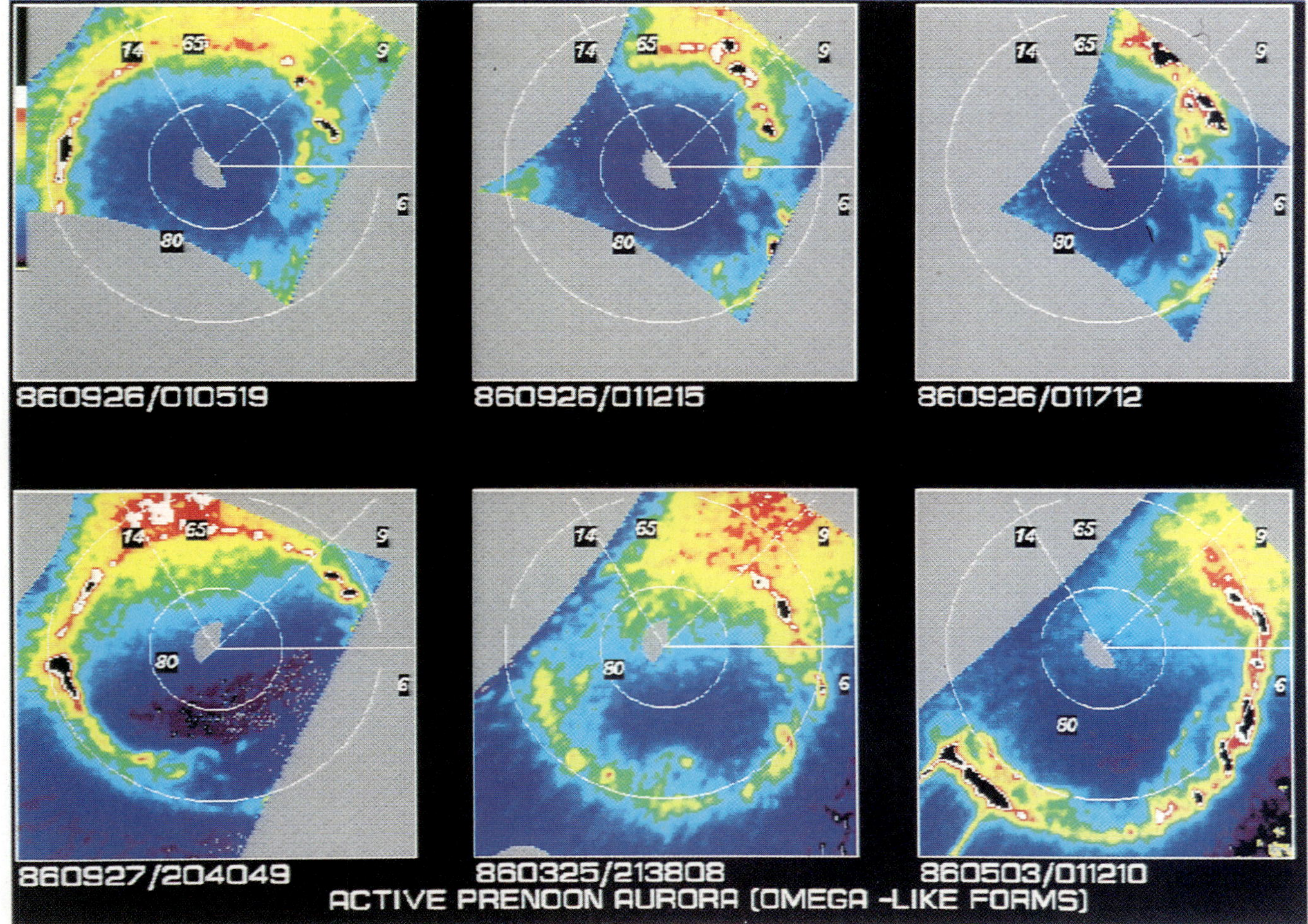

Plate 4. Four examples of disturbed morning sector aurora. The top row shows the temporal development of forms within the double oval on September 26, 1986 between 0105 UT and 0117 UT. The bottom row illustrates three other examples of such a morphology.

differ from previously reported events as seen from the ground which are usually closer to local noon. Three phases were identified from the study of one of these events. A magnetic impulse occurs first, then an equatorward motion and brightening of the UV auroras, and finally an auroral transient which moves poleward and antisunward into the dark region poleward of the main UV auroral distribution.

2) Changes to the region of weak noon sector aurora associated with an IMF transition from B_y positive to negative have been described. The correspondence with what one might expect for convection patterns is striking. It appears that asymmetries in the dayside aurora may help to understand temporal changes in the large scale convection. The transition from one configuration to the other took approximately 20 minutes.

3) High latitude dayside systems occur at the interface between polar rain and the plasma mantle (as defined by DMSP particle data). This observation strongly supports a mantle current system separate from the region 1 and 2 systems. It is also perhaps the best evidence that on occasion, UV auroral emissions as seen by Viking may occur on open field lines. (This auroral morphology should be strongly distinguished from high latitude UV polar arcs). These systems exist dominantly when IMF B_z is positive and IMF B_y is positive consistent with merging poleward of the cusp on the front surface of the magnetotail during this interval.

4) A new auroral morphology termed the "impulsive low latitude auroral form" has been described and associated with ground magnetic pulsations (Y component) with a period of about 3 minutes (corresponding to a

frequency of 5.2 mHz). The dynamics of this form may be closely linked to Alfvén transit times. Faster transit times at lower latitudes than at higher ones may cause the auroral form to appear first at the lower latitudes. An impulse in the solar wind density appears to be directly related to the onset of this auroral transient and variations in the IMF B_z component match the frequency of the ground magnetic pulsations. This may be an observation of a surface wave driving Alfvén waves inside the magnetosphere which manifest themselves in the ionosphere as this auroral morphology. This morphology occurs dominantly when IMF B_y is negative and does not appear to depend on the sign of B_z.

5) Magnetic pulsations precede structured UV auroral activations by a couple of cycle times. This supports the hypothesis of bouncing Alfvén waves with current initially carried by low energy electrons ($< .5$ keV) which are subsequently accelerated to higher energies.

6) The dynamics of "fan arcs" in the morning sector may be a result of a disturbance establishing a voltage source in the dayside boundary layers. These systems appear dominantly when the IMF B_y is negative and IMF B_z is positive. This is consistent with the high latitude morning sector being on closed field lines during these times. As the entire boundary layer region is activated a continuum of Alfvén waves modulate currents in the ionosphere at the natural resonant frequency of each field line. The result is a system of auroral arcs parallel to one another with the spacing between them increasing with latitude. As observed these arcs would propagate westward and poleward and should be associated with magnetic pulsations.

7) The split between the dayside and nightside auroral distribution between 13 and 17 MLT indicates that the dayside boundary layer may extend over a limited local time range and that the nightside system may penetrate to local times as early as 13 MLT. These systems appear distinct from one another when the IMF B_y component is negative during magnetically disturbed time intervals.

8) Spiral forms observed with spatial scales greater than about 100 km consistently have the most equatorward arc westward of the main spiral and the most poleward arc on the eastward side. The resulting anticlockwise sense of the spiral when viewed in the direction of the magnetic field is opposite to what we could expect from a KHI vortex generated in the Earth's equatorial plane. Movement of forms within the spiral however have the expected clockwise sense.

9) Omega-like bands in the late morning sector occur on the most equatorward portion of the double oval configuration indicating that they have a near Earth source. The wavelength associated with these forms is about 0.6 MLT. They occur preferentially during IMF $B_y < 0$.

Acknowledgements. The authors would like to thank Ann Marie Morris in helping with production of this paper and V. Petrov from IZMIRAN for supplying a valuable magnetometer data base. We would also like to thank one of the referees for pointing out the differences between the observations seen on the ground and those seen from space. The Viking UV imager was built as a project of the National Research Council of Canada and this work was supported under a grant by the Natural Science and Engineering Research Council of Canada.

REFERENCES

Akasofu, S.-I., Auroral arcs and auroral potential structure, *in Physics of Auroral Arc Formation*, Geophys. Mono. 26, edited by S.-I. Akasofu and J. R. Kan, American Geophysical Union, Washington, DC, p. 1, 1981.

Anger, C. D., S. K. Babey, A. L. Broadfoot, R. G. Brown, L. L. Cogger, R. Gattinger, J. W. Haslett, R. A. King, D. J. McEwen, J. S. Murphree, E. H. Richardson, B. R. Sandel, K. Smith and A. Vallance Jones, An ultraviolet auroral imager for the Viking spacecraft, *Geophys. Res. Lett., 14*, 387-390, 1987a.

Burch, J. L., Precipitation of low-energy electrons at high latitudes: Effects of interplanetary magnetic field and dipole tilt angle, *J. Geophys. Res., 77*, 6696, 1972.

Bythrow, P. F., T. A. Potemra, R. E. Erlandson, L. J. Zanetti, and D. M. Klumpar, Birkeland currents and charged particles in the high-latitude prenoon region: A new interpretation, *J. Geophys. Res., 93*, 9791-9803, 1988.

Cogger, L. L., J. S. Murphree, S. Ismail, and C. D. Anger, Characteristics of dayside 5577 Å and 3914 Å aurora, *Geophys. Res. Lett., 4*, 413-416, 1977.

Cowley, S. W. H., and W. S. Hughes, Observations of an IMF sector effect in the Y magnetic field or geostationary orbit, *Planet. Space Sci., 31*, 73-90, 1983.

Cowley, S. W. H., Magnetospheric asymmetries associated with the Y-component of the IMF, *Planet. Space Sci., 29*, 79-96, 1981.

Craven, J. D., J. S. Murphree, L. A. Frank, and L. L. Cogger, Simultaneous optical observations of transpolar arcs in the two polar caps, *Geophys. Res. Lett., 18*, 2297-2300, 1991.

Dandekar, B. S., and C. P. Pike, The midday discrete auroral gap, *J. Geophys. Res., 83*, 4227-4236, 1978.

Eather, R. H., Dayside auroral dynamics, *J. Geophys. Res., 89*, 1695-1700, 1984.

Eliasson, L., R. Lundin, and J. S. Murphree, Polar cap arcs observed by the Viking spacecraft, *Geophys. Res. Lett., 14*, 451-454, 1987.

Elphinstone, R. D., K. Jankowska, J. S. Murphree, and L. L. Cogger, The configuration of the northern auroral distribution for interplanetary magnetic field B_z northward, 1, IMF B_x and B_y dependencies as observed by the Viking satellite, *J. Geophys. Res., 95*, 5791-5804, 1990.

Elphinstone, R. D., J. S. Murphree, L. L. Cogger, D. Hearn, M. G. Henderson and R. Lundin, Observations of the auroral distribution prior to substorm onset, in: *Magnetospheric Substorms*, Geophys. Mono., American Geophysical Union, Washington, D.C., 257, 1991a.

Elphinstone, R. D., D. Hearn, J. S. Murphree, and L. L. Cogger, Mapping using the Tsyganenko long magnetospheric model and its relationship to Viking auroral images, *J. Geophys. Res., 96*, 1467-1480, 1991b.

Elphinstone, R. D., J. S. Murphree, D. J. Hearn, L. L. Cogger, P. T. Newell, and H. Vo, Observations of the UV dayside aurora and their relationship to DMSP particle boundary definitions, *Ann. Geophys., 19*, 815-826, 1992.

Elphinstone, R. D. and D. J. Hearn, The auroral distribution and its relation to magnetospheric processes, *Adv. Space Res., 13*, 17-27, 1993.

Elphinstone, R. D., J. S. Murphree, D. J. Hearn, W. Heikkila, M. G. Henderson, and L. L. Cogger, The auroral distribution and its mapping according to substorm phase, in press *J. Atmos. Terr. Phys.*, 1993.

Fairfield, D. H., On the average configuration of the geomagnetic tail, *J. Geophys. Res., 84*, 1950-1958, 1979.

Feldstein, Ya. I., and Yu. I. Galperin, The auroral luminosity of the high-latitude upper atmosphere: Its dynamics and relationship to the large-scale structure of the Earth's magnetosphere, *Rev. Geophys., 23*, 217-275, 1985.

Greenwald, R. A., K. B. Baker, J. M. Ruohoniemi, J. R. Dudeney, M. Pinnock, N. Mattin, J. M. Leonard, and R. P. Lepping, Simultaneous conjugate observations of dynamic variations in high latitude dayside convection due to changes in IMF B_y, *J. Geophys. Res., 95*, 8057-8071, 1991.

Heikkila, W. J., and J. D. Winningham, Penetration of magnetosheath plasma to low altitudes through the dayside magnetospheric cusps, *J. Geophys. Res., 76*, 883-891, 1971.

Heikkila, W. J., Definition of the cusp in:*The polar cusp*, D. Reidel Publishers, Hingham, Mass., 387-395, 1985.

Jacobsen, B., P. E. Sandholt, B. Lybekk, and A. Egeland, Transient auroral events near midday: Relationship with solar wind/magnetosheath plasma and magnetic field conditions, *J. Geophys. Res., 96*, 1327-1336, 1991.

Jankowska, K., R. D. Elphinstone, J. S. Murphree, L. L. Cogger, and D. Hearn, The configuration of the auroral distribution for interplanetary magnetic field B_z northward, 2, ionospheric convection consistent with Viking observations, *J. Geophys. Res., 95*, 5805-5816, 1990.

Jones, A. V., R. L. Gattinger, F. Creutzberg, R. A. King, P. Prikryl, L. L. Cogger, D. J. McEwen, F. R. Harris, C. D. Anger, J. S. Murphree and R. A. Heelis, A comparison of Canopus ground optical data with images from the Viking UV camera, *Geophys. Res. Lett., 14*, 391-394, 1987.

Kremser, G., and R. Lundin, Average spatial distributions of energetic particles in the mid-altitude cusp/cleft region observed by Viking, *J. Geophys. Res., 95*, 5733-5766, 1990.

Lockwood, M., Radar observations of the plasma flow signatures of the cusp in: *Geospace Environmental Modelling*, GEM WSR-3, National Science Foundation, Washington, D.C., 23-25, 1990.

Meng, C.-I., and R. Lundin, Auroral Morphology of the Midday Oval, *J. Geophys. Res., 91*, 1572-1584, 1986.

Murphree, J. S., and R. D. Elphinstone, Correlative studies using the Viking imagery, *Adv. Space Res., 8*, 9-19, 1988.

Murphree, J. S., R. D. Elphinstone, L. L. Cogger, and D. D. Wallis, Short-term dynamics of the high latitude auroral distribution, *J. Geophys. Res., 94*, 6969-6974, 1989.

Murphree, J. S., R. D. Elphinstone, D. Hearn and L. L. Cogger, Large-scale high-latitude dayside auroral emissions, *J. Geophys. Res., 95*, 2345-2354, 1990.

Murphree, J. S., and R. D. Elphinstone, *Geospace Environment Modelling: Workshop report on Intercalibrating cusp signatures*, GEM WSR-3, National Science Foundation, Washington, D.C., 14-17, 1990.

Newell, P. T., and C.-I. Meng, On quantifying the distinctions between the cusp and the cleft/LLBL, in *Electromagnetic coupling in the polar clefts and caps*, Kluwer Academic Publishers, Dordrecht, 87-101, 1989.

Newell, P. T., and C.-I. Meng, Intense keV energy polar rain, *J. Geophys. Res., 95*, 7869-7879, 1990.

Newell, P. T., S. Wing, C.-I. Meng, and V. Sigillito, A neural-network-based system for monitoring the aurora, *John Hopkins APL Technical Digest, 11*, 291-299, 1990.

Newell, P. T., W. J. Burke, C.-I. Meng, E. R. Sanchez, and M. E. Greenspan, Identification and observations of the plasma mantle at low altitude, *J. Geophys. Res., 96*, 35-45, 1991a.

Newell, P. T., S. Wing, C.-I. Meng, and V. Sigillito, The auroral oval position, structures, and intensity of precipitation from 1984 onward: An automated online data base, *J. Geophys. Res., 96*, 5877-5882, 1991b.

Reiff, P. H., J. L. Burch, and R. A. Heelis, Dayside auroral arcs and convection, *Geophys. Res. Lett., 5*, 391-394, 1978.

Reiff, P. H., Some comments on all three questions, in: *Geospace Environmental Modelling*, GEM WSR-3, National Science Foundation, Washington, D.C., p. 98-99, 1990.

Rosenbauer, H., G. Grunwaldt, M. D. Montgomery, G. Paschmann, and N. Sckopke, Heos 2 Plasma observations in the distant polar magnetosphere: The plasma mantle, *J. Geophys. Res., 80*, 2723-2737, 1975.

Sanchez, E. R., C.-I. Meng, and P. T. Newell, Observations of solar wind penetration into the Earth's magnetosphere: The plasma mantle, *Johns Hopkins APL Technical Digest, 11*, 272-278, 1990.

Sandahl, I., and P. Lindqvist, Electron populations above the nightside auroral oval during magnetic quiet times, *Planet. Space Sci., 38*, 1031-1049, 1990.

Sandford, B. P., Aurora and airglow intensity variations with time and magnetic activity at southern high latitudes, *J. Atmos. Terr. Phys., 26*, 749-769, 1964.

Sandholt, P. E., A. Egeland, C. S. Deehr, V. G. Vorobjev, G. V. Starkov, V. L. Zverev and Ya.I. Feldstein, Dayside auroral luminosity in relation to the IMF and magnetospheric substorm activity, in: *Electrodynamical Processes in High Latitudes*, Proc. of International Symposium Polar Geomagnetic Phenomena, International Association of Geomagnetism and Aeronomy, 116-145, 1988.

Sandholt, P. E., M. Lockwood, B. Lybekk and A. D. Farmer, Auroral bright spot sequence near 1400 MLT: Coordinated optical and ion drift observations, *J. Geophys. Res., 95*, 21,095-21,109, 1990a.

Sandholt, P. E., M. Lockwood, T. Oguti, S. W. H. Cowley, K. S. C. Freeman, B. Lybekk, A. Egeland, and D. M. Willis, Midday auroral breakup events and related energy and momentum transfer from the magnetosheath, *J. Geophys. Res., 95*, 1039-1060, 1990b.

Sandholt, P. E., J. Moen, A. Rudland, D. Opsvik, W. F. Denig, and T. Hansen, Auroral event sequences at the dayside polar cap boundary for positive and negative IMF B_y, *J. Geophys. Res.*, in press, 1992a.

Sandholt, P. E., M. Lockwood, W. E. Denig, R. C. Elphic, and S. Leontjev, Dynamical auroral structure in the vicinity of the polar cusp: multipoint observations during southward and northward IMF, *Ann. Geophys., 10*, 483-497, 1992b.

Shepherd, G. G., Dayside cleft aurora and its ionospheric effects, *Rev. Geophys., 17*, 2017-2033, 1979.

Tsyganenko, N. A., Global quantitative models of the geomagnetic field in the cislunar magnetosphere for different disturbance levels, *Planet. Space Sci., 35*, 1347-1358, 1987.

Vorobjev, V. G., G. V. Starkov, and Y.-I. Feldstein, The auroral oval during the substorm development, *Planet. Space Sci., 24*, 955, 1976.

Vo, H., Ph.D. Thesis, Implications of Viking dayside auroral measurements, *Univ. of Calgary*, 1992.

Wolfe, A., and R. L. Kaufmann, MHD wave transmission and production near the magnetopause, *J. Geophys. Res., 80*, 1764, 1975.

R.D. Elphinstone, D.J. Hearn, J.S. Murphree, L.L. Cogger, M.L. Johnson and H.B. Vo, Department of Physics and Astronomy, University of Calgary, 2500 University Drive N.W., Calgary, Alberta, Canada T2N 1N4.

Auroral Expansion Into the Dayside Polar Cap: Ground and Satellite Observations in the Prenoon Sector

P. E. Sandholt, J. Moen, and D. Opsvik

Department of Physics, University of Oslo, Norway

W. F. Denig and W.J. Burke

Geophysics Directorate, Phillips Laboratory, Hanscom AFB, Massachusetts

A specific category of auroral events taking place at the dayside polar cap boundary during southward-directed interplanetary magnetic field (IMF) is exemplified by one case of combined ground and satellite observations in the prenoon sector. The detailed temporal and spatial evolution of these optical events were monitored by all-sky TV imagery mapped onto a geographic coordinate system. Latitude profiles (along the satellite track) of particle precipitation and ionospheric convection in the vicinity of the optical events were important for the determination of the solar wind - magnetosphere coupling mode involved. The observed north-westward event motion in the prenoon sector is typical for periods of positive $IMFB_Y$. The optical events appear as a brightening of the 630.0 nm aurora at the latitude of cleft/LLBL precipitation at the eastern boundary of the field-of-view of the TV camera, near 09 MLT. A continuous motion into the polar cap, the region of mantle precipitation, was observed before the optical emission faded out after 10-15 minutes, ~ 500 km poleward of the continuous cleft arc. The auroral forms typically cover ~ 500 km in the east - west direction and ~ 50-200 km in the north - south direction. The optical, particle and ion drift observations in combination with the available IMF and solar wind plasma data for this category of auroral events indicate that they represent dynamical structures of Dungey cell convection over the polar magnetosphere, possibly initiated by pulses of enhanced merging rate ($B_n \neq 0$) at the dayside magnetopause.

INTRODUCTION

The question of signatures in the dayside aurora related to solar wind - magnetosphere coupling modes has been in the focus in recent years. Auroral precipitation signatures of dayside magnetosphere boundary layers have been studied extensively by using data from polar orbiting satellites (Newell and Meng, 1992). Newell and Meng distinguish between the following precipitation regions: polar rain, mantle, cusp, cleft/LLBL, traditional CPS and traditional BPS. However, the mapping of the ionospheric regions to the magnetosphere is still controversial.

Auroral Plasma Dynamics
Geophysical Monograph 80

Information on the spatial and temporal evolutions of transient boundary layer events can be obtained by continuous observations of the ionospheric signatures from the ground. The combination of ground-based imaging techniques and "snapshot" latitude profiles of e.g. particle precipitation measured from polar satellites is an interesting approach to the problem of time-dependent boundary layer processes and their coupling to the ionosphere, which will be applied in this study.

Time-varying magnetopause merging/reconnection, Kelvin-Helmholtz instabilities excited by boundary layer flow shears, and upstream plasma irregularities (e.g. dynamic pressure pulses) are among the phenomena which are expected to give rise to auroral footprints in the cusp/cleft ionosphere.

According to recent numerical simulation studies (Wei and Lee, 1992) counterclockwise plasma vorticity in the post-noon sector low-latitude boundary layer can develop as a result of Kelvin - Helmholtz instabilities at the magnetopause. The associated coupling to the ionosphere, involving field-aligned currents and parallel electric fields, may give rise to a series of bright auroral spots in the cleft ionosphere, predominantly in the post-noon sector since the current is directed out of the ionosphere there. Observational evidence of such auroral signatures has been presented by e.g. Lui et al. (1989).

Friis-Christensen et al. (1988) identified the magnetic signatures of travelling double-vortex systems whose speeds and directions of interior motion do not meet the requirements expected for merging events. Kivelson and Southwood (1991) showed that such twin vortices can however, be explained as effects of pressure pulses on the magnetopause that launch fast-mode MHD waves into the magnetosphere. In the presence of inhomogeneities in the magnetospheric plasma, fast waves couple to shear Alfvén waves. These, in turn, carry bipolar currents into/out of the ionosphere. Electric fields associated with the Alfvén waves produce the observed ionospheric plasma convection.

Transient events of discrete auroral forms are observed around the time of sharp changes in the IMF orientation and/or the solar wind dynamic pressure (Heikkila et al., 1989; Jacobsen et al., 1991). A study of the detailed relationship between the phenomenon of short-lived twin vortical flow patterns in the cleft ionosphere (cf. Friis-Christensen et al. (1988)) and the associated optical auroral activity is underway (H. Lühr, private communication, 1992). The preliminary analysis of a Jan. 6, 1992 event shows a patch of enhanced 630.0 nm emission moving in the same direction as the magnetic signature. Discrete auroral structures were observed at the trailing edge of this prenoon sector event, in the

region of inferred upward directed field-aligned current (convergent electric field).

The search for ionospheric signatures of time-dependent magnetopause merging and the associated flux transfer events (FTEs) have focused on a specific category of quasi-periodic optical events occurring at cusp/cleft and mantle latitudes within ~09 - 15 MLT during intervals of southward-directed IMF (Sandholt et al., 1989, 1990, 1992). In this study we present one example of combined ground and satellite observations of this category auroral events. The purpose is to document the characteristic event motion in relation to the particle precipitation environment and ionospheric ion flow (zonal component) obtained from polar satellites. The satellite pass (DMSP F9) of the present case occurred close to the magnetic meridian through the ground site at Ny Ålesund, Svalbard, near the leading edge of a westward-moving optical event in the prenoon sector. This optical event is an individual distinct case within a series of similar auroral activity. The motion pattern in the prenoon sector is typical for positive IMF B_Y conditions. The auroral form faded out near the 08 MLT meridian, within the field of view of the all-sky TV camera in Ny Ålesund, allowing the determination of its spatial dimension.

OBSERVATIONS

Figure 1 shows a quick-look display of the photometer scans for the period 0501 - 0616 UT, Jan. 12, 1992. The wavelength is 630.0 nm and zenith (Ny Ålesund) corresponds to 75,4° MLAT. Black dots mark the zenith angle location of maximum intensity during intervals of little auroral activity whereas intensity maxima of dynamical structures are marked by yellow colour.
Four major events of poleward expanding auroral events are seen in this interval, each event lasting typically 10 min. An exceptionally long-lived event is observed within 0520 - 0535 UT. In this paper we focus on the 0510 - 0520 UT event, since a close DMSP F9 pass occurred along the Ny Ålesund meridian, covering the latitude range 71.8° - 78.1° MLAT within 0510 -0512 UT.

Information on polar cap convection obtained from dawn-dusk passes of satellite DMSP F8 within the intervals 0455 - 0515 UT (southern hemisphere) and 0550 - 0610 UT (northern hemisphere) indicates that IMF B_Z was negative and B_Y positive during the time of the auroral activity displayed in figure 1.

The all-sky TV image sequence in figure 2 shows the evolution of auroral activity within 0510 -0520 UT. Each picture represents a 2.5 sec. integration of the 630.0 nm video-recording in Ny Ålesund. These images are mapped onto a geographic coordinate system at 300 km altitude. It is noted that this projection technique can be

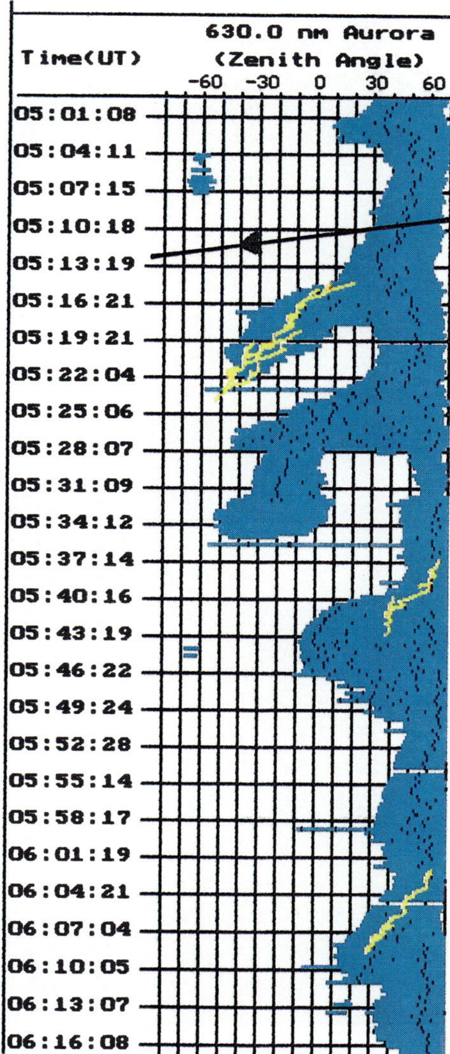

Fig. 1. Ny Ålesund meridian scanning photometer observations for the period 0501-0616 UT, Jan. 12, 1992. The location in zenith angle of the 630.0 nm auroral luminosity is marked by blue color. North is to the left. Intensity maxima are shown by black dots and transient auroral forms are indicated by yellow color. Satellite pass along the Ny Ålesund meridian (0510-0512 UT) is marked by a straight line in the plot.

somewhat misleading in case of long auroral rays, which are frequently observed in such cusp/cleft auroral events (e.g. Sandholt et al., 1989, 1990, Lockwood et al., 1980 b). This effect (projection of rays) is seen in the 0512 and 0514 UT images in figure 2. Being aware of this problem, the technique is adequate for the purpose of determining the general motion pattern of the transient auroral forms considered in this paper.

The satellite pass has been mapped along the geomagnetic field from 800 to 300 km (typical auroral altitude) and marked in the 0510, 0512 and 0514 UT images in figure 2. The location of the satellite at these three times is indicated by red circular dots in the figure. The locations of north-Norway, Svalbard (island West-Spitsbergen) and Greenland are shown. The F9 satellite track was close to the magnetic meridian through Ny Ålesund (the ground site) in this case.

The 0510 UT picture shows two bright auroral "spots", one on each side of the satellite trajectory. At 0512 UT the eastern event has become brighter and it has expanded westward. The leading edge is very close to the satellite track/the Ny Ålesund meridian. The following images show the northwestward motion of this event. At 0518 UT it is located at $\sim$76 - 77° MLAT whereas the continuous cleft emission is located at 72 - 73° MLAT. The last picture (0520 UT) marks the fading of this event, well within the field of view of the TV camera in Ny Ålesund, as well as the brightening of the next event (cf. figure 1) to the east of Svalbard. The new event initially follows a very similar evolution as the previous one (cf. figure 1).

Since the whole auroral form is within the camera field of view during the interval 0516 - 0520 UT it is possible to estimate its spatial extent. In the bright phase around 0516 UT it covers $\sim$400 km in the east-west direction and $\sim$200 km north-south. In the fading phase the north-south dimension seems to increase, according to the 0520 UT image.

The main features of the satellite data (figure 3) are the following: A channel of strong eastward (sunward) flow is detected within 72.5-73.5° MLAT with enhanced BPS electron precipitation in this region. This observation fits with the bright auroral structure seen at this latitude in the 0512 UT image. This is the eastern edge of an auroral arc extending outside the field of view in the westward (tailward) direction (figure 2). Enhanced westward ion drift ($\sim$1.5-3kms^{-1}) is detected within 74.5-76.5° MLAT, which is the latitude range of the northwestward moving auroral event (cf. the 0514-0516 UT images), within the region of mantle precipitation. An interesting triple reversal of the east-west flow component is observed within $\sim$73.5-74.5°MLAT. This phenomenon requires further consideration, which will be reported elsewhere.

SUMMARY AND DISCUSSIONS

Figure 4 summarizes essential features of the ground and satellite observations. The observed auroral events occurred during positive IMF B$_Y$ and negative IMF B$_Z$,

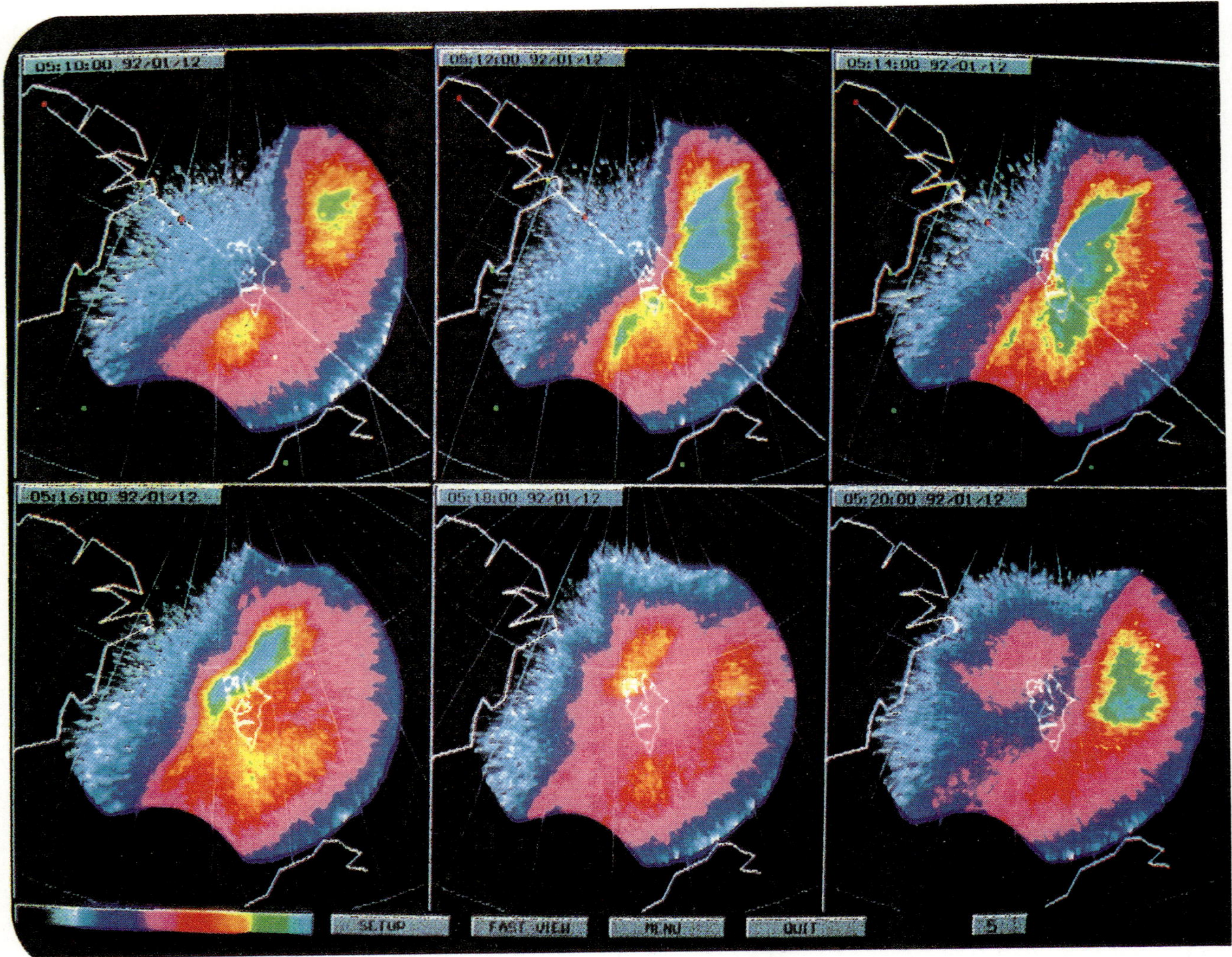

Fig. 2. 630.0 nm auroral image sequence for the period 0512-0520 UT, Jan. 12, 1992. All-sky pictures (2,5 sec. integration time) have been mapped onto a geographical coordinate system at 300 km altitude. The satellite track (projected down to 300 km altitude) within 0510-0514 UT is marked in the upper panel images.

as documented indirectly by ion drift data from polar orbiting satellites.

The preliminary evidence suggests that similar auroral events are moving eastward (duskward) around noon and in the postnoon sector during negative IMF B_Y (Sandholt et al., 1992). Prenoon events have been selected in this study because the complementary information from polar satellites is limited to the ~09 MLT region. This information allows the study of the optical events in relation to the particle precipitation environment and the ionospheric ion drift pattern.

The northwestward motion of the auroral events is illustrated by an arrow in figure 4. The different zones of particle precipitation are also indicated. Bright 630 nm auroral forms in the southwestern part of the field of view correspond to BPS particles and eastward plasma flow (cf. also Newell et al., 1991), whereas the main optical events occur further north, associated with westward convection (zonal component) and LLBL or mantle precipitation. The optical events expand northwestward, possibly originating within the zone of cleft/LLBL precipitation. The most severe limitation of

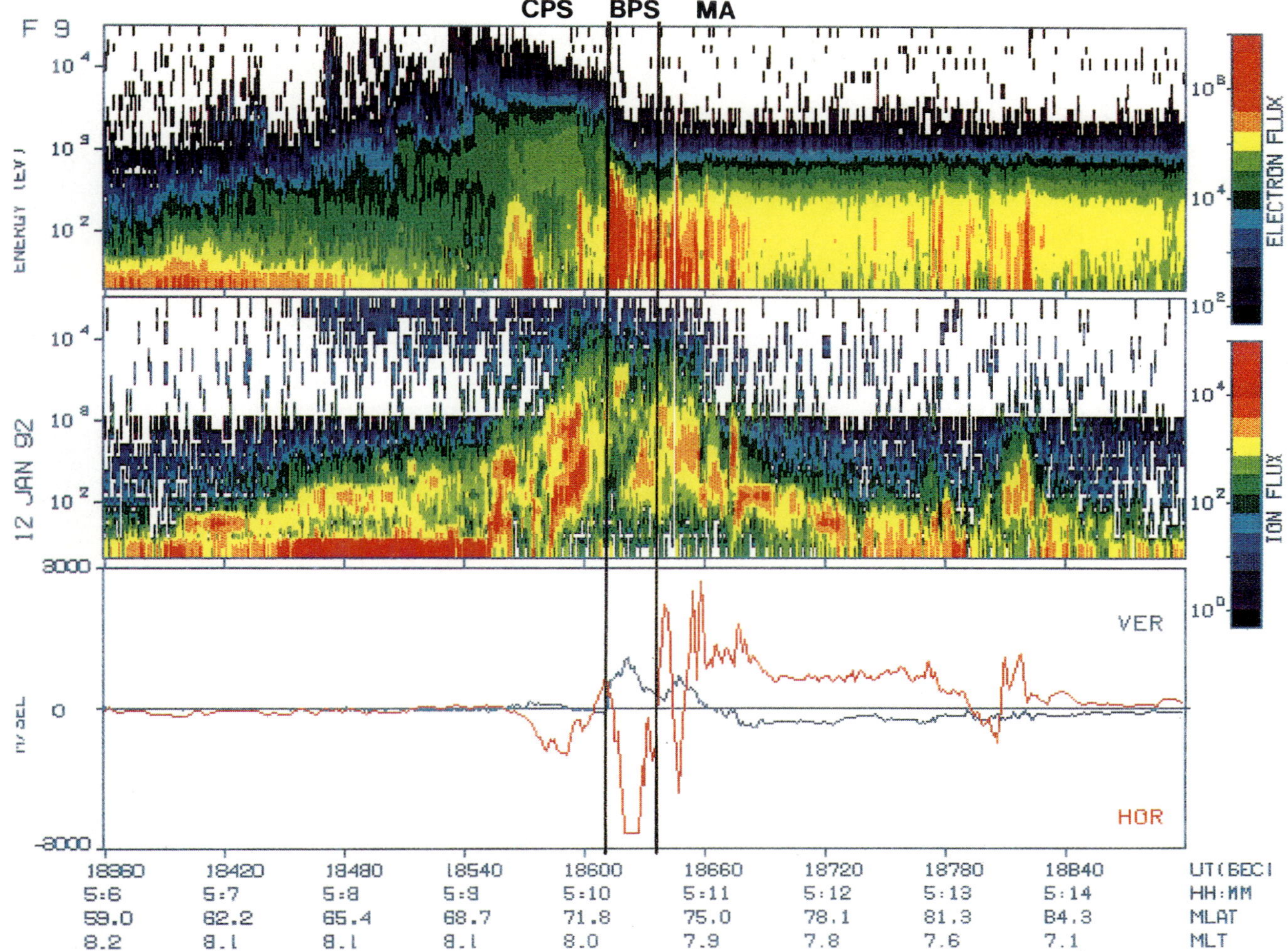

Fig. 3. Panels 1 and 2 show electron and ion precipitation spectra along the ~08 MLT meridian for the period 0506-0514 UT, Jan. 12, 1992. Horizontal (east-west) and vertical (blue) ion drifts are plotted in the lower panel.

this kind of study is due to the field of view of the TV camera, which is ~1000 km in diameter for F-layer auroral emissions.

The main characteristics of the auroral phenomenon in focus here may be summarized as follows:

i)   a periodic sequence of moving auroral forms (often appearing as rayed sheets)at the polar cap boundary, within ~08-15MLT

ii)   the recurrence time is typically within 5-15 min.

iii)   occurrence during southward-directed IMF (the continuous cusp/cleft auroral emission is located south of 75° MLAT)

iv)   spatial scale of red line aurora: ~50-200 km (N-S) times 300-1000km (E-W)

v)   westward (dawnward) or eastward (duskward) motion during positive and negative $IMFB_Y$, respectively (Sandholt et al., 1992)

vi)   optically fade out after 10-15 min., within the region of mantle precipitation, ~500 km poleward of the continuous cleft emission.

vii)   event occurrence is not correlated with major solar wind dynamic pressure changes (Sandholt et al., 1992)

The motion pattern of the auroral forms is illustrated by the image sequence in figure 2. They appear within

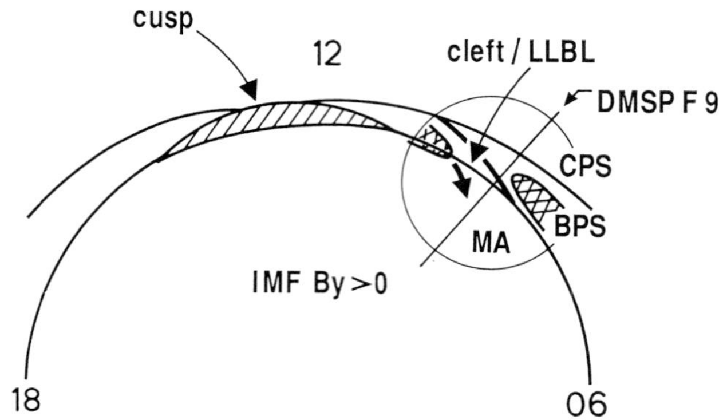

Fig. 4. Schematical illustration of particle precipitation regions and typical motion path of optical events (arrow) in the prenoon sector for positive IMFB$_Y$ condition. The satellite track and the field of view of the TV camera in Ny Ålesund are also indicated.

the field of view as a brightening of the continuous cleft auroral arc and subsequently move into the polar cap (mantle precipitation) where they fade out (cf. also the scanning photometer data in figure 1). An enhanced ion flow speed (zonal component) is observed at the latitude of the optical events. In the case presented here a plateau of constant, but smaller, westward ion flow velocity is observed poleward of the optical events, within ~78-81° MLAT.

These observations are consistent with essential features of the predicted ionospheric signatures of sequences of transient and patchy magnetopause merging (Southwood, 1985; Lockwood et al., 1990; Cowley et al. 1991).

Due to the present limited understanding of ionosphere - magnetosphere field line mapping there is some uncertainty on whether the auroral events in the fading phase, associated with "mantle" precipitation poleward of the continuous cleft arc, are coupled to the high-latitude boundary layer or the low-latitude boundary layer. This is a critical point for the identification of the solar wind - magnetosphere coupling mode involved, i.e. merging or viscous process. One of the most specific properties of the merging process is the associated zonal ionospheric ion drift related to the IMF B$_Y$ polarity (Heppner and Maynard, 1987; Cowley et al., 1991 a, b). The presently available data indicate that a similar motion pattern exists for the optical events. Thus, better statistics on the east-west motion of the auroral events in relation to IMF B$_Y$ will be a crucial test of the present interpretation in terms of magnetic merging/reconnection.

The field-aligned current configuration associated with auroral events similar to those presented here may be inferred from the satellite ion drift observations, as demonstrated by Denig et al. (1992).

Acknowledgments. The optical observation campaign in Ny Ålesund was supported by The Norwegian Polar Research Institute and by The Norwegian Research Council for Science and the Humanities. This study has been supported by NATO grant RG 85/05 21 and grant AFOSR - 90 - 0082.

REFERENCES

Cowley, S.W.H., M.P. Freeman, M. Lockwood, and M.F. Smith, The ionospheric signature of flux transfer events, in "Cluster: Dayside Polar Cusp", ESA SP - 330, pp. 105-112, Nordwijk, The Netherlands, 1991 a.

Cowley, S.W.H., J.P. Morelli and M. Lockwood, Dependence of convective flows and particle precipitation in the high latitude dayside ionosphere on the X and Y components of the interplanetary magnetic field, J. Geophys. Res., 96, 5557-5564, 1991, b.

Denig, W.F., W.J. Burke, N.C. Maynard, F.J. Rich, B. Jacobsen, P.E. Sandholt, A. Egeland, S. Leontjev and V.G. Vorobjev, Ionospheric signatures of dayside magnetopause transient: A case study using satellite and ground measurements, J. Geophys. Res, in press (1992).

Friis-Christensen, E., M.A. Mc Henry, C.R. Clauer, and S. Vennerström, Ionospheric travelling convection vortices near the polar cleft: A triggered response to sudden changes in the solar wind, Geophys. Res. Lett., 15, 253-256, 1988.

Heikkila, W.J., T.S. Jörgensen, L.J. Lanzerotti and C.G. Maclennan, A transient auroral event on the dayside, J. Geophys. Res., 94, 15291-15305, 1989.

Heppner J.P., and N.C. Maynard, Empirical high-latitude electric field models, J. Geophys. Res., 92, 4467-4489, 1987.

Jacobsen, B., P.E. Sandholt, B. Lybekk and A. Egeland, Transient auroral events near midday: Relationships with solar wind/magnetosheath plasma and magnetic field conditions, J. Geophys. Res., 96, 1327-1336, 1991.

Kivelson, M.G., and D.J. Southwood, Ionospheric travelling vortex generation by solar wind buffeting of the magnetosphere, J. Geophys. Res., 96, 1661-1667, 1991.

Lockwood, M., S.W.H. Cowley and P.E. Sandholt, Transient reconnection: search for ionospheric signatures, EOS, Trans. Am. Geophys. Union, 71 (20), 709-720, 1990 a.

Lockwood, M., P.E. Sandholt, A.D. Farmer, S.W.H. Cowley, B. Lybekk and V.N. Davda, Auroral and plasma flow transients at magnetic noon, Planet. Space Sci., 38, 973-993, 1990 b.

Lui, A.T.Y.,D. Venkatesan, and J.S. Murphree, Auroral bright spots on the dayside J. Geophys. Res., 94, 5515-5522, 1989.

Mc Henry, M.A., and C.R. Clauer, Modeled ground magnetic signatures of flux transfer events, J. Geophys. Res., 92, 11231-11240, 1987.

Newell, P.T.,W.J. Burke, E.R. Sanches, C.-I. Meng, M.E. Greenspan and C.R. Clauer, The low-latitude boundary layer and the boundary plasma sheet at low altitude: Prenoon precipitation regions and convection reversal boundaries, J. Geophys. Res., 96, 21013-21023, 1991.

Newell, P.T., S. Wing, C.-I. Meng and V. Sagillito, The auroral oval position, structure, and intensity of precipitation from 1984 onward: an automated on-line data base, J. Geophys. Res., 96, 5877-5882, 1991.

Newell, P.T. and C.-I. Meng, Mapping the dayside ionosphere to the magnetosphere according to particle precipitation characteristics, Geophys. Res. Lett. 19, 609-612, 1992.

Sandholt, P.E.,B. Lybekk, A. Egeland, R. Nakamura and T. Oguti, Midday auroral breakups, J. Geomagn. Geoelectr., 41, 371-387, 1989.

Sandholt, P.E.,M. Lockwood, T. Oguti, S.W.H. Cowley, K.S.C. Freeman, B. Lybekk, A. Egeland and D.M. Willis, Midday auroral breakup events and related energy and momentum transfer from the magnetoheath, J. Geophys. Res., 95, 1039-1060, 1990.

Sandholt, P.E., J. Moen, W.F. Denig, A. Rudland, D. Opsvik and T. Hansen, Auroral event sequences at the dayside polar cap boundary, for positive and negative IMF B_Y J. Geophys. Res., in press (1992).

Southwood, D.J., Theoretical aspects of ionosphere - magnetosphere - solar wind coupling, Adv. Space Res., 5, 7-14, 1985.

Wei, C.Q. and L.C. Lee, Coupling of magnetopause boundary layer to the polar ionosphere, J. Geophys. Res., in press (1992).

J. Moen, D. Opsvik and P.E. Sandholt, Department of Physics, University of Oslo, P.O. Box 1048, Blindern, 0316 Oslo, Norway

W.J. Burke and W.F. Denig, Geophysics Directorate, Phillips Laboratory, Hanscom AFB, MA 01731, USA.

High-Latitude Electrodynamics and Aurorae
During Northward IMF

L. G. BLOMBERG AND G. T. MARKLUND

Department of Plasma Physics, Alfvén Laboratory, Royal Institute of Technology, Stockholm

The large-scale auroral morphology and its associated electrodynamics for northward interplanetary magnetic field (IMF) is characteristically different from that for southward IMF. For northward IMF significant auroral activity is often present poleward of the "normal" auroral oval, and a number of different polar auroral configurations can occur. Two extreme situations which are considered here are either one where one or more discrete arcs separated from the "normal" oval are present in the polar region, or one where diffuse auroral activity is found in a large fraction of this region. The latter case might be associated with so-called NBZ (field-aligned) currents. We briefly review the typical signatures in terms of optical emissions, field-aligned currents, electric fields, and plasma convection usually encountered in the polar region when IMF $B_z > 0$, and their interrelationships. This is the starting point for a more detailed discussion, driven by modeling results, of the relationships between electric field, field-aligned current, and conductivity in the two distinct situations mentioned. In particular we discuss the influence the polar field-aligned currents have on the convection pattern, on the small as well as on the large scale. As expected, currents associated with discrete isolated arcs give rise mainly to small-scale modifications, whereas currents related to auroral activity in an expanded oval can modify the overall convection pattern substantially.

1. INTRODUCTION

The global distribution of the aurora shows a qualitative as well as a quantitative dependence on the interplanetary magnetic field (IMF). For southward IMF the intense auroral activity is usually confined to the region known as the auroral oval while for northward IMF considerable auroral activity is often encountered poleward of this region [e.g., Davis, 1963; Berkey et al., 1976; Lassen and Danielsen, 1978]. The existence of auroral arcs at very high latitudes has been known for a long time [cf. e.g., Davis, 1960; Denholm, 1961; Denholm and Bond, 1961; Eather and Akasofu, 1969]. These arcs, variably referred to as "Sun-aligned arcs," "polar cap arcs," "high-latitude arcs," and "polar arcs," have been observed for many years using all-sky cameras [e.g., Lassen and Danielsen, 1978], and photometers on board polar-orbiting spacecraft, such as ISIS and DMSP [e.g., Gussenhoven, 1982; Ismail and Meng, 1982; Hoffman et al., 1985]. However, due to the limited view of all-sky cameras and photometers on low-altitude satellites these arcs were considered as relatively isolated features. It was not until the advent of imaging instruments on high-altitude

spacecraft, such as Dynamics Explorer 1 [e.g., Frank et al., 1982, 1986] and Viking [e.g., Murphree et al., 1987], that the existence of so-called transpolar arcs, luminous features extending across the "polar cap" linking the dayside aurora to the nightside oval, was established.

Another manifestation of the activity in the "polar cap" during periods of northward IMF is the presence of a large-scale field-aligned current system [e.g., Iijima et al., 1984; Araki et al., 1984; Iijima and Shibaji, 1987] occupying a large fraction of the region poleward of region 1 [cf. Iijima and Potemra, 1976, 1978]. This may be related to widespread diffuse aurora, but this relation is not firmly established. If sufficiently strong, these currents are capable of driving a large-scale convection system comprising three or four cells, in contrast to the two-cell system typically found for southward IMF.

High-latitude observations of electric fields, magnetic fields, and ion drifts have been interpreted in terms of multi-cell convection patterns in a number of studies [e.g., Burke et al., 1979; Zanetti et al., 1984; Potemra et al., 1984; Heelis et al., 1986]. Numerical modeling [e.g., Rasmussen and Schunk, 1987, 1988] has also indicated consistency between the large-scale field-aligned currents typically observed and multi-cell convection. Theoretical arguments based on magnetic merging have also been advanced to explain four-cell convection [e.g., Russell, 1972]. Some authors, however, favor a picture where the sunward convection occurring at high latitudes is not related to a multi-cell convection pattern but rather to a distorted two-cell pattern [e.g., Heppner, 1977; Friis-Christensen et al., 1985; Heppner and Maynard, 1987].

Several theories regarding the origin of polar arcs [e.g., Burke et al., 1979; Akasofu and Roederer, 1983; Chiu et al., 1985; Kan and Burke, 1985; Lyons, 1985], and regarding the cause of large-scale sunward convection [e.g., Russell, 1972; Reiff, 1982; Reiff and Burch, 1985] have been developed. It is not the purpose of the present paper to discuss these models and theories in detail, although reference to some of the features in these models will be made in relation to our own modeling results.

We will briefly overview some of the observations made of the high-latitude ionosphere during periods of northward IMF. This is followed by a discussion of some of our own recent modeling results of the electrodynamics pertaining to these phenomena.

2.   OVERVIEW OF OBSERVATIONS

Observations of auroral activity far poleward of the "normal" auroral oval have been recorded for many years. Our knowledge of this phenomenon has increased rapidly in recent years, particularly since imaging devices were included in the instrumentation flown on high-altitude orbiting spacecraft. In spite of this, many fundamental questions concerning polar arcs and their associated electrodynamics on the small (arc) scale as well as on the global scale remain to be answered.

From optical observations it is possible to distinguish two distinct transpolar arc situations. One where the arc is an isolated feature surrounded by what from particle data appears to be polar cap (i.e., open) field lines, while the particle distribution found on the arc-associated field lines usually have typical plasma sheet properties. However, the inferred acceleration potential is normally lower than that typical for auroral oval arcs. The other situation is that of an expanded (or rather contracted) oval, where open field lines are seen on one side of the transpolar arc, while on the other side the region between the transpolar arc and the oval is filled with diffuse aurora. Note, however, that the distinction made between these two situations might be quite arbitrary since it depends on the sensitivity threshold of the optical instrument, and on the identification of open or closed field lines. (Actually, the fact that polar arcs appear to be more isolated from Dynamics Explorer images than from Viking images, may well be due to the different sensitivities of the two instruments.) Whether such an identification is possible from particle data with any certainty is at present uncertain. Nevertheless, in this paper we focus our attention on these two distinct situations, although in many cases reality is better described as a situation somewhere in-between these extremes.

This section summarizes part of the overwhelming observational material that has been published on northward IMF electrodynamics and polar arcs.

2.1. *Optical Signatures and Classifications*

Gussenhoven [1982], based on images from 1600 DMSP polar passes, divided polar arcs into four typical categories: morning sector arcs, evening sector arcs, single Sun-aligned arcs, and midnight sector arcs. All but the single arcs seemed to be associated with an expanded oval. All except the last category occur for similar conditions: northward IMF, high solar wind velocity, moderate K_p and Dst levels, but rather low AE. The last type appears during substorm conditions. The location of the arcs depend somewhat on B_y, morning arcs are more common for $B_y < 0$ (in the northern hemisphere) and conversely. Morning sector arcs are, however, by far the most common type.

Ismail and Meng [1982] gave another classification of polar arcs. Their study was also based on DMSP imaging data; however, they divided the arcs into three types: 1) distinctly Sun-aligned arcs, 2) evening/morning arcs expanded from but still connected to the oval, and 3) hook shaped arcs connecting to the oval. They found that all three types occur predominantly for $B_z > 0$, type 1 for low AE, type 2 during substorm recovery, while type 3 is not seen during substorm recovery. Type 1 arcs occur mostly in the northern hemisphere for $B_x > 0$ and in the southern hemisphere for $B_x < 0$. This suggests that this arc type does not appear conjugately in both hemispheres.

Statistics on the distribution of discrete arcs as seen by all-sky cameras during periods of northward IMF were compiled by Lassen and Danielsen [1978] (Figure 1). They distinguish between two patterns of discrete arcs. The first pattern is when the arcs are ordered along the oval, or in other words the usual oval arcs. The other pattern (the polar cap pattern) consists of Sun-aligned arcs as well as oval-aligned arcs at high latitude (78-80 degrees). As expected, the oval pattern dominates for southward B_z. As the IMF turns northward the auroral configuration moves polewardly and turns gradually into the polar cap pattern. For $B_y > 0$ the location of the morning sector oval remains essentially unchanged while the evening oval expands in the poleward direction. For $B_y < 0$ the morning oval contracts while the evening oval fades away. Polar arcs are mainly seen on the morning side of the "polar cap." They are slightly more commonly observed when IMF B_y is positive.

Ismail et al. [1977] used Isis 2 photometer data to study the characteristics of polar arcs. They recorded a 2-1 asymmetry in the occurrence frequency between morning and evening side in favor of the morning side. The arcs were most frequently observed when both K_p and AE were low. The intensity of the optical emissions was found to vary considerably along the arc, but the intensity ratio of 557.7 nm to 391.4 nm emissions remained relatively constant, indicating excitation by particles of roughly the same energy. This energy was found from particle data to be rather low, usually below 1 keV. Particle data also indicated that the proton fluxes were negligible. They also concluded that the magnetic activity was usually higher prior to the formation of polar arcs than after.

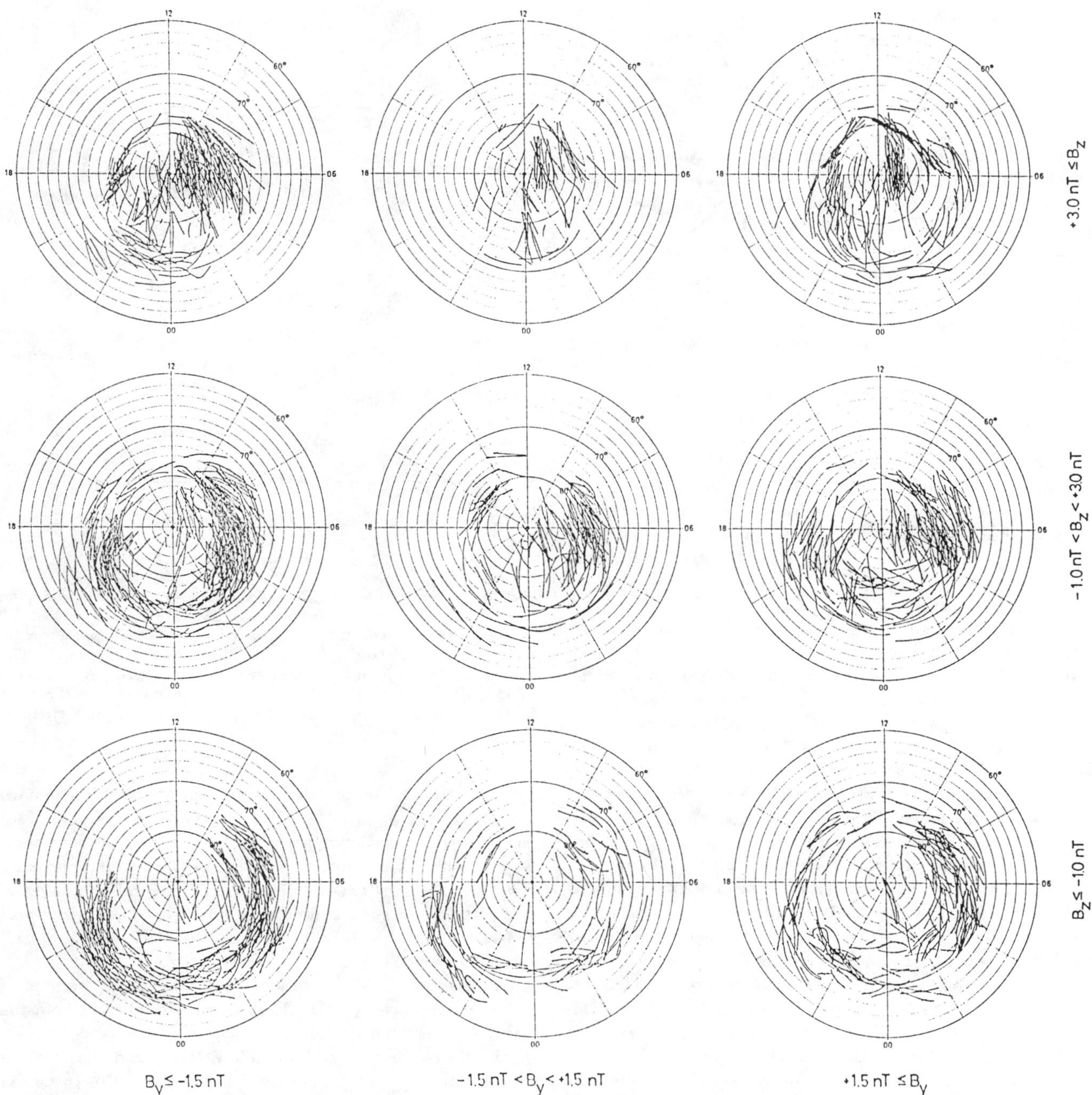

Fig. 1. All-sky camera observations from Greenland binned according to the prevailing IMF orientation. The fact that arcs are observed at substantially higher latitudes for northward IMF is apparent in this figure. The predominance for morning sector polar arcs for IMF $B_z > 0$ (top row) is also clearly seen. (After Lassen and Danielsen [1978], their Figure 3.)

2.2. *Magnetic Signatures and Field-Aligned Currents*

Burke et al. [1982] found from S3-2 data a 1-1 correspondence between transverse components of the electric and the residual magnetic fields above polar arcs, implying field-aligned currents closing via ionospheric Pedersen currents [cf. Sugiura et al., 1982; Sugiura 1984]. They also observed that the field-aligned current was embedded in a region of negative space charge, and that the charge carrier of upward current was weakly field-aligned electron fluxes with a temperature of a few hundred eV, accelerated by a potential drop of about 1 kV. They further proposed a model of polar arcs having their "generator" at the magnetopause.

Menietti and Burch [1987] studied DE 1 electron and ion data for transpolar events. They concluded that there was a downward field-aligned current on the dawnside of the arc, and a narrower region of upward field-aligned current on the duskside. This seems to be in conflict with many other observations, and, as discussed below, it is also inconsistent with sunward convection in the vicinity of the arc. This makes these cases an interesting special class.

Satellite magnetometer data often indicate the presence of an additional large-scale system of field-aligned currents at polar latitudes when IMF $B_z > 0$. This additional current system typically consists of two regions of oppositely flowing current in turn opposite to the adjacent region 1. It is usually occupying the major part of the region poleward of the auroral oval (i.e., poleward of region 1). This current system is sometimes observed to be associated with an extended region of diffuse aurora, in some cases with discrete aurora appearing at its poleward edge. However, the relation between these large-scale currents and diffuse polar aurora is at present not clear, since no statistical study of simultaneous magnetometer and imager data has been carried out to date. The relationship between aurora in the oval and field-aligned current has been studied by, for example, Kamide and Akasofu, [1976].

Many observations of this current system have been reported [e.g., McDiarmid et al., 1977; Araki et al., 1984; Iijima et al., 1984; Iijima and Shibaji, 1987]. Iijima et al. [1984] introduced the term "NBZ currents" indicating their association with northward IMF (Figure 2). It was shown in the studies by Iijima et al. [1984] and Iijima and Shibaji [1987] that the NBZ currents are typically most pronounced on the dayside and in the summer hemisphere, or in other words where the ionospheric conductivity is high. On magnetic field lines connected to the dark ionosphere the current pattern is typically more turbulent. The (Sun-aligned) dividing line between the upward and the downward flowing parts of the current system is displaced towards dawn or dusk depending on IMF B_y. In the northern hemisphere the displacement is in the direction of B_y.

Horwitz and Akasofu [1979] studied the horizontal current system in the polar cap during northward B_z, from Resolute Bay magnetic variations. They found that whereas the nightside current pattern is qualitatively the same irrespective of B_z, the dayside pattern for $B_z > 0$ is considerably different from that for $B_z < 0$. For northward IMF, the dayside pattern can be thought of as two juxtaposed standard two-cell patterns, such that antisunward current is flowing in the central portion of the polar cap. This implies associated sunward convection and is in agreement with the observations by Burke et al. [1979] (see below).

2.3. Electric Field and Plasma Convection Signatures

For southward IMF the convection pattern is usually of the two-cell type with mainly antisunward flow at polar latitudes. For northward IMF the convection pattern is

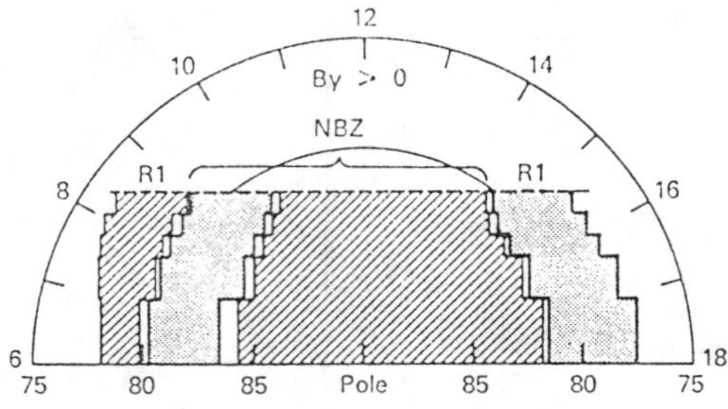

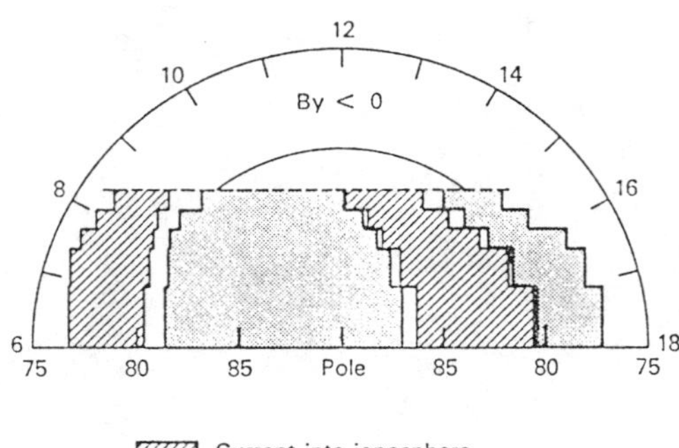

Fig. 2. The spatial distribution of the large-scale NBZ and Region 1 field-aligned currents from a statistical study of MAGSAT magnetometer data. This plot is from the south polar region during $B_y > 0$ (top) and during $B_y < 0$ (bottom). The characteristic shift in the geometry in accordance with the sign of IMF B_y is very clear. This shift is in the opposite direction in the northern hemisphere. After Iijima et al. [1984], their Figure 9.

typically more complicated [e.g., Heppner and Maynard 1987]. Burke et al. [1979] (Figure 3) found that S3-2 electric field data from periods of $B_z > 0$ were consistent with a four-cell convection pattern in the sunlit polar cap. Two of these cells were confined to the polar cap, and had circulation such that the plasma flow in-between was sunward. In the dark polar cap the flow was typically confused.

Heelis et al. [1986] studied DE 2 data from periods of prolonged northward IMF. They frequently observed plasma flow consistent with a four-cell pattern on the sunward side of the dawn-dusk meridian. They too found a turbulent "non-pattern" on the nightside. The geometry of the polar cells was found to depend on B_y.

Zanetti et al. [1984] derived the ionospheric flow pattern from MAGSAT magnetometer data. Their observations suggest three- and four-cell patterns, with a transition between them in response to a changing IMF B_y.

Sunward convection at polar latitudes was also reported by Reiff [1982].

2.4. Particle Signatures and Ionospheric Conductivity

Burch et al. [1979] studied the properties of polar cap electron acceleration regions using AE-D data. They found

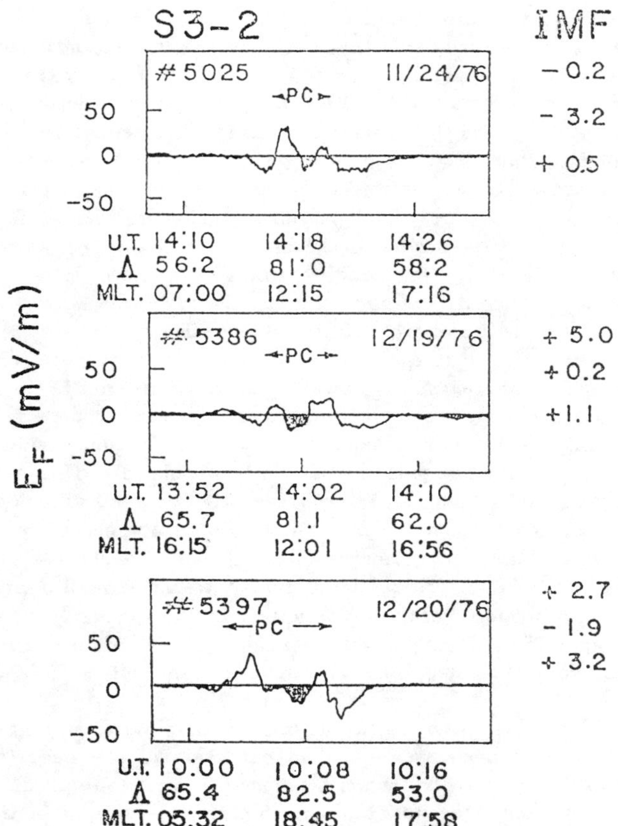

Fig. 3. Electric field data from S3-2. PC denotes the polar cap, and the shaded areas in the panels indicate regions where the electric field points from dusk to dawn corresponding to sunward plasma convection. In all three instances the IMF was pointing northward. After Burke et al. [1979], their Figure 3.

that these acceleration regions appear with almost equal probability at all MLT within $80° < \Lambda < 85°$. The electron spectra and angular distribution were generally consistent with acceleration by parallel electric fields. The existence of these acceleration regions was clearly correlated with $B_z > 0$ and $B_x < 0$ (in the northern hemisphere). The observed ion convection was usually consistent with converging ionospheric Pedersen currents, implying upward field-aligned current. They concluded that their observations support a picture of the source as being due to a localized magnetic interconnection between the solar wind and polar cap field lines, leading to a current system similar to the substorm current wedge, but occurring for northward IMF and mapping to the magnetopause rather than to the neutral sheet. (A model where the same mechanism is operating but actually connecting to the neutral sheet, in complete analogy to the substorm current wedge concept, has been put forward by Israelevich [1988].)

Frank et al. [1982] concluded from DE data that the plasma properties above transpolar arcs are similar to those found above the poleward section of the auroral oval.

Hoffman et al. [1985] studied a polar arc event using AE-C data and DMSP photographs. Their data indicated a source at 5-8 R_E on field lines threading the plasma sheet boundary layer. Contrary to many other observations, their electron spectra did not show signs of electrostatic acceleration.

Eliasson et al. [1987] examined Viking particle data and found that polar arc events are caused by accelerated particle distributions similar to plasma sheet boundary layer. The electron angular distribution indicated closed field lines. They concluded that many transpolar arcs are better described as the poleward edge of an expanded (contracted) oval during quiet conditions.

Evidence for the expanded oval configuration was collected by Meng [1981] who from DMSP electron data and simultaneous optical observations found that multiple polar arcs typically coincide with broad regions of low-energy (< 1 keV) electron precipitation. The arcs were often embedded in a widened soft-precipitation part of the normal oval, and usually occurred between substorms during periods of magnetospheric quiesence. The oval widening occurred conjugately in both hemispheres.

Mizera et al. [1987] examined an interesting event where data were available from DMSP F6 in the northern hemisphere and from NOAA 7 in the southern. Polar arcs were seen conjugately. However, whereas the arc in the north appeared to be surrounded by open field lines, the southern arc was embedded in a broad region of low level precipitation, presumably connected to the plasma sheet. Both arcs had similar source regions.

Simultaneous data from the two polar caps were studied by Reiff [1982] who found reversed (i.e., sunward) convection in both hemispheres. The field-aligned currents were significantly weaker in the winter hemisphere, which implies that the ionospheric conductivity plays an important role.

Heelis et al. [1986] examined particle data from periods of prolonged northward IMF. They found that the polar convection cells typically occurring during these interplanetary conditions could be on both open and closed field lines. Occasionally entire cells were found on open field lines. Sunward convection on closed field lines was also apparent.

2.5. Summary of Observations

The vast observational material on auroral arcs, field-aligned currents, and plasma convection at very high latitudes gathered over several decades indicate that the polar auroras show up in a variety of different configurations. In spite of their many-sided appearance some typical properties can be found. These can be summarized as follows:

1) Two basic situations: one or more bright distinct polar (or transpolar) arcs, or large regions of diffuse aurora presumably related to a large-scale current system (the NBZ currents), are seen poleward of the classical oval. In the former case the arcs are single features (isolated arcs) that may be surrounded on both sides by open field lines. In the latter case they occur at the poleward edge of a polewardly expanded oval (i.e., the arc appears to be bounded by open

field lines on one side and by closed field lines on the other).

2) These phenomena occur predominantly (almost exclusively) in association with northward IMF (i.e., during $B_z > 0$ or shortly after a southward turning). The occurrence of isolated arcs also shows different dependence on IMF B_x in the two hemispheres, implying that this arc type does not appear conjugately in both hemispheres.

3) NBZ currents are typically well ordered on the dayside but turbulent on the nightside, and the convection pattern is much steadier in the sunlit part of the ionosphere than in the dark part, indicating a dependence on the ionospheric conductivity.

4) The emission intensity may vary a lot along the arc, but the shape of the spectrum is essentially independent of the intensity, indicating excitation by particles of roughly the same characteristic energy. This characteristic energy is typically quite low (< 1 keV). The particles exciting the aurora are typically electrons, whose spectrum and angular distribution resemble those of central or boundary plasma sheet electrons.

5) Sunward convection is often (but not always) seen in conjunction with bright polar arcs. In this case it is typically localized to the immediate vicinity of the arc. Sunward convection is often seen also in association with NBZ currents. In this case it extends over a larger region. The sunward convection normally occurs on closed field lines, but it may in some cases occur (partly) on open field lines.

6) The modifications to the standard two-cell global convection pattern to account for sunward convection at extreme latitudes can take two basic forms: either the two-cell pattern is severely distorted, or one or two additional cells appear in the polar cap. In the latter case the global pattern has three or four cells.

Even though these properties are statistically characteristic of the high-latitude ionosphere, it should be kept in mind that significant deviations are not uncommon.

3.   Some Possible Configurations of the Magnetosphere

Simultaneous in situ measurements in the entire magnetosphere would greatly enhance our understanding of the magnetosphere-ionosphere interaction processes giving rise to polar aurora during periods of northward IMF. Since such measurements are hard to come by (to say the very least), we are restricted to theory, extrapolation, and conjectures based on single (or few) point measurements, when building ourselves a picture of the magnetospheric configuration responsible for the formation of these arcs (the situation is, of course, basically the same for any IMF orientation).

Numerous theories and/or models describing the magnetospheric topology and the interaction processes have been advanced. Without going into details we briefly review and comment on some of the fundamental concepts in these theories.

For southward IMF the plasma sheet is normally thought of as a region extending across the tail from dawn to dusk but relatively narrow in the north-south direction, as seen when looking towards the Sun. It typically extends from some ten Earth radii tailward of the Earth to many tens of Earth radii into the tail. When tracing the magnetic field lines emanating from this extended region one ends up in the auroral zone. Thus, the entire plasma sheet is topologically connected to the latitudinally limited ionospheric region where the auroral displays normally occur. It is often argued that the auroral oval "maps" to a "band" in the tail with a limited extent in the Sun-tail direction. However, such an "inverse" argument is usually not valid, and thus conclusions drawn from it are often misleading.

For northward IMF considerable complication arises. The magnetospheric topology is often described in terms of either a tilted plasma sheet, mapping to an expanded (contracted) oval, or a bifurcated tail lobe, where the region dividing the lobe into two parts maps to a transpolar arc. However, the mapping along field lines is sensitive to small variations in the magnetic field, and it is therefore not necessary to assume that the geometry of the plasma sheet needs to be modified in order for it to map to an expanded oval or a transpolar arc. Asymmetries in the magnetospheric magnetic field, might well be sufficient to account for the topological connections. This issue is at present the subject for debate.

The NBZ currents (occurring for northward B_z) have sometimes been viewed as an extension of the cusp (or mantle) currents which are frequently observed for any IMF orientation. One possible way to close this current in the magnetosphere is that the NBZ currents arise as a partial diversion of the tail current (i.e., the current flowing from dusk to dawn along the flanks of the magnetosphere). In this picture the region 2 current close in the ring current region, and the region 1 current at the inside of the magnetopause boundary layer.

4.   Modeling of Auroral Electrodynamics

In order to gain a more thorough understanding of the relationships between electric fields, ionospheric conductivity, and field-aligned currents in a clear-cut way, without obscuring the basic principles by taking too much of the fine structure which forms a natural and unavoidable part of reality into account, we have performed a series of numerical modeling studies [Marklund et al., 1987; 1988; 1991; Blomberg and Marklund, 1988; 1991a, b; Marklund and Blomberg, 1991].

In all these studies, the field-aligned current and the conductivity, sometimes assumed on a purely theoretical basis and sometimes inferred from and calibrated against measurements, have been used as model input. The numerical model is then used to calculate the ionospheric potential distribution (or convection pattern). The reasons for choosing the current as input rather than the electric field are several. First, we view the field-aligned current as the fundamental mediator of electrodynamical interaction of the ionosphere and the magnetosphere. This does not imply any particular cause-and-effect relationship between current and electric field. Secondly, the field-aligned current is easier to model than the electric field (e.g., the upward field-aligned current is clearly related to the precipitating particles causing the auroral emissions).

4.1. General Aspects of the Modeling

The different modeling studies that we have performed for northward IMF situations can be divided into three groups: 1) idealized models of transpolar arc-associated large-scale electrodynamics, 2) idealized models of the large-scale electrodynamics associated with NBZ current systems, and 3) event modeling. A common problem in these studies is the distribution of the (downward) return current. It is reasonable to assume that upward currents coincide with discrete aurora. However, there is no simple way to infer the location of downward current from the auroral distribution. The only possible way to determine it is by in situ measurements, which are never available more than at a few points at a time.

4.2. Transpolar Arc Modeling

In our idealized modeling study [Marklund and Blomberg, 1991] we have considered three basic transpolar arc associated field-aligned current systems. One where the polar arc is represented by a single sheet of upward current, one where the upward (arc) current sheet is locally balanced by a downward current just duskward of the upward current, and one where the downward (return) current is located just dawnward of the arc current (Figure 4).

The convection pattern for the first case has either two or three cells, depending on the relative magnitudes of the field-aligned current in the dusk sector region 1/2 system and that in the transpolar arc sheet as well as the conductivity distribution. In both cases one morning cell is seen. The evening cell is much more affected by the introduction of the polar current. As the transpolar arc current is increased the evening cell expands polewardly, while the electric field in its central part decreases in magnitude. If the polar current is sufficiently strong, the convection reversal in the evening cell will move to very high latitudes, almost to the location of the transpolar arc. This means that sunward plasma convection is present in a large part of what is normally the polar cap. Whether this region is on open or closed field lines is, however, an open question. Admittedly, the morning cell geometry changes from that found for a reference case where the polar region is void of current, but this change is not qualitatively important. A nice example of this case was found in Viking orbit 213, where the UV imager registered a transpolar arc, and the electric field data indicated sunward convection throughout the region between the oval and the polar arc (Figure 5, top right).

In the second model case, with two balancing transpolar current sheets, the convection pattern shows additional local structure in the vicinity of the polar currents, while the remainder of the pattern is essentially unchanged from the reference case. If the polar arc currents are strong enough two small cells will emerge in the arc region, so that sunward plasma flow appears in-between. Even for weaker arc currents sunward flow is present. In this case it is not, however, associated with additional convection cells. Rather, the convection pattern has an inverted S-shape, when viewed from dusk. It is remarkable how well the effect of the polar current system on the convection pattern is confined to a small region. This indicates that closure of the additional currents occurs almost completely locally. Also this case is exemplified (Figure 5, bottom right) by Viking data (from orbit 379).

The third model case, similar to the second but with the polar arc currents interchanged, again demonstrates the locality of the modifications to the convection pattern. In this case the local effect on the convection is, as expected, the opposite. That is, the electric field (or plasma flow speed) does not reverse, but is enhanced. No examples of this case have been found so far in Viking data, but the DE observations of Menietti and Burch [1987] as well as recent observations by Akebono [Obara et al., 1992] indicate such a polarity of the arc associated field-aligned currents.

4.3. NBZ Current Situations

Our modeling of the convection system in the presence of polar so-called NBZ currents [Blomberg and Marklund, 1991a] considers a number of different current configurations as model input. By varying the region 2 to region 1, and the NBZ to region 1 current ratios we obtained convection patterns representative of different IMF orientations (Figure 6). Typically, the region 2 to region 1 ratio decreases as IMF turns northward. At the same time the NBZ to region 1 ratio increases. It is not clear what determines the proportion between these ratios, but observations indicate that large differences between different events exist. Therefore, whereas it is relatively clear that the importance of the region 2 current decreases and that an NBZ current system often evolves in response to a northward turning of the IMF, the details of the current system evolution are not well understood. Another dependence is found on the y-component of the IMF. For $B_y \approx 0$ the dawn (upward) and the dusk (downward) NBZ current regions are typically of equal size. However, for $B_y \not\approx 0$ the dividing line between the two regions is often displaced. In the northern hemisphere this displacement is in the direction of B_y. The (integrated) magnitudes of the two currents are typically inversely dependent on the size of the region. The dependence is, however, not linear.

Assuming a conductivity distribution typical of summer solstice (reasonable since large-scale NBZ type currents are usually found in the sunlit portion of the polar region), the convection patterns derived for the case with equal (but opposite) NBZ currents are either of a two-cell or a four-cell type. The one which results depends on the ratio of NBZ current to current in the region 1/2 system. Choosing a conductivity distribution representative of a different season (with a smaller night-day gradient) results in the additional cells being shifted toward the dayside.

However, modeling a $B_y \not\approx 0$ case, where the upward and downward NBZ currents differ in magnitude, results in much more interesting convection patterns. Here, three-cell

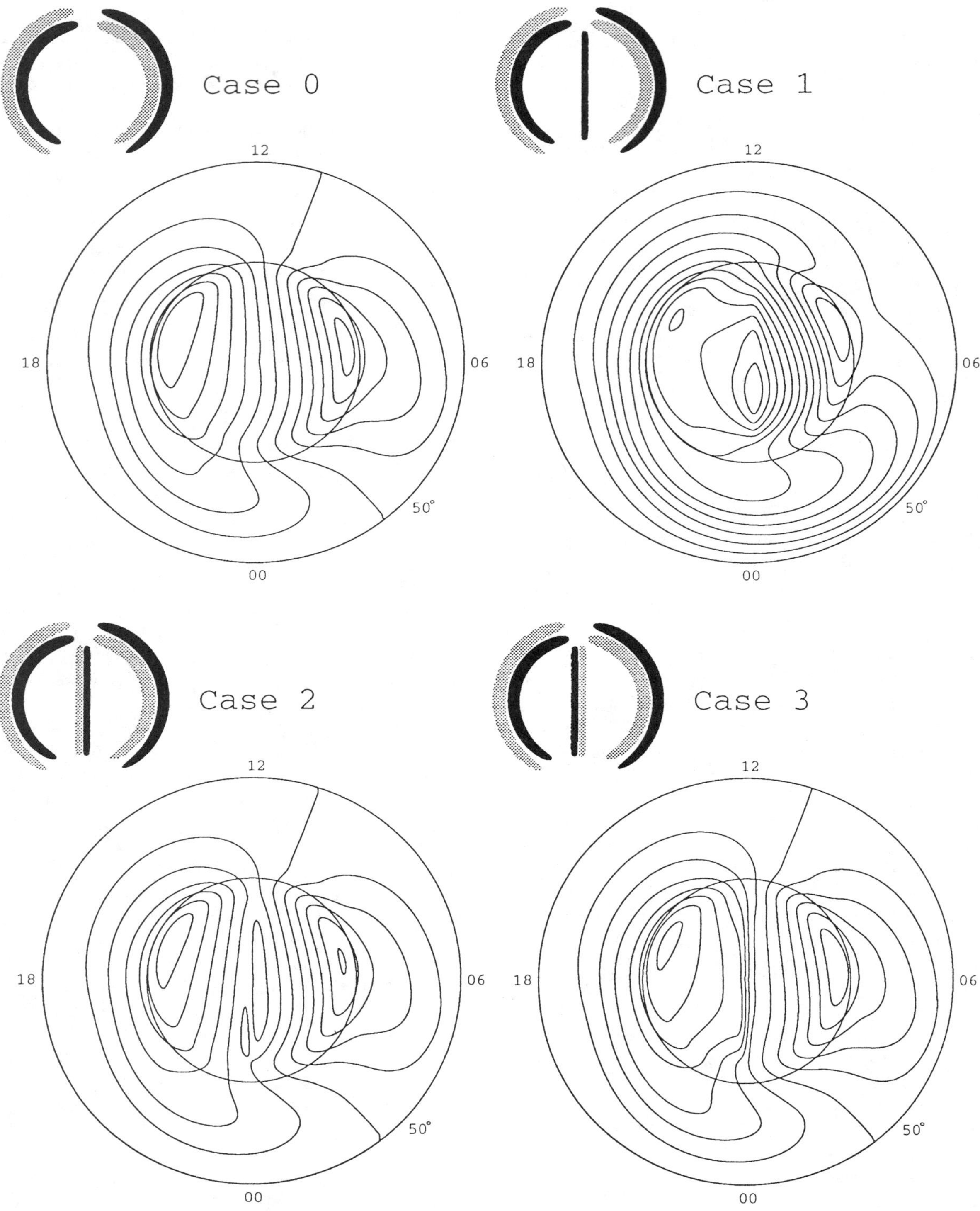

Fig. 4. Model results by Marklund and Blomberg [1991]. Case 0 is a reference case where field-aligned currents are absent in the polar region. In Case 1 a transpolar arc current system is introduced where the arc is represented by a single sheet of upward current. In Case 2 the arc is represented by a pair of oppositely directed currents, the downward return current being located duskward of the upward arc current. Case 3 is the opposite of Case 2.

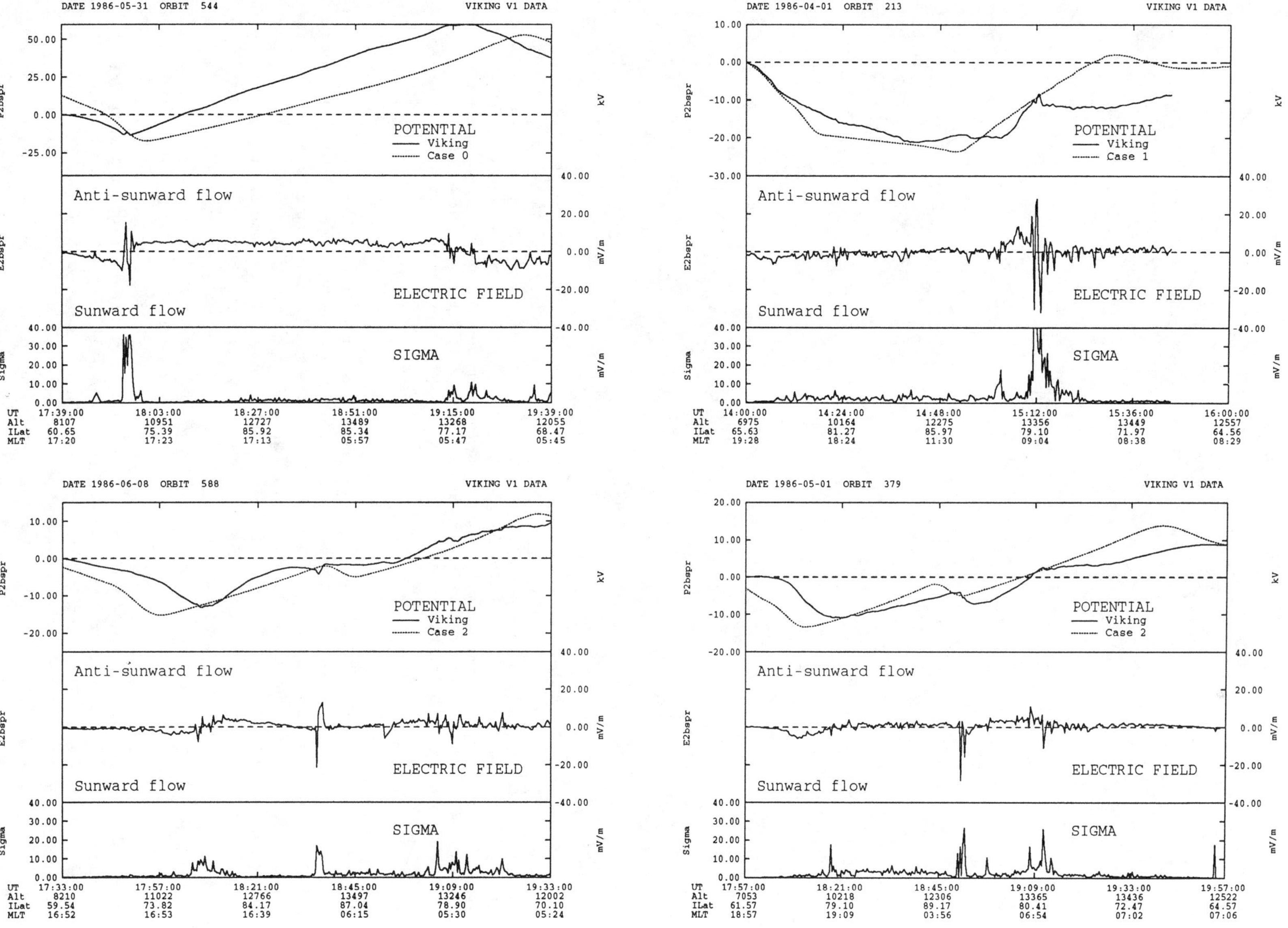

Fig. 5. Comparisons of the model results in Figure 4 to Viking data. The qualitative agreement is seen to be good.

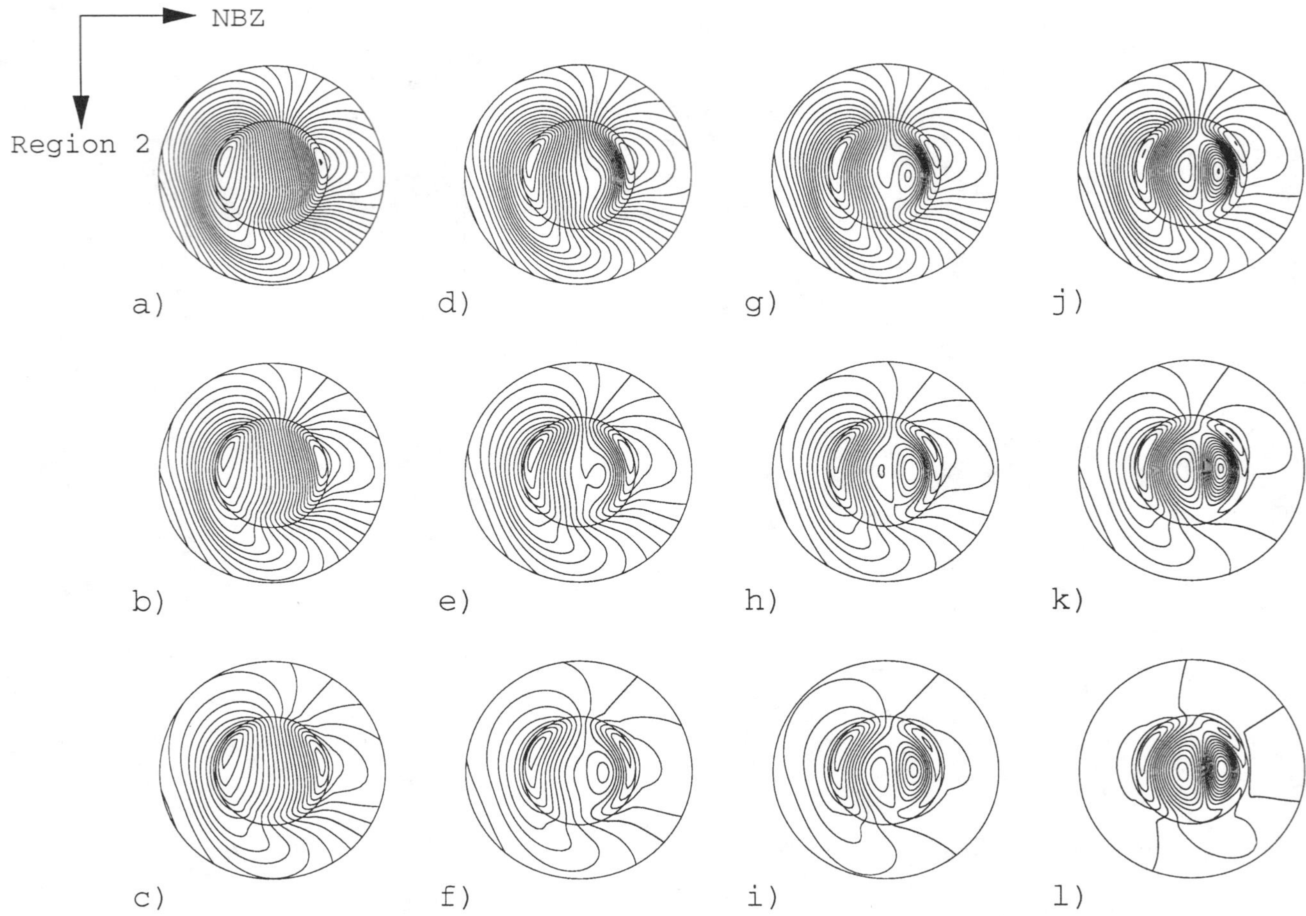

Fig. 6. Model convection patterns by Blomberg and Marklund [1991]. Going from left to right in each row the NBZ current increases. Going from top to bottom in each column the Region 2 current increases. The top left pattern thus represents a case where only a Region 1 current is present. During a turning from southward to northward IMF the Region 2 current typically decreases while the NBZ current typically increases. The evolution of the convection pattern during such a turning can be pictured by going from bottom left to top right in the figure.

patterns enter as a third fundamental configuration. It is concluded from the model results that during a transition from southward to northward IMF a three-cell pattern might well show up as an intermediate state, even though the initial pattern is of a two-cell type, and the final state normally a four-cell pattern (Figures 6 and 7). Knipp et al. [1991] reached similar conclusions from modeling using the AMIE technique [cf. Richmond and Kamide, 1988].

4.4. Event Studies

Besides the model studies using idealized input data, we have used our numerical model to "reconstruct" the global electrodynamics pertaining to specific events. In this "event modeling," auroral images acquired by the ultraviolet imager on Viking, are used to determine the "auroral geography" and also to infer upward field-aligned currents and associated conductivity enhancements. The distribution of downward field-aligned currents is based on statistical current distributions, calibrated against magnetometer measurements from satellites crossing high-latitude field lines during the event. From these input data (supplemented with the conductivity contribution from solar EUV) the potential distribution is calculated. As a check on the results they can be compared with satellite electric field data.

In one such event study (from September 25, 1986) [Marklund et al., 1991] a dawn side transpolar arc case was modeled. In this case a convection pattern deviating considerably from the idealized situations discussed above was found. This convection pattern could be described as being of a distorted two-cell type, with a clear kink in the flow pattern at the location of the transpolar arc. Even though diffuse aurora was seen in the image between the polar arc

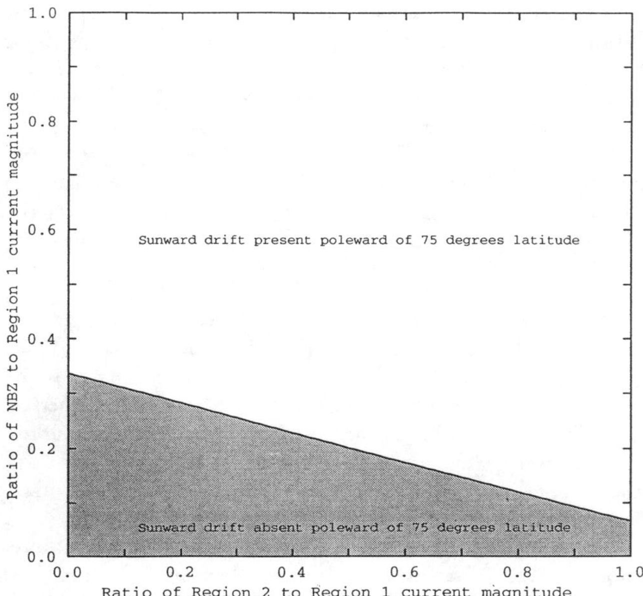

Fig. 7. Area in Region 2 to Region 1 and NBZ to Region 1 current magnitude parameter space where sunward convection appears in the polar region for the modeled convection patterns of Figure 6.

and the dawn oval, the transpolar arc current was modeled as a single sheet. The reason for this was that magnetometer data, surprisingly enough, did not indicate any field-aligned currents in this region.

5. Discussion

The characteristic features of the high-latitude ionosphere could loosely be grouped into three categories: small-scale phenomena, large-scale phenomena, and global (i.e., pertaining to the magnetosphere-ionosphere system as a whole) phenomena. Roughly speaking one could say that small-scale phenomena are observed in relation to polar arcs, large-scale phenomena are related to situations where the more extended NBZ current systems are present, while the global situation is important in both cases.

5.1. General Considerations

In spite of the many observations of the high-latitude ionosphere's electrodynamical properties during periods of prolonged northward IMF, several fundamental issues remain unclear. One such issue is whether the large-scale NBZ currents appear in association with auroral activity at very high latitudes. Another is whether transpolar arcs can appear as isolated structures surrounded by polar cap (i.e., open) field lines, or if they always occur at the poleward edge of a widened oval. A third issue is the topology of the magnetotail and the magnetospheric current system under these conditions. The latter is at present a rather speculative area.

Irrespective of these problems we find it worthwhile to model the electrodynamics of the high-latitude ionosphere assuming the presence of additional field-aligned current systems at polar latitudes. We have considered idealized situations with localized transpolar arc-associated currents as well as large-scale NBZ-type currents. These idealized model situations shed light on fundamental relations between the large-scale distributions of current, conductivity, and electric fields, as well as on the importance of the (ambient) large-scale electric field to the signatures of a localized field-aligned current system. These relations are examined in more detail below. In addition to the idealized modeling we have also performed event modeling. The results from such studies contain, of course, more of reality than any idealized picture. However, at the same time more complication is introduced, and the fundamental relationships between the different parameters of interest are therefore somewhat obscured. For this reason both kinds of studies are important, and in particular much is to be gained from a combination of the two.

5.2. Large-Scale Effects

Generally speaking, large-scale modifications to the polar convection pattern can be created by polar field-aligned currents in two basic ways; either by a spatially extended (NBZ type) current system, or by a spatially small (arc related) current system carrying a significant net current. The effects of these two configurations are quite different. NBZ currents tend to "drive" additional large-scale convection cells at high latitudes, while smaller-scale transpolar arc-related net currents have the effect of expanding either the dusk cell (if the net current is upward) or the dawn cell (if the net current is downward) so that it reaches to very high latitudes. In both cases, however, sunward convection is present in the "polar cap." The overall pattern is quite different though. When the dawn or dusk cell expands to extreme latitudes the pattern is still of a two-cell type and a polar-orbiting satellite would only record one (field or convection) reversal as it traverses "polar cap" field lines, in contrast to the multi-cell pattern situations arising from NBZ currents.

Whether a multi-cell convection pattern is "produced" by a given polar current depends also on the conductivity. If the conductivity in the "polar cap" is high, a significant current is needed to "drive" additional convection cells. Observations indicate that NBZ currents are most pronounced in regions of high conductivity, mostly on the dayside of the summer hemisphere. This is also where evidence of additional large-scale convection cells is found, if at all.

5.3. Small-Scale Effects

Polar arc-related field-aligned currents introduce local modifications to the convection pattern. This is true irrespective of whether these additional currents are locally balanced. However, if they are indeed balanced the influence on the convection pattern is only local, whereas if not, they influence also the global convection pattern, as discussed above.

Depending on the magnitude of the currents the local modifications could take different forms; either reversing the convection so that it is sunward locally, or introducing kinks and bends in the convection streamlines. If the polarity of the arc-related currents is such that the downward return current is located dawnward of the upward current, convection will not be reversed but rather enhanced in the anti-sunward direction. There is little observational evidence of this situation, but it could not be ruled out as a possibility.

It is quite clear that local conductivity gradients, which are not included in our modeling, can have a significant influence on the convection signatures of a polar arc-associated field-aligned current system. Such gradients introduce local distortions in the flow pattern in and around the arc. The same effect also arises from local variations along the arc in the field-aligned current density. However, arc-associated conductivity gradients are believed to be less important for polar arcs than for oval arcs, since the characteristic energy of the electrons exciting polar auroras are generally lower than for oval auroras. This means that ionization is produced mainly at F region altitudes. Ionization at these altitudes contributes relatively little to the height-integrated conductivity.

5.4. Coupling to the Magnetosphere

Of fundamental importance to a complete understanding of the electrodynamics of the high-latitude ionosphere is a picture of the global (three-dimensional) current system. In our view the field-aligned current is the primary mediator of electrodynamical interaction of the ionosphere and the magnetosphere. These currents transfer not only electric charge, but also momentum, stress, and energy between different parts of the magnetosphere-ionosphere system.

$$j_\parallel = \nabla \cdot \mathbf{J}_\perp = \Sigma_P \nabla \cdot \mathbf{E}_\perp + \mathbf{E}_\perp \cdot \nabla \Sigma_P + \hat{\mathbf{B}} \times \mathbf{E}_\perp \cdot \nabla \Sigma_H \quad (1)$$

The equation of continuity (1) relating the field-aligned current to the ionospheric electric field and conductivity distribution consists of three terms, one including the divergence of the electric field, the other two including gradients of the conductivity. These terms are often thought of as describing field-aligned current of magnetospheric and ionospheric origin, respectively [e.g., Boström, 1974]. However, such a description is usually an oversimplification, since the magnetosphere-ionosphere system is characterized not only by its strong electrodynamical interaction, but also by strong feedback. Redistribution of ionospheric charge in response to an imposed magnetospheric electric field could change the electric field drastically. Likewise, particle precipitation driven by a magnetospheric generator will lead to an enhanced ionospheric conductivity, which in turn modifies the relationship between electric field and current. Because of this intimate relationship between the magnetosphere and the ionosphere, the magnetosphere-ionosphere system must be regarded as a whole, in order to be understood in full.

It should also be pointed out that the standard form of the "coupling equation" applies to steady-state only. Variations on a short time scale in the current or the electric field may lead to local accumulation of charge which invalidates the steady-state relation.

Whereas the ionospheric closure of these currents is reasonably well understood, the magnetospheric closure is much more of an open question. In our opinion this is one of the outstanding problems of magnetospheric physics today.

6.   Concluding Views and Remarks

The task of understanding the electrodynamics of the high-latitude ionosphere during periods of northward IMF is a difficult one. One approach which has proven to be very fruitful is a combination of observations with numerical modeling, and idealized modeling in relation to these studies.

Much more work is needed to obtain a complete picture of the magnetosphere-ionosphere system and its electrodynamics. No doubt, future space probe missions will contribute to this picture, and if the computer development continues to proceed as it has over the last few decades (which is most likely) the level of sophistication of numerical models will increase continuously.

However, we believe that recent modeling studies, where efforts to combine the modeling with observational material as well as theoretical investigations have been made, demonstrate that this kind of combined studies is a most valuable complement to more traditional approaches. It is also our firm belief that this will remain true also in the future, since simultaneous measurements throughout the magnetosphere will never be carried out.

These recent modeling studies have focussed on the ionosphere, mainly because of the greater availability of observational material, compared to the magnetosphere. However, a most worthwhile continuation of these studies would be to extend them to address some outstanding problems in the physics of the magnetosphere. Among the most urgent outstanding questions in magnetospheric electrodynamics (particularly under northward IMF conditions) of today we can identify the following: the closure of field-aligned currents (particularly in the magnetosphere, the closure in the ionosphere is today reasonably well understood), the magnetospheric current system in general, the magnetic topology (particularly of the magnetotail), the energy transfer mechanisms operating at the magnetopause, the importance of magnetic merging versus viscous interaction for producing the "polar cap potential drop," and whether the magnetospheric generators act as current or voltage sources. Another issue of great importance is the auroral acceleration process. However, the details of this complicated process are outside the scope of the present context.

Our wish to see modeling efforts extended to the magnetosphere is not to be interpreted as if we consider the ionosphere as being understood in full. Preferably, the ionospheric modeling should proceed in parallel.

Finally, we wish to stress that although we in this paper have discussed two distinct basic situations, reality is, of course, immensely more complicated. Furthermore, dividing the discussion into large-scale and small-scale effects is, admittedly, quite arbitrary.

Acknowledgments. The authors are grateful to the two referees for their constructive comments.

The Viking Project was managed and operated by the Swedish Space Corporation under contract from the Swedish National Space Board.

References

Akasofu S.-I. and M. Roederer, Polar cap arcs and the open regions, *Planet. Space Sci.*, *31*, 195, 1983.

Araki, T., T. Kamei, and T. Iyemori, Polar cap vertical currents associated with northward interplanetary magnetic field, *Geophys. Res. Lett.*, *11*, 23, 1984.

Berkey, F. T., L. L. Cogger, and S. Ismail, Evidence for a correlation between Sun-aligned arcs and the interplanetary magnetic field direction, *Geophys. Res. Lett.*, *3*, 145, 1976.

Blomberg, L. G. and G. T. Marklund, The influence of conductivities consistent with field-aligned currents on high-latitude convection patterns, *J. Geophys. Res.*, *93*, 14493, 1988.

Blomberg, L. G. and G. T. Marklund, High-latitude convection patterns for various large-scale field-aligned current configurations, *Geophys. Res. Lett.*, *18*, 717, 1991a.

Blomberg, L. G. and G. T. Marklund, A numerical model of ionospheric convection derived from field-aligned currents and the corresponding conductivity, *Rep. TRITA-EPP-91-03*, Royal Inst. of Technol., Stockholm, 1991b.

Boström, R., Ionosphere-magnetosphere coupling, in *Magnetospheric Physics*, pp. 45, edited by B. M. McCormac, D. Reidel, Hingham, Mass., 1974.

Burch, J. L., S. A. Fields, and R. A. Heelis, Polar cap electron acceleration regions, *J. Geophys. Res.*, *84*, 5863, 1979.

Burke, W. J., M. C. Kelley, R. C. Sagalyn, M. Smiddy, and S. T. Lai, Polar cap electric field structures with a northward interplanetary magnetic field, *Geophys. Res. Lett.*, *6*, 21, 1979.

Burke, W. J., M. S. Gussenhoven, M. C. Kelley, D. A. Hardy, and F. J. Rich, Electric and magnetic field characteristics of discrete arcs in the polar cap, *J. Geophys. Res.*, *87*, 2431, 1982.

Chiu, Y. T., N. C. Crooker, and D. J. Gorney, Model of oval and polar cap arc configurations, *J. Geophys. Res.*, *90*, 5153, 1985.

Craven, J. D., J. S. Murphree, L. A. Frank, and L. L. Cogger, Simultaneous optical observations of transpolar arce in the two polar caps, *Geophys. Res. Lett.*, *18*, 2297, 1991.

Davis, T. N., The morphology of the polar aurora, *J. Geophys. Res.*, *65*, 3497, 1960.

Davis, T. N., Negative correlation between polar cap visual activity and magnetic activity, *J. Geophys. Res.*, *68*, 4447, 1963.

Denholm, J. V., Some auroral observations inside the southern auroral zone, *J. Geophys. Res.*, *66*, 2105, 1961.

Denholm, J. V. and F. R. Bond, Orientation of polar auroras, *Aust. J. Phys.*, *14*, 193, 1961.

Eather, R. H. and S.-I. Akasofu, Characteristics of polar cap auroras, *J. Geophys. Res.*, *74*, 4794, 1969.

Eliasson, L., R. Lundin, and J. S. Murphree, Polar cap arcs observed by the Viking satellite, *Geophys. Res. Lett.*, *14*, 451, 1987.

Frank, L. A., J. D. Craven, J. L. Burch, and J. D. Winningham, Polar views of the Earth's aurora with Dynamics Explorer, *Geophys. Res. Lett.*, *9*, 1001, 1982.

Frank, L. A., J. D. Craven, D. A. Gurnett, S. D. Shawhan, D. R. Weimer, J. L. Burch, J. D. Winningham, C. R. Chappell, J. H. Waite, R. A. Heelis, N. C. Maynard, M. Sugiura, W. K. Peterson, and E. G. Shelley, The theta aurora, *J. Geophys. Res.*, *91*, 3177, 1986.

Friis-Christensen, E., Y. Kamide, A. D. Richmond, and S. Matsushita, Interplanetary magnetic field control of high-latitude electric fields and currents determined from Greenland magnetometer data, *J. Geophys. Res.*, *90*, 1325, 1985.

Gussenhoven, M. S., Extremely high latitude auroras, *J. Geophys. Res.*, *87*, 2401, 1982.

Heelis, R. A., P. H. Reiff, J. D. Winningham, and W. B. Hanson, Ionospheric convection signatures observed by DE 2 during northward interplanetary magnetic field, *J. Geophys. Res.*, *91*, 5817, 1986.

Heppner, J. P., Empirical models of high-latitude electric fields, *J. Geophys. Res.*, *82*, 1115, 1977.

Heppner, J. P., and N. C. Maynard, Empirical high-latitude electric field models, *J. Geophys. Res.*, *92*, 4467, 1987.

Hoffman, R. A., R. A. Heelis, and J. S. Prasad, A Sun-aligned arc observed by DMSP and AE-C, *J. Geophys. Res.*, *90*, 9697, 1985.

Horwitz, J. L. and S.-I. Akasofu, On the relationship of the polar cap current system to the north-south component of the interplanetary magnetic field, *J. Geophys. Res.*, *84*, 2567, 1979.

Iijima, T., and T. A. Potemra, The amplitude distribution of field-aligned currents at northern high latitudes observed by Triad, *J. Geophys. Res.*, *81*, 2165, 1976.

Iijima, T., and T. A. Potemra, Large-scale characteristics of field-aligned currents associated with substorms, *J. Geophys. Res.*, *83*, 599, 1978.

Iijima, T., T. A. Potemra, L. J. Zanetti, and P. F. Bythrow, Large-scale Birkeland currents in the dayside polar region during strongly northward IMF: A new Birkeland current system, *J. Geophys. Res.*, *89*, 7441, 1984.

Iijima, T. and T. Shibaji, Global characteristics of northward IMF-associated (NBZ) field-aligned currents, *J. Geophys. Res.*, *92*, 2408, 1987.

Ismail, S., D. D. Wallis, and L. L. Cogger, Characteristics of polar cap Sun-aligned arcs, *J. Geophys. Res.*, *82*, 4741, 1977.

Ismail, S., and C.-I. Meng, A classification of polar cap auroral arcs, *Planet. Space Sci.*, *30*, 319, 1982.

Israelevich, P. L., The change of the magnetospheric configuration due to cross-tail current disruption and its possible relation to the appearance of Θ-aurora, preprint, Space Research Institute, U. S. S. R. Academy of Sciences, 1988.

Kamide, Y. and S.-I. Akasofu, The location of the field-aligned currents with respect to discrete auroral arcs, *J. Geophys. Res.*, *81*, 3999, 1976.

Kan, J. R., and W. J. Burke, A theoretical model of polar cap auroral arcs, *J. Geophys. Res.*, *90*, 4171, 1985.

Knipp, D. J., A. D. Richmond, B. Emery, N. U. Crooker, O. de la Beaujardière, D. Evans, and H. Kroehl, Ionospheric convection response to changing IMF direction, *Geophys. Res. Lett.*, *18*, 721-724, 1991.

Lassen, K. and C. Danielsen, Quiet-time pattern of auroral arcs for different directions of the interplanetary magnetic field in the $Y - Z$ plane, *J. Geophys. Res. 83*, 5277, 1978.

Lyons, L. R., A simple model for polar cap convection patterns and generation of θ auroras, *J. Geophys. Res.*, *90*, 1561, 1985.

Marklund, G. T., L. G. Blomberg, T. A. Potemra, J. S. Murphree, F. J. Rich, and K. Stasiewicz, A new method to derive "instantaneous" high-latitude potential distributions from satellite measurements including auroral imager data, *Geophys. Res. Lett.*, *14*, 439, 1987.

Marklund, G. T., L. G. Blomberg, K. Stasiewicz, J. S. Murphree, R. Pottelette, L. J. Zanetti, T. A. Potemra, D. A. Hardy, and F. J. Rich, Snapshots of high-latitude electrodynamics using Viking and DMSP F7 observations, *J. Geophys. Res.*, *93*, 14479, 1988.

Marklund, G. T., L. G. Blomberg, J. S. Murphree, R. D. Elphinstone, L. J. Zanetti, R. E. Erlandson, I. Sandahl, O. de la Beaujardière, H. Opgenoorth, and F. J. Rich, On the electrodynamical state of the auroral ionosphere during northward IMF: A transpolar arc case study, *J. Geophys. Res.*, *96*, 9567, 1991.

Marklund, G. T. and L. G. Blomberg, On the influence of localized electric fields and field-aligned currents associated with polar arcs on the global potential distribution, *J. Geophys. Res.*, *96*, 13977, 1991.

McDiarmid, I. B., E. E. Budzinski, M. D. Wilson, and J. R. Burrows, Reverse polarity field-aligned currents at high latitudes, *J. Geophys. Res.*, *82*, 1513, 1977.

Meng, C.-I., Polar cap arcs and the plasma sheet, *Geophys. Res. Lett.*, *8*, 273, 1981.

Menietti, J. D., and J. L. Burch, DE 1 observations of theta aurora plasma source regions and Birkeland current charge carriers, *J. Geophys. Res.*, *92*, 7503, 1987.

Mizera, P. F., D. J. Gorney, and D. S. Evans, On the conjugacy of the aurora, high and low latitude, *Geophys. Res. Lett.*, *14*, 190, 1987.

Murphree, J. S., L. L. Cogger, C. D. Anger, D. D. Wallis, and G. G. Shepherd, Oval intensifications associated with polar arcs, *Geophys. Res. Lett.*, *14*, 403, 1987.

Obara, T., T. Mukai, H. Hayakawa, A. Nishida, K. Tsuruda, S. Machida, and H. Fukunishi, Akebono (EXOS-D) observation of small scale electromagnetic signature relating to the polar cap precipitation, *J. Geophys. Res., in press*, 1992.

Potemra, T. A., L. J. Zanetti, P. F. Bythrow, A. T. Y. Lui, and T. Iijima, B_y-dependent convection patterns during northward interplanetary magnetic field, *J. Geophys. Res.*, *89*, 9753, 1984.

Rasmussen, C. E., and R. W. Schunk, Ionospheric convection driven by NBZ currents, *J. Geophys. Res.*, *92*, 4491, 1987.

Rasmussen, C. E., and R. W. Schunk, Ionospheric convection inferred from interplanetary magnetic field-dependent Birkeland currents, *J. Geophys. Res.*, *93*, 1909, 1988.

Reiff, P. H., Sunward convection in both polar caps, *J. Geophys. Res.*, *87*, 5976, 1982.

Reiff, P. H., and J. L. Burch, IMF B_y-dependent plasma flow and Birkeland currents in the dayside magnetosphere, 2. A global model for northward and southward IMF, *J. Geophys. Res.*, *90*, 1595, 1985.

Richmond, A. D., and Y. Kamide, Mapping electrodynamic features of the high-latitude ionosphere from localized observations: Technique, *J. Geophys. Res.*, *93*, 5741, 1988.

Russell, C. T., The configuration of the magnetosphere, in *Critical problems of Magnetospheric Physics*, p. 1, edited by E. R. Dyer, Jr., National Academy of Science, Washington, D. C., 1972.

Sugiura, M., N. C. Maynard, W. H. Farthing, J. P. Heppner, B. G. Ledley, and L. J. Cahill, Jr., Initial results on the correlation between the magnetic and electric fields observed from the DE-2 satellite in the field-aligned current regions, *Geophys. Res. Lett.*, *9*, 985, 1982.

Sugiura, M., A fundamental magnetosphere-ionosphere coupling mode involving field-aligned currents as deduced from DE-2 observations, *Geophys. Res. Lett.*, *11*, 877, 1984.

Zanetti, L. J., T. A. Potemra, T. Iijima, W. Baumjohann, and P. F. Bythrow, Ionospheric and Birkeland current distributions for northward interplanetary magnetic field: Inferred polar convection, *J. Geophys. Res.*, *89*, 7453, 1984.

L. G. Blomberg and G. T. Marklund, Department of Plasma Physics, Alfvén Laboratory, Royal Institute of Technology, S-100 44 Stockholm, Sweden.

Convection and Electrodynamic Signatures in the Vicinity of a Sun-aligned Arc: Results from the Polar Acceleration Regions and Convection Study (Polar ARCS)

L. A. Weiss,[1] E. J. Weber,[2] P. H. Reiff,[1] J. R. Sharber,[3] J. D. Winningham,[3]
F. Primdahl,[4] I. S. Mikkelsen,[5] C. Seifring,[6] and E. M. Wescott[7]

An experimental campaign designed to study high-latitude auroral arcs was conducted in Sondre Stromfjord, Greenland, on February 26, 1987. The Polar Acceleration Regions and Convection Study (Polar ARCS) consisted of a coordinated set of ground-based, airborne, and sounding rocket measurements of a weak, sun-aligned arc system within the duskside polar cap. A rocket-borne barium release experiment, two DMSP satellite overflights, all-sky photography, and incoherent scatter radar measurements provided information on the large-scale plasma convection over the polar cap region while a second rocket instrumented with a DC magnetometer, Langmuir and electric field probes, and an electron spectrometer provided measurements of small-scale electrodynamics. The large-scale data indicate that small, sun-aligned precipitation events formed within a region of antisunward convection between the duskside auroral oval and a large sun-aligned arc further poleward. This convection signature, used to assess the relationship of the sun-aligned arc to the large-scale magnetospheric configuration, is found to be consistent with either a model in which the arc formed on open field lines on the dusk side of a bifurcated polar cap or on closed field lines threading an expanded low-latitude boundary layer, but not a model in which the polar cap arc field lines map to an expanded plasma sheet. The antisunward convection signature may also be explained by a model in which the polar cap arc formed on long field lines recently reconnected through a highly skewed plasma sheet. The small-scale measurements indicate the rocket passed through three narrow (≤ 20 km) regions of low-energy (≤ 100 eV) electron precipitation in which the electric and magnetic field perturbations were well correlated. These precipitation events are shown to be associated with regions of downward Poynting flux and small-scale upward and downward field-aligned currents of 1-2 $\mu A/m^2$. The paired field-aligned currents are associated with velocity shears (higher and lower speed streams) embedded in the region of anti-sunward flow.

1. Introduction

The "polar cap boundary," the boundary between the bright, approximately circular, oval of auroral emission and the darker, interior polar region, has been variously defined on the basis of magnetospheric topology (the mapping of an assumed open/closed field line boundary) [Birn et al., 1991; Elphinstone et al., 1991], the locus of the electric field (convection) reversal [Heelis et al., 1980; Torbert et al., 1981], and the poleward extent of plasma sheet particle precipitation [Winningham and Heikkila, 1974; Meng, 1981; Makita et al., 1988; Rich et al., 1990]. Although these boundaries may be located near one another, they often do not coincide and become especially difficult to interpret during times of northward Interplanetary Magnetic Field (IMF $B_z > 0$) when sun-aligned arcs, additional field-aligned current systems, and highly structured electric fields can extend over the entire high-latitude region [Burke et al., 1979, 1982; Iijima et al., 1984; Hardy, 1984; Zanetti et al., 1990].

Polar cap aurorae are observed in the polar cap during quiet magnetospheric times when the auroral oval is least active, (i.e., when the IMF B_z component is northward), and are predominantly aligned along the earth-sun line [e.g., Lassen and Danielsen, 1978; 1989]. A common type of polar cap aurora is created by precipitating electrons with average energies between 50 and 300 eV, resulting in weak (usually subvisual) multiple, F-layer arcs [Weber and Buchau, 1981; Hardy, 1984]. A less common type of polar cap aurora is one which stretches across the entire polar cap from noon to midnight, forming what appears from high altitude spacecraft as a 'theta' aurora [Frank et al., 1986; Nielsen et al., 1990]. Bright sun-aligned arcs are also often observed at the poleward

[1]Department of Space Physics and Astronomy, Rice University, Houston, TX
[2]Phillips Laboratory, Hanscom AFB, Bedford, MA
[3]Southwest Research Institute, San Antonio, TX
[4]Danish Space Research Institute, Lyngby, Denmark
[5]Danish Meteorological Institute, Copenhagen, Denmark
[6]Naval Research Laboratory, Washington, DC
[7]Geophysical Institute, University of Alaska, Fairbanks, AK

Auroral Plasma Dynamics
Geophysical Monograph 80

boundary of a region of soft electron precipitation extending poleward from one or both edges of the auroral oval, resulting in a 'D' or 'teardrop' shaped polar cap, respectively [Murphree and Cogger, 1981; Elphinstone et al., 1990; Hones et al., 1989]. Recent attempts at mapping the open/closed field line boundary to the ionosphere as a function of IMF have had moderate success in reproducing the observed changes in polar cap shapes [Birn et al., 1991; Elphinstone et al., 1991; Toffoletto and Hill, 1990; 1992].

Like arcs in the auroral oval, polar cap aurorae are believed to be the optical signature of upward field-aligned currents where $\nabla \cdot \Sigma E < 0$ [Reiff et al., 1978; Lyons, 1980; Burke et al., 1982; Chiu and Gorney, 1983; Chiu, 1989; Valladares and Carlson, 1991]. On the dayside, the electrons carrying the majority of this upward current have either been interpreted as precipitating on closed field lines threading the LLBL [Lassen and Danielsen, 1989; Lundin et al., 1990] or on open field lines threading the plasma mantle [Newell et al., 1991]. At other local times, it has been suggested that weak, polar cap arcs result from the precipitation of accelerated polar rain electrons on open field lines [Hardy et al., 1982; Burke et al., 1982; Gussenhoven and Mullen, 1989], or from plasma sheet or low latitude boundary layer electrons on closed field lines [Meng, 1981; Murphree et al., 1982; Lundin and Evans, 1985; Lundin et al., 1990].

The Polar Acceleration Regions and Convection Study (Polar ARCS) was designed to provide the simultaneous convection and electrodynamic information needed to investigate the magnetospheric topology and generation mechanisms of high-latitude polar cap arcs. This paper summarizes the primary measurements and results of this study, which can be divided into two categories: large-scale phenomena, which lead to the examination of different models of polar cap morphology; and smaller-scale phenomena, which allow the investigation of the electrodynamic structure of the arcs. More detailed descriptions of the instruments, measurements, and results of Polar ARCS can be found in Weiss [1991].

2. EXPERIMENT DESCRIPTION

Polar ARCS consisted of a coordinated set of ground-based, airborne, and in situ measurements of high-latitude sun-aligned arcs over Sondrestrom, Greenland, on February 26, 1987. Three charges of barium were injected into the F-region ionosphere immediately prior to the launch of a Black Brant IX instrumented with a DC magnetometer, Langmuir and electric field probes, and an electron spectrometer. Simultaneous electron density and line-of-sight velocity measurements were made by the Sondrestrom Incoherent Scatter Radar, and all-sky images of auroral activity were recorded using the All-Sky Imaging Photometer (ASIP) aboard the Airborne Ionospheric Observatory (AIO).

Figure 1 (a) summarizes, in geographical coordinates, the relative orientations of a large, sun-aligned arc observed by the ASIP, the ground trajectories of two DMSP satellite passes occurring within 30 minutes of launch, and the position of the auroral oval (as inferred from the DMSP electron measurements and ground-based all-sky images from Dye 2). The fields-of-view of the ASIP and the radar are denoted by the circular and semi-circular regions, respectively. The rocket's ground trajectory is shown as a straight line stretching from Sondrestrom to the northeast, and the three ionized barium tracks (resulting from rocket-borne releases just prior to launch of the instrumented rocket) are shown as irregular lines at the end of the rocket trajectory. The sun-aligned arc is oriented along a line 50° west of north, between the magnetic meridian (39° west of north) and the sun-aligned direction (~60° west of north).

The 6300 Å ASIP images indicated the presence of diffuse, sun-aligned precipitation throughout the region over Sondrestrom between the sun-aligned arc and the auroral oval. The dawn - dusk F6 number flux spectrograms [Weiss, 1991] show a clearly defined region of enhanced electron and ion precipitation over the middle of the polar cap, indicative of the presence of a transpolar (theta) aurora [Frank et al., 1986] at the time of the pass; the theta aurora would appear just off the scale of Figure 1 (a), to the upper right. The theta aurora, auroral oval, and sun-aligned arc are shown in MLT/IL coordinates in Figure 1 (b).

The instrumented rocket was launched at 2349:10 UT (~22 MLT) to the northeast, reaching an apogee of 380 km and covering a ground range of 166 km. Shortly after launch, the nosecone portion of the rocket separated from the main payload in order to provide a magnetically clean environment for the DC science magnetometer aboard this separated section. Data were obtained from both payloads for nearly 600 seconds after launch. Electron precipitation measurements indicate that the rocket passed through three narrow ($\leq$ 20 km) regions of low-energy ($\leq$ 100 eV) electron precipitation. These events are interpreted as small-scale structures that lie within the region of weak, sun-aligned precipitation in the region between the auroral oval and the large sun-aligned arc in the all-sky images. Shortly after the decision to launch was made, the large arc began to fade from a pre-launch brightness of about 300 R (subvisual) to less than 100 R by 2358 UT. This type of weak, F-region sun-aligned arc is typical of those identified by Weber and Buchau [1981] as resulting from the structured precipitation of electrons with average energies of only a few hundred eV.

Although IMP-8 was in the magnetosheath at the time of the experiment (and thus no IMF measurements are available), the presence of sun-aligned arcs, the precipitation over the center of the polar cap, and the contracted state of the auroral oval (Figure 1 (b)) argue that the IMF B_z component was northward during this period. Moreover, the geomagnetic three-hourly average K_p index was 1 for the six hours prior to and three hours after the flight, confirming that the magnetosphere was in a very quiet state.

3. OBSERVATIONS

3.1. Large-scale measurements

The large-scale Polar ARCS measurements consist of the all-sky images, incoherent scatter radar measurements, DMSP

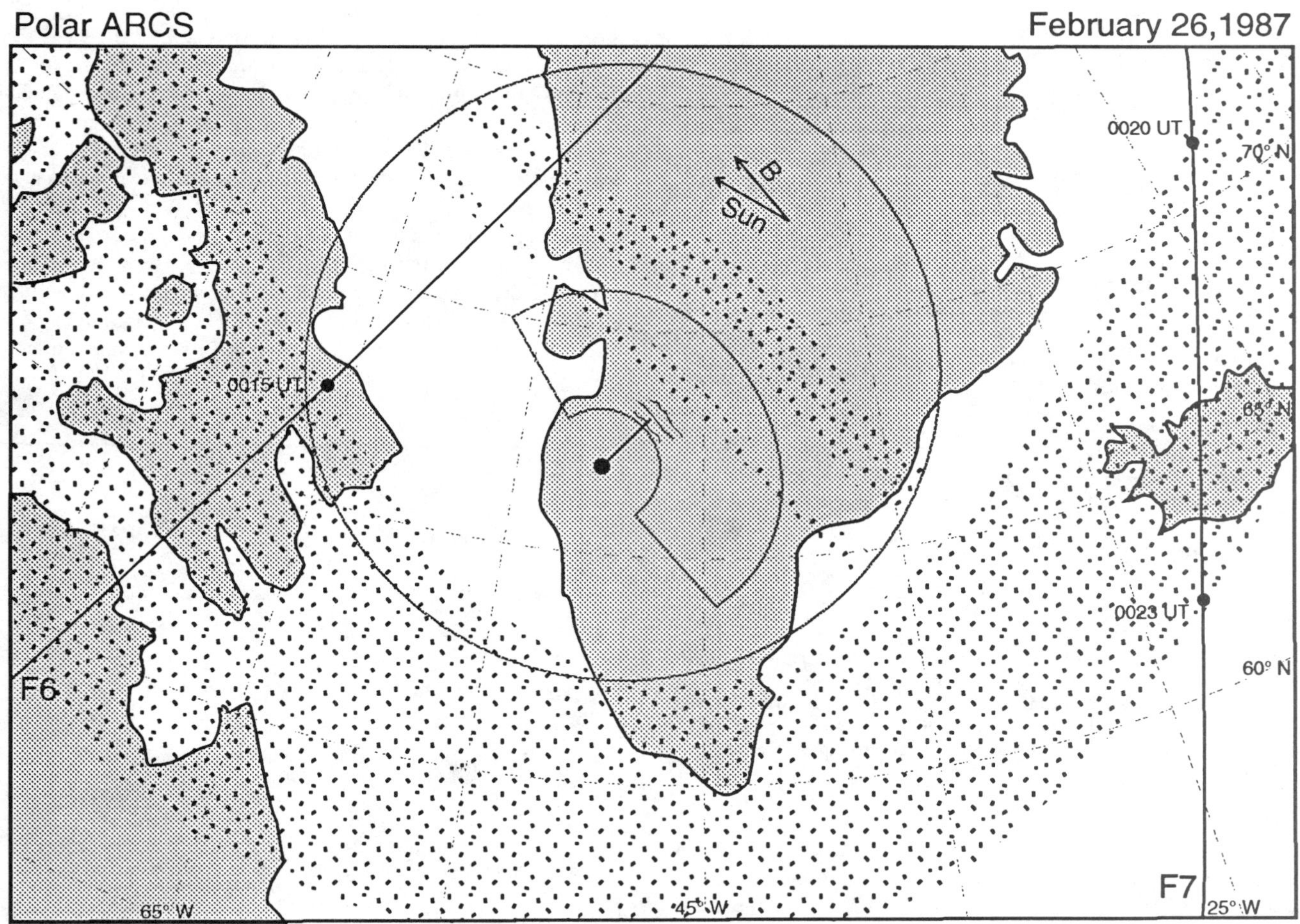

Fig. 1(a). A geographic diagram showing the location and orientation of the sun-aligned arc and auroral oval, and the spatial relationship between the different measurement regions. The large circle represents the ASIP field-of-view and the smaller partial circle is the radar field-of-view. The instrumented rocket track is denoted by the line stretching from Sondrestrom to the northeast. The motion of the three barium tracks are shown as irregular lines at the end of the rocket trajectory.

particle measurements, barium convection data, and DC convection electric field measurements. The all-sky images and DMSP overflight data are used to establish the scale and relative orientation of the sun-aligned arc and auroral oval, as shown in Figures 1 (a) and (b). The ionospheric convection signature in the region between the sun-aligned arc and the dusk-side auroral oval is inferred from the remaining independent measurements: the motion of the three barium jets, the radar l.o.s. velocities, and the rocket electric field data.

A detailed description of the barium release experiment is given by Mikkelsen [1987]. The horizontal motion of the three barium jets (projected along the field line to 100 km altitude) is shown in geographic coordinates in Figure 2. The rocket's ground trajectory, the three observation sites, and the rocket range east of the radar have also been noted. The horizontal drift speed and direction was determined by plotting the position of the three ionized jets (labeled with time markers in minutes after 2300 UT) in planes perpendicular and parallel to the local magnetic meridian plane; the result was found to be an average plasma drift speed of 440 m/s in a direction 50° east of south (i.e., the antisunward direction).

The plasma convection between the precipitation events and the large sun-aligned arc is corroborated by the radar l.o.s. velocity measurements. Prior to and during the instrumented rocket flight, the Sondrestrom radar operated in a mode combining both azimuth and elevation scans, directly measuring the electron number density, line-of-sight (l.o.s.) plasma velocity, and ion and electron temperatures in the region northwest of the rocket trajectory (see Figure 1(a)). Figure 3 shows l.o.s. velocity measurements for two combined azimuth scans from 2333:07 - 2340:45 UT. The l.o.s. vectors indicate antisunward flow at approximately 500 m/s in a direction slightly east of the magnetic meridian, or about 50° east of south. Figures 4 (a) and (b) show electron number densities and l.o.s. velocities from an elevation scan from 2346:17 -

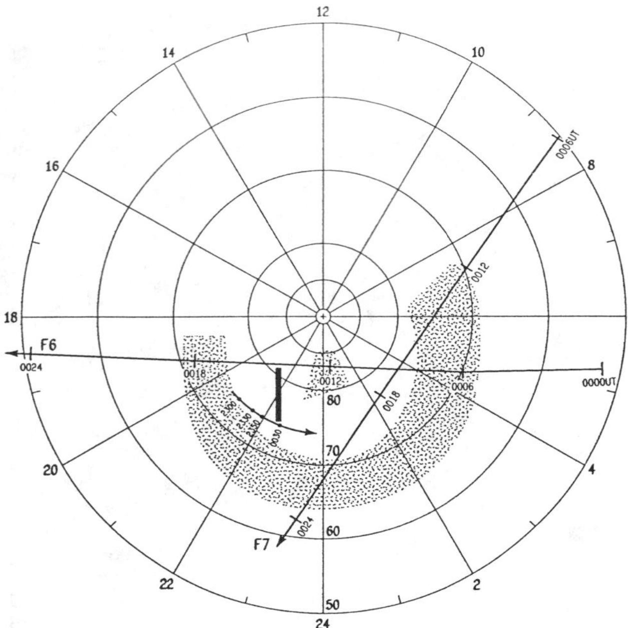

Fig. 1(b). The location of the sun-aligned arc (dark gray bar) and the tracks of the Sondrestrom ground station (from 2300 - 0030 UT) and DMSP F-6 and F-7 satellites in Magnetic Local Time / Invariant Latitude (MLT / IL) format. The observed portion of the transpolar arc and the auroral oval (as inferred from the DMSP spectrograms) are shown in light gray.

2349:11 UT in a plane perpendicular to the magnetic meridian. The region of enhanced density at a range of 600 km is associated with the large, sun-aligned arc seen in the ASIP images. Negligible l.o.s. velocity measurements indicate that the convection is either negligible or is nearly perpendicular to this plane, i.e. along the magnetic meridian. If the latter interpretation is true, the measurements agree with and extend the region of anti-sunward convection determined by the motion of the barium jets.

Finally, the rocket-borne electric field measurements indicate that the large-scale (convection) electric field in the region traversed by the rocket had an average westward component of 10 mV/m and average southward component of 12 mV/m, corresponding to plasma convection in a direction ~50° east of south at an average velocity of 300 m/s. Superposed on these large-scale components (not shown) are deviations occurring on time scales of 40 - 60 seconds associated with the precipitation events and small scale current structures traversed by the rocket (see Figures 5 and 6).

3.2. Smaller-scale measurements

The smaller scale data consist of measurements made by the electron spectrometer, electric field detector, magnetometer, and Langmuir probe during the instrumented rocket flight. Three electron precipitation events were detected during the flight by a tophat electrostatic analyzer [Sharber et al., 1988]. The events have the appearance of high latitude arcs but are

very narrow in extent, the widest being about 20 km. The precipitating energy flux profiles in the energy range 1.0 eV - 1.0 keV integrated over the downcoming hemisphere for the first two precipitation events are shown in the top panel of Figures 5 and 6. The events are very weak, with an average electron energy ≤ 100 eV and a maximum energy flux of 0.12 erg / cm^2 s. Since the third precipitation event was weaker yet, the electrodynamics study will be restricted to the first two events. The horizontal scale of Figures 5 and 6 corresponds to 22 km and 27 km, respectively.

During the course of the flight, measurements of the ambient electric and magnetic fields were made by a double-probe electric field detector and a three-axis fluxgate magnetometer

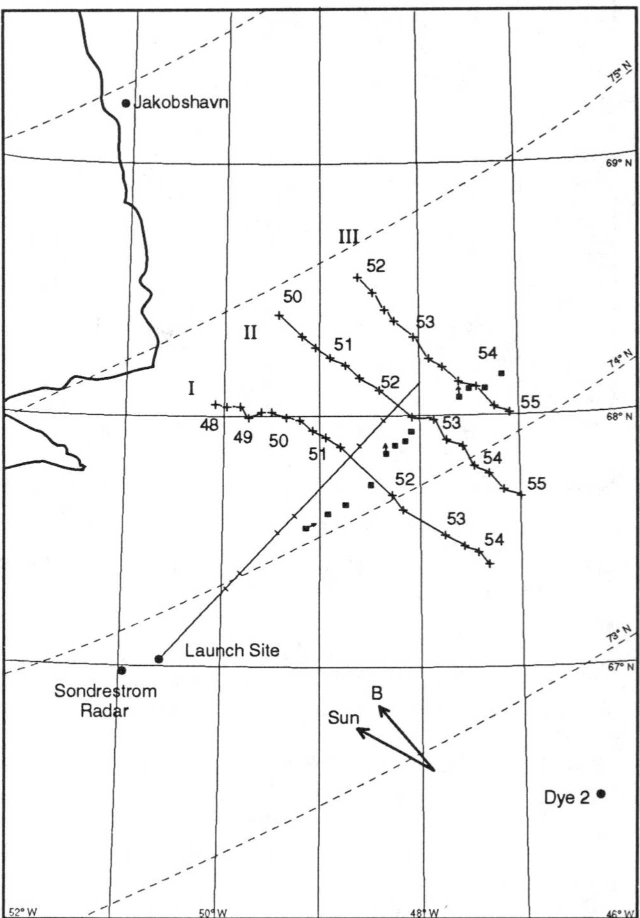

Fig. 2. The horizontal motion of the three barium jets plotted in geographic coordinates at a reference altitude of 100 km. Time markers along each trajectory are in minutes after 2300 UT. The tracks of the neutral barium / strontium clouds are indicated by squares and the firing directions of the three shaped charges are indicated by arrows. Note that the direction of magnetic north (azimuth of the line perpendicular to invariant latitude, -27°) differs from that of the magnetic meridian plane (azimuth of the plane passing through the local magnetic field line and local zenith, -39°)

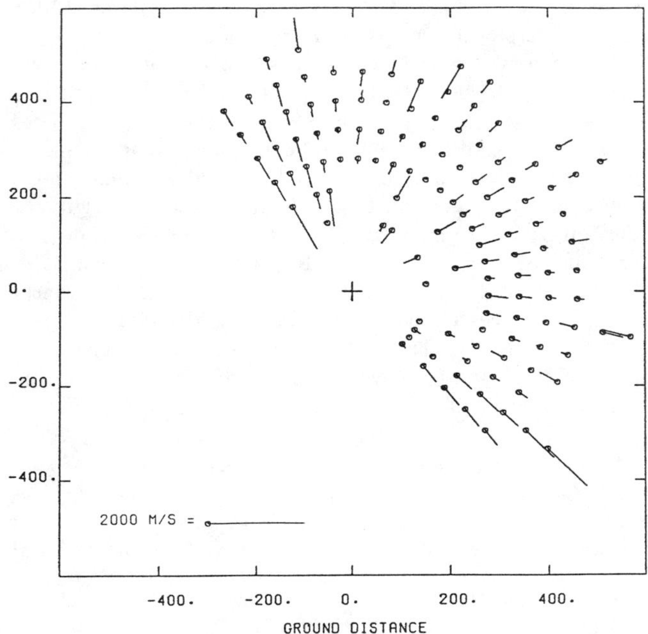

Fig. 3. Line-of-sight velocity measurements from two combined azimuth scans (at 35° elevation angle) from 2333:07 - 2340:45 UT. The sunward direction is ~60° west of north, which is at the top of the figure. Ground distances are in km.

[Primdahl, 1979]. After the appropriate data reduction, the perturbation electric and magnetic field components were rotated into a coordinate system in which x is perpendicular to the arc (positive eastwards), y is parallel to the arc (positive northward), and z is directed upward antiparallel to the magnetic field. The cross-arc electric field perturbation (δE_x) and arc-aligned magnetic field perturbation (δB_y) are shown in the second panel of Figures 5 and 6. Both sets of data have been smoothed to remove the spin period oscillation, and the average convection electric field (-15 mV/m) was subtracted from E_x for the purpose of highlighting the anticorrelation between E_x and δB_y (discussed in Section 4.2). The electric field data are consistent with arc-aligned ionospheric shear flows embedded in a region of anti-sunward convection. No significant cross-arc component of the flow (i.e., arc-aligned electric field) remained after the rotation.

The Langmuir probe measured the thermal electron density and temperature throughout the flight. No significant enhancement in electron density was detected in association with the precipitation events, a result consistent with the softness of the precipitation [e.g., Roble and Rees, 1977]. In contrast, significant increases in the electron temperature were observed in the precipitation regions due to the short time constant for heating the electron gas at F-layer altitudes. The average background Pedersen conductivity in the region traversed by the rocket was found to be 0.6 mho using the measured electron number density profile.

4. DISCUSSION

4.1. Morphology of the Polar Cap

The large-scale measurements presented in Section 3.1 are used in this section to explore the feasibility of four different models of polar cap morphology that have been used to explain the presence of weak sun-aligned arcs. These models are here called the 'open' [Hardy et al., 1982; Burke et al., 1982; Chiu, 1989; Gussenhoven and Mullen, 1989], 'bifurcated tail' [Frank et al., 1982; Kan and Burke, 1985; Frank and Craven, 1988; Toffoletto and Hill, 1990], 'expanded plasma sheet' [Meng, 1981; 1988; Murphree et al., 1982; Makita et al., 1991], and 'expanded low-latitude boundary layer (LLBL)' [Lundin and Evans, 1985; Lundin et al., 1990] models. A fifth model, which relies on the rotation of the tail x-line, is also introduced and discussed briefly [Reiff et al. 1992].

Figure 7 schematically summarizes these five different models. The left side of the figure shows distant ($x \geq 10\ R_E$) cross-sections of the northern magnetotail lobe (y-z plane, looking from the distant tail sunward), while the right-hand column displays the corresponding topology of the high-latitude ionosphere, including the predicted ionospheric plasma flow (dotted lines). Regions of antisunward plasma

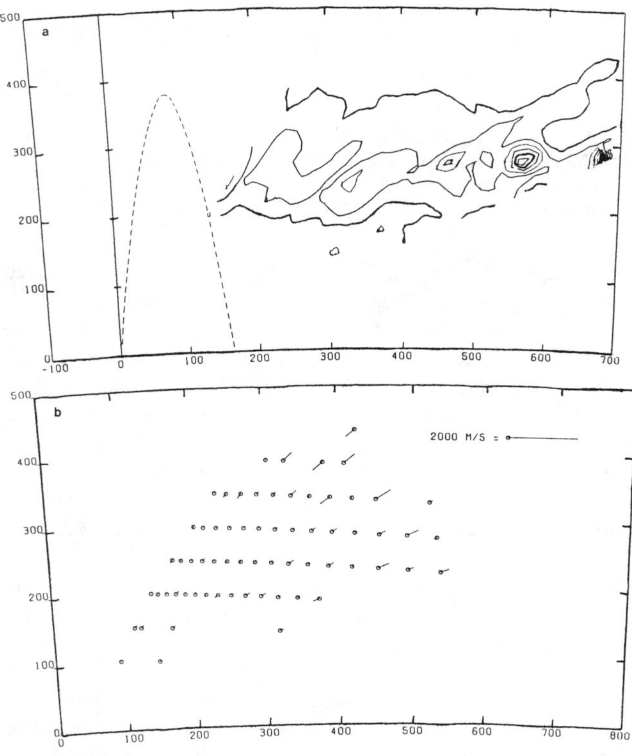

Fig. 4. (a) Electron number density and (b) line-of-sight velocity measurements from an elevation scan from 2346:17 - 2349:11 UT in a plane perpendicular to the magnetic meridian. The rocket's trajectory is shown in the top panel. The density contours are plotted in increments of 0.5×10^5, reaching a peak density of 3.0×10^5 cm^{-3}.

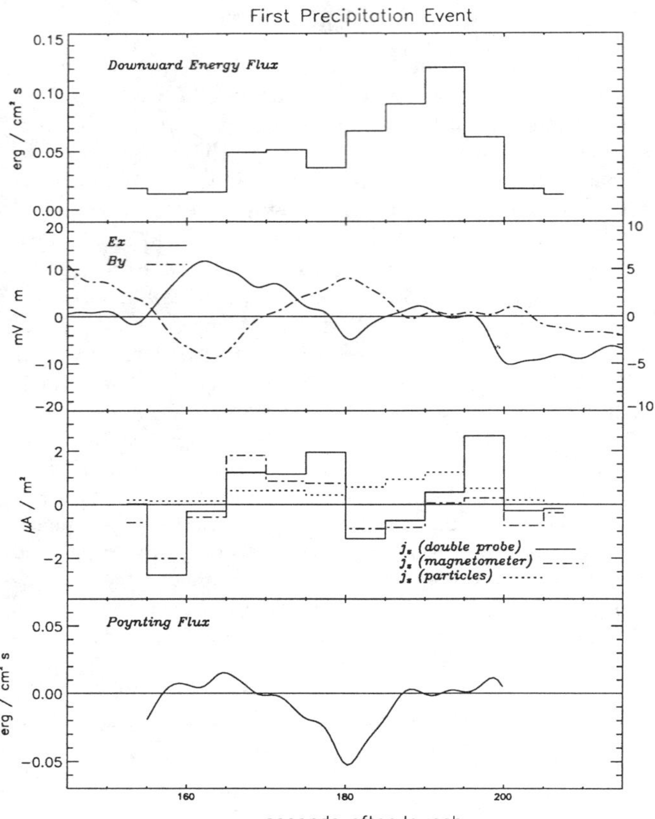

Fig. 5. Relationship between (*a*) particle precipitation, (*b*) cross-arc and parallel electric and magnetic field perturbations, δE_x and δB_y, (*c*) field-aligned current density, and (*d*) Poynting flux for the first precipitation event.

convection in the tail (LLBL and lobe) are labeled with circled dots, while regions of sunward convection (PSBL and plasma sheet) are denoted by circled crosses. In both the magnetotail and ionosphere, closed field line regions are shown in light gray (LLBL), medium gray (plasma sheet), or dark gray (PSBL), while open field line regions are shown in white. We have denoted the ionospheric projection of the PSBL as the boundary region separating open and closed field lines. Counter-streaming particle distributions and velocity dispersed ions, on the other hand, would only be observed in that portion of the PSBL in which there is a merging outflow. The (observed) sun-aligned arc is designated by the short solid line at ~22 MLT and the (inferred) theta aurora is shown as a solid line extending from noon to midnight across the polar cap.

Polar cap arcs have been argued to occur on open field lines on the basis of two types of observations: the measurement of electron spectra within polar cap arcs with the appearance of an accelerated polar rain (magnetosheath) distribution [Hardy et al., 1982; Burke et al., 1982] and, secondly, observations of sun-aligned arcs imbedded in regions containing simultaneous solar flare (> 100 keV) and polar rain electron precipi-

tation [Gussenhoven and Mullen, 1989]. If the Polar ARCS data are interpreted in terms of an open model, shown in Figure 7 (a), the weak, sun-aligned arc would map along open field lines to a position in the tail lobe denoted by the black dot. Antisunward convection of plasma along these open field lines maps directly to the polar ionosphere, where the arc is shown imbedded in a region of antisunward convection. Although the ionospheric convection in Figure 7 (a) has been depicted as a four-cell pattern with an extended region of sunward flow over the center of the polar cap (i.e., $B_y \approx 0$), alternate versions of the open model for northward IMF with a strong B_y component consist of three-cell convection patterns [Reiff and Burch, 1985; Chiu, 1985]. The (inferred) theta aurora is depicted at the central convection reversal of the dawn NBZ Birkeland current region [Reiff and Burch, 1985; Lyons, 1985; Zanetti et al., 1990].

In a bifurcated tail interpretation, Figure 7 (b), the sun-aligned arc also maps to the open tail lobe, but the transpolar aurora instead maps to a region of closed field lines extending upward from the central plasma sheet. The bifurcated tail model was originally developed to explain observations of

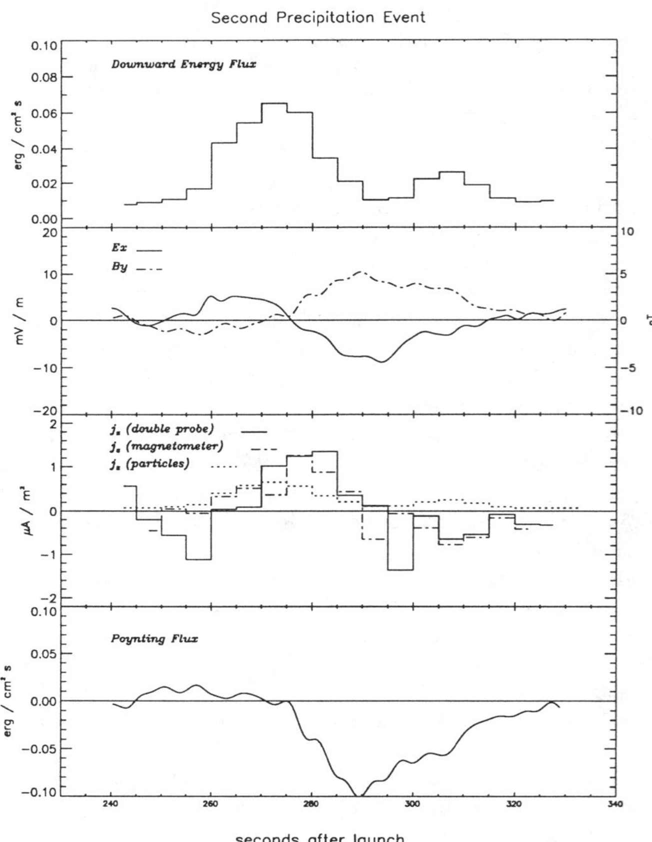

Fig. 6. Relationship between (*a*) particle precipitation, (*b*) cross-arc and parallel electric and magnetic field perturbations, δE_x and δB_y, (*c*) field-aligned current density, and (*d*) Poynting flux for the second precipitation event.

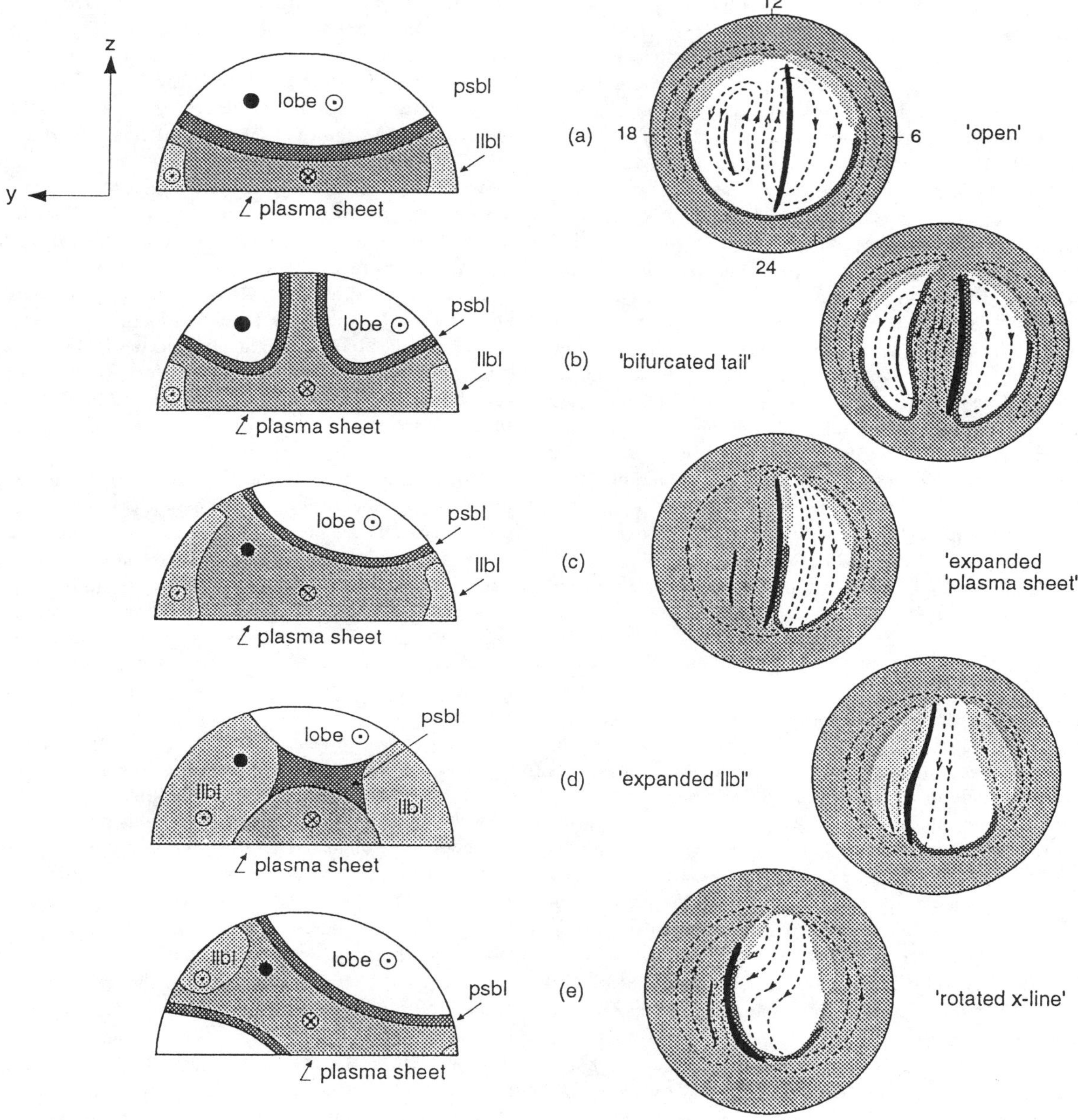

Fig. 7. Schematic diagram of the mapping of the high-latitude ionosphere to the distant magnetotail for the (a) 'open', (b) 'bifurcated tail', (c) 'expanded plasma sheet', (d) 'expanded LLBL', and (e) 'rotated x-line' models of polar cap morphology. A cross-section of the northern magnetotail lobe (y-z plane, looking from the tail sunward) is shown on the left, and the corresponding ionospheric convection pattern (dotted lines) and location of the sun-aligned and theta arc (solid lines), is shown on the right. Closed field line regions are shown in light gray (LLBL), medium gray (plasma sheet), and dark gray (PSBL), while open field line regions are shown in white. The solid dot in each figure on the left represents a field line threading the weak, sun-aligned arc at 22 MLT.

~keV ion precipitation and sunward plasma convection within arcs which appeared to be flanked by open field lines [e.g., Kan and Burke, 1985; Frank et al., 1988; Frank and Craven, 1988]. This model has been supported by magnetotail observations of plasma sheet protrusions into the lobe [Huang et al., 1987, 1989], and antiparallel merging models suggesting a split dayside merging line [Crooker, 1979; Toffoletto and Hill, 1990]. In a bifurcated tail scenario, weak, sun-aligned arcs observed on either side of a theta aurora occur on open field lines.

Both the expanded plasma sheet model and the expanded LLBL model (Figures 7 (c) and (d), respectively) propose that sun-aligned arcs occur on closed field lines threading a widened auroral oval. A closed field line source region of polar cap arcs has been supported by observations of trapped-particle signatures and the 'filled-in' nature of optical, electron, and ion measurements in the region between the average auroral oval and the polewardmost bright arc [Peterson and Shelley, 1984; Eliasson et al., 1987; Makita et al., 1991; Hones et al., 1989]. The difference between the expanded plasma sheet and LLBL models is really an issue of basic magnetospheric configuration; i.e., where the field lines in the widened auroral oval map to in the magnetosphere. In an expanded plasma sheet model, Figure 7 (c), the sun-aligned arc maps on closed field lines to an asymmetrically expanded plasma sheet in the tail. In the ionosphere, the region of open field lines is much smaller and shifted toward dawn to accommodate the expansion of the duskside oval; the theta aurora occurs at the poleward boundary of the expanded auroral oval, and the sun-aligned arc maps to the (sunward convecting) plasma sheet. This model was originally based on the continuous nature of simultaneous optical and electron precipitation measurements across the polar cap during quiet geomagnetic conditions [Meng, 1981, 1988; Murphree and Cogger, 1981; Murphree et al., 1982] and has been supported by observations of the poleward expansion of the soft (< 500 eV) electron precipitation region during northward IMF conditions [Makita et al., 1988; Rich et al., 1990].

In the expanded LLBL model, Figure 7 (d), the low-latitude boundary layer is significantly larger in the tail cross-section with the result that the sun-aligned arc field line maps to the antisunward convection region within the LLBL. The theta aurora, which maps to the boundary between open and closed field lines, is connected to the LLBL on the dayside and the plasma sheet boundary layer (PSBL) on the nightside [Lundin et al., 1990]. This model relies on an alternate view of solar wind – magnetosphere interaction in which the transfer of solar wind energy to the magnetosphere is primarily accomplished not through magnetic merging but through viscous interactions and actual mass transfer in the LLBL [Axford and Hines, 1961; Eastman et al., 1976; Sonnerup, 1980; Zhu and Kivelson, 1988; Stasiewicz, 1989]. Specifically, sun-aligned arcs map to polarization features resulting from finger-like injections of plasma into the LLBL [Lundin and Evans, 1985], which has been shown to widen during quiet times [Williams et al., 1985; Mitchell et al., 1987]. Vasyliunas [1979] argued that the LLBL should map to the ionosphere as a crescent shaped region near local noon; this projection would seem to be confirmed by studies of auroral oval-magnetosphere mapping using low-altitude particle measurements [e.g., Newell and Meng, 1992]. Studies using the Tsyganeko '87 and '89 magnetic field models have shown, however, that the ionospheric footprint of the LLBL (closed field lines near the dawn and dusk flanks of the magnetotail) may be significantly larger during quiet times, filling the region between a teardrop shaped polar cap and the auroral oval [Elphinstone et al., 1991; Birn et al.,1991].

The weak, sun-aligned arc system of February 27, 1987 can be examined in light of these models of polar cap arc forma-

tion using the convection information presented in Section 3.1. Though arguably limited in spatial extent, the barium motion, radar l.o.s. velocities, and rocket electric field measurements indicate an extended region (~400 - 500 km) of antisunward convection in the region between the large sun-aligned arc and the duskside auroral oval. Furthermore, the DMSP F-6 electron spectrogram shows an extended region of soft, sporadic electron precipitation poleward of the duskside auroral oval up to a magnetic latitude of ~82°, where strong electron and ion precipitation signatures have been identified as a theta aurora over the center of the polar cap. The sun-aligned arc identified by ASIP and the soft precipitation events encountered by the rocket lie within this precipitation zone (though further into the nightside than the DMSP pass). These observations may be accounted for by sun-aligned arc formation on closed field lines threading an expanded (antisunward convecting) LLBL, but are also consistent with a model in which they formed on open field lines on the dusk side of a bifurcated tail lobe (Figure 7 (b)). A model in which the polar cap is entirely open (Figure 7 (a)) appears to be ruled out, since it is inconsistent with the strong ion precipitation observed over the center of the polar cap (unless you invoke the Gussenhoven and Mullen [1989] scenario of an earthward-streaming lobe ion component). The antisunward convection signature appears least consistent with a model in which the polar cap arc field lines map to an expanded plasma sheet (Figure 7 (c)), since one would expect sunward convection in this region.

Although it would seem clear (from both modeling and magnetotail measurements) that the ionospheric footprint of the LLBL should expand during quiet times, it is still questionable whether arcs deep within the nightside polar cap can be magnetically connected to this source. We are beginning to explore a different model of polar cap arc formation, shown in Figure 7 (e), which accounts for sun-aligned arcs (as well as their IMF B_y dependence) at the poleward boundary of a closed field line region [Reiff et al., 1992]. In this model, a transpolar arc occurs as the result of the conductivity gradient across the ionospheric projection of the tail x-line, which is both sun-aligned and separated from the auroral oval for near-northward IMF B_z [Toffoletto and Hill, 1992]. In this skewed plasma sheet geometry, the region between the sun-aligned arc (open-closed boundary) and the auroral oval (convection reversal boundary) maps to very long, plasma sheet field lines recently reconnected through the rotated x-line. In this sense the model is topologically an 'expanded plasma sheet' model, with the significant difference that the region between the x-line and the auroral oval is still convecting antisunward in the polar cap. Flow irregularities of plasma sheet or ionospheric origin could account for the presence of weak sun-aligned arcs in this closed region.

4.2. Electrodynamic Structure

Figures 5 and 6 summarize the relationship between particle precipitation, cross-arc and parallel electric and magnetic field perturbations (δE_x and δB_y), field-aligned current density, and Poynting flux for the first and second precipitation events, respectively. Note the high degree of anti-correlation between δB_y and δE_x in Figures 5 and 6. Smiddy et al. [1980] and Sugiura et al. [1982] have shown that anti-correlation between the perturbation magnetic field parallel to an infinite field-aligned current sheet and the cross-sheet component of the electric field results from a combination of current continuity, Ohm's law, and Ampere's law such that $\Delta E_x/\Delta \delta B_y = -(\mu_0 \Sigma_p)^{-1}$, where Σ_p, the height-integrated Pedersen conductivity, is assumed to be uniform over the interval Δx. This relationship has been demonstrated with electric and magnetic field data acquired by the DE, HILAT and VIKING satellites [Sugiura et al., 1982; Sugiura, 1984; Vickrey et al., 1986; Bythrow et al., 1984] and several rocket experiments [Primdahl and Marklund, 1986; Primdahl et al., 1987]. The Polar ARCS measurements extend those results, showing that the relationship holds for δB perturbations of magnitude ≤ 10 nT. Using the measured values of δB_y and E_x, the calculated height-integrated Pedersen conductivities in the two events fall between 0.4 - 0.5 mho. As one would expect for such weak events, these values of Σ_p are comparable (within experimental uncertainty) to the average background Pedersen conductivity (0.6 mho) calculated from the Langmuir probe number density profile.

The current density in each event, shown in the third panel of Figures 5 and 6, respectively, was calculated using three different measurements: (1) magnetic field perturbations in the arc coordinate system ($j_z = \partial B_y/\mu_0 \, \partial x$); (2) spatial variations in the cross-arc electric field and/or conductivity ($-j_z = \Sigma_p \, \partial E_x/\partial x + E_x \, \partial \Sigma_p/\partial x$); and (3) precipitating electron fluxes ($j_z = env_d$, where nv_d is the electron flux over the downcoming hemisphere). The first two calculations yield well-correlated, paired, upward and downward field-aligned currents of densities between 1-2 $\mu A / m^2$, with the upward currents associated with regions of $\nabla \cdot \mathbf{E} < 0$. The currents computed using the number flux measurements are upward throughout both events due to the fact that a significant portion of the downward current is carried by thermal electrons with energies less than 10 eV [Burch et al., 1983]; with the addition of an onboard noise source, low-energy upflowing electrons and thus downward current regions were undetectable [Weiss, 1991]. Magnetosphere-ionosphere coupling models have shown that locally-closed, paired, upward and downward field-aligned currents in polar cap arcs can result from either mesoscale velocity shear structures in the ionosphere [Chiu, 1989], or from multiply-bouncing Alfvén waves originating at mesoscale velocity shear structures in the magnetosphere [Zhu et al., 1993]. The latter model predicts a paired current structure scale size of tens of kilometers, with a spacing between arcs on the order of 50 km.

The energy flux into the ionosphere due to particle precipitation is plotted in the top panels of Figures 5 and 6 to show the relationship between particle precipitation and the field-aligned current density. One would normally expect the maximum particle energy flux to be coincident with an upward field-aligned current. In both events, however, the maximum energy flux is displaced by ~5 s from the region of peak upward current, a situation not observationally uncommon in auroral arcs [Lu et al., 1991]. This displacement may be explained by either an asymmetric parallel potential drop or by mapping to a nonuniform density structure in the magnetosphere. A second form of energy input into the high-latitude ionosphere is through the Joule dissipation of electromagnetic energy. Kelley et al. [1991] have shown that a reliable measure of electromagnetic energy flux into the ionosphere is the vertical component of the Poynting flux, in this case, $P_z = E_x \delta B_y / \mu_0$, plotted in the bottom panels of Figures 5 and 6 in units of erg / cm^2-s. In both events the Poynting flux is downward, indicating a transfer of energy from the magnetosphere to the ionosphere. The maximum electromagnetic energy input is comparable to the kinetic input due to particle precipitation (but both are significantly smaller than the 1-10 erg / cm^2-s typically observed in auroral oval arcs). Note also that the Poynting flux maximizes in the region between upward and downward currents (where the product of the quantities E_x and δB_y have their largest values), while the maximum energy flux due to electron precipitation is displaced toward the upward current region, a situation expected in a paired current sheet geometry.

5. SUMMARY AND CONCLUSIONS

Radar, optical, and in situ (rocket and satellite) data from the Polar ARCS campaign have been presented and used to analyze the electrodynamic structure of a weak, sun-aligned arc within the duskside polar cap as well as its relationship to large-scale magnetospheric boundaries. Electron precipitation measurements indicate that the rocket passed through three narrow ($\leq$ 20 km) regions of low-energy ($\leq$ 100 eV) electron precipitation. These events are interpreted as small-scale structures that lie within a region of weak, sun-aligned precipitation in the region between the auroral oval and the brightest sun-aligned arc in the all-sky images. DMSP F-6 measurements of enhanced ion and electron precipitation over the center of the polar cap further suggests the presence of a transpolar ('theta') aurora at this time.

The relationship between the high latitude ionosphere and the magnetosphere, especially the question of whether field lines on which polar cap arcs occur are open or closed, has been widely debated and remains one of the outstanding topics of magnetospheric physics today. In this paper we have used one aspect of the Polar ARCS measurements, namely the anti-sunward convection signature in the region between the sun-aligned arc and the dusk-side auroral oval, as a means of evaluating different models of polar cap arc formation. The observed convection (as determined by the radar l.o.s. velocity measurements, barium convection data, and electric field double probe measurements) is least consistent with a model in which the auroral field lines map to an expanded plasma sheet [Meng, 1981; Murphree et al., 1982], since one would expect sunward convection on these reconnected field lines. An exception to the expected sunward convection signature on plasma sheet field lines might arise if the polar cap arc maps to field lines recently reconnected through a highly skewed x-line [Reiff et al., 1992]; in this case, the ionospheric end of the field line may still be convecting antisunward even though it is moving earthward in the plasma sheet. The antisunward convection signature is also consistent with a model in which the weak, sun-aligned arc system formed on open field lines on the dusk side of a bifurcated tail lobe [e.g., Kan and Burke, 1985; Frank and Craven, 1988; Toffoletto and Hill, 1990] or a model in which it formed on closed field lines threading the low-latitude boundary layer [Lundin and Evans, 1985; Lundin et al. 1990]. In the latter case, however, we note that while the teardrop shape of the quiet-time auroral oval can be accounted for by the expansion of the LLBL away from the dayside cleft, it seems unlikely that arcs deep within the nightside polar cap can be topologically attributed to this source. To date, the authors are unaware of an extended (statistical) study of ionospheric convection in the region between sun-aligned arcs and the nearby auroral oval using satellite data. As demonstrated in this paper, such a study may help evaluate competing models of polar cap arc formation and thus determine the source region(s) of these arcs.

Very weak but highly correlated electric and magnetic field perturbations made aboard the Polar ARCS instrumented rocket indicate that the weak, F-layer precipitation events crossed by the rocket are associated with small-scale, paired, field-aligned current sheets. The correlation of δB_y and E_x imply uniform Σ_p and provides a measure of that quantity, in this case ~0.4 - 0.5 mho, values consistent with that calculated from the Langmuir probe measurements. The current density in each event was calculated using the magnetic field (Ampere's Law), electric field (current divergence plus Ohm's law), and particle precipitation measurements, yielding well correlated field-aligned currents of densities between 1-2 $\mu A/m^2$. The upward and downward current regions can be associated with velocity shears (higher and lower speed streams) embedded in a region of anti-sunward flow. Such mesoscale shear flows in association with polar cap arcs have been confirmed by many observations [Carlson et al., 1984; Hoffman et al., 1985; Mende et al., 1988; Weber et al., 1989 Valladares and Carlson, 1991]. These velocity shears appear to be of magnetospheric origin since the Poynting flux is predominately downward in the vicinity of the arcs. If the weak, sun-aligned arcs formed on open field lines in a bifurcated tail geometry, the velocity shears map to regions of the magnetosheath where nonuniform flow may be easily accounted for. If the arcs map to the LLBL, the velocity shears may result from finger-like injections of magnetosheath plasma onto closed LLBL field lines, or, if they map to a highly skewed plasma sheet, the velocity shears may originate as bursty flows in the tail. In all cases, the visual emissions are produced in the upward current regions by precipitating electrons carrying the outgoing current from the electric field divergence.

Acknowledgments. The authors would like to acknowledge the Danish Commission for Scientific Research in Greenland for permission to conduct the Polar ARCS experiments. We also thank Jim Vickrey of SRI, Bob Robinson of LPARL, Rudy Frahm of SwRI, and the two referees for their help in improving and evaluating this paper. Information critical to the analysis of the magnetometer and electron spectrometer data was supplied by Poul Jensen, Carl Bargainer, John Scherrer, and Nick Eaker of SwRI. We thank DuWayne Bostow of the Geophysical Institute for managing the high explosive shaped charges and machining the barium metal cones, and Gale Weeding of the Denver Research Institute for fabricating the high explosive Octol charges. We would also like to thank the personnel of NASA Wallops Flight Facility, and in particular Jay Brown, the payload manager for payload fabrication and launch support. This research was supported at Southwest Research Institute under Phillips Laboratory grant AFOSR-87-0203, at Rice University by NASA under grant NAGW 1655 and by the National Science Foundation under grant ATM-91-03440, at Phillips Laboratory by AFOSR task 2310G9, and at the University of Alaska at Fairbanks by NASA under grant NAG6-1.

REFERENCES

Axford, W. I., and C. O. Hines, A unifying theory of high-latitude geophysical phenomena and geomagnetic storms, *Can. J. Phys., 39*, 1433, 1961.

Birn, J., E. W. Hones, Jr., J. D. Craven, L. A. Frank, R. D. Elphinstone, and D. P. Stern, On open and closed field line regions in Tsyganenko's field model and their possible associations with horse collar auroras, *J. Geophys. Res., 96*, 3811, 1991.

Burch, J. L., P. H. Reiff, M. Sugiura, Upward electron beams measured by DE-1: A primary source of dayside region-1 Birkeland currents, *Geophys. Res. Lett., 10*, 753, 1983.

Burke, W. J., M. C. Kelley, R. C. Sagalyn, M. Smiddy, and S. T. Lai, Polar cap electric field structures with a northward interplanetary magnetic field, *Geophys. Res. Lett., 6*, 21, 1979.

Burke, W. J., M. S. Gussenhoven, M. C. Kelley, D. A. Hardy, and F. J. Rich, Electric and magnetic field characteristics and discrete arcs in the polar cap, *J. Geophys. Res., 87*, 2431, 1982.

Bythrow, P. F., T. A. Potemra, W. B. Hanson, L. J. Zanetti, C.-I. Meng, R. E. Huffman, F. J. Rich, and D. A. Hardy, Earthward directed high-density Birkeland currents observed by HILAT, . *Geophys. Res., 89*, 9114, 1984.

Carlson, H. C., V. B. Wickwar, E. J. Weber, J. Buchau, J. G. Moore, and W. Whiting, Plasma characteristics of polar-cap F-layer arcs, *Geophys. Res. Lett., 9*, 895, 1984.

Chiu, Y. T., Formation of polar cap arcs, *Geophys. Res. Lett., 16*, 125, 1989.

Chiu, Y. T., and D. J. Gorney, Eddy intrusion of hot plasma into the polar cap and formation of polar cap arcs, *J. Geophys. Res., 85*, 543, 1980.

Chiu, Y. T., and D. J. Gorney, Eddy intrusion of hot plasma into the polar cap and formation of polar cap arcs, *Geophys. Res. Lett., 10*, 463, 1983.

Chiu, Y. T., N. U. Crooker, and D. J. Gorney, Model of oval and polar cap arc configurations, *J. Geophys. Res., 90*, 5153, 1985.

Eastman, T. E., E. W. Hones, Jr., S. J. Bame, and J. R. Asbridge, The magnetospheric boundary layer: site of plasma, momentum and energy transfer from the magnetosheath into the magnetosphere, *Geophys. Res. Lett., 3*, 685, 1976.

Eliasson, L., R. Lundin, and J. S. Murphree, Polar cap arcs observed by the Viking satellite, *Geophys. Res. Lett., 14*, 451, 1987.

Elphinstone, R. D., K. Jankowska, J. S. Murphree, and L. L. Cogger, The configuration of the auroral distribution for interplanetary magnetic field B_z northward: 1. IMF B_x and B_y dependencies as observed by the Viking satellite, *J. Geophys. Res., 95*, 5791, 1990.

Elphinstone, R. D., D. Hearn, J. S. Murphree, and L. L. Cogger, Mapping using the Tsyganenko long magnetospheric model and its relationship to Viking Auroral Images, *J. Geophys. Res., 96*, 1467, 1991.

Frank, L. A., J. D., Craven, J. L. Burch, and J. D. Winningham, Polar views of the Earth's aurora with Dynamics Explorer, *Geophys. Res. Lett., 9*, 1001, 1982.

Frank, L. A., J. D. Craven, D. A. Gurnett, S. D. Shawhan, D. R. Weimer, J. L. Burch, J. D. Winningham, C. R. Chappell, J. H. Waite, R. A. Heelis, N. C. Maynard, M. Sugiura, W. K. Peterson, and E. G. Shelley, The theta aurora, *J. Geophys. Res., 91*, 3177, 1986.

Frank, L. A., and J. D. Craven, Imaging results from Dynamics Explorer 1, *Rev. Geophys., 26*, 249, 1988.

Gussenhoven, M. S., and E. G. Mullen, Simultaneous relativistic electron and auroral particle access to the polar caps during interplanetary magnetic field B_z northward: A scenario for an open field line source of auroral particles, *J. Geophys. Res.*, *94*, 17,121, 1989.

Hardy, D. A., W. J. Burke, and M. S. Gussenhoven, DMSP optical and electron measurements in the vicinity of polar cap arcs, *J. Geophys. Res.*, *87*, 2413, 1982.

Hardy, D. A, Intense fluxes of low-energy electrons at geomagnetic latitudes above 85°, *J. Geophys. Res.*, *89*, 3883, 1984.

Heelis, R. A., J. D. Winningham, W. B. Hanson, and J. L. Burch, The relationships between high-latitude convection reversals and the energetic particle morphology observed by Atmospheric Explorer, *J. Geophys. Res.*, *85*, 3315, 1980.

Hoffman, R. A., R. A. Heelis, and J. S. Prasad, A sun-aligned arc observed by DMSP and AE-C, *J. Geophys. Res.*, *90*, 9697, 1985.

Hones, E. W., Jr, J. D. Craven, L. A. Frank, D. S. Evans, and P. T. Newell, The horse-collar aurora: A frequent pattern of the aurora in quiet times, *Geophys. Res. Lett.*, *16*, 37, 1989.

Huang, C. Y., L. A. Frank, W. K. Peterson, D. J. Williams, W. Lennartsson, D. G. Mitchell, R. C. Elphic, and C. T. Russell, Filamentary structures in the magnetotail lobes, *J. Geophys. Res.*, *92*, 2349,1987.

Huang, C. Y., L. A. Frank, J. D. Craven, Simultaneous observations of a theta aurora and associated magnetotail plasmas, *J. Geophys. Res.*, *94*, 10137,1989.

Iijima, T., T. A. Potemra, L. J. Zanetti, and P. F. Bythrow, Large-scale Birkeland in the dayside polar region during strongly northward IMF: A new Birkeland current system, *J. Geophys. Res.*, *89*, 7441, 1984.

Kan, J. R., and W. J. Burke, A theoretical model of polar cap auroral arcs, *J. Geophys. Res.*, *90*, 4171, 1985.

Kelley, M. C., D. J. Knudsen, and J. F. Vickrey, Poynting flux measurements on a satellite: a diagnostic tool for space research, *J. Geophys. Res.*, *96*, 201, 1991.

Lassen, K., and C. Danielsen, Quiet time pattern of auroral arcs for different directions of the interplanetary magnetic field in the Y-Z plane, *J. Geophys. Res.*, *83*, 5277, 1978.

Lassen, K., and C. Danielsen, Distribution of auroral arcs during quiet geomagnetic conditions, *J. Geophys. Res.*, *94*, 2587, 1989.

Lu, G., P. H. Reiff, J. L. Burch, and J. D. Winningham, On the auroral current-voltage relationship, *J. Geophys. Res.*, *96*, 3523, 1991.

Lundin, R., and D. S. Evans, Boundary layer plasmas as a source for high-latitude, early afternoon, auroral arcs, *Planet. Space Sci.*, *32*, 1389, 1985.

Lundin, R., L. Eliasson, J. S. Murphree, The quiet-time aurora and the magnetosphere configuration, in *Auroral Physics*, ed. C.-I. Meng, M. J. Rycroft, and L. A. Frank, p.159, Cambridge University Press, Cambridge, U.K., 1990.

Lyons, L. R., Generation of large-scale regions of auroral currents, electric potentials, and precipitation by the divergence of the convection electric field, *J. Geophys. Res.*, *85*, 17, 1980.

Makita, K., C.-I. Meng, and S.-I. Akasofu, Latitudinal electron precipitation patterns during large and small IMF magnitudes for northward IMF conditions, *J. Geophys. Res.*, *93*, 97, 1988.

Makita, K., C.-I. Meng, and S.-I. Akasofu, Transpolar auroras, their particle precipitation and IMF B_y component, *J. Geophys. Res.*, *96*, 14085, 1991.

Mende, S. B., J. H. Doolittle, R. M. Robinson, R. R. Vondrak, and F. J. Rich, Plasma drifts associated with a system of sun-aligned arcs in the polar cap, *J. Geophys. Res.*, *93*, 256, 1988.

Meng, C.-I., Polar cap arcs and the plasma sheet, *Geophys. Res. Lett.*, *8*, 273, 1981.

Meng, C.-I., Auroral oval configuration during the quiet condition, in *Electromagnetic Coupling in the Polar Clefts and Caps*, ed. R. E. Sandholt and A. Egeland, p.61, Kluwer Academic Publishers, Dordrecht, Netherlands, 1988.

Mikkelsen, I. S., Drift of the Ba-jets released over Greenland on February 26 and March 5, 1987, *Progress Report for Grant AFOSR-87-0203*, Danish Meteorological Institute, Copenhagen, Denmark, 1987.

Mitchell, D. G., F. Kutchko, D. J. Williams, T. E. Eastman, L. A. Frank, and C. T. Russell, An extended study of the low-latitude boundary layer on the dawn and dusk flanks of the magnetosphere, *J. Geophys. Res.*, *92*, 7394, 1987.

Murphree, J. S., and L. L. Cogger, Observed connections between apparent polar cap features and the instantaneous diffuse auroral oval, *Planet. Space Sci.*, *29*, 1143, 1981.

Murphree, J. S., C. D. Anger, and L. L. Cogger, The instantaneous relationship between polar cap and oval auroras at times of northern interplanetary magnetic field, *Can. J. Phys.*, *60*, 349, 1982.

Newell, P. T., W. J. Burke, C.-I. Meng, E. R. Sanchez, and M. E. Greenspan, Identification and observations of the plasma mantle at low altitude, *J. Geophys. Res.*, *96*, 35, 1991.

Newell, P. T., and C.-I. Meng, Mapping the dayside ionosphere to the magnetosphere according to particle precipitation characteristics, *Geophys. Res. Lett.*, *19*, 609, 1992.

Nielsen, E., J. D. Craven, L. A. Frank, and R. A. Heelis, Ionospheric flows associated with a transpolar arc, *J. Geophys. Res.*, *95*, 21169, 1990.

Peterson, W. K., and E. G. Shelley, Origin of the plasma in a cross-polar cap auroral feature (theta aurora), *J. Geophys. Res.*, *89*, 6729, 1984.

Primdahl, F., The fluxgate magnetometer, *J. Phys. E: Sci. Instrum.*, *12*, 241, 1979.

Primdahl, F. and G. Marklund, Birkeland currents correlated with direct-current electric fields observed during the CENTAUR Black Brant X rocket experiment, *Can. J. Phys.*, 64, 1412, 1986.

Primdahl, F., G. Marklund, and I. Sandahl, Rocket observations of E-B field correlations showing up- and down-going Poynting flux during an auroral breakup event, *Planet. Space Sci.*, 35, 1287, 1987.

Reiff, P. H., J. L. Burch, and R. A. Heelis, Dayside auroral arcs and convection, *Geophys. Res. Lett.*, *5*, 391, 1978.

Reiff, P. H. and J. L. Burch, IMF By-dependent plasma flow and birkeland currents in the dayside magnetosphere 2. A global model for northward and southward IMF, *J. Geophys. Res.*, *90*, 1595, 1985.

Reiff, P. H., F. R. Toffoletto, L. A. Weiss, and J. S. Murphree, A new model of polar cap convection and auroral arcs during northward IMF, *EOS, Transactions, 73*, p.466, 1992.

Rich, F. J., D. A. Hardy, R. H. Redus, and M. S. Gussenhoven, Northward IMF and patterns of high-latitude precipitation and field-aligned currents: The February 1986 Storm, *J. Geophys. Res.*, *95*, 9893, 1990.

Roble, R. G., and M. H. Rees, Time-dependent studies of the aurora: effects of particle precipitation on the dynamic morphology of ionospheric and atmospheric properties, *Planet. Space Sci.*, *25*, 991, 1977.

Sharber, J. R., J. D. Winningham, J. R. Scherrer, M. J. Sablik, C. A. Bargainer, P. A. Jensen, B. J. Mask, and N. Eaker, Angle resolving energy analyzer (AREA): A versatile "top-hat" instrument for space plasma measurement, *IEEE Transactions on Geosciences and Remote Sensing*, 474, 1988.

Smiddy, M., W. J. Burke, M. C. Kelley, N. A. Saflekos, M. S. Gussenhoven, D. A. Hardy, and F. J. Rich, Effects of high-latitude conductivity on observed convection electric fields and Birkeland currents, *J. Geophys. Res.*, *85*, 6811, 1980.

Sonnerup, B. U. O., Theory of the low-latitude boundary layer, *J. Geophys. Res.*, *85*, 2017, 1980.

Stasiewicz, K., A fluid finite ion larmor radius model of the magnetopause layer, *J. Geophys. Res.*, *94*, 8827, 1989.

Sugiura, M., N. C. Maynard, W. H. Farthing, J. P. Heppner, and B. G. Ledley, Initial results on the correlation between the magnetic and electric fields observed from the DE-2 satellite in the field-aligned current regions, *Geophys. Res. Lett.*, *9*, 985, 1982.

Sugiura, M., A fundamental magnetosphere-ionosphere coupling mode involving field-aligned currents as deduced from DE-2 observations, *Geophys. Res. Lett.*, *11*, 877, 1984.

Toffoletto, F. R., and T. W. Hill, Magnetic field configuration of the theta aurora, *Geophys. Res. Lett.*, *17*, 595, 1990.

Toffoletto, F. R., and T. W. Hill, A non-singular model of the open magnetosphere, accepted for publication, *J. Geophys. Res.*, Sept., 1992.

Torbert, R. B., C. A. Cattell, F. S. Mozer, and C.-I. Meng, The boundary of the polar cap and its relation to electric fields, field-aligned currents, and auroral particle precipitation, in *Physics of Auroral Arc Formation, Geophys, Monogr. Ser.*, vol. 25, edited by S.-I. Akasofu and J. R. Kan, p.143, Washington D.C., AGU, 1981.

Valladares, C. E., and H. C. Carlson, Jr., The electrodynamic, thermal, and energetic character of intense sun-aligned arcs in the polar cap, *J. Geophys. Res., 96,* 1379, 1991.

Vasyliunas, V. M., Interaction between the magnetospheric boundary layers and the ionosphere, in *Proceedings of Magnetospheric Boundary Layers Conference,* ESA SP-148, p.387, 1979.

Vickrey, J. F., R. C. Livingston, N. B. Walker, T. A. Potemra, R. A. Heelis, M. C. Kelley, and F. J. Rich, On the current-voltage relationship on the magnetospheric generator at intermediate spatial scales, *Geophys. Res. Lett., 13,* 495, 1986.

Weber, E. J., and J. Buchau, Polar cap F layer auroras, *Geophys. Res. Lett., 8,* 125, 1981.

Weber, E. J., M. C. Kelley, J. O. Ballenthin, S. Basu, H. C. Carlson, J. R. Fleischman, D. A. Hardy, N. C. Maynard, R. F. Pfaff, P. Rodriguez, R. E. Sheehan, and M. Smiddy, Rocket measurements within a polar cap arc: Plasma, particle, and electric field circuit parameters, *J. Geophys. Res., 94,* 6692, 1989.

Weiss, L. A., A study of high-latitude auroral arcs using radar, optical, and *in situ* techniques, *Ph.D. Thesis,* Rice University, Houston, TX, October, 1991.

Williams, D. J., D. G. Mitchell, T. E. Eastman, and L. A. Frank, Energetic particle observations in the low-latitude boundary layer, *J. Geophys. Res., 90,* 5097, 1985.

Winningham, J. D., and W. J. Heikkila, Polar cap auroral electron fluxes observed with ISIS 1, *J. Geophys. Res., 79,* 949, 1974.

Zanetti, T. A. Potemra, R. E. Erlandson, F. F. Bythrow, B. J. Anderson, J. S. Murphree, and G. T. Marklund, Polar region Birkeland current, convection, and aurora for northward interplanetary magnetic field, *J. Geophys. Res., 95,* 5825, 1990.

Zhu, L., J. J. Sojka, R. W. Schunk, and D. J. Crain, A time-dependent model of polar cap arcs, in press, *J. Geophys. Res.,* 1993.

Zhu, X., and M. G. Kivelson, Analytic formulation and quantitative solutions of the coupled ULF wave problem, *J. Geophys. Res., 93,* 8602, 1988.

Plasma Flows Associated With an Auroral Arc at the Polar Cap Boundary

H. A. GALLAGHER JR. AND R. L. CAROVILLANO

Boston College, Department of Physics, Chestnut Hill, Massachusetts

E. J. WEBER

Phillips Laboratory, Geophysics Directorate/GPI, Hanscom Air Force Base, Bedford, Massachusetts

J. F. VICKREY

Geoscience and Engineering Center, SRI International, Menlo Park, California

A specialized incoherent scatter radar (ISR) mode is used together with an all sky imaging photometer (ASIP) to obtain high spatial and temporal resolution measurements of auroral arcs. We discuss the observations obtained in this manner of an L shell aligned arc located in the post midnight sector (2 MLT). A detailed comparison of the 6300 Å and 4278 Å intensities measured by the ASIP and the electron densities and ion velocities measured by the ISR indicates that the optical auroral arc is collocated with an E region electron density enhancement and just equatorward of a latitudinally narrow region of enhanced eastward plasma flow. The line of sight (LOS) velocities indicate that the total plasma convection consists of an enhanced eastward plasma flow and the mainly antisunward background plasma flow. By subtracting the line of sight component of an assumed background ionospheric convection from the total measured line of sight velocity we are able to isolate the enhanced eastward plasma flow. The subtracted background flow was chosen to be consistent with the measured ion velocity of 350 m/s across the arc. In this way we determine the location of the eastward flow enhancement to be at the poleward edge of the auroral arc. The relative motion between the ionospheric plasma and the auroral arc is suggestive of reconnection processes occurring in the magnetosphere.

INTRODUCTION

Studies of the electrodynamics of auroras have been made at many local times with an incoherent scatter radar (ISR) using many different antenna modes in conjunction with other ground and spaced based instrumentation. Many of these studies have been summarized by Vondrak and Robinson [1989] and Marklund [1984]. De la Beaujardiere et al. [1977] employed the Chatanika ISR and an all sky camera to study auroral electrodynamics in the context of auroral location and morphology. More recently, cooperative all sky imager and ISR investigations of auroral electrodynamics have been conducted at Sondre Stromfjord, Greenland [Doolittle et al. 1990, Robinson and Mende 1990, and Weber et al. 1991]. These studies have featured ISR modes with increased spatial coverage to determine better the local ionospheric convection associated with auroral arcs. Additionally, the all sky imagers are used to suggest possible behavior of the magnetospheric source region as well as give an accurate location and morphological classification of the auroral arc. Because of its high magnetic latitude, Sondre Stromfjord is an ideal location for studying auroral arcs at or near the polar cap boundary. Robinson and Mende [1990] used azimuthal scans of the ISR to analyze auroral arcs in the dusk sector during a period of magnetic quiescence. They found the arc to be located near a reversal in the ionospheric convection with a small component of plasma flow across the auroral arc. Doolittle et al. [1990] examined an evening sector aurora using two perpendicular elevation scans. This arc was determined to be located near a shear in the component of plasma drift parallel to the arc with a significant component of flow across the arc. Weber et al. [1991] discussed two auroras differing in local times, morphologies and associated F region velocities observed at the polar cap boundary simultaneously by the Sondre Stromfjord ISR and the Phillips Laboratory all sky imaging photometer (ASIP). The first arc was an evening sector arc on 12 December 1988 located along an equipotential just equatorward of a shear in the F

Auroral Plasma Dynamics
Geophysical Monograph 80

region ion velocity. Such an arc might be generated by a gradient in the magnetospheric convection electric field [Lyons 1980]. Lyons has suggested that a convergence of the magnetospheric convection electric field, as is frequently observed as one traverses the convection reversal in the dusk sector of the magnetosphere, maps downward to the ionosphere. Under certain conditions the current required to satisfy the current continuity equation in the ionosphere can not be supplied by the ionosphere and must therefore be supplied by the magnetosphere through energetic particle precipitation. The second arc was in the midnight sector on 30 October 1989 and was also located equatorward of a shear in the F region velocity. However, superposed on the shear was a uniform equatorward plasma velocity of 500 m/s perpendicular to the arc.

In this paper we discuss a third auroral event in the post midnight sector (2 MLT) on 10 February 1991, observed using the same techniques as Weber et al. [1991]. The aurora in this study has associated magnetometer and optical signatures that are characteristic of auroras that occur during substorms [de la Beaujardiere et al. 1991, Robinson and Vondrak 1990, Robinson et al. 1990, Pudovkin et al. 1991]. After a steady poleward advance of the aurora from the southern horizon, the auroral arc, although dynamic, is relatively stationary near the Sondre Stromfjord zenith. The observed plasma flow associated with the arc is determined to consist of a latitudinally narrow region of enhanced eastward flow and a background plasma convection with a component of 350 m/s equatorward across the auroral arc. The relative motion between the intense auroral arc and the ionospheric plasma may be a signature that indicates the presence of magnetospheric reconnection [Vasyluinus 1975]. Detailed ionospheric observations of such processes may have important consequences for reconnection modelling efforts. Observations of plasma flow across the polar cap boundary are also important for ionospheric convection models which simulate reconnection in the models with a prescribed plasma flow across the polar cap boundary in the ionosphere [Moses et al. 1989 and Lockwood et al. 1990] .

OBSERVATIONS

In our experiment the ASIP recorded the 4278 Å, 5577 Å, and 6300 Å intensities with a 180° field of view lens. One image at each wavelength was obtained every 68 seconds. The ISR measured the ionospheric plasma parameters, N_e, V_i, T_e, T_i , using a specialized scanning mode described by Weber et al. [1991]. The radar scan cycle consists of three "parallel" scans each contained in a plane. Two of the planes form 28 degree angles with a vertical center plane and are symmetric about this center plane. The three planes intersect in a line that is tangent to the earth at the location of the radar. Using the real time optical images, the line of intersection is orientated so that the three planes are approximately perpendicular to the arc. Eleven minutes is required to complete the three scan cycle of the ISR. The auroral arc discussed in this paper is nearly aligned along an L-shell. In this case, the line of intersection of the three scan planes is along the magnetic meridian and the center scan is an elevation scan in the plane of the magnetic meridian. The

approximate path of the radar scans is indicated by the lines superposed on the images at 03:52:39 UT and 03:52:09 UT, which are transformed to a geographic coordinate system at 250 km and 110 km altitude, respectively (Figure 1). The radar cycle begins with the east scan, slewing north to south. Next the elevation scan is made northward along the magnetic meridian followed by the west scan which slews north to south. This mode of operation allows us to make radar measurements slightly displaced in time and in azimuth to obtain high resolution line of sight (LOS) plasma velocities both parallel and perpendicular to the arc. The simultaneous operation of the instruments enables us to make comparisons between the measured radar parameters and the observed optical boundaries.

At approximately 03:00 UT the polar cap boundary is located near the southern horizon in the 6300 Å image. The boundary is identified as the intensity gradient between the dark polar cap and the luminous aurora. The boundary advances steadily poleward in time until the arc is at the radar zenith at approximately 03:45 UT, where it remains stationary for more than 10 minutes. The Sondre Stromfjord magnetometer indicates that a small substorm occurs from 03:15 UT to approximately 04:00 UT. During this interval the magnetometer shows a negative bay in the H component of the magnetic field with a maximum negative excursion of approximately 100 nT occurring at 03:35 UT. Kp is 2 during this interval and no IMF data was available for this period. A detailed evalutation of the 4278 Å images shows that the poleward advance of the optical emissions occurs systematically as new auroral arcs form poleward of existing or previous aurora. These newly formed arcs in general are L-shell aligned, have durations of 2 to 10 minutes, show a large amount of spatial structure and move equatorward after they form. Similar behavior of auroral arcs has been reported by Pudovkin et al. [1991] and associated with auroral substorm phenomena. Although the aurora is active, the images indicate that the arc or precipitation region is quite stationary during the radar cycle from 03:47:27 UT to 03:57:59 UT. Figure 1 shows the 6300 Å and 4278 Å images corresponding to each radar scan.

The first row in Figure 2 shows the contours of electron density determined by the three radar scans as a function of ground range and altitude. Superposed on the electron density contours are the corresponding 6300 Å (bold faced) and 4278 Å relative intensities plotted as a function of ground range. In order to determine the ground range, emission heights of 250 km and 110 km were assumed for the 6300 Å and 4278 Å emissions, respectively. The 6300 Å intensities are scaled in Figure 2 (rows 1 and 2) so that zero intensity is at 250 km altitude and the maximum intensity is at 400 km altitude, whereas the 4278 Å scale has zero intensity at 110 km altitude and maximum intensity at 260 km altitude. These are not absolute but relative intensities emphasizing the character of the optical emissions corresponding to the radar scan. All three radar scans show similar characteristics in the E region electron density. Poleward of the zero km ground distance (50° azimuth) the electron density at 125 km altitude is extremely low. At 125 km altitude there is a sharp change in the electron density at a ground range of approximately zero. This gradient, which is the poleward edge of the auroral E region density enhancement, is

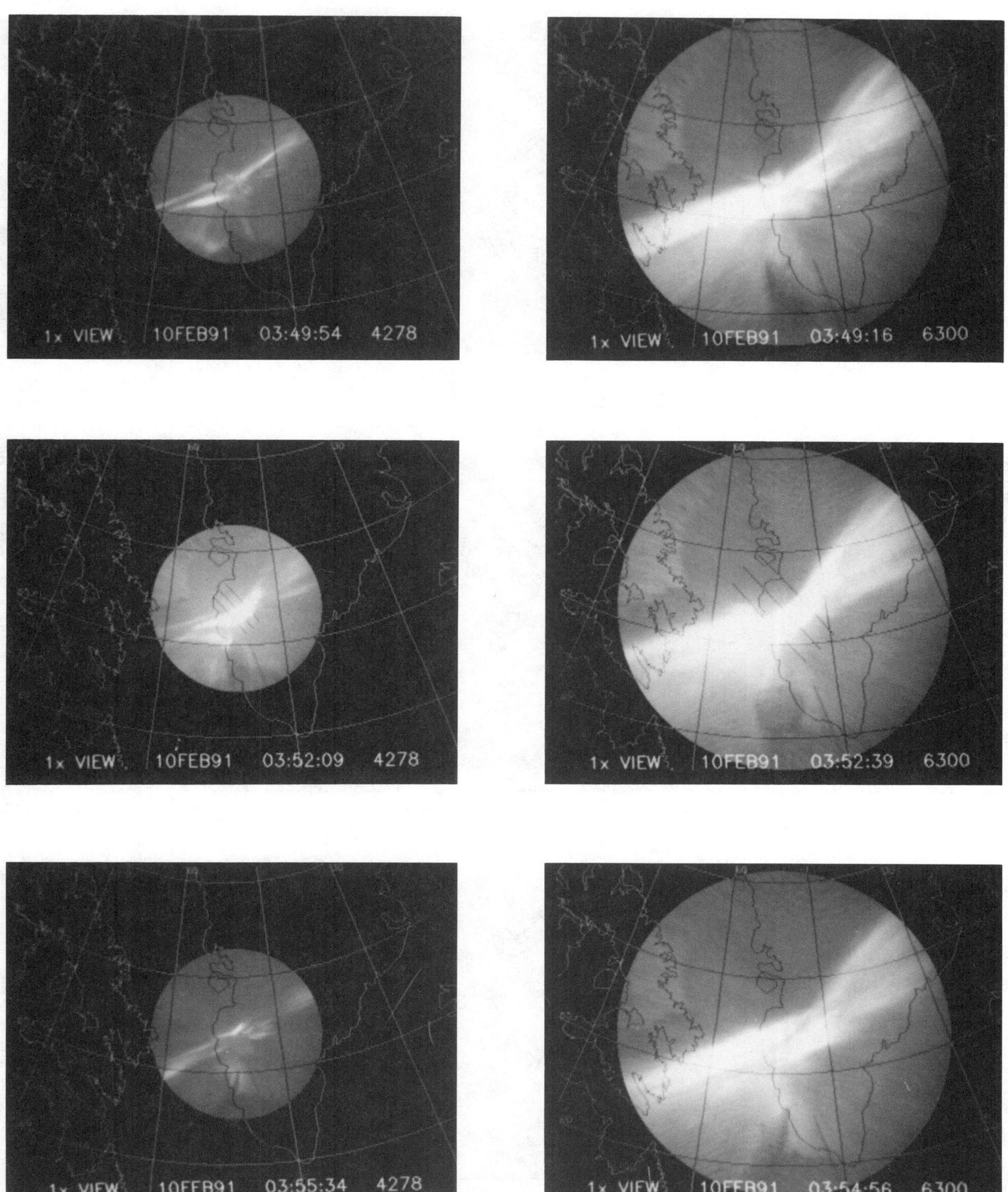

Fig. 1. The 4278 Å (left) and 6300 Å (right) images for 10 February 91 are shown in a geographical coordinate system.   Each 4278 Å and 6300 Å image pair corresponds in time to one of the three radar scans. The locations of the radar scans at 140 km and 250 km are indicated by three lines in the images at 03:52:09 UT and 03:52:39 UT respectively.

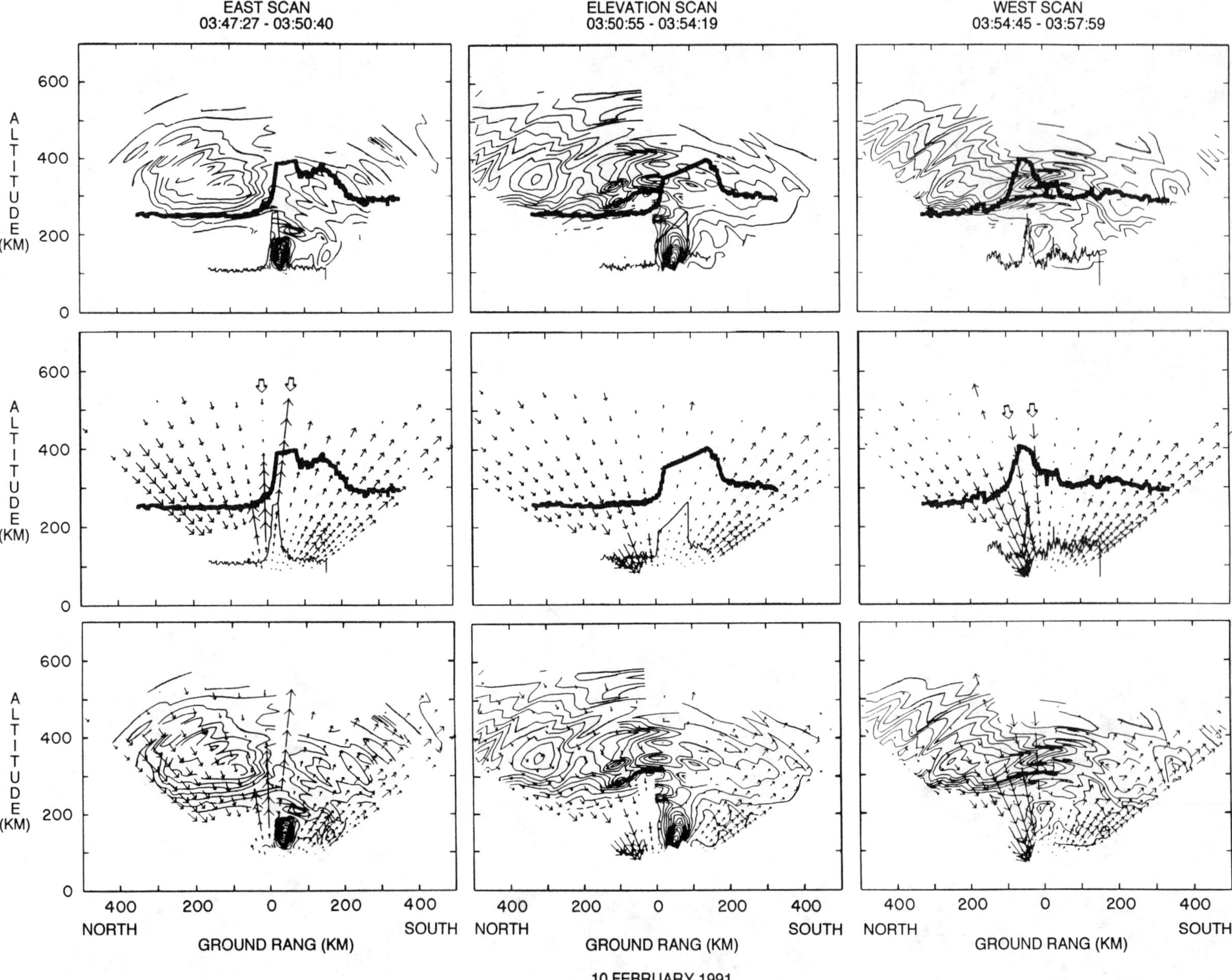

Fig. 2. Electron density contours (row 1), LOS ion velocities (row 2), and the superposition of LOS ion velocities onto electron density contours (row 3) for the radar cycle from 03:47:27 UT to 03:54:45 UT. The relative intensities of the 6300 Å (bold) and 4278 Å emissions are superposed on the electron density contours and the LOS ion velocities. An altitude of 250 km and 140 km was assumed for the 6300 Å and 4278 Å emissions respectively.

the suggested location of the separatrix mapped from the magnetosphere to the ionosphere [de la Beaujardierre 1991]. Equatorward of the arc there exist E region electron densities suggestive of softer electron precipitation. The 6300 Å intensities show similar and coincident behavior. Poleward of the arc, the 6300 Å emissions are extremely low. In the east and elevation scans the sharp gradient in the optical intensity near zenith is almost exactly aligned with the electron density gradient, and clearly delineates the polar cap boundary. The gradient in the optical intensity in the west scan is not exactly collocated with the electron density gradient. During the time interval of the west scan, a new arc formed slightly poleward of the prior location of the previous aurora. The formation of this aurora could account for the slight offset between the optical intensity gradient and the gradient in the electron density at 125 km. Although the 6300 Å intensity decreases equatorward of the arc, where the electron precipitation is softer, it does not decrease to the low level found in the polar cap.

The second row of Figure 2 shows the LOS velocities for the three scans, with the 6300 Å and 4278 Å intensities again superposed. The LOS velocities in both the east and west scans, which are directed toward the radar site poleward of the arc and away from the radar site equatorward of the arc, suggest the presence of an equatorward flow across the arc. The elevation scan which measures the ion velocity in the plane of the magnetic meridian verifies the plasma drifts across the arc. The LOS velocities measured in the magnetic zenith determine the plasma velocity parallel to the magnetic field to be approximately zero, and we assume in our analysis that the vertical velocity is zero everywhere within the field of view of the radar. The F region ion velocities measured by this scan imply a uniform horizontal component of 350 m/s southward directed at 141° azimuth. Weber et al. [1991] and de la Beajardierre et al. [1991] have also observed relatively uniform flow perpendicular to the auroral signature at the polar cap boundary.

The three scans demonstrate the presence of a strong, latitudinally narrow eastward flow about 50 km north of the zero ground range. Thus the east and west scans measure strong enhancements in the flow away and towards the radar respectively near this location. F region line of sight velocities that show evidence of this enhanced flow (indicated by the outline arrows in the east and west scans of Figure 2 row 2) have maginitudes that are significantly greater than adjacent LOS velocity measurements. Additionally, in the most poleward line of sight signature in the east scan, the LOS velocity decreases with increasing altitude and range finally reversing at 500 km altitude. The significant change in the LOS velocity between adjacent LOS measurements is inconsitent with the expected change if the plasma flow were uniform. A more realistic interpretation is that the measurements span a narrow spatial inhomogeneity in the plasma flow with a large component of velocity parallel to the arc. From its apparent location in the east and west scan we take the feature to be nearly L-shell aligned and passing near zenith with an 1150 m/s component of velocity directed eastward along the arc. The elevation scan further verifies the presence of the strong eastward velocity

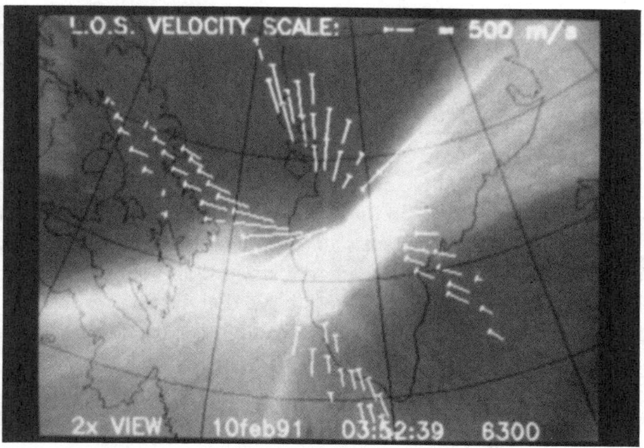

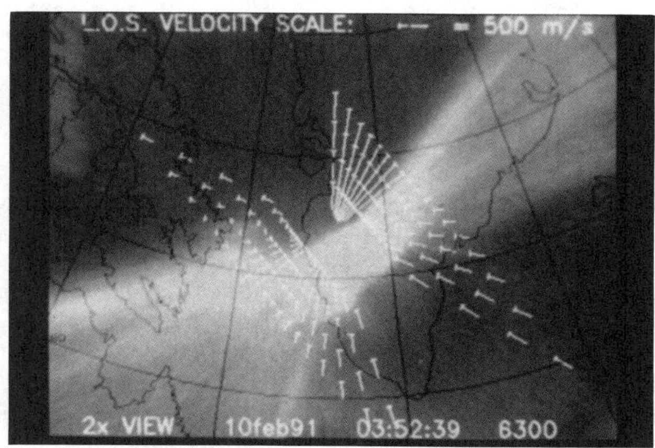

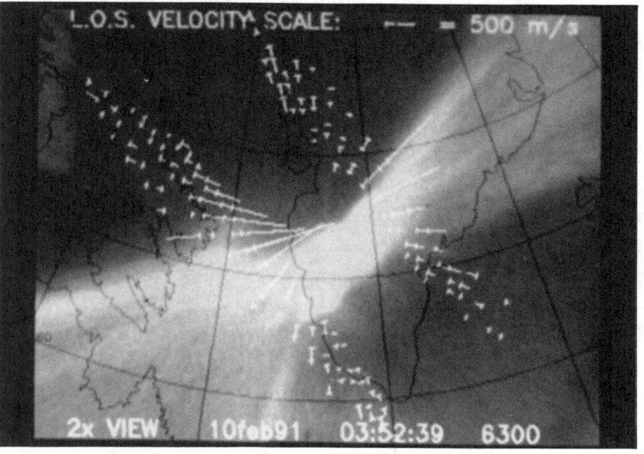

Fig. 3. (Top) The horizontal LOS plasma velocities for the east and west scans are superposed on a 2x enlargement of the 6300 Å image at 03:52:39 UT. (Middle) LOS plasma velocities for an estimated background velocity field consisting of a horizontal velocity of 457 m/s along 180° azimuth poleward of the arc and 350 m/s along 141° equatorward of the arc. (Bottom) Residual velocities determined by subtracting the background velocity field from the measured velocities.

enhancement. The large equatorward velocities near 50 km north in the E region are at an altitude where the ions are expected to Pederson-drift. The Pederson drifting ions would be driven by an equatorward component of the electric field. This electric field would map upwards to the F region where it would cause the plasma to ExB drift towards the east, which is consistent with the drift measured by the east and west radar scans. The LOS velocities are superposed on the electron density contours for all three scans in the third row of Figure 2. This high resolution comparison of velocities and densities demonstrates that the enhanced F region plasma flow is located at the poleward edge of the E region density enhancement which is associated with the arc.

The LOS velocities measured by the west scan during the previous cycle (03:43:19-03:46:33 UT, not shown) are quite similar to those shown in the west scan in Figure 2 row 2. Additionally, the east scan from the subsequent cycle (03:58:43-04:02:07 UT, not shown) has features similar to those in shown in the east scan in Figure 2 row 2. Consequently, despite the active nature of the aurora, the assumption that a steady state condition holds throughout the radar field of view from 03:47:27 UT to 03:57:59 UT seems valid. Assuming that steady state conditions prevail for the duration of the radar cycle, we may isolate the strong eastward flow at the poleward edge of the arc by subtracting an estimate of the background plasma flow. This technique is the same technique employed by Weber et al. [1991] to isolate an arc associated shear in the plasma flow from an antisunward component of plasma flow. In this case, our background flow will consist of regions of uniform flow poleward and equatorward of the arc. Robinson and Mende [1990] similarly used the two component flow to interpret LOS velocities measured at multiple azimuths. In Figure 3a the horizontal projection of the LOS velocities is superposed on a 2x enlargement of a 6300 Å image in a geographical coordinate system. Using a trial and error method, the background horizontal plasma velocity is estimated to have uniform components of 457 m/s along 180° azimuth poleward of the arc and 350 m/s along 141° azimuth equatorward of the arc. The LOS components of the estimated background ionospheric flow (Figure 3b) are subtracted from the horizontal projections of the east and west scan. The residual velocities are superposed on the 6300 Å image (Figure 3c). The residual velocities clearly show a strong eastward flow at the poleward edge of the optical aurora. Other background velocity fields were considered, and the pattern described above gave the best and most systematic results.

CONCLUSIONS

The importance of accurately locating the separatrix is essential when discussing likely effects of the reconnection process [Weber et al. 1991, de la Beaujardiere et al. 1991, Moses et al. 1989]. Plasma flow across a boundary separating different magnetic topologies is a primary signature of reconnection [Vasyluinas 1975]. The flow across the separatrix would be driven by an electric field lying in the plane of the separatrix and would be proportional to the reconnection rate [de la Beaujardiere et al. 1991, Vasyluinas 1975]. For the post midnight sector arc

discussed herein, we measure a plasma flow of 350 m/s across the most poleward arc that would represent the boundary between open and closed field lines. A second important feature is the strong eastward flow observed along the auroral arc at its poleward edge. Preliminary analysis shows that this enhanced eastward flow is consistent with a current sheet pair, downward poleward of the arc and upward within the arc. Both the cross arc plasma flow and the strong eastward flow poleward of the arc are features common to the midnight sector auroral arc on 30 October 1989 analyzed by Weber et al. [1991]. However, in the present case the eastward flow is much stronger. The observational technique described in this paper is designed and well suited for making high temporal and spatial resolution comparisons between the plasma flows and the optical emissions associated with an auroral arc. A data base containing approximately twenty-five stable auroral arcs, many located at the polar cap boundary, has been acquired using the technique and will be used to investigate the plasma flows associated with auroral arcs at a variety of local times between dusk and dawn.

Acknowledgments. We are grateful to the Danish Commission for Scientific Research in Greenland for approval to conduct these experiments, to the Sondre Stromfjord site crew for their assistance in gathering this data, to M. Mcready for processing the radar data, to P. Ning and M. Colerico for processing the optical data, and to E.Friis-Christensen for providing the magnetometer data. The research at Boston College was supported by the Air Force Office of Scientific Research and by the National Aeronautics and Space Administration's Graduate Student Researcher's Program. The research at the Air Force Phillips Laboratory was partially supported by the Air Force Office of Scientific Research task 2310G9, and at SRI by AFGL F19628-90-K-0036 and NSF cooperative agreement ATM-8822560.

REFERENCES

de la Beaujardiere, O. et al., Radar observation of electric fields and currents associated with auroral arcs, *J. Geophys. Res.*, 82, 5051, 1977.

de la Beaujardiere, O., et al., Sondrestrom radar measurements of the reconnection electric field, *J. Geophys. Res.*, 96, 13907, 1991.

Doolittle, J.H. et al., An observation of ionospheric convection and auroral arc motion, *J. Geophys. Res.*, 95, 19123, 1990.

Lockwood, M., S. W. H. Cowley, and M.P. Freeman, The excitation of plama convection in the high latitude ionosphere, *J. Geophys. Res.*, 95,7961, 1990.

Lyons, L.R., Generation of large-scale regions of auroral currents, electric potentials, and precipitation by the divergence of the convection electric field, *J. Geophys. Res.*, 85, 17, 1980.

Marklund, G. Auroral arc classification scheme based on the observed arc-associated electric field pattern, *Planet. Space Sci.*, 32, 192, 1984.

Moses, J.J. et al., Polar cap deflation during magnetospheric substorms, *J. Geophys. Res.*,94, 3785, 1989.

Pudovkin, M.I. et al., On separation of the potential and vortex parts of the magnetotail electric field, *Planet. Space Sci.*, 39, 563, 1991.

Robinson, R.M., and R.R. Vondrak, Electrodynamic properties of auroral surges, *J. Geophys. Res.*, 95, 7819, 1990.

Robinson, R.M., and S.B. Mende, Ionization and electric field properties of auroral arcs during magnetic quiescence, *J. Geophys. Res.*, 95, 21111, 1990.

Robinson, R.M. et al., Sondrestrom and EISCAT radar observations of poleward-moving auroral forms, *J. Atmos. Terr. Phys.*, 52, 1990.

Vasyliunas, V.M., Theoretical models of magnetic field line merging, I, *Rev. Geophys.*, 13, 303, 1975.

Vondrak, R.R. and R.M. Robinson, Electrodynamics of auroral and polar cap arcs at very high latitudes, in Electromagnetic Coupling in the Polar Clefts and Caps, edited by P.E. Sandholt and S. Egeland, pp. 269-284, Kluwer Academic, Boston, Mass., 1989.

Weber, E.J. et al., Coordinated radar and optical measurements of stable auroral arcs at the polar cap boundary, *J. Geophys. Res.*, 96, 17847, 1991.

R. L. Carovillano and H. A. Gallagher Jr., Department of Physics, Boston College, Chestnut Hill, MA 02167.

J.F. Vickrey, Geoscience and Engineering Center, SRI International, 33 Ravenswood Avenue, Menlo Park, CA 94025.

E. J. Weber, Phillips Laboratory, Geophysics Directorate/GPI, Hanscom Air Force Base, MA 01731.

Plasma Convection and Currents in the Auroral Zone

LING ZHANG AND ROBERT L. CAROVILLANO

Boston College, Department of Physics, Chestnut Hill, Massachusetts

The work we present is based upon an analytical model of the global configuration of ionospheric plasma convection, height-integrated currents, electric fields and potentials, with emphasis upon auroral zone effects. Sheet-like field-aligned-currents (FACs), located at the high- and low-latitude boundaries of the auroral zone, are used to represent the region I and region II currents. Utilizing the cross-cap driving potential and the region II FACs as driving mechanisms, the familiar two-cell convection pattern and the dawn to dusk electric field within the polar cap results, along with auroral electrojets and low latitude shielding. The dependence of the ionospheric convection upon the relative phase and intensity of the driving FACs is discussed, including properties such as the convection rotation and twisting in the auroral zone, and potential penetration to the sub-auroral latitudes. The effects of the Hall to Pedersen conductivity ratio and the degree of the auroral zone conductivity enhancement on the twisting of the convection pattern and on low latitude electrical shielding are also demonstrated.

INTRODUCTION

The importance of field-aligned currents(FACs) dates back to *Kristian Birkeland*'s [1908] suggestion on the origin of auroral currents. The direct observation of FACs is more recent [*Zmuda and Armstrong*, 1974a, b; *Sugiura*, 1975; *Iijima and Potemra*, 1976a, b]. Today we realize that FACs are associated with a wide variety of auroral phenomena including visual and radar forms, ionospheric currents, and ionospheric plasma instabilities.

Considerable effort in recent years has determined the statistical characteristics of FACs, such as their location, flow direction, and intensity [e.g., *Akasofu and Ahn*, 1981; *Iijima et al.*, 1984; *Zanetti et al.*, 1990; *Burch et al.*, 1985]. The FACs can be categorized in at least five different regions [*Saflekos et al.*, 1982]. These regions include (1) the region I system, (2) the region II system, (3) the dayside "cusp" system, (4) the polar cap "NBZ" system, and (5) the "Harang discontinuity" midnight region of overlapping and multi-sheet FACs [*Potemra*, 1983]. In this paper we shall examine the potential distribution, electric field, and current in the ionosphere associated with simple representations of the driving FACs with emphasis on the auroral zone.

During the last decade significant research has been conducted on high latitude convection. Empirical studies include those that rely on information from satellite crossings of the polar regions [e.g., *Heppner and Maynard*, 1987; *Burch et al.*, 1985,

Heelis et al., 1986], or on statistical analyses of large collections of ground-based measurements [e.g., *Wand and Evans*, 1981; *Foster*, 1983; *Foster et al.*, 1986; *de la Beaujardiere et al.*, 1986; *Alcayde et al.*, 1986]. The convection pattern most commonly displays a two-cell pattern [review by *Richmond*, 1983], highlighted by the anti-sunward flow across the polar cap. The basic pattern of the auroral zone convection is generally westward in the evening sector and eastward in the morning sector. *Vasyliunas* [1970; 1972] developed a fundamental mathematical model for ionospheric convection and magnetospheric-ionospheric coupling. Early analytical work on convection also includes that of *Wolf* [1974], *Atkinson and Hutchison* [1978] and *Nopper and Carovillano* [1979]. Numerical convection studies have also been conducted by many people, e.g., *Nopper and Carovillano* [1978]; *Kamide and Matsushita* [1979a,b]; *Blomberg and Marklund* [1991]. The approach in this work is to develop a simple model and formalism for ionospheric convection, to obtain analytical solutions at high latitudes, and to study the effect on convection associated with each individual physical parameter introduced. The objective is to establish a better and more secure relationship between proposed theoretical ideas (such as driving mechanisms) and characteristics of ionospheric convection observed or deduced empirically from measurements.

THE MODEL

Since our objective is to study global scale ionospheric electrodynamics, the ionosphere is assumed to be a very thin, spherical conducting shell; electrically neutral and highly conductive; and described by the MHD approximation,

Auroral Plasma Dynamics
Geophysical Monograph 80

$E+v\times B/c=0$. Here E and B are the electric and magnetic field, respectively, v is the convection velocity of the ionospheric plasma, and c is the speed of light. To obtain the equipotential contours and the plasma convection pattern, and to relate the FACs to the horizontal currents, the geomagnetic field model is needed. Because our interests here are confined to high latitudes, the geomagnetic field is assumed to be radially inwards and of constant magnitude at high latitudes in the northern hemisphere.

The observed pattern of FACs appears to have a narrow latitudinal extension in statistical representations [*Potemra*, 1983; *Bythrow et al.*, 1981]. It is adequate here, however, to represent the region I and region II FACs by two sets of sheet currents that intersect the ionosphere at the constant colatitudes θ_1 and θ_2, respectively. The region between θ_1 and θ_2 corresponds to the auroral zone and may be given enhanced electrical conductivity. For numerical evaluations we use $\theta_1=18°$ (the polar cap boundary) and $\theta_2=26°$ (the equatorial boundary of the aurora). Because our main concern is in high latitudes in this paper, $\theta=40°$ is about the maximum colatitude of interest in the calculations made.

Following the model and work of *Vasyliunas* [1970; 1972], we use height-integrated Pedersen (Σ_P) and Hall (Σ_H) conductivities. We keep the ratio Σ_H/Σ_P constant and are able to provide for an auroral zone of enhanced conductivity [*Hardy et al.*, 1987; *Ahn et al.*, 1989]. We denote Σ as the Hall-to-Pedersen conductivity ratio which is usually larger than unity,

$$\Sigma=\frac{\Sigma_H}{\Sigma_P},\qquad(1)$$

and define the ratios of the Pedersen conductivities in the different zones by

$$p_1=\frac{\Sigma_P^I}{\Sigma_P^{II}},\qquad p_3=\frac{\Sigma_P^{III}}{\Sigma_P^{II}},\qquad(2)$$

where Σ_P^I, Σ_P^{II} and Σ_P^{III} are the height-integrated Pedersen conductivities in zone I (the polar cap), zone II (the auroral zone) and zone III (the sub-auroral region), respectively. An enhanced auroral conductivity is represented by $p_1<1$ and $p_3<1$.

FORMALISM

Current continuity applies throughout the ionosphere,

$$J_{\|}=\nabla\cdot J,\qquad(3)$$

and we require that the electrical potential V satisfies the boundary condition

$$V(\theta=\theta_1)=V_1\sin\varphi\qquad(4)$$

to account for the dawn-to-dusk E field inside the polar cap. Here $J_{\|}$ and J in (3) represent the FAC and the height-integrated current, respectively, V_1 in (4) corresponds to half of the cross-cap driving potential which is chosen to fit suggested observational values [e.g., *Reiff et al.*, 1981], and φ is the azimuthal angle measured counterclockwise from the midnight meridian. We represent the region II FACs (RIICs) by

$$J_{\|}^{II}=-J_2\sin(\varphi+\delta_2),\qquad(5)$$

where J_2 and δ_2 are given constants, representing the intensity and the phase shift of the RIICs, respectively. The "$-$" sign is used in Eq. (5) to emphasize that the RIICs essentially have the opposite polarity of the region I FACs (RICs), whose derived form at θ_1 is $J_{\|}^{II}=-J_2\sin(\varphi+\delta_2)$. Both the RICs and the RIICs satisfy Eq. (3) along with the height-integrated current in the ionosphere. The height-integrated ionospheric current J is related to the ionospheric E field through the Ohm's law relationship,

$$J=\Sigma_P E+\Sigma_H \hat{B}\times E.\qquad(6)$$

The potential satisfies Laplace's equation in each zone. The boundary conditions are that the potential is continuous and the currents are conserved at θ_1 and θ_2. The electrostatic potential distribution, the electric field, the height-integrated ionospheric currents and the region I currents at θ_1 are determined analytically in this way. Specifically, the values of Σ, p_1 and p_3 (Eqs. (1) and (2)) parameterize the results.

SELECTED RESULTS AND DISCUSSIONS

Convection patterns and zero equipotential

Typical effects of region I and region II FACs are shown in Figure 1, where the convection patterns in the ionosphere are presented in polar plots. The format of the figures is: local noon is at the top of the plot, and dusk is at the left; the dotted circles are centered at the geomagnetic pole and form a background coordinate grid at magnetic latitudes of 80°, 70°, 60° and 50°; the dot-dashed circles locate the boundaries of the auroral zone at θ_1 and θ_2. The input parameters are listed at the top of the figure, and the output (calculated) parameters are given inside the parentheses. I_2 and I_1 are respectively the total intensity of the RIIC and the RIC, and V_2 and φ_2 are the maximum value and the phase shift of the potential distribution at θ_2. The properties of the FACs are also specified in the plots. The symbol $\otimes$/O represents down/upward FACs, and $\lozenge$ represents the location of nodes where the FACs switch polarity and have zero intensity. The plot is quantitative in that between every pair of adjacent $\otimes$s or Os, the current is the same.

Figure 1a is the elementary example showing equipotentials for the case of uniform conductivity and zero RIIC. The 2-cell convection pattern is evident, with the uniform anti-sunward flow across the polar cap and the return flow penetrating to lower latitudes. This pattern is almost identical to the one presented by others [e.g., *Vasyliunas*, 1970; *Kamide and Matsushita*, 1979] under similar conditions. In Figure 1b, RIIC is present and the auroral conductivity is enhanced by a factor 5. For the same value of the cross cap potential ($2V_1$) as in Figure 1a the potential at θ_2 is decreased by about 30% re-

CONVECTION ELECTRIC POTENTIAL IN THE IONOSPHERE

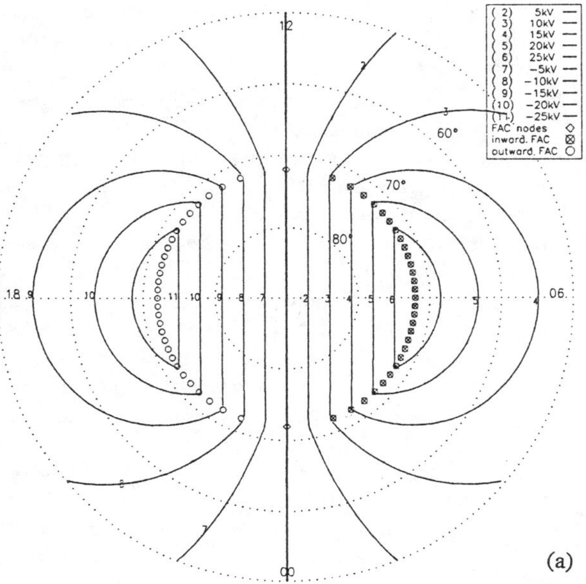

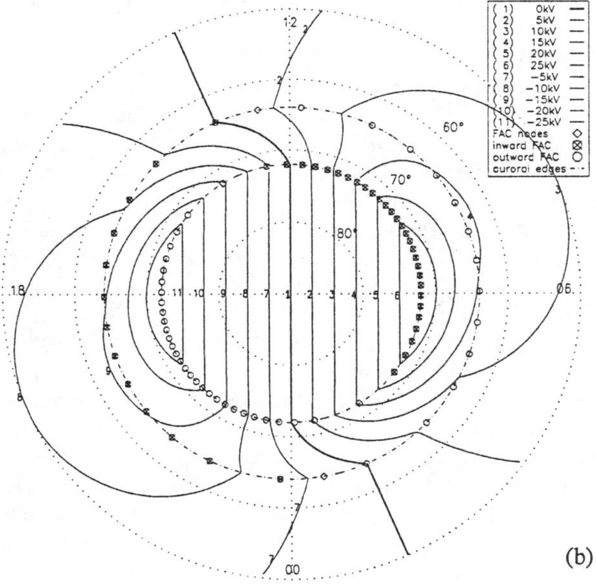

Fig. 1. 1a, the equipotential lines and the convection pattern driven by the cross-cap potential V_1 for the case of uniform conductivity. Lines are almost equally spaced and parallel in the polar cap which indicates a dawn-to-dusk electric field. Plasma flow is anti-sunward inside the polar cap and returns in the opposite direction, and the familiar 2-cell pattern forms. 1b, the convection pattern in the presence of RIC and RIIC. Input parameters are listed at the top of the figure, and the calculated quantities are given in the parentheses.

sulting in low latitude electrical shielding ($\theta > \theta_2$). The shielding is also manifested by the azimuthal twist of the equipotential lines (and the convection) in the auroral zone. The basic 2-cell convection pattern remains in Figure 1b, though its symmetry is altered and the flow is confined to higher latitudes because of the auroral twist.

The convection pattern in Figure 1b is quite similar in both the auroral and sub-auroral zones to those given by *Barbosa* [1984] who used a variational method to minimize Joule dissipation. Overall features are also strikingly similar to the modeled studies of *Siscoe and Maynard* [1991] who used distributed (and not sheet) FAC representations.

Of special interest is the contour of zero potential which separates the cells of positive and negative potentials. The positive/negative equipotentials coincide with the dawn/dusk side convection cell where the circulation is counterclockwise/clockwise. In Figure 1a the zero-potential line is along the noon-midnight meridian. In Figure 1b the zero-potential extends across the polar cap along the noon-midnight meridian and rotates from midnight/noon through the auroral zone by an angle $-\varphi_2$ into the predawn/dusk sector. Very similar zero-potential rotation can be seen in the empirical convection pattern presented by *Foster* [1983; his Figure 2d]. Equatorward of the auroral zone, the zero-potential is again a straight line and lies along a meridian.

In this paper a uniform anti-sunward convection always exists in the polar cap. This will no longer persist under other physical conditions. For example, (1) the introduction of cusp currents leads to a quite structured polar cap convection and results in east-west flow components[*Zhang and Carovillano*, 1990]; (2) if the RIC is taken as input rather than the cross-cap potential, a RIC of ideal polarity leads to a clockwise-tilted convection sweeping from late morning towards late evening within the polar cap [*Zhang*, 1993]; (3) The inclusion of a day-night conductivity gradient squeezes the polar cap equipotentials towards dawn and effectively rotates the nightside convection lines a bit clockwise [*Atkinson and Hutchison*, 1978].

The zero potential line can best indicate the extent of the distortion and rotation of the overall convection pattern. The value of $-\varphi_2$ determines the extent of the zero potential rotation from midnight at θ_2, and the analysis of the rotation simplifies to the study of the variation of $-\varphi_2$ with the input parameters. The parameter V_2 indicates how much of the polar cap potential penetrates to the sub-auroral region from the higher latitude driver(s). In the equipotential contour plot the line with $V=V_2$, where $V_2<V_1$, is the one which comes from high latitudes and just touches the circle $\theta=\theta_2$ (tangentially) at the azimuth $\pi/2-\varphi_2$. Equipotential contours with V larger than V_2 are closed within the region $\theta \leq \theta_2$, and the other contours with V smaller than V_2 enter the region $\theta \leq \theta_2$ and penetrate to the sub-auroral region.

The Degree of Twist

The azimuthal twist in the convection pattern $-\varphi_2$ can be very large. It can be separated into independent contributions from the conductivities $-\varphi_{20}$ and from the RIICs $-\Delta\varphi_2$:

$$\varphi_2 = \varphi_{20}(\Sigma, p_3) + \Delta\varphi_2(I_2, \delta_2). \qquad (7)$$

For either $\delta_2=0$ or $I_2=0$, we get $\varphi_2=\varphi_{20}$ and the sub-auroral rotation is determined only by the Hall-to-Pedersen conductivity ratio Σ and the meridional conductivity ratio p_3. Only if $p_3=1$ or $\Sigma=0$ does $\varphi_{20}=0$ so that there is no rotation of the convection pattern. The RIIC may result in an increase or a decrease from φ_{20} depending on the values of I_2 and δ_2. We have also proved that the angle $-\varphi_{20}$ is always in the first quadrant ($0°$ to $90°$), so that the twist due to the conductivities is always counterclockwise. This is shown in Figure 2a, where $-\varphi_{20}$ is seen

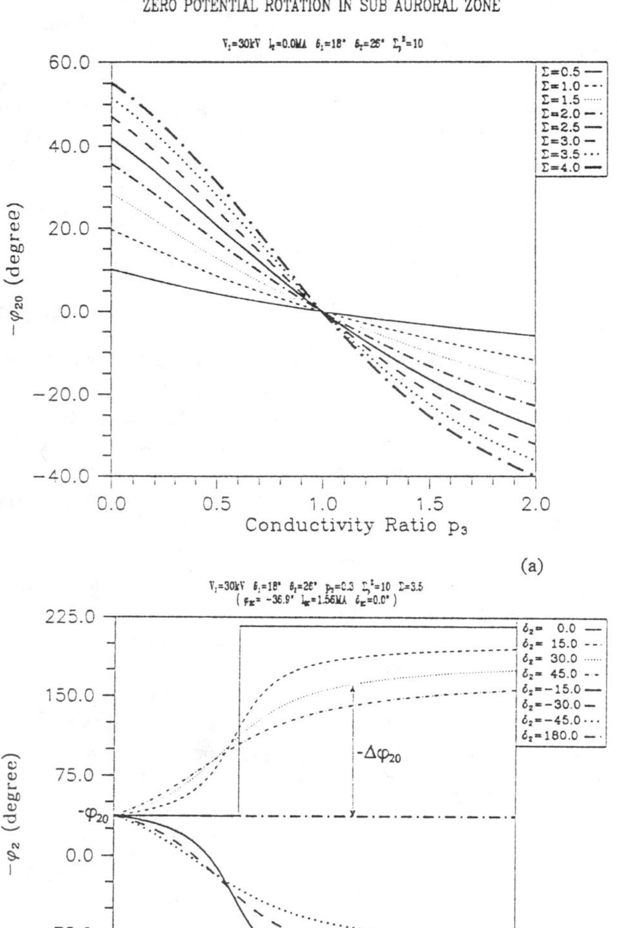

Fig. 2.   Properties of the zero-equipotential line ($V=0$). 2a shows the dependence of the zero-equipotential rotation on the Hall to Pedersen conductivity ratio and the auroral zone conductivity enhancement. 2b, the dependence of the zero-equipotential rotation on the intensity and phase of the region II currents is shown.

to increase monotonically with both Σ at fixed p_3 and with enhanced auroral zone conductivity (p_3 decreasing from 1) at fixed Σ. Therefore both the increase in the Hall to Pedersen conductivity ratio and the auroral zone conductivity enhancement lead to a greater rotation of the zero-potential in the counterclockwise direction. (If p_3 were to be larger than 1, the rotation would reverse and be clockwise).

Additional rotation, independent of the conductivity parameters Σ and p_3, results from the RIIC as shown in Figure 2b where $-\varphi_2$ is plotted versus I_2 with δ_2 as a parameter. The vertical offset in the plot gives the value of $-\varphi_{20}$ which is determined by the conductivity parameters, and $-\Delta\varphi_2$ is the additional shift which is equal to the difference between $-\varphi_2$ and $-\varphi_{20}$. Note that $\Delta\varphi_2\equiv0$ for $\delta_2=180°$.

Interesting features are demonstrated in Figure 2b: (1) $-\Delta\varphi_2$ is in the range $\pm(0°$ to $180°)$ if δ_2 is in the range $\pm(0°$ to $180°)$, i.e., a clockwise rotation of the RIIC induces a counterclockwise rotation in the convection pattern. (2) $-\Delta\varphi_2(-\delta_2) = \Delta\varphi_2(\delta_2)$: the convection pattern rotation reverses direction while keeping the same magnitude if the RIIC phase shift reverses sign. (3) For $\delta_2\neq0°$ or $180°$, $|-\Delta\varphi_2|$ increases monotonically with I_2 and saturates asymptotically. (4) The additional rotation is zero for $I_2 \leq I_{20}$ and jumps to $180°$ for $I_2 > I_{20}$ at $\delta_2=0°$. This step-function feature can be understood as follows. While $\delta_2=0$ and I_2 is relatively small, the overall potential distribution is dominated by the cross-cap driving potential V_1 and the potential is positive/negative at the dawn/dusk side, and $\Delta\varphi_2=0$. But when the RIIC is large enough it will reverse the sign of the potential at θ_2 from that of the cross-cap driver V_1, so that $\Delta\varphi_2=180°$.

If there is no RIIC or if the RIIC is not significant when properly compared to the cross-cap driving potential, the rotation of the convection pattern in the auroral zone is not very large and the convection cells essentially circulate about the RIC (as in Figure 1). If the RIIC is strong, the auroral twist in the convection pattern rotation can be quite large, as demonstrated in Figure 3. In Figure 3, the auroral conductivity enhancement is large with $\varphi_{20}\approx-32°$ in Figure 3a and $\varphi_{20}\approx-48°$ in Figure 3b. The phase of the RIIC is of opposite signs in 3a and 3b, and the intensity I_2 is 2.5 MA in 3b verses 1 MA in 3a. As in Figure 1b, the anti-sunward convection across the central polar cap twists through the auroral zone, extends into the sub auroral region to a degree dependent in the conductivity enhancement, and returns to the polar cap. In Figure 3a, there is a 2-cell convection pattern, counterclockwise on the dawnside and clockwise on the duskside, controlled by the RIC in the usual manner. In Figure 3b, however, where the total RIIC is very large, two additional cells form at θ_2 and significant electrical penetration to sub auroral latitudes occurs. Extra convection cells of this type can also be seen in the convection patterns given by *Zi and Shen* [1986], and may be inferred from the equivalent current patterns presented by *Crooker and Siscoe* [1981], though the effect is not identified by the authors. In another approach utilizing radar measurements, *Alcayde et al.* [1986] found that the potential can be as high as 36 kV at mid-latitudes (verses 62 kV across the polar cap) at high magnetic activities,

CONVECTION ELECTRIC POTENTIAL IN THE IONOSPHERE

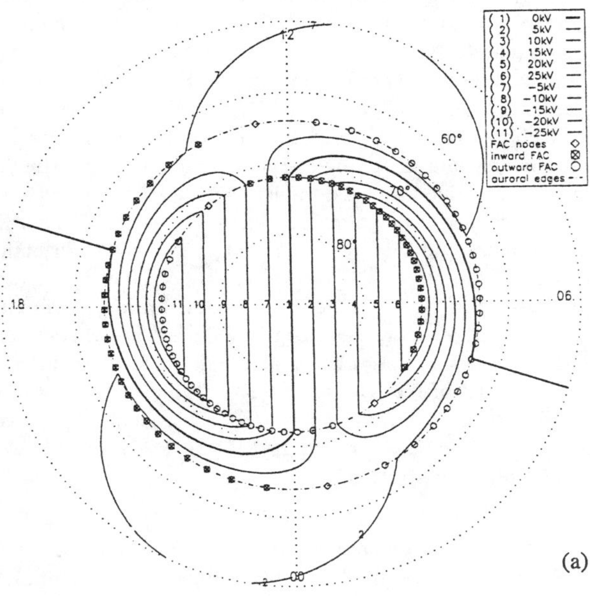

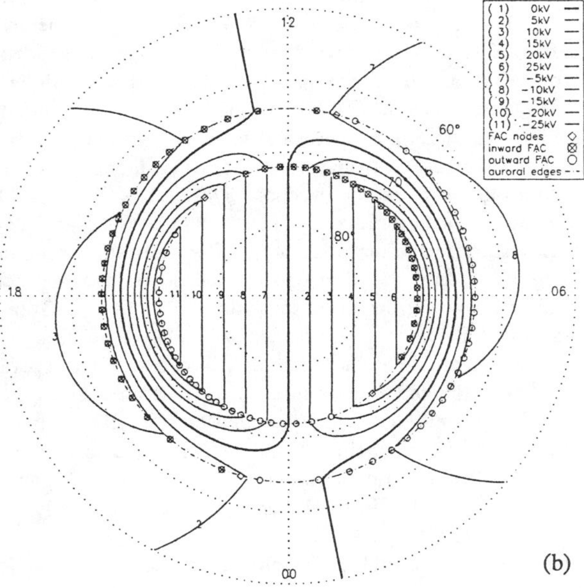

Fig. 3.   Special effects of the RIIC. 3a gives a case with a large "rotation" of the convection pattern in the auroral zone and a large shielding. 3b has strong RIIC which in addition to the large rotation produces two additional lower latitude convection cells and diminished electrical shielding.

and hence conclude that a large electric field can indeed penetrate to low latitudes under such conditions.

Potential Penetration to Sub-Auroral latitudes

The dependence of the potential amplitude V_2 at θ_2 upon the conductivity ratios p_3 and Σ (see Eq. (2)) are shown in Figure 4a and upon the intensity and phase of the RIICs is shown in

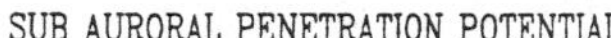

SUB AURORAL PENETRATION POTENTIAL

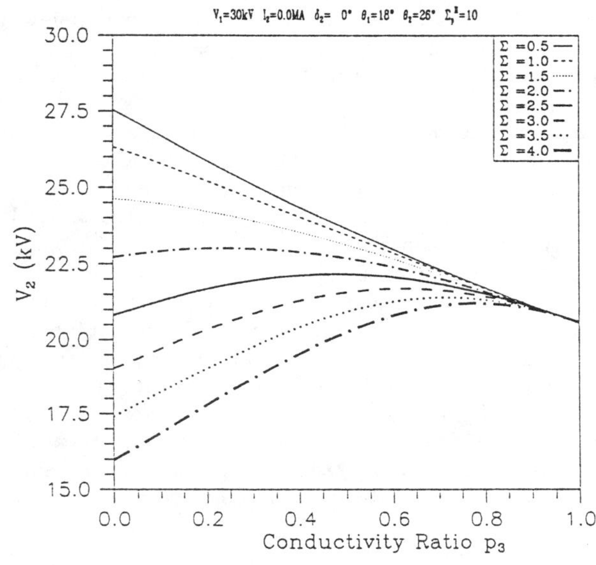

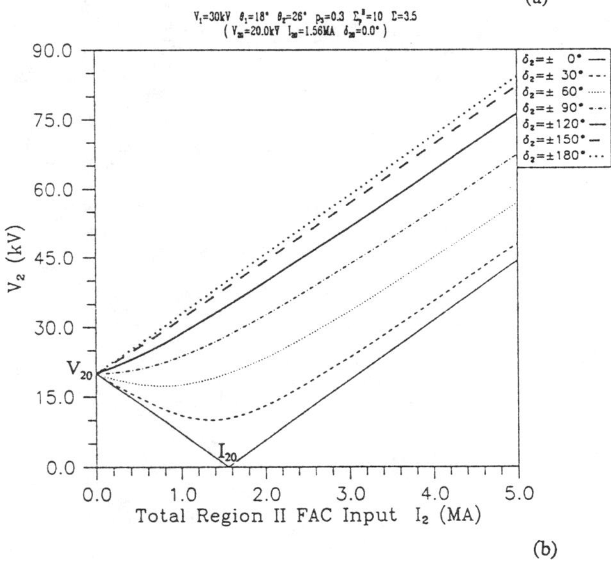

Fig. 4.   Lower latitude ($\theta > \theta_2$) electrical shielding conditions. 4a, the dependence of the potential penetration to the sub auroral region on the Hall to Pedersen conductivity ratio and the auroral zone conductivity enhancement is given. 4b, the dependence on the intensity and phase of the region II currents is given.

Figure 4b. In Figure 4a (where $I_2=0$), V_2 is seen to decrease monotonically with Σ for given p_3. If the conductivity at θ_2 is continuous ($p_3=1$), V_2 is independent of Σ; otherwise a higher Hall to Pedersen conductivity ratio always results in greater shielding of the potential in the sub-auroral region. For realistic values of Σ (>1) and p_3 (<0.2) [e.g., *Hardy et al.*, 1987] a more enhanced conductivity in the auroral zone also results in greater shielding.

In the numerical convection study by *Nopper and Carovillano* [1978], an example was given which demonstrated that the RIIC can greatly restrict the potential penetrating to low latitudes. We now describe our analytical results on this same effect. Figure 4b shows the dependence of the potential amplitude at θ_2, i.e., V_2, upon the total intensity of the RIIC I_2. Along each curve in Figure 4b, the phase δ_2 takes on the values noted in the legend increasing from 0° to 180° from the bottom to the top of the figure. These curves form an open parallelogram-like figure. We denote by V_{20} the value of V_2 for $I_2=0$. V_{20} depends upon the values of Σ and p_3 and scales linearly with the cross-cap potential V_1. Similarly we denote as I_{20} the value of I_2 for which $V_2=0$.

If $V_2\neq0$, there is electrical penetration to sub-auroral latitudes. The significance of $V_2=0$, which occurs at I_{20}, is that it requires perfect electrical shielding, i.e., none of the potential penetrates into the sub-auroral region and $E=0$ there. This case of perfect shielding was studied by *Crooker and Siscoe* [1981] and *Barbosa* [1984] and can be seen in the representative empirical convection patterns by *Heppner and Maynard* [1987]. Figure 5 is an example of the convection pattern for the case of perfect

shielding. The zero equipotential coincides with the noon-midnight meridian for $\theta<\theta_2$ and with the auroral boundary at $\theta=\theta_2$; all of the convection is confined to the polar cap and the auroral zone; and the potential in the sub-auroral region is identically zero. In addition, the convection cells are symmetric with respect to the noon-midnight meridian for any conductivity parameters.

As shown by the lower bound curve in Figure 4b, at $\delta_2=0$, V_2 decreases linearly with I_2 when $I_2\leq I_{20}$, and then increases linearly with I_2 when $I_2\geq I_{20}$ with the same slope. This interesting result comes about as follows. The superimposed contributions to the potential due to V_1 and the RIIC are 180° out of phase and therefore of opposite sign at θ_2. When $I_2\leq I_{20}$, the V_1 contribution dominates and requires $-\varphi_2=0$. But the two contributions cancel more and more as I_2 increases, until $V_2=0$ at $I_2=I_{20}$ (as in Figure 5). For $I_2>I_{20}$, the RIIC contribution dominates, $-\varphi_2=180°$, and V_2 increases linearly with I_2. In general, better shielding occurs if the polarity of the RIIC is close to being opposite to that of the RIC. For a RIIC typical of the observations (small $|\delta_2|$ and $I_2\leq I_{20}$) strong RIIC or small $|\delta_2|$ leads to better shielding; otherwise, strong RIIC or large $|\delta_2|$ would reduce the shielding effect.

The Electric Field E and the ionospheric currents J

The derived electric field **E** at two points at the same latitude and 180° apart in longitude appears to have the same magnitude and the same direction (in the polar plot format). This analytical result agrees with the observational behavior of the average electric field patterns summarized by *Heelis* [1989] that the dominant variation in the average electric field is diurnal. Other analytical properties of **E** are as follows. (1) Inside the polar cap: **E** is directed dawn-to-dusk with magnitude independent of azimuth and approximately constant ($E\approx15$ mV/m for $V_1=30$kV and $\theta_1=18°$). (2) In the sub-auroral region: The magnitude of **E** is directly proportional to V_2, is independent of azimuth, and decreases from high to low latitudes (e.g., from $\theta=\theta_2$ to $\theta=40°$ it decreases by about 57%). The direction of **E** is independent of latitude and makes an angle $\varphi+\varphi_2-90°$ counterclockwise from the meridian at φ. Therefore the variation of the magnitude and direction of the electric field can be constructed or inferred from the discussion of V_2 and φ_2. (3) In the auroral zone: **E** varies significantly with both latitude and azimuth. Its magnitude at each latitude has two maxima separated 180° in azimuth and two minima also separated 180° in azimuth. The magnitude normally decreases from high latitude to low latitude along any azimuth, although it can also increase if I_2 is large. The electric field can be quite large in the auroral zone compared to that in the other two zones, especially if the auroral conductivity is enhanced.

Figure 6a shows a typical **E** field vector plot. The **E** field is very strong in the auroral zone, its average magnitude there being about twice as large as in the polar cap. Also we see significant shielding of the **E** field below the auroral zone. The magnitude of **E** in zone III is about 1/20th compared with that in the auroral zone. The **E** field has a positive divergence centered at local dawn and a negative divergence centered at local

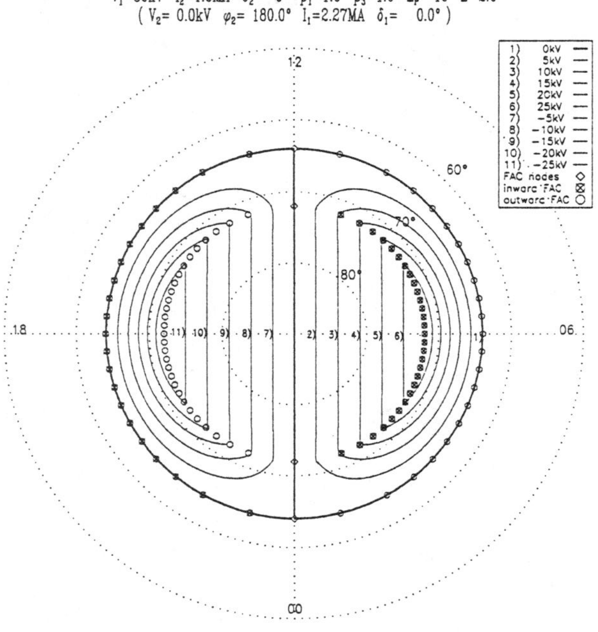

Fig. 5.   The equipotential convection pattern with perfect shielding.

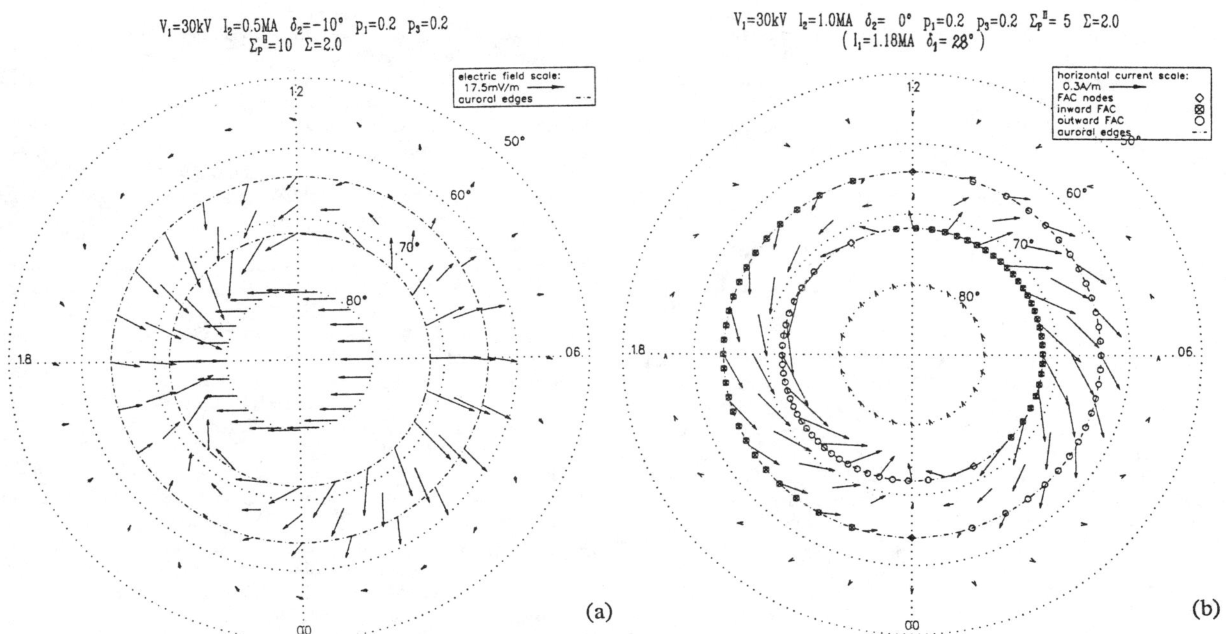

Fig. 6.   Vector plots of the electric field (6a) and height-integrated ionospheric current (6b). In the polar cap, **E** is dawn-to-dusk and **J** is rotated towards noon by the large Hall current. In 6a, it is apparent that the combination of the RIIC and the conductivity enhancement in auroral zone results in significant shielding below auroral zone. In 6b, the electrojets are westward in early morning and eastward in the afternoon and are very intense because of the combined effects of the RIIC and the conductivity enhancement in auroral zone. The trapping of the current in the auroral zone and the electrical shielding of low latitudes go together.

dusk along the boundary at θ_1; the polarity of divergence reverses at boundary θ_2. A positive/negative divergence in **E** corresponds to inward/outward directed FACs. In the auroral zone (at all latitudes) the magnitude reaches a maximum in the vicinity of local dawn and dusk and gradually decreases in azimuth with its weakest intensity near local noon and midnight. The **E** field in zone III has a very small intensity and a dipole-like shape.

The magnitude of the height-integrated ionospheric current **J** in the three zones varies with azimuth φ in the same fashion as the electric field. The direction of the horizontal current can be found by simply rotating the **E** field direction clockwise by the angle $\cos^{-1}\left(1+\Sigma^2\right)^{-1/2}$.

As with the **E** field, the current appears to be the same (both magnitude and direction) at two locations 180° apart along the same latitude. Figure 6b shows a vector plot for the horizontal current along with the FAC representations. (1) Inside the polar cap: the current is approximately constant in magnitude and has a uniform flow direction making an angle $\cos^{-1}\left(1+\Sigma^2\right)^{-1/2}$ with the dawn-dusk meridian. Small/large Hall to Pedersen ratio Σ leads to a more/less dawn-dusk oriented current flow. (2) In the auroral zone: we see clearly a very intense westward current (electrojet) on the dawn side and an eastward electrojet on the dusk side. We can understand these electrojets as follows.

With the extreme assumption that there is no Hall current, the total ionospheric current would be a Pedersen current parallel to **E** (like in the pattern shown in Figure 6a). The Pedersen current connects the RICs and RIICs. It is more intense both at the dawn and dusk sectors and is directed northwards or southwards. In fact, however, there is a stronger Hall current which flows perpendicular to the Pedersen current. It is the Hall current, flowing eastwards or westwards in the auroral zone, that is principally responsible for the electrojets.

Although it may not be obvious in Figure 6b, the electrojet has a 180° rotational balance, i.e., the total auroral electrojet current flowing, say, eastward at one MLT near the dusk sector equals the total auroral electrojet flowing westward 12 hours away in the dawn sector.

Acknowledgment. This research was supported in part by the Air Force Systems Command under Contract AF19628-86-C-0029.

REFERENCES

Ahn, B.-H., H. W. Kroehl, Y. Kamide, and D. J. Gorney, Estimation of ionospheric electrodynamic parameters using ionospheric conductance deduced from bremsstrahlung X-ray image data, J. Geophys. Res., 94, 2565, 1989.

Akasofu, S.-I., and B.-H. Ahn, Distribution of the field-aligned currents, ionospheric currents, and electric fields in the polar region on a very quiet day and a moderately disturbed day, J. Geophys. Res., 86, 753, 1981.

Alcayde, D., G. Caudal, and J. Fontanari, Convection electric fields and electrostatic potential over $61°<\Lambda<72°$ invariant latitude observed with the European Incoherent Scatter facility, 1, Initial results, J. Geophys. Res., 91, 233, 1986.

Atkinson, G., and D. Hutchison, Effect of the day-night ionospheric conductivity gradient on polar cap convective flow, J. Geophys. Res., 83, 726, 1978.

Barbosa, D. D., An energy principle for high-latitude electrodynamics, J. Geophys. Res., 89, 2881-2890, 1984.

Birkeland, K., The Norwegian Polaris Expedition 1902-1903, Vol. 1, Sect. 1, Aschhoug, Oslo, 1908.

Blomberg, L. G. and G. T. Marklund, High-latitude convection patterns for various large-scale field-aligned current configurations, Geophys. Res. Lett., 18, 717, 1991.

Burch, J. L., P. H. Reiff, J. D. Menietti, R. A. Heelis, W. B. Hanson, S. D. Shawhan, E. G. Shelley, M. Sugiura, D. R. Weimer, J. D. Winningham, IMF By-dependent plasma flow and Birkeland currents in the dayside magnetosphere 1. Dynamics Explorer Observations, J. Geophys. Res., 90, 1577, 1985.

Bythrow, P. F., R. A. Heelis, W. B. Hanson, and R. A. Power, Observational evidence for a boundary layer source of dayside region I field-aligned currents, J. Geophys. Res., 86, 5577, 1981.

Crooker, N. U., and G. L. Siscoe, Birkeland currents as the cause of the low-latitude asymmetric disturbance field, J. Geophys. Res., 86, 11201, 1981.

de la Beaujardiere, V. W. Wickwar and J. H. King, Sondrestrom radar observations of the effect of the IMF B_y-component on polar cap convection, in Solar Wind-Magnetosphere Coupling, edited by Y. Kamide and J. A. Slavin, Terra Scientific Publishing Company, 495, 1986.

Foster, J. C., An empirical electric field model derived from Chatanika radar data, J. Geophys Res., 88, 981, 1983.

Foster, J. C., J. M. Holt, R. G. Musgrove, and D. S. Evans, Ionospheric convection associated with discrete levels of particle precipitation, Geophys. Res. Lett., 13, 656, 1986

Hardy, D.C., M.S. Gussenhoven, R. Raistrick and W.J. McNeil, Statistical and functional representations of the pattern of auroral energy flux, number flux, and conductivity, J. Geophys. Res., 92, 12275, 1987

Heelis, R. A., in The Earth's Ionosphere—Plasma Physics and Electrodynamics, by M. C. Kelley, Academic Press Inc., San Diego, California, 1989.

Heelis, R. A., P. H. Reiff, J. D. Winningham and W. B. Hanson, Ionospheric convection signatures observed by DE 2 during northward interplanetary magnetic field, J. Geophys. Res., 91, 5817, 1986.

Heppner, J. P. and N. C. Maynard, Empirical high-latitude electric field models, J. Geophys Res., 92, 4467, 1987.

Iijima, T., and T. A. Potemra, Field-aligned currents in the dayside cusp observed by TRIAD, J. Geophys. Res., 81, 5971, 1976b.

Iijima, T., and T. A. Potemra, The amplitude distribution of field-aligned currents at northern high latitudes observed by TRIAD, J. Geophys. Res., 81, 2165, 1976a.

Iijima, T., T. A. Potemra, L. J. Zanetti, and P. F. Bythrow, Large scale Birkeland currents in the dayside polar region during strongly northward IMF: A new Birkeland current system, J. Geophys. Res., 89, 7441, 1984.

Kamide, Y., and S. Matsusita, Simulation studies of ionospheric electric fields and currents in relation to field-aligned currents, 1, Quiet periods, J. Geophys. Res., 84, 4084, 1979a.

Nopper, R.W., and R.L. Carovillano, Polar-equatorial Coupling During Magnetically Active Periods, Geophys. Res. Lett., 699, 1978.

Nopper, R. W., and R. L. Carovillano, Ionospheric Electric Fields Driven by Field-aligned Currents, Geophys. Monograph 21, 1979.

Potemra, T. A., Birkeland currents: Present Understanding and Some Remaining Questions, High-Latitude Space Plasma Physics, ed. Dengt Hultqvist and Tor Hagfors, 335, Plenum Pub. Co., 1983

Reiff, P. H., R. W. Spiro, and T. W. Hill, Dependence of polar cap potential drop on interplanetary parameters, J. Geophys Res., 86, 7639, 1981.

Richmond, A. D., Ionospheric electrodynamics and irregularities: a review of contributions by US scientists from 1979 to 1982, Revs. Geophys. Space Phys., 21, 234, 1983

Saflekos, N. A., R. E. Sheehan, and R. L. Carovillano, Global nature of field-aligned currents and their relation to auroral phenomena, Revs. Geophys. Space Phys., 1982.

Siscoe, G. L. and N. Maynard, Distributed two-dimensional region 1 and region 2 currents: model results and data comparisons, J. Geophys. Res., 96, 21071, 1991.

Sugiura, M., Identifications of the polar cap boundary and the auroral belt in the high-altitude magnetosphere: A model for field-aligned currents, J. Geophys. Res., 80, 2057, 1975.

Vasyliunas, V.M., Mathematical Models of Magnetospheric Convection and Its Coupling to the Ionosphere. in B.M. McCormac (Ed.), Particle and Fields in the Magnetosphere (Reidel 1970).

Vasyliunas, V.M., The Interrelationship of Magnetospheric Processes. in B.M. McCormac (Ed.), Earth's Magnetospheric Processes (Reidel 1972).

Wand, R. H., and J. V. Evans, The penetration of convection electric fields to the latitude of Millstone Hill (Λ = 56 deg), J. Geophys Res., 86, 5809, 1981.

Wolf, R. A., Calculations of magnetospheric electric fields. in Magnetospheric Physics, edit by B. M. McCormac, D. Reidel, Hingham Mass., 1974.

Zanetti, L.J., T.A. Potemra, R.E. Erlandson, P.F. Bythrow and B.J. Anderson, Polar region Birkeland current, convection, and aurora for northward interplanetary magnetic field, J. Geophys. Res., 95, 5825, 1990.

Zhang, L., Analytical Study of High Latitude Plasma Convection in the Ionosphere, Ph. D. Thesis, Boston College, 1993.

Zhang, L., and R.L. Carovillano, A simple analytical model for ionospheric convection with region I and DPZ currents (abstract), EOS transaction, AGU Fall Mtg. supp., 1990

Zi, M., and C. Shen, A Time-dependent Analytical Model with Day-Night Ionospheric Conductivity Gradient for Magnetospheric Convection, Planet. Space Sci., 34, No. 4, p353-362, 1986.

Zmuda, A. J., and J. C. Armstrong, The diurnal flow pattern of field-aligned currents, J. Geophys. Res., 79, 4611, 1974b.

Zmuda, A. J., and J. C. Armstrong, The diurnal variation of the region with vector magnetic field changes associated with field-aligned currents, J. Geophys. Res., 79, 2501, 1974a.

L. Zhang and R. L. Carovillano, Boston College, Department of Physics, Chestnut Hill, MA 02167.

Auroral Weak Double Layers: A Critical Assessment

HANNU E. J. KOSKINEN AND ANSSI M. MÄLKKI

Finnish Meteorological Institute, Department of Geophysics,
Helsinki, Finland

Weak double layers (WDLs) were first observed in the mid-altitude auroral magnetosphere in 1976 by the S3–3 satellite. The observations were confirmed by Viking in 1986, when more detailed information of these small-scale plasma structures became available. WDLs are upward moving rarefactive solitary structures with negative electric potential. The potential drop over a WDL is typically 0–1 V with electric field pointing predominantly upward. The structures are usually found in relatively weak (≤ 2 kV) auroral acceleration regions where the field-aligned current is upward, but sometimes very small. The observations suggest that WDLs exist in regions of cool electron and ion background. Most likely the potential structures are embedded in the background ion population that may drift slowly upward. There have been several attempts for plasma physical explanation of WDLs but so far the success has not been very good. Computer simulations have been able to produce similar structures, but usually for somewhat unrealistic plasma parameters. A satisfactory understanding of the phenomenon requires consideration of the role of WDLs in the magnetosphere-ionosphere (M–I) coupling, including the large-scale electric fields, both parallel and perpendicular to the magnetic field, and the Alfvén waves mediating the coupling. In this report we give a critical review of our present understanding of WDLs. We try to find out what can be safely deduced from the observations, what are just educated guesses, and where we may go wrong.

INTRODUCTION

The question of the formation of a potential difference between the ionosphere and magnetosphere belongs to the most important outstanding problems of auroral plasma dynamics. At present there is a wealth of evidence that the plasma on the auroral field lines is capable of maintaining magnetic field-aligned potential drops of the order of ten thousand volts. The most convincing evidence is based on particle observations below, within, and above the so called auroral acceleration region, where the parallel potential energy is converted to field-aligned energy of electrons and ions. Especially valuable have been the rare occasions with two spacecraft on nearly the same field line, as e.g., Dynamics Explorer 1 and 2 [Reiff et al., 1988]. To some extent it has also been possible to use the Viking electric field instrument to estimate the parallel electric field [Block and Fälthammar, 1990; see also Mozer and Torbert, 1980].

Although the majority of auroral plasma physicists now accepts the existence of parallel potential drop at altitudes of 1–2 R_E above the auroral oval, their internal structure and the underlying physical processes have remained largely un-

known. A few years ago Block [1984] argued that the observed electron spectra below the acceleration region cannot result either from one single electric double layer or from uniformly distributed parallel electric field. Instead, there must be irregularities in the potential structure to allow for low energy electrons to drift into the acceleration region from the sides of the flux tube. Block suggested that the weak double layers (WDLs) observed in 1976 by S3-3 [Temerin et al., 1982] were consistent with the known facts.

An electric double layer is a localized space charge region with a net electric potential drop over it [Block, 1973]. The weak double layers are characterized by the smallness of the potential drop. The net potential drop over a single WDL is at most of the order of the thermal energy of the ambient electrons ($e\,\Delta\phi \leq T_e$). Although WDLs are small, they have been directly observed on auroral field lines, and when observed, they often occur in large numbers near each other [Temerin et al., 1982; Boström et al., 1988]. Thus it is natural to ask, whether WDLs could act as building blocks of a larger parallel potential drop. More generally, we should like to understand their final role in the global electrodynamic processes coupling the magnetosphere and ionosphere together.

The investigation of WDLs experienced a renaissance after the Viking satellite had confirmed their existence in 1986 and considerably extended the observational information on the small-scale potential structures [Boström et al., 1988; Koskinen et al., 1989]. Furthermore, Viking yielded new insight into the plasma environment where WDLs were observed [Koskinen et al., 1990]. In this presentation we discuss the status of our knowledge of WDLs about five years after the Viking observations. We review the main observational characteristics including some previously unpublished statistical information. We consider critically the latest theoretical and simulation approaches for the physical explanation of the structures. Finally, we discuss the consequences of the observed potential structures on particle energization and on the large-scale M–I coupling.

OBSERVATIONS OF WEAK DOUBLE LAYERS AND THE PLASMA ENVIRONMENT WHERE THEY ARE FOUND

In this section we review the Viking WDL observations together with other supporting observational data. We do not dwell deeply into the methods of observations that have been described elsewhere [Boström et al., 1988; Koskinen et al., 1989, 1990]. However, it is important to note some of the instrumental constraints in order to avoid too optimistic an interpretation of the experimental data.

The direct observation of WDLs was made by the low-frequency wave instrument (V4L) of the Uppsala Division of the Swedish Institute of Space Physics. Four spherical electric probes of diameter 10 cm in the tips of 40 m long wire booms were used. The probes formed two orthogonal 80 m electric field antennas in the spin plane of the spacecraft (i.e., nearly the orbital plane, as well as the plane of the ambient magnetic field). The spin rate of the spacecraft was 3 rpm. Two of the probes on opposite sides of the spacecraft could alternatively be used as Langmuir probes with a positively biased voltage to measure plasma density and density fluctuations ($\Delta n/n$). In order to have enough dynamics for wave fluctuations the signals were high-pass filtered at 0.5 Hz. The maximum sampling rate of the waveform was 853.3 s^{-1}. Due to telemetry limitations only two waveforms could be sampled simultaneously. Three main observation modes were used, measuring simultaneously:

1)   two components of wave electric field

2)   one electric field component and density fluctuations at one Langmuir probe

3)   two measurements of density fluctuations 80 m apart from each other

The third option was the greatest improvement as compared with S3–3. It allowed good estimates for the speed and size of the WDL structures. Unfortunately, one of the dual-purpose probes failed after three and half months of operations in June 1986, after which mode 2) above was mostly used.

In Figure 1 we show some representative Viking observations at an altitude of 7670 km. The magnetic latitude was 76.2° and the magnetic local time 0130. In the lower panel $\Delta n/n$ data features a number of small-scale density depletions. The upper panel shows the electric field component along the 80 m boom directed in this case at an angle of 26° with respect to the ambient magnetic field. Assuming that the density depletions were moving upward, the electric field points toward

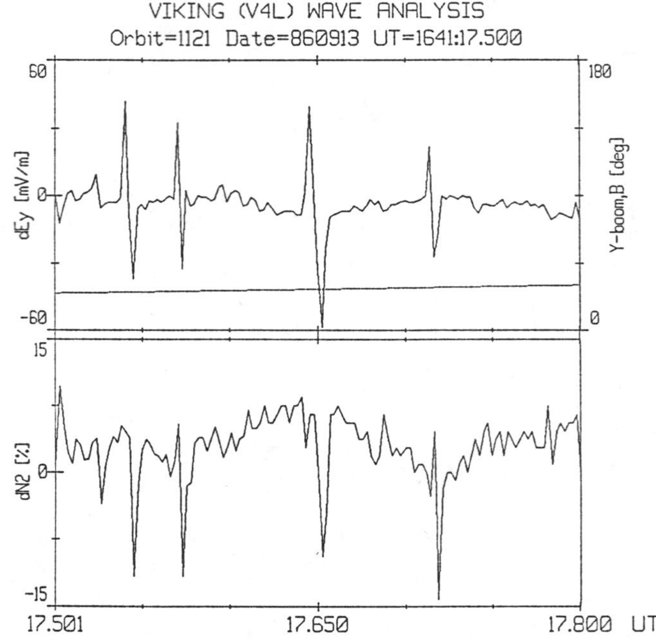

Fig. 1.   Example of WDL observations. The upper panel shows the AC-coupled electric field along the 80 m long boom. The dashed line indicates the boom angle with respect to the ambient magnetic field. In the lower panel the $\Delta n/n$ is shown. Time is in seconds, i.e., the entire plot is over 0.3 s.

the depletions, and consequently, the potential associated with the depletions is negative.

After analyzing a large number of similar data in all three operational modes of the instrument listed above, the following main conclusions were reached. The data from the $\Delta n/n$ probes indicated that ion-scale density depletions up to about 50% of the ambient cool plasma density were observed. With a pair of $\Delta n/n$ probes it was possible to determine the speed of the structures and thereafter their size. It was found that in most cases the structures were propagating upward along the magnetic field (in the spacecraft frame of reference). Typical speeds were 5–30 km/s but speeds up to the maximum observational resolution (68 km/s) were observed. The size of the structures along the magnetic field was typically 100–300 m. The perpendicular size is more difficult to estimate but considering the asymmetry of the depth of the structures, when the boom was nearly perpendicular to the magnetic field, one can deduce that the perpendicular scales were somewhat larger but perhaps not much. Using the mode 2) above the density depletions were found to be associated with negative electric potential. From S3–3 it had not been possible to determine the polarity unambiguously since there were no independent means of determining the direction of propagation. The conclusion by Temerin et al. [1982] was that either the structures had a negative potential and moved up, or a positive potential and moved down. There were good theoretical reasons to believe that the potential was negative [Lotko and Kennel, 1983], but it could not be proven prior to Viking observations. Based on this information the observed structures have been interpreted as depletions in the ambient ion population, "ion holes". Notice that although the holes were observed in the electron density, there is no contradiction. One can easily show that a simple BGK-equilibrium for a negative potential and ion depletion results in electron depletion of about the same depth [Mälkki et al., 1989]. As a matter of fact the solitary ion hole property seems to be a more fundamental feature of the observed structures than the potential drop, and we would prefer to call the structures solitary waves, ion holes, or solitary structures. The auroral plasma community seems to prefer the name WDL.

Sometimes the electric field signature over the ion holes is asymmetric indicating a net potential drop over the structure. When a potential drop was observed, it preferentially implied an upward directed electric field. In the early reports [Boström et al., 1988; Koskinen et al., 1989] the tendency for the upward polarity seems to have been slightly exaggerated due to limited statistics and biased selection of the most clear examples to be studied in detail. A recent, more thorough, study confirms this overall tendency but shows also that a large number of WDLs had a downward polarity. Large majority of the observed ion holes exhibited a very small potential drop, but in some few cases the potential reached up to 2–3 V [A. M. Mälkki et al., A statistical survey of auroral solitary waves and weak double layers: Occurrence and net voltage, manuscript in preparation]. The Langmuir probe estimates for the background temperature were of the order of a few electron volts. This is consistent with the relationship between maximum density depletion and potential variations assuming Boltzmannian electrons with temperatures in the range 2–20 eV. Thus the observed double layers qualify to the category of weak double layers.

In addition to the characteristics of the double layers themselves we need information about the plasma environment where they were observed. Examples of the plasma environment were presented by Koskinen et al. [1990]. The most consequent feature is that, so far, we have not found WDLs without upflowing ions (UFI). Usually the energetic ion distribution was a field-aligned proton beam with a typical characteristic energy 0.1–2 keV, but some cases with considerably higher ion energies have been identified. In the small subset of all Viking data, that have been scanned for WDLs, there are occasionally similar plasma structures in regions of upflowing ion conics. Note however, that UFIs are much more commonly observed than WDLs.

In regions with WDLs a cool background ion population appears to coexist with the beam population. Its detection is difficult because the observations were made in relatively thin plasma where the satellite was usually charged up to a few volts by photoelectron emission, keeping the cool ions away from the ion detectors. However, in a few cases when the spacecraft potential was 2 V, or less, background protons have been found moving up at speeds of a few tens of km/s, i.e., the same speed as the observed WDLs [Koskinen et al., 1990]. The density of such a population was a few ions per cm^3, i.e., about an order of magnitude more than the density of a typical ion beam.

The identification of the cool background ions was important in several respects. First of all there must be some medium to dig the holes into. Due to their slow speed, which is perhaps the most certain piece of information we have, it is quite difficult to think that the holes would be in the energetic UFI population, which passes the satellite much faster. Second, the plasma density depletions were observed by Langmuir probes, which clearly indicated the existence of a cool electron background. In about 20% of the analysed WDL events the cool electron density was less than 1 cm^{-3}, otherwise somewhat higher but less than 10 cm^{-3}. Thus the local plasma density was low but not so low that it would consist of energetic electrons and ions only. Consequently, the speed of the ion holes was of the order of the thermal ion speed of the background plasma, i.e., in the velocity space the holes were not very far from the center of the background ion distribution function (Figure 2). Their size along the magnetic field was several Debye lengths of the background.

WDLs are not the only ion-scale fluctuations in the auroral acceleration region. The auroral acceleration region is rich in various wave phenomena and it is not at all clear, which wave modes are related with WDLs. We know from Viking observations that electrostatic ion cyclotron (EIC) waves are commonly (more than 80% of cases) seen in regions of WDLs. And if we do not find the waves exactly at the same place, we do not usually have to look far from the structures to find them. Of course, this is not so surprising because the EIC waves are frequently found in ion beam regions in any case. The role of the EIC waves is an intriguing question because they may be associated with the perpendicular dynamics of the ion holes and WDLs.

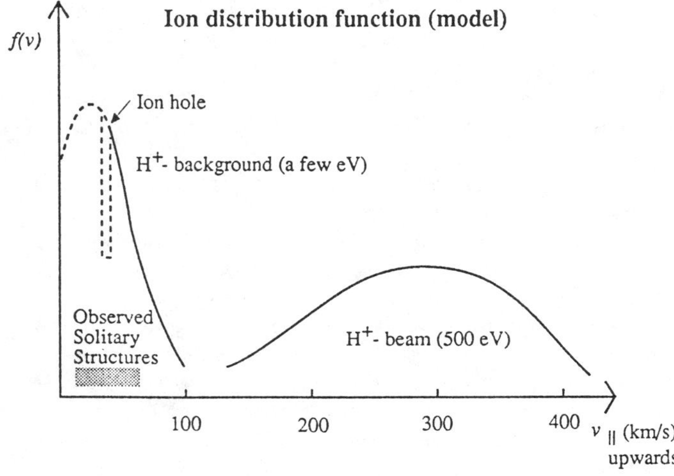

Fig. 2. Model of dominating ion distributions according to Koskinen et al. [1990]. The observed ion hole velocities are indicated by shading.

Another important question is the relationship between WDLs and field-aligned currents (FACs). For theoretical work this is especially critical since several of the proposed theories and computer simulations assume a current, sometimes even a relatively large current. Because the large-scale FAC is typically upward in upward ion beam regions, it is not surprising that it was typically upward also in association with WDLs. However, FAC as estimated from the Viking magnetometer in WDL regions was relatively weak, typically below the critical current for linear ion-acoustic or EIC instabilities [c.f. Mälkki et al., 1989]. The more we have studied the data, the less important the existence of a net current seems to have become. We have even found a few cases where the net current is weakly downward, although here again the signatures are not very clear and the identification of the structures as WDLs may be controversial.

It is important to realize that most supporting data of the plasma environment was acquired with poorer temporal resolution than WDLs themselves. For example, the parallel current is estimated from the deviations in the magnetic field by one sample per ≈ 100 m and several samples are required for a reliable current estimate. Thus the current may have been more structured and locally higher than can be inferred from the magnetometer data. Only the cool electron density was measured with a comparable time resolution by Langmuir probes, whereas a complete ion or energetic electron distribution required half a spin of the spacecraft (i.e. 10 s, during which the spacecraft traveled about 50 km).

WHERE TO LOOK FOR WEAK DOUBLE LAYERS

In general the easiest way of finding WDLs from Viking data is to look for well-defined UFI beams at energies of the order of 1 keV. This does not guarantee that you would find anything, but on the other hand we have not found many cases of WDLs in other environments. There are some exceptions where the ion distribution is conical, and some isolated WDL-like signatures have been found in association with a 10 kV potential drop and a strong density depletion (from ≈ 5 cm^{-3} to 0.1 cm^{-3}). However, these are very special cases, which have not been analyzed in full detail. On the other hand, we see no a priori reason why there would exist only one type of WDL-like structures in auroral plasmas of widely varying parameter ranges.

In Figure 3 we show some preliminary results of an ongoing statistical study of the occurrence of WDLs. Panel 3a) shows where, in invariant latitude and magnetic local time, we have found WDLs. The figure is based on 1017 identified WDLs. The distribution is somewhat biased by the Viking orbit. The velocity determination was possible only during the first few months of the mission when both $\Delta n/n$ probes were operational. For this reason more events from that period have been selected for detailed analysis. At that time the orbit favored the dayside half of the northern hemisphere. On the other hand, we have specifically looked for WDLs from several nightside passes and still there is an almost void region in pre-midnight-midnight sector. Another striking feature is that a ma-

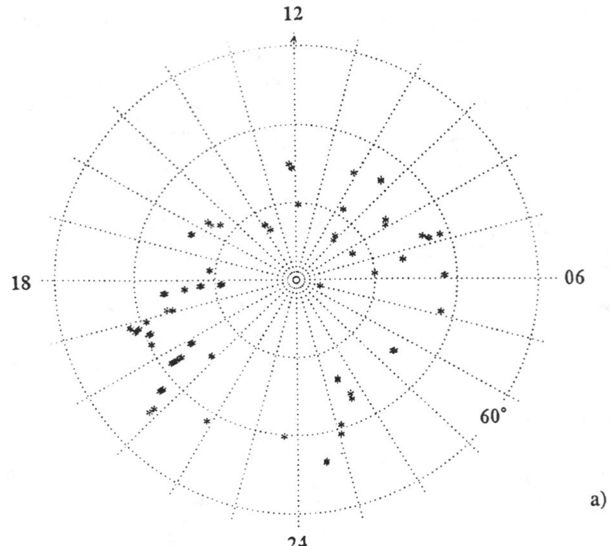

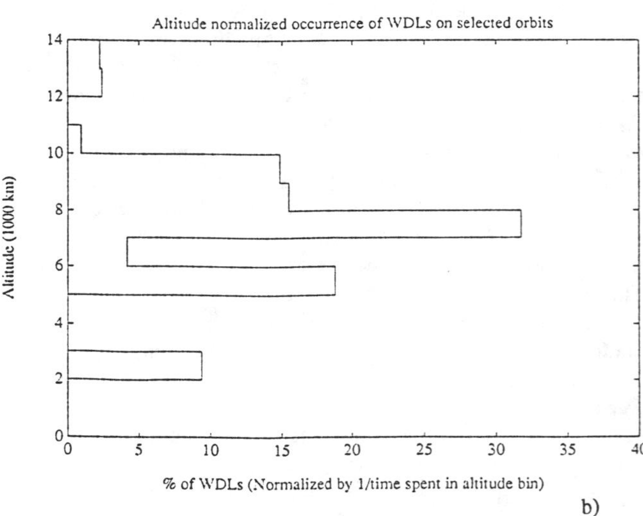

Fig. 3. a) Magnetic latitude vs. local time distribution of 1017 identified WDLs. b) Altitude distribution of WDLs normalized by the time spent in altitude bins of 1000 km.

jority of WDL observations is from relatively high latitudes. This means that WDLs are often observed within high-latitude ion beams that typically are relatively weak and associated with weak high-latitude (polar) auroral arcs [e.g., Eliasson et al., 1987].

Panel 3b) shows altitude distribution of WDLs normalized by the time spent in altitude bins of 1000 km. These statistics are based only on a subset of panel 3a) observations, because in order to diminish subjective bias, we have in 3b) used spacecraft passes which have been investigated throughout and not concentrated on ion beam regions. Most WDLs have been found at altitudes between 7000 and 11000 km, which is consistent with our general experience from WDL observations which are not included in this statistics. In 3b) there are a few WDLs from altitudes 2000–3000 km. Our normalization by the time spent in the altitude interval strongly exaggerates the weight of these observations. Furthermore, the classification of these particular structures as WDLs is somewhat controversial as they do not belong to the clearest cases in the data. We are currently doing a more thorough study on the statistical distribution [A. M. Mälkki et al., manuscript in preparation, referred to above].

One reason for the limited extent of the available statistics is that we have not found a reliable way of automating the analysis. In the spatial and temporal scales we are interested in, the data show a lot of turbulence, irregularities, etc. The WDLs themselves are sometimes well-defined but often one has to judge subjectively whether to include the signature or not as a WDL. Thus all data are scanned manually with the full time resolution.

From a physics point of view it may be more informative to relate the observations with the auroral acceleration region and large-scale plasma cavities. Concluding from energetic particle observations, the WDLs were found either within or above the parallel acceleration region. Since we practically always observed the field-aligned ion beam in connection with WDLs, we may never find them below the entire parallel potential drop, perhaps not even at its lower part. It has been suggested that WDLs form at the edges of large-scale depletions, caused by the parallel electric field removing background plasma from the flux tube, which would allow cool plasma to drift into the acceleration region [Block, 1984], but we cannot confirm this by Viking data. Instead WDLs have been found inside, at the edges, and outside of large-scale density depletions. One should, however, be careful with this conclusion; We do not know the actual shape of the cavities since their depth was measured only along the trajectory of the satellite.

STATUS OF PLASMA PHYSICAL THEORIES ATTEMPTING
TO EXPLAIN WEAK DOUBLE LAYERS

As mentioned above the concept of an ion hole can be described in terms of BGK-equilibrium in one dimension. In such an equilibrium the ion hole is the primary feature. If electrons are assumed to be Boltzmannian, they follow the formation of the hole in such a way that a net negative potential remains. The potential together with electrons and ions forms an equilibrium in the ion trapping time scale ($L / \sqrt{(e\,|\phi|/m_i)}$) , where L is the length of the hole and ϕ its potential). It evolves in ion time scale (ω_{pi}^{-1} , inverse of the ion plasma frequency), and is static for electrons. If the electrons are drifting with respect to the hole, some are reflected, which results in a charge imbalance and thus leads to a potential difference over the hole. This may be the reason why some of the ion holes have the double layer characteristics. However, the BGK picture is highly idealized and does not explain what creates the holes in the first place. This is yet an unsolved question.

In the level of plasma kinetic theory, the free energy for the formation of an ion hole comes from the streaming of the charged particles, regardless of the way in which this streaming is forced by the M–I coupling. Whether the process should be described as current-driven, beam-plasma, or beam-beam interaction, depends on the actual mechanism that converts the streaming to the observed dielectric response. Both the excitations of linear modes as well as the nonlinear response must be taken into account. The first attempts for a theoretical description were based on ion-acoustic (IAC) waves in the nonlinear regime, driven either by electron current or by an ion beam. Such an approach is attractive for theoretical studies because the investigation can be carried relatively far in an analytic form. Although the early studies could not explain the observed structures very well, they contributed to understanding of many basic issues of, e.g., negatively charged ion holes. We will not discuss the various theories in any detail [see, e.g., Mälkki et al., 1989] but concentrate on a few most recent and promising theoretical developments.

We think that a successful theory of WDLs must explain their relationship (if any) with UFI beams and EIC waves. As discussed above (Figure 2) the ion holes move relatively slowly with respect to the spacecraft. Consequently, if a beam-plasma instability drives the holes (e.g., nonlinear development of IAC waves), they cannot be riding on the tail of the beam ions, which would be the first order result. Even an oxygen (O^+) beam is too fast, and its role seems to be minor in any case [Koskinen et al., 1990]. Thus the existence of the background population is a critical issue. It is natural to ask why the parallel potential drop would not remove all cool plasma from the acceleration region. If the WDLs are irregularities in the potential drop, they may allow for cool plasma to penetrate into the region of net parallel potential drop. On the other hand, we know from particle data that some parallel potential is always below the spacecraft (ion beam) and often also above (field-aligned characteristics of energetic electrons), but this does not necessarily imply that the parallel electric field is locally strong in the regions where WDLs are observed. This is a crucial question which we do not know the answer to. If we can identify WDLs in regions of strong local parallel electric field, this problem will become even more acute. The second question, concerning EIC waves, is related to the perpendicular dynamics of the ion holes: Are the holes directly connected with the wave or is their simultaneous appearance only coincidental?

A general problem with current-driven theories has been that they often require stronger currents than have been observed. A theory that avoids this problem, is the nonlinear phase-space ion hole theory originally developed by Dupree [1982] and later applied to Viking observations by Tetreault [1988, 1991]. It predicts that a nonlinear interaction between background electrons and ions results in the growth of the holes for any finite drift between the populations, even below the thresholds for linear IAC or EIC instabilities. As a seed for a hole thermal fluctuations are already sufficient. This theory has been investigated also using computer simulations [Berman et al., 1986], but only in one spatial dimension. In addition to the requirement of weak electric current, one of the strongest arguments in favor of this theory is that it predicts speeds for the ion holes comparable to those observed. Tetreault [1991] listed several other arguments in support of the theory. We think that some arguments were quite optimistically selected from published analyses of Viking data and a careful experimentalist might have been more reserved. Tetreault [1991] also seems to have fully adopted the suggestion that the entire parallel electric field is built up by WDLs and that the observed ion beam is a consequence of acceleration by these small potential jumps. We cannot prove this suggestion, nor entirely disprove it either, with our data. However, the consequent existence of an ion beam created below the spacecraft casts some doubts on this hypothesis. In any case a thorough theory of electric fields in the auroral acceleration region is clearly a wider issue than the microphysics of WDLs. Furthermore, the one-dimensional ion hole theory does not give any direct hints about the role of EIC waves. Tetreault [1991] investigated this problem by adding a perpendicular term in the dielectric function. He came up with an interesting result that the ion holes might be a source for the waves.

When the paper by Mälkki et al.[1989] was written, we found the nonlinear ion hole theory as the most promising approach to the WDL problems. We still think that it belongs to the most serious alternatives, but meanwhile other approaches have also made positive progress. For some time, an argument against IAC wave processes has been the high velocity of the produced structures. This was challenged by recent simulation studies [Gray et al., 1991; Hudson et al., talk at the Chapman Conference on Auroral Plasma Dynamics, 1991], involving background electrons and ions and an ion beam. Gray et al. [1991] found that although the slow beam mode on the positive slope of the beam distribution function is destabilized first, the ion hole grows in the background population due to nonlinear wave coupling between the fast and slow beam acoustic modes and the background acoustic mode. They were able to produce perhaps the most realistic looking ion holes so far. The plasma had to be forced with somewhat unrealistically strong beams but also here the progress is in the right direction [Hudson et al., talk at the Chapman Conference on Auroral Plasma Dynamics, 1991]. Another problem is that the perpendicular dynamics is not taken into account. Preliminary results show that the holes evolve also in two spatial dimensions, but the limitations in the applicable number of particles in the simulation system introduces difficulties with numerical noise [M. K. Hudson, private communication, 1991].

Another recent simulation was performed by Mottez et al. [1992]. They also included background electrons and ions and an ion beam. The simulation was quite similar to the beam-background study of Gray et al. [1991], but it was performed in two spatial dimensions. The higher dimensionality makes the interpretation quite different, however. The free energy comes from the ion beam. First the beam excites an oblique mode in the ion cyclotron regime. This wave is either a beam-background two-stream mode or a kinetic EIC wave, or, perhaps, some hybrid of these [Mottez et al., 1992]. The wave grows to a finite amplitude and its positive and negative frequency components beat, producing a purely perpendicular zero-frequency fluctuation. This is effectively a field-aligned striation of the plasma. Simultaneously, the beam ions drag the electrons along their direction, and finally the ion holes are produced within the striations by electron-ion interaction. Also this work is still in a preliminary phase and more studies are required before a final evaluation can be given. However, the first results are very promising. It is especially interesting to notice the natural role of the ion cyclotron waves. Because the waves are primarily needed to create the field-aligned striations, they are not necessarily observed in the same place as fully developed ion holes or WDLs.

It is noteworthy that the interpretation of all the three approaches described above depends critically on the nonlinear properties of the plasma. The modern simulation models have become so complicated that the mere analysis of the simulation runs is almost as difficult as interpretation of data observed in space. At the same time when more efficient computer codes are developed, much more effort must be paid to the diagnostics methods and physical interpretation of the results. Evidently, the final words have not been told yet. So far the simulations have been electrostatic. The formation of field-aligned striations in the simulation by Mottez et al. [1992] may cause filamentation of the FAC as well. Here we may encounter the limit how far electrostatic studies can be carried. The smallness of the plasma beta does not guarantee that the process stays electrostatic in the nonlinear regime. The excited modes may couple, e.g., with Alfvén waves that mediate the M–I coupling.

Finally there is the problem of appropriate boundary conditions. In most particle simulations periodic boundary conditions are used, which simplifies the numerical codes. On the other hand, the particle populations get heated in the system and loose the free energy, whereas in nature new free energy is continuously fed into the region of interest because the particle streaming is forced by the entire magnetospheric circuitry. It is not easy to find physically relevant nonperiodic boundary conditions. A new approach to the M–I coupling was presented by Goertz et al. [1991] who described the global M–I coupling in the framework of an Alfvén wave model and used it to determine the boundary conditions for kinetic simulations. Also this model is in a premature phase and has not been used to produce WDLs. Goertz believed that it would become useful for this problem as well [private communication, 1991]. Our opinion is that somehow the small-scale plasma dynamics must be tied together with the global M–I coupling.

THE SIGNIFICANCE OF AURORAL WEAK DOUBLE LAYERS

We have already indicated above that we are not quite confident with the idea that the WDLs would be a general solution to the problem of formation of large-scale parallel electric fields in the auroral acceleration region. The estimate by Koskinen et al. [1989], that WDLs might contribute an average parallel electric field of one volt over a few kilometers, has sometimes been interpreted that it was really suggested that a large potential drop would be possible to build-up with WDLs. Although WDLs have been observed within an altitude range 4000–13000 km, we do not know how high the WDL region is within any specific flux tube. Neither do we have any direct information on the parallel electric field between two consecutive WDLs. Viking made perpendicular cuts through the region where it observed the small-scale potential structures passing by the spacecraft along the magnetic field. Of course, we cannot disprove that suggestion either, but we consider the evidence in favor of it rather vague. The question how the parallel potential difference is maintained is, in general, an open question, as well as is the question of how much WDLs contribute to it.

Another question about the significance of WDLs is their contribution to the energization of charged particles. Koskinen et al. [1990] discussed the problem and concluded that energetic electrons of plasma sheet origin go through the region and are adiabatically accelerated downward by the amount of the integrated potential drop of all WDLs they encounter [see also, Lotko, 1986]. The question of energetic ions, i.e., the ion beam, is more tricky. Koskinen et al. [1990] stated that the ions are adiabatically accelerated upward and perpendicularly heated by the perpendicular dynamics. The statement concerning the upward acceleration was based on the assumption that the beam ions are like single test particles, experiencing a potential drop created entirely by the background plasma. However, if the beam more or less directly participates in the creation of the ion holes, the question is not so simple. The fact the macroscopic current and electric field are both locally upward ($\mathbf{j} \cdot \mathbf{E} > 0$) is not enough to conclude that the beam ions are accelerated. All participating particle species must be considered, taking into account the negative energy of the holes, the streaming directions of particles and the hole, etc. This is evidently an extremely difficult problem.

Also the perpendicular ion dynamics is complicated. We have made three dimensional test particle simulations of ion evolution in a static ion hole model, where each ion was scattered by a large number of randomly distributed holes on its way upward [Mälkki, 1991]. The holes were assumed spatially symmetric, conserving the total energy. In Figure 4 we show results of a simulation including a homogeneous accelerating potential along the magnetic field. The initial distribution (at the bottom of 4a) had a zero perpendicular temperature. The other distributions were computed for altitudes 200, 1000, and 1800 km. When the ion drift energy was comparable with the hole depth, i.e., roughly the first 200 km, the ion population was rapidly heated in perpendicular direction. Thus the holes redistributed a part of the parallel energy. When we added more parallel energy, the final perpendicular temperature remained the same. This can be explained by the softness of the collisions between particles and holes when the energy is large with respect to the hole depth. Thus we conclude that the ion holes and WDLs contribute only weakly to the heating of ion beams. For initial heating of cool ions the holes seem to be efficient but then one encounters the problem of particle trapping by the dynamically evolving holes and there is a need for self-consistent treatment of the problem.

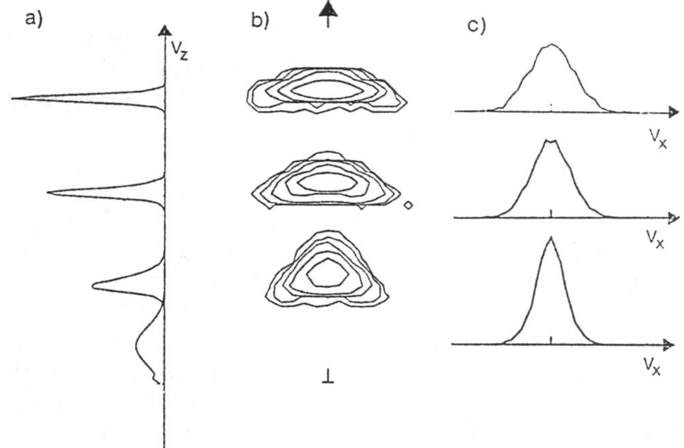

Fig. 4. Evolution of a test particle population in a simulation of interaction between ion hole and the particles including a field-aligned accelerating potential (z-direction). a) shows the distribution function along the magnetic field, b) is the contour plot in v_z–v_x space, and c) the perpendicular velocity distribution. The distributions are computed for 200, 1000, and 1800 km above the initial distribution which was Maxwellian along the magnetic field with zero perpendicular temperature (shown only in a)).

We conclude that the significance of the WDLs is not very well understood at present. We can safely say that they are a part of the local auroral plasma response to the global M–I coupling. They seem to contribute to the irregular, or turbulent, characteristics of the auroral acceleration region and the parallel potential drop. Being electric potential structures, they also participate in the processes that redistribute energy between particle populations. However, an assessment of how important they are as compared with various wave modes cannot be given until we understand the local plasma dynamics much better than today.

Summary

We should like to conclude this review with an attempt to evaluate the various pieces of information we have on WDLs. It is clear that such an evaluation is highly subjective and will certainly be subject to change within the near future.

Concerning the observations we are most certain of the propagation direction and speed with respect to the satellite, and of the size along the magnetic field. Note, however, that this conclusion applies only for the typical cases, where we can identify well-formed structures without hesitation. We are also quite certain that we have observed density holes with negative potential. That the holes often form a WDL-like potential drop is also well-documented, as well as their association with UFI beams. Somewhat less certain is the association with EIC waves and with the large-scale parallel potential drops because here more assumptions are needed to interpret the data. The same applies to the existence of the cool background plasma which is not always directly observable. The existence and necessity of a net upward FAC is even more uncertain. We have some indication of the perpendicular structure of the ion holes but are quite uncertain of the actual scale length in that direction. And finally, every now and then one observes similar plasma structures that do not fit well into the realm of the general trends discussed above. It is quite possible that our typical ion holes and WDLs are not the only solitary structures that auroral plasmas can create.

On the theoretical front we have learned a lot since the time of the S3–3 observations. The plasma structures clearly require kinetic treatment. Ion dynamics is quite certainly three-dimensional whereas electrons can probably be treated as one dimensional but perhaps not as a fluid. Also it is evident that we have to deal with complicated nonlinear plasma physics when looking for physical explanations. Although it is probable that the observed EIC waves play an important role in the dynamics of ion holes, we do not know what the role is. The waves may be important in the creation of the holes, or the holes may emit the waves.

The most uncertain part is the geophysical role of the WDLs. Of course, somehow they must be associated with the M–I coupling. They may build-up some parallel potential drop and contribute to heating of cool ions. Also they may be coupled with the Alfvén waves mediating the M–I coupling but it is not known how. We have gradually become quite uncertain whether the WDLs can build up large parallel potential drops and we do not think that they can be major contributors to the ion heating. Both these highly important questions of auroral plasma dynamics remain unsolved for the time being.

In conclusion we can state that there still are quite a few open matters in the physics of auroral weak double layers. Our somewhat negative attitude to their grand role in the M–I coupling should not be understood as an understatement of the importance of the work trying to explain the physics of WDLs. After all, they are directly observed plasma phenomena in a highly interesting geophysical region. In space physics, studies must often be based on quantities derived from relatively uncertain observations. It is important to try to understand especially those phenomena which have been observed directly.

Acknowledgements. This report is based on results obtained through a collaboration within the Viking low-frequency wave experiment (V4L) team and with other Viking experimenters. Discussions with R. Boström, G. Chanteur, P. Gray, B. Holback, G. Holmgren, M. Hudson, R. Lundin, F. Mottez, A. Roux, and D. Tetreault are gratefully acknowledged.

REFERENCES

Berman, R. H., D. J. Tetreault, and T. H. Dupree, Growth of nonlinear intermittent fluctuations in linearly stable and unstable simulation plasma, *Phys. Fluids.*, *29*, 2860, 1986.

Block, L. P., Potential double layers in the ionosphere, *Cosmic Electrodynamics*, *3*, 349, 1972.

Block, L. P., Three-dimensional potential structure associated with Birkeland currents, in *Magnetospheric Currents, Geophys. Monogr. Ser., vol 28*, T. A. Potemra, ed. (AGU, Washington D.C., 1984) p. 315.

Block, L. P., and C.-G. Fälthammar, The role of magnetic-field-aligned electric fields in auroral acceleration, *J. Geophys. Res.*, *95*, 5877, 1990.

Boström, R., G. Gustafsson, B. Holback, G. Holmgren, H. Koskinen, and P. Kintner, Characteristics of solitary waves and weak double layers in the magnetospheric plasma, *Phys. Rev. Lett.*, *61*, 82, 1988.

Dupree, T. H., Theory of phase-space density holes, *Phys. Fluids*, *25*, 277, 1982.

Eliasson, L., R. Lundin, and J. S. Murphree, Polar cap arcs observed by the Viking satellite, *Geophys. Res. Lett.*, *14*, 451, 1987.

Goertz, C. K., T. Whelan, and K.-I. Nishikawa, A new numerical code for simulating current-driven instabilities on auroral field lines, *J. Geophys. Res.*, *96*, 9579, 1991.

Gray, P. C., M. K. Hudson, W. Lotko, and R. Bergmann, Decay of ion beam driven acoustic waves into ion holes, *Geophys. Res. Lett.*, *18*, 1675, 1991.

Koskinen, H., R. Boström, and B. Holback, Viking observations of solitary waves and weak double layers on auroral field lines, in *Ionosphere-Magnetosphere-Solar Wind Coupling Processes*, edited by T. Chang, G. B. Crew, and J. R. Jasperse, p. 147, Scientific, Cambridge, Mass., 1989.

Koskinen, H. E. J., R. Lundin, and B. Holback, On the plasma environment of solitary waves and weak double layers, *J. Geophys. Res.*, *95*, 5921, 1990.

Lotko, W., Diffusive acceleration of auroral primaries, *J. Geophys. Res.*, *91*, 191, 1986.

Lotko, W., and C. F. Kennel, Spiky ion acoustic waves in collisionless auroral plasma, *J. Geophys. Res.*, *88*, 381, 1983.

Mälkki, A., Numerical simulation of perpendicular ion heating by ion holes in the auroral magnetosphere, in the proceedings of the 4th International School for Space Simulation, p. 194, Kyoto and Nara, Japan, 1991.

Mälkki, A., H. Koskinen, R. Boström, and B. Holback, On theories attempting to explain observations of solitary waves and weak double layers in the auroral magnetosphere, *Phys. Scr.*, *39*, 787, 1989.

Mottez, F., G. Chanteur, and A. Roux, Filamentation of the plasma in the auroral region, by an ion-ion instability: A process for the formation of bi-dimensional potential structures, *J. Geophys. Res.*, *97*, 10801, 1992.

Mozer, F. S., and R. B. Torbert, An average parallel electric field deduced from the latitude and altitude variations of the perpendicular electric field below 8000 kilometers, *geophys. Res. Lett.*, *7*, 219, 1980.

Reiff, P. H., H. L. Collin, J. D. Craven, J. L. Burch, J. D. Winningham, E. G. Shelley, L. A. Frank, and M. A. Friedman, Determination of auroral electrostatic potentials using high- and low-altitude particle distributions, *J. Geophys. Res.*, *93*, 7441, 1988.

Temerin, M., K. Cerny, W. Lotko, and F. S. Mozer, Observations of double layers and solitary waves in the auroral plasma, *Phys. Rev. Lett.*, *48*, 1175, 1982.

Tetreault, D. J., Growing ion holes as the cause of auroral double layers, *Geophys. Res. Lett.*, *15*, 164, 1988.

Tetreault, D. J, Theory of electric fields in the auroral acceleration region, *J. Geophys. Res.*, *96*, 3549, 1991.

H. E. J. Koskinen and A. M. Mälkki, Finnish Meteorological Institute, Department of Geophysics, P. O. Box 503, SF-00101 Helsinki, Finland

Are Weak Double Layers Important for Auroral Particle Acceleration?

Anders I. Eriksson and Rolf Boström

Swedish Institute of Space Physics, Uppsala Division, S-755 91 Uppsala, Sweden

The acceleration of auroral particles is often attributed to quasi-static electric fields with a component along geomagnetic field lines. Such fields may be sustained by weak double layers (WDLs). In this model, the energy for the particle acceleration is provided by the energy flow associated with the field-aligned currents mediating the stationary magnetosphere-ionosphere coupling, and the use of WDLs as particle accelerators is therefore dependent on their inclusion in the global magnetospheric circuit pattern. To answer the question of the importance of WDLs for particle acceleration, one must therefore consider the circuit as a whole, including the generator part. We conclude that parallel electric fields sustained by WDLs contribute to particle acceleration, but it is difficult to evaluate from satellite observations how important they are.

Introduction

The currents that flow along the geomagnetic field between the ionosphere and magnetosphere (Birkeland currents) transport huge amounts of energy. They do so in their attempt to harmonize the convection of the ionospheric plasma, which couples to the neutral atmosphere and thus possesses a large inertia, to the plasma motions in the vast regions of the magnetosphere. The acceleration of auroral particles takes place in the regions of space where these currents flow, which suggests that in the acceleration processes, some energy is tapped from the electromagnetic energy flux and converted to particle kinetic energy. For this mechanism to work, we must have $\mathbf{j} \cdot \mathbf{E} > 0$ somewhere along a field line in the acceleration region, so that work is done on the plasma by the electromagnetic field. We thus need an electric field with a component along the geomagnetic field, henceforward called a parallel electric field. To study the energetics of the situation, it is suitable to represent the electromagnetic energy flux as the Poynting flux caused by the magnetic field of the field-aligned current, and the electric field associated with the closure of these currents through the resistive ionosphere. Where $\mathbf{j} \cdot \mathbf{E} > 0$, the Poynting flux has a negative divergence, and energy balance is maintained.

The existence and nature of this quasi-static parallel electric field is still a subject of debate. As the electric field in the perpendicular direction almost always dominates over the weak parallel field, the measurement of the latter is intricate. Most evidence for the large-scale quasi-static parallel electric field is therefore indirect, derived from measurements of accelerated auroral particles [Burch, 1991], although there are some direct measurements of such fields (Mozer et al. [1980], and references therein; see also Lindqvist and Marklund [1990] and Block and Fälthammar [1990]).

A part of this parallel electric field may be continuously distributed in space (see Chiu and Schultz [1978] and references therein) , but from simulations and laboratory experiments (reviewed by Schamel [1986]), there is reason to expect that a significant part is localized in a series of small steps, called weak double layers (WDLs). The S3-3 and Viking satellites had the possibility of measuring electric fields with high time resolution, and solitary structures that could be interpreted as WDLs were indeed observed [Temerin et al., 1982; Boström et al., 1988]. This paper is a tutorial discussion of WDLs, with emphasis on the title question – are they important for the acceleration of auroral particles? The detailed microscopic theories for WDLs are in this context of minor interest (a review of these matters is given by Mälkki et al. [1989], updated by Koskinen and Mälkki [this volume]). Instead, we give a picture of how the WDLs fit into the general context of quasi-static magnetosphere-ionosphere coupling. Discussions and criticism against theories where quasi-static parallel electric fields are used for explaining particle acceleration [Bryant et al., 1992] have shown that there is a need to clarify these issues and provide a qualitative picture of what a basically electrostatic model of the magnetospheric system might look like. We do not claim that the real magnetosphere is an electrostatic system, but it is interesting to explore the possibilities of such a model, as the time-independent case is the simplest possible description.

The formation of weak double layers

The general idea of how a weak double layer is formed may be illustrated as follows [Hasegawa and Sato, 1982; Tetreault, 1988]. We consider a one-dimensional plasma with an upward field-aligned current, carried by thermal electrons. Let us assume that somewhere in this plasma there exists a localized minimum of the electrostatic potential (a potential well) with amplitude (in volts) less than the electron temperature (in eV). We assume this structure to be governed by the ion dynamics, so that its lifetime is so long as to seem stationary for the electrons. The electrons will then see the electrostatic field as conservative, and electrons with too low kinetic energy to pass the potential well will be reflected (Figure 1a). As the electrons carry an upward current, this reflection will create an excess negative charge above the potential well, and a net positive charge below it. The resulting asymmetric potential structure is a WDL. Electrons with energy sufficiently high to pass the well will be accelerated downwards by the electric fields associated with the excess charge built up by the reflecting electrons. These electric fields, marked as P and R in Figure 1a, are in principle free to propagate, while the potential well stays confined at Q as it is governed by the ion dynamics. We may thus have situations where the mainly upwardly directed field at Q is the only part of the field that stays in the vicinity of the original potential well (Figure 1b). It is then meaningful to speak of a net voltage drop over the WDL, in the sense that the integral of the parallel electric field over the length of the original potential well is non-zero.

If we consider an ensemble of WDLs along a field line, and assume that the properties of the original potential wells are

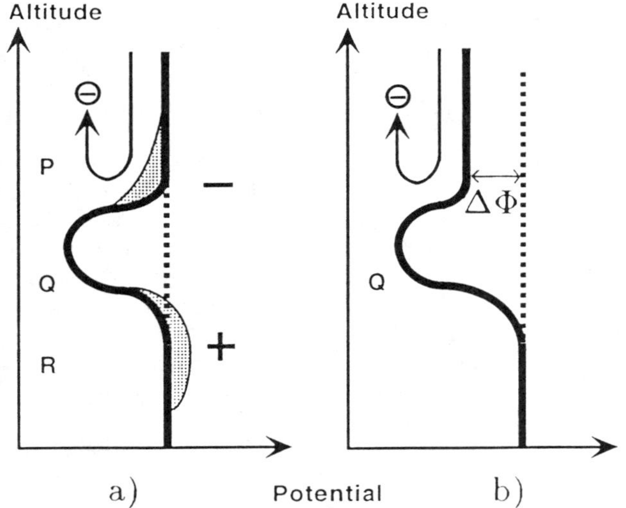

Fig. 1. The formation of a weak double layer from a symmetric potential well (left) in the presence of an upward current, carried by downstreaming electrons. The potentials from the reflecting electrons at P and R (shaded areas) are free to propagate in space, and thus an asymmetric potential structure (right) with a net voltage drop $\Delta\Phi$ over the region of the original potential well is formed.

the result of current driven instabilities associated with the field-aligned current, the average effect of the WDLs can be described in terms of an anomalous resistivity. However, the use of this concept is somewhat misleading, as the relationship between current and electric field does not have to be linear. Neither should this resistivity be thought of as implying heat losses. In the simple one-dimensional picture presented above, there is no heating, and all electromagnetic energy is converted to energy of the accelerated particles. In a two- or three-dimensional model, some pitch angle scattering will occur [Rowland and Palmadesso, 1983], and thus some energy goes to heat. Still, the load on the electric circuit imposed by the WDLs should rather be compared to the load of an electric motor, where a part of the electromagnetic energy used goes to useful work, than to a simple resistor, where all energy goes to heat. In the case of a WDL, the work is done on the electrons that pass through it and get accelerated.

As long as the electron gyroradius r_{ce} is much less than the dimension of the potential well perpendicular to the magnetic field, the electron motion is essentially one-dimensional. As any structure governed by the ion dynamics is expected to have perpendicular dimensions at least on the ion gyroradius r_{ci} scale, the one-dimensional picture above should be applicable for the electrons. The simple one-dimensional model of WDL formation given above should then be valid for any potential well on a spatial scale set by the ions. Thus, coherent or turbulent low-frequency (ion acoustic, electrostatic ion cyclotron) waves should in principle be as important for the formation of weak parallel potential as soliton-like phenomena (ion acoustic solitons, ion phase space vortices). Rowland and Palmadesso [1983] found in simulations that the potential drop over a current carrying region with electrostatic ion waves was localized in space as a series of small voltage drops, one at every minimum of the wave electrostatic potential. However, there are two facts that make the solitary phenomena particularly interesting. First, numerical studies show that large-amplitude potential wells associated with long-lived (several ion plasma and cyclotron periods) solitary ion wave structures (ion phase space vortices, ion holes) are often formed in ion wave turbulence [e.g., Kofoed-Hansen et al., 1989], and the potential drops and WDL structures thus tend to be associated with solitary potential wells [Rowland and Palmadesso, 1987]. Second, it is possible to measure the net voltage drop over a solitary double layer by a satellite double probe experiment. This is difficult for periodic or turbulent wave fields, but the anomalous resistivity of such fields may still be important for the total parallel electric field.

Observations of solitary potential structures and weak double layers

The previously published observations of solitary structures from S3-3 [Temerin et al., 1982] and Viking [Boström et al., 1988; Koskinen et al., 1990; Boström, 1992; Mälkki et al., 1992], which are also discussed elsewhere in this volume [Koskinen and Mälkki, this volume], may be summarized as follows:

- Localized solitary electron density depletions of up to 50 % exist in what is known as the auroral acceleration region. We call these solitary structures.

- The structures are local minima of the electrostatic potential with an amplitude of up to 5 V.

- The structures are propagating upwards with typical speeds of 10 – 30 km/s. This is a realistic range of the ion acoustic velocity c_s, and thus indicates the presence of a cold ion population.

- The density and potential signals relate approximately as predicted by a Boltzmann relation with electron temperature T_e about 2 – 20 eV, which indicates that a population of cold electrons is present.

- Their dimension parallel to the geomagnetic field is typically 50 – 250 m, which is of the order of 5 – 50 Debye lengths λ_D.

- The perpendicular size of the structures is difficult to estimate, but is at least of the same order as the parallel. Assuming a cold ion population around 1 eV, the proton gyroradius r_{cp} is 20 – 100 m.

- The solitary structures are observed in regions of beams of upflowing ions with energies of typically 0.5 – 1 keV.

- The large-scale field-aligned currents in the regions of observation are weak, $\sim 10^{-7}$ A/m^2. The current is generally upward, but there are indications of a downward current in a few cases. The current structure on a smaller scale (less than a few kilometers) is not known.

- The observed linear packing density (fraction of time in data occupied by solitary structures) is typically 5 – 10 % in the regions where we observe them. For shorter time periods, up to 20 % packing density is observed.

Examples of Viking data are shown in Figure 2. Presently, we are in the process of completing a statistical study of some 2,000 solitary structures. The detailed results are not yet complete, but one result may be added to the list above:

- The characteristic evolution time for the structures (e-folding time for growth or decay) is on the order of 10 ms or more.

This estimate is based on the difference between the amplitudes of the density fluctuation recorded by two Langmuir probes separated by 80 m along a field line (Figure 2b).

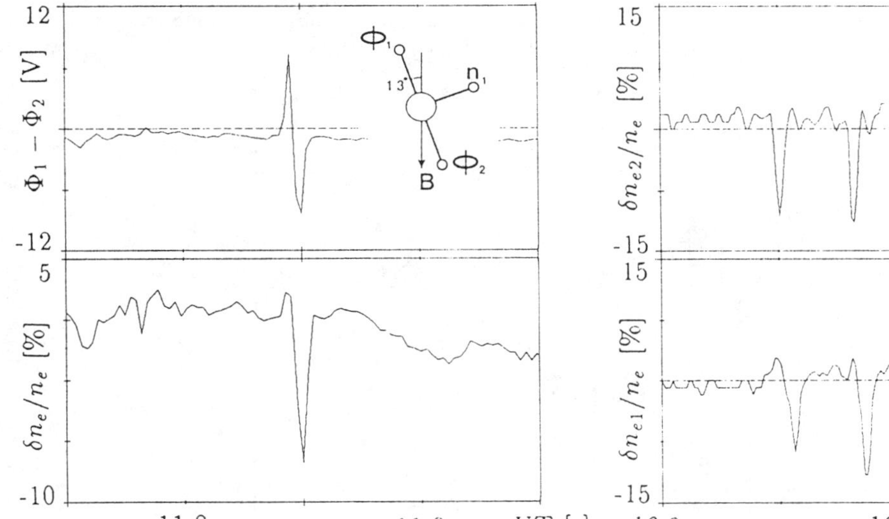

Fig. 2. Viking observations of solitary structures. The insets show the satellite orientation relative to the geomagnetic field. a) Potential and density fluctuation signature of a passing rarefactive solitary structure with a negative potential. b) Two point measurements of density fluctuations, separated by 80 m. Two solitary structures are seen first by the lower, then by the upper probe. If the structures move strictly upward, the difference in amplitude in the two channels can be used to estimate the time evolution.

If this characteristic time is compared to the typical ion and electron cyclotron and plasma frequencies of $f_{cp} = 50 - 200$ Hz, $f_{ce} = 100 - 500$ kHz, $f_{pi} = 1 - 4$ kHz, $f_{pe} = 50 - 200$ kHz, we find that the lifetime of the structures seems to lie on the order of several ion plasma or cyclotron periods. Typical plasma parameters relevant for the Viking observations are $T_e = T_i = 10$ eV, and $n_e = 5$ cm^{-3}. The electron thermal speed and drift speed, assuming they carry currents as stated above, should typically be at least 10^5 m/s. Combining this information on speed, frequencies, and characteristic lengths, we find that the structures are indeed seen as stationary on an electron time scale. According to the theory outlined in the previous section, such structures should develop WDL characteristics if located in a current carrying region. Using plasma parameters as above, the average potential drop over a WDL formed from an ion hole can be calculated [Mälkki et al., 1992] to be of the order of 0.2 V. As was stated above, this voltage drop can be measured by satellites, and examples of voltage drop estimates for individual WDLs are found in the results from S3-3 and Viking [Temerin et al., 1982; Boström et al., 1988]. Using statistics of some 600 Viking measurements of solitary structures, Mälkki et al. [1992] have found a preliminary estimate of the average potential drop over a WDL to be around 0.1 V with the lower potential above the layer. We emphasize that this is a very preliminary result of the statistical investigation mentioned above. Combined with the observed linear packing density of some 10 %, this gives a potential drop of the order of 0.4 keV if the acceleration region is taken to be 6,000 km along a field line, and the parallel extent of a WDL is 150 m. The immediate interpretation of this is that the WDLs should be of minor importance for the acceleration of auroral particles to the highest energies. However, the material does not allow this conclusion. First, the spread in

net potential over the structures is considerable. Structures with net potential drops of some volts are observed [Temerin et al., 1982; Boström et al., 1988]. When taking the average value above, we are likely to mix structures from different regions. A lot of structures with zero net voltage drop are included, and they may be associated with regions of no particle acceleration or field-aligned current. Second, the method of determining the net potential drop is sensitive to the time evolution of the structures, and a careful analysis of this effect is needed. Third, the figure above refers only to solitary WDLs that can be clearly identified in the data. We cannot say anything about the contributions from more disordered wave fields. From the observations, we can neither conclude nor exclude that the WDLs sustain potential drops of keV order along the field lines.

WEAK DOUBLE LAYERS AS CIRCUIT ELEMENTS

The interest in WDLs as possible particle accelerators lies in their possibility to sustain a parallel electric field stationary on a typical electron time scale. This possibility is completely dependent of the connection of the WDL to a circuit system with a generator part. A single WDL left on its own could not give any net acceleration of auroral particles by a stationary electrostatic process. In fact, such a WDL could not even exist. If there is no generator which accepts the load on the circuit caused by the WDL, the current will decay. When the current decays, the solitary structure resumes its symmetric shape, and its WDL property vanishes.

Figure 3 illustrates an idealized model of how the potential pattern of a newly formed WDL is connected to the magnetospheric circuit. Figure 3a shows the situation at outset. The magnetic field is assumed to be pointing downwards, and a current flows along it. The contours show the

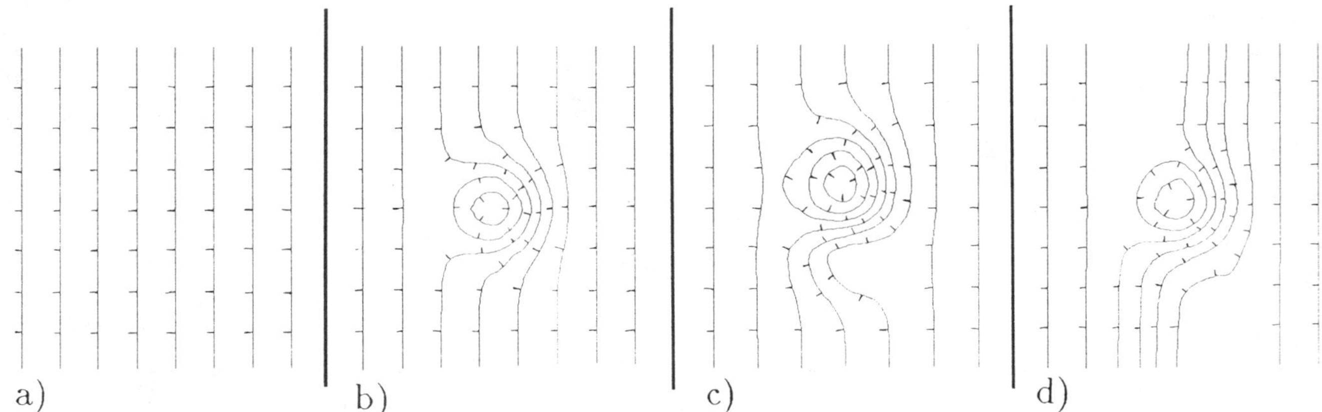

a)   b)   c)   d)

Fig. 3. Idealized model of the evolution of the potential pattern during the formation of a WDL and its connection to the global circuit. The magnetic field is directed from top to bottom in the figures, and the lines shown are electrostatic equipotentials. The small bars on the equipotentials point in the direction of the electric field. a) The potential pattern at outset with a perpendicular electric field. b) A localized potential well is formed. c) The potential well develops into a WDL by reflection of low-energy electrons. d) The potential caused by electron reflection propagate along the field lines.

equipotentials of a large scale perpendicular electric field. In Figure 3b, a potential well is formed in the center of the figure. This develops into a WDL by the reflection mechanism discussed above. The asymmetric potential distribution is shown in Figure 3c. When the excess charges from the electron reflection propagate upwards and downwards, we have a situation as in Figure 3d, where the net potential drop is established locally over the WDL. Comparing Figures 3c and 3d, the equipotentials have moved slightly to the right (left) above (below) the WDL when the electrostatic field (P and R in Figure 1a) of the reflected electrons has passed.

The crucial feature of this qualitative model is the propagation of the fields P and R. If these fields should stay in the vicinity of the WDL, no net voltage drop will form. Electrons will be retarded and accelerated locally. To couple the local potential fluctuation to a global circuit, we must move the regions P and R to the ionosphere and generator regions, respectively. This is possible, as these fields are not tied to the potential well region Q, but may propagate freely in space. When P propagates downward, its excess positive charge will give it the signature of a positive pulse of field-aligned current. Similarly, R will be a negative pulse of field-aligned current. Changes in field-aligned currents are generally transported as Alfvén waves, as this is the only low frequency mode with a current component along the magnetic field. The net voltage drop over the structure depends on the amplitude of the potential well, and thus the characteristic time for the parallel potential drop will be the same as for the solitary potential well itself. According to the discussion of the Viking measurements above, this means that the characteristic frequencies of the wave signals carrying the information about the WDL should be well below the ion gyrofrequency, where Alfvén waves propagate. This naive picture of a microscopic coupling between WDLs and Alfvén waves needs to be clarified. The importance of the coupling of these two phenomena has been emphasized by Lysak and Hudson [1987] , but quantitative studies taking care of the microphysics are still lacking. An obvious reason for this is the problems encountered in the study of the coupling between phenomena of so disparate scale sizes ($\sim 10^6$ m for the Alfvén waves, and $\sim 10^2$ m for the solitary structures).

The WDL will have decayed long before the information on the new load imposed by it on the circuit has reached the generator region, but in a turbulent region, others will have formed instead and taken its place. For the global dynamics, it is the average anomalous resistivity and parallel electric field of all WDLs that matters. The dynamical coupling of the ionospheric and magnetospheric parts of the circuit by Alfvén waves in the presence of a region of current dependent anomalous resistivity has been studied by Lysak and Dum [1983] (see also review by Lysak [1990]). They show how the information about a region of anomalous resistivity is transmitted to the generator region, and how a new global potential pattern, including a parallel electric field, is set up. This should be an accurate description of the large-

scale properties of the effect of many double layers. Thus, there is no problem with the connection of the WDLs to the global circuit, although the interesting microscopic coupling between WDLs and Alfvén waves deserves more study.

Parallel electric fields associated with Alfvén waves present an alternative mechanism for auroral particle acceleration [e. g., Hasegawa, 1976; Seyler, 1990]. A detailed discussion of this alternative is outside the scope of this paper, but we want to point out that the anomalous resistivity of WDLs is important for such theories. The parallel field-aligned current in the Alfvén waves should act as any other field-aligned current and produce weak double layers from electrostatic potential wells. Also, the Alfvén waves are partially reflected by this anomalous resistivity, which changes their propagation characteristics and the current and potential patterns they set up [Lysak and Dum, 1983].

MAGNETOSPHERIC GENERATORS

If we consider a completely electrostatic system, the total energy, kinetic plus potential, of any charged particle in a collisionless plasma is conserved. This is used in our description of the acceleration of particles by double layers above, where potential energy is converted to kinetic energy by the parallel electric field. It is also instructive to look at the generator part of the circuit in this single-particle picture. Generator models are generally presented in MHD terms [e.g. Boström, 1975], although Vlasov methods have also been used in discussions of field-aligned currents [e.g. Birmingham, 1992]. It is important to note that there is no contradiction between collisionless particle motion theory and MHD, as the normally used MHD relations can be deduced from the single particle drift motions. However, MHD does not tell us much about the particle dynamics.

MHD generators can be driven by convection or thermal energy [Boström, 1975]. Figure 4 illustrates the particle picture of the convection driven dynamo of Rostoker and Boström [1976]. A plasma streams through a region (shaded) where the convection velocity and electric field decreases in the direction of the flow (the geometry of this situation is illustrated by Rostoker and Boström [1976, Fig. 6]). In this region of changing electric field, the particles are affected by the polarization drift. This is proportional to particle mass, so the heavy ions drift to the right in Figure 4, while the light electrons essentially continue straight ahead. This drives a perpendicular current through the region, and thus creates field-aligned currents at the region boundaries. The perpendicular current is directed so as to have $\mathbf{j} \cdot \mathbf{E} < 0$, and thus the region is a generator. In this simple model, we get an excess of ions at the right and electrons at the left. If the field-aligned current at the left is carried by electrons and at right by ions, the model preserves continuity for both species.

Magnetospheric current circuits are often illustrated by a two-dimensional figure like Figure 5a. As the energy of any single particle is conserved also in the generator region G, no individual electron can move from S to B (or ion from B

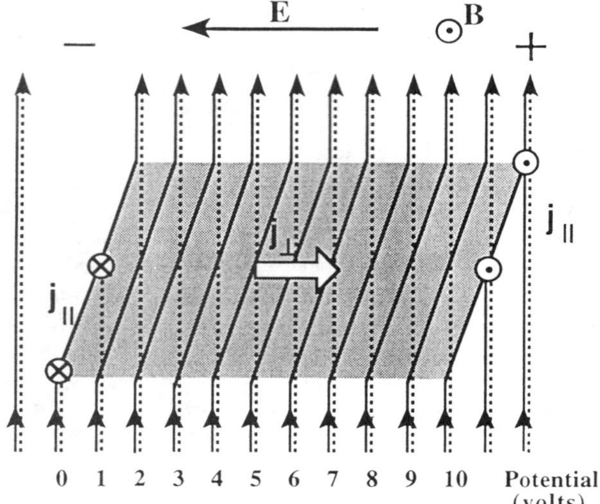

Fig. 4. A schematic single-particle picture of a MHD dynamo. Plasma streams into the shaded generator region, where it slows down and loses kinetic energy (in a realistic geometry, this requires diverging streamlines). Solid and dashed lines mark ion and electron trajectories, respectively. The dashed lines are also equipotential contours, as the polarization drift is small for the electrons. In this generator, ten ions lose 2 eV of kinetic energy when they climb to correspondingly higher potential under influence of the polarization drift $\mathbf{v}_p = (\mathbf{v}\cdot\nabla)\mathbf{E}/(\omega_{ci}B)$. At the edges, two ion-electron pairs at 10 eV become available. If there is a conductive region, e.g. the ionosphere, in a plane above the paper, this charge separation will be removed by the flow of field-aligned currents as indicated.

to S) unless the particle's initial kinetic energy is at least as high as the difference in potential energy. However, it is not necessary for the generator to move any particle across all equipotentials in a limited region of space. The key point is that a real generator is three-dimensional, so the bulk particle motion is not confined to the plane of Figure 5a. In the schematic generator of Figure 4, ten ions give up 2 eV of energy each, and two ion-electron pairs become available at 10 V potential. The generator yields a few particles with high potential energy to carry the field-aligned current by taking a small amount of kinetic energy from many particles. Most of the particles stream through the generator region with only a slight change in their velocity, and in fact no particle goes across all equipotentials. Current closure does not imply that any single particle moves all the way between the poles of the generator. In this respect, the generator behaves like an ordinary battery, where ions move across equipotentials in each cell, but no individual ion moves across the whole battery. Also, the currents in the circuit connected to the battery will in general be carried by electrons and not by ions as in the battery. This is the case also in the generator schematically illustrated above – the current is carried by different particle populations in different regions.

The generator schematically described above works by converting particle kinetic energy associated with the bulk motion of the plasma to electromagnetic energy. The streamlines of this convection are the equipotential contours in the plane perpendicular to the magnetic field. If we consider how the equipotential surfaces in three dimensions are closed, we realize that there is a finite amount of convection energy to draw on. Also, the generator of Figure 4 loses particles all the time, as both currents out of it are carried by particles that leave the generator region. These questions of energy and particle continuity have been discussed by Block [1984]. In general, the implication is that the generator will be time-dependent, unless it utilizes an effectively infinite flow like the solar wind. However, it is theoretically possible to construct a generator that is stationary on any finite time scale by choosing the equipotentials so as to enclose a plasma of sufficiently high total convection energy. This is all we are interested in, unless we desire a global model which is stationary on an infinite time scale. In this case, we would have to use the energy of a stationary solar wind directly in the generator. To close the equipotentials, we then must follow them into the collisional solar plasma, where the energy of single particles is no longer conserved. However, this is clearly outside the scope of any magnetospheric

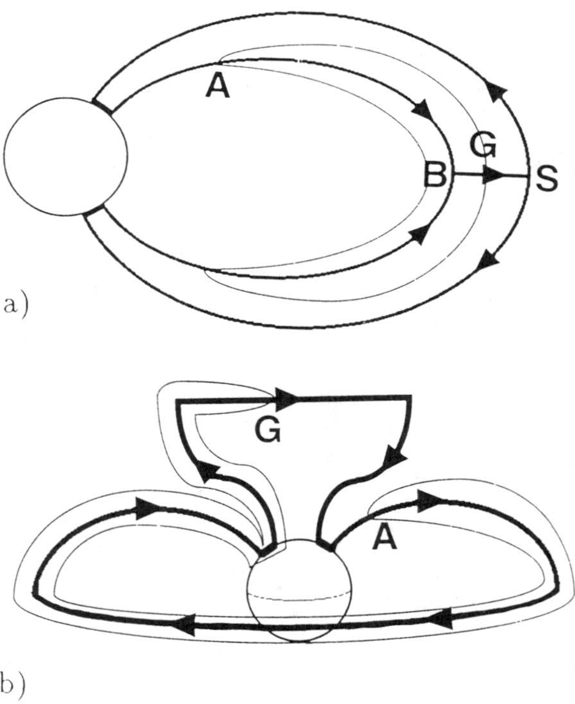

Fig. 5. Two schematic pictures of magnetospheric current configurations. Arrows indicate currents, and the thin lines are selected equipotentials. a) In a two-dimensional illustration, the generator G is located along the same field line as the load A. b) In three dimensions, the generator may be located far from the load. In this picture, the earth is viewed from a night-side position, and the generator G is placed out in the solar wind. Electromagnetic energy to the load at A is transmitted by currents closing through the ionosphere and a partial ring current.

model. We are not looking for an electrostatic model of the entire solar system.

It has been suggested [Bryant et al., 1992; see also comment by Borovsky ,1992] that stationary electrostatic fields cannot be used for particle acceleration, as they conserve particle energy. It is argued that any particle in such a field may be accelerated in one part of the circuit, but it has to be retarded in another part, and so there is no net acceleration. We think this is an improper use of the term 'acceleration'. A particle falling down to lower potential energy *is* accelerated, even if it has been retarded by an equal potential hill somewhere else in the system at some earlier time. And it is not necessary that it shall ever have climbed a potential hill anywhere in the magnetospheric system. As was pointed out in the preceding section, current closure in an open system does not imply that any single particle has to go around all the circuit. It is only if we insist on a global model, describing all the universe at all times, that we have to close the system. Schematic pictures like Figure 5a can sometimes be misleading. The generator does not have to be located on the same field line as the acceleration region. In Figure 5b, we show an example of a current configuration where there is no generator on the same field line as the acceleration region A. Instead, the current closes through a partial ring current, and the generator G is located on open field lines in the solar wind. Also, the current passes through the collisional ionosphere, where the energy of any single particle is no longer conserved, even though the system is electrostatic on a macroscopic level. There is, for example, no problem in supplying the acceleration region A with electrons of sufficient potential energy in this model. We do not propose that Figure 5b is necessarily a correct representation of the real current system in the magnetosphere. We just want to point out that the closure of currents and equipotential surfaces in the ionosphere-magnetosphere-solar wind system may be much more complicated than a simple two-dimensional picture like Figure 5a can show, and that it is possible to construct magnetospheric circuits that fulfill any reasonable demand on energy and particle continuity.

In the above, we have only discussed conservative processes. To what extent processes not conserving particle energy are important for the global magnetosphere-ionosphere circuit we do not know. Wave-particle interactions could be important in the generator as well as in the acceleration region. However, it is hard to find any source of the enormous energy amounts required to drive the magnetosphere-ionosphere current system other than large scale plasma motion (magnetospheric convection or solar wind motion on open field lines) or pressure gradients. Thus it should be possible to describe the generators mainly in MHD terms, but the precise generator mechanisms and configurations deserve more study.

CONCLUDING DISCUSSION

In this paper, we have given a tutorial discussion of some concepts of interest for the acceleration of auroral particles by static electrostatic fields. The reason for our interest in such models is the fact that large scale electric fields and currents are undoubtedly present in the auroral regions, and as stated in the introduction, the energy flow associated with these is a natural candidate as an energy supply for particle acceleration. The WDLs provide a general mechanism for how parallel electric fields are formed in a turbulent current carrying plasma. Such a mechanism is necessary if a part of the electromagnetic energy flux is to be used for particle acceleration. Thus WDLs, when present, must contribute to particle acceleration, but it does not exclude the possibility that other acceleration processes are present and at least sometimes dominate. Of other models for auroral particle acceleration, mechanisms relying on the parallel electric field of Alfvén waves [Hasegawa, 1976; Seyler, 1990] share one property of the electrostatic model discussed in this paper, i.e., that the acceleration is a natural consequence of the fields or waves that mediate the magnetosphere-ionosphere coupling. Other theories, like the mechanism for electron acceleration by lower-hybrid waves suggested by Bryant et al. [1991], do not have this property of directly utilizing a part of the electromagnetic energy flow of the magnetosphere-ionosphere circuit. This is no argument against other mechanisms as such; they may very well work, but to provide the energy necessary for particle acceleration, they must rely on energy flows other than the electromagnetic energy flux primarily associated with the magnetosphere-ionosphere coupling.

When we have discussed particle acceleration in this paper, we have concentrated on electrons. The average electric field sustained by an ensemble of WDLs along a field line can be seen as quasi-static, i.e., energy conserving, for both ions and electrons (see the discussion by Block and Fälthammar [1990]). However, the microstructure of the parallel field in the model above is governed by the WDL dynamics, which takes place on a time scale set by the thermal ions. Hence, the field is not stationary for these ions, and their interaction with the WDLs is more complicated. The electrons and the observed keV ions should, however, be accelerated by the parallel electric field.

Our main points can be summarized as follows:

1. From theory and simulations, weak double layers are associated with potential wells in a plasma with current carrying electrons. The effects of an ensemble of WDLs can be described as an anomalous resistivity which sustains a large-scale parallel electric field in a region of field-aligned current.

2. The notion of a weak double layer does not necessarily apply only to localized solitary structures, but it is only for such structures we have satellite measurements of net voltage drops.

3. From satellite observations, we can neither conclude nor exclude that the observed solitary double layers alone are capable of sustaining potential drops along

the field lines of the kV order. We have no observational knowledge of voltage drops sustained by less ordered (non-solitary) forms of anomalous resistivity and double layers.

4. The mechanisms for how a parallel electric field is sustained in the acceleration region are more studied and better understood than the details of the magnetospheric generators and the global circuit pattern.

5. The relative importance of electrostatic acceleration by quasi-static parallel electric fields and processes relying on wave-particle interactions is not established, and is likely to be different in different situations. Quantitative evaluation of occurrence frequencies of each phenomenon is needed, as well as reliable estimates of their efficiency. These problems are difficult to address experimentally, but estimates can and must be done.

6. Parallel electric fields sustained by weak double layers are still, ten years after the first report of satellite observations, a candidate for explanations of auroral particle acceleration.

Acknowledgments. The analysis of the data from the Viking wave experiment is the result of team work at the Uppsala Division of the Swedish Institute of Space Physics, involving also co-investigators from the Finnish Meteorological Institute and Cornell University. In particular, we thank P.-O. Dovner, G. Holmgren, and A. Mälkki for discussions and comments. The figures were drawn by D. Forss.

References

Birmingham, T. J., Birkeland currents in an anisotropic, magnetostatic plasma, *J. Geophys. Res.*, *97*, 3907–3917, 1992.

Block, L. P., and C.-G. Fälthammar, The role of magnetic-field-aligned electric fields in auroral acceleration, *J. Geophys. Res.*, *95*, 5877–5888, 1990.

Block, L. P., Three-dimensional potential structure associated with Birkeland currents, in *Magnetospheric currents*, edited by T. A. Potemra, volume 28 of *Geophys. Monogr. Ser.*, pp. 315–324, AGU, Washington D. C., 1984.

Borovsky, J. E., Double layers do accelerate particles in the auroral zone, *Phys. Rev. Lett.*, *69*, 1054–1056, 1992.

Boström, R., G. Gustafsson, B. Holback, G. Holmgren, H. Koskinen, and P. Kintner, Characteristics of solitary waves and weak double layers in the magnetospheric plasma, *Phys. Rev. Lett.*, *61*, 82–85, 1988.

Boström, R., Mechanisms for driving Birkeland currents, in *Physics of the hot plasma in the magnetosphere*, edited by B. Hultqvist and L. Stenflo, pp. 341–362, Plenum Press, New York, 1975.

Boström, R., Observations of weak double layers on auroral field lines, *IEEE Trans. Plasma Sci.*, in press, 1992.

Bryant, D. A., A. C. Cook, Z.-S. Wang, U. de Angelis, and C. H. Perry, Turbulent acceleration of auroral electrons, *J. Geophys. Res.*, *96*, 13829–13839, 1991.

Bryant, D. A., R. Bingham, and U. de Angelis, Double layers are not particle accelerators, *Phys. Rev. Lett.*, *68*, 37–39, 1992.

Burch, J. L., Diagnosis of auroral acceleration mechanisms by particle measurements, in *Auroral physics*, edited by C-I Meng and M J Rycroft and L A Frank, pp. 97–106, Cambridge University Press, 1991.

Chiu, Y. T., and M. Schulz, Self-consistent particle and parallel electrostatic field distributions in the magnetospheric-ionospheric auroral region, *J. Geophys. Res.*, *83*, 629–642, 1978.

Hasegawa, A., and T. Sato, Existence of a negative potential solitary-wave structure and formation of a double layer, *Phys. Fluids*, *25*, 632–635, 1982.

Hasegawa, A., Particle acceleration by MHD surface wave and formation of aurora, *J. Geophys. Res.*, *81*, 5083–5090, 1976.

Kofoed-Hansen, O., H. L. Pecseli, and J. Trulsen, Coherent structures in numerically simulated plasma turbulence, *Physica Scripta*, *40*, 280–294, 1989.

Koskinen, H. E. J., and A. Mälkki, Auroral weak double layers: a critical assessment, this volume.

Koskinen, H., R. Lundin, and B. Holback, On the plasma environment of solitary waves and weak double layers, *J. Geophys. Res.*, *95*, 5921–5930, 1990.

Lindqvist, P. A., and G. Marklund, A statistical study of high-altitude electric fields measured on the Viking satellite, *J. Geophys. Res.*, *95*, 5867–5876, 1990.

Lysak, R. L., and C. T. Dum, Dynamics of magnetosphere-ionosphere coupling including turbulent transport, *J. Geophys. Res.*, *88*, 365–380, 1983.

Lysak, R. L., and M. K. Hudson, Effect of double layers on magnetosphere-ionosphere coupling, *Laser and Particle Beams*, *5*, 351–366, 1987.

Lysak, R. L., Electrodynamic coupling of the magnetosphere and ionosphere, *Space Sci. Rev.*, *52*, 33–87, 1990.

Mälkki, A., H. Koskinen, R. Boström, and B. Holback, On theories attempting to explain observations of solitary waves and weak double layers in the auroral magnetosphere, *Physica Scripta*, *39*, 787–793, 1989.

Mälkki, A., R. Boström, and G. Holmgren, Viking observations of weak double layers and solitary structures in the auroral acceleration region: latest results on WDL polarities, in *Proceedings of the third symposium on plasma double layers and related results, Innsbruck, Austria*, 1992.

Mozer, F. S., C. A. Cattell, M. K. Hudson, R. L. Lysak, M. Temerin, and R. B. Torbert, Satellite measurements and theories of low altitude auroral particle acceleration, *Space Sci. Rev.*, *27*, 155–213, 1980.

Rostoker, G., and R. Boström, A mechanism for driving the gross Birkeland current configuration in the auroral oval, *J. Geophys. Res.*, *81*, 235–244, 1976.

Rowland, H. L., and P. J. Palmadesso, Anomalous resistivity due to low-frequency turbulence, *J. Geophys. Res.*, *88*, 7997–8002, 1983.

Rowland, H. L., and P. J. Palmadesso, Spiky parallel dc electric fields in the aurora, *J. Geophys. Res.*, *92*, 299–303, 1987.

Schamel, H., Electron holes, ion holes, and double layers, *Physics Reports*, *140*, 161–191, 1986.

Seyler, C. E., A mathematical model of the structure and evolution of small-scale discrete auroral arcs, *J. Geophys. Res.*, *95*, 17199–17215, 1990.

Temerin, M., K. Cerny, W. Lotko, and F. S. Mozer, Observations of double layers and solitary waves in the auroral plasma, *Phys. Rev. Lett.*, *48*, 1175–1179, 1982.

Tetreault, D. J., Growing ion holes as the cause of auroral double layers, *Geophys. Res. Lett.*, *15*, 164–167, 1988.

R. Boström and A. I. Eriksson, Swedish Institute of Space Physics, Uppsala Division, S-755 91 Uppsala, Sweden.

The Strong-Double-Layer Model of Auroral Arcs: An Assessment

JOSEPH E. BOROVSKY

Space and Atmospheric Sciences Group, Los Alamos National Laboratory

The strong-double-layer model for auroral arcs utilizes a magnetized double-layer structure to accelerate magnetospheric electrons downward. The strong point of the model is that it explains many auroral phenomena in a simple and cohesive manner. The weak point of the model is that the electric-field strengths can be larger than those observed by satellites. The standing predictions of the model are that (1) the magnitudes of the strong electric fields in the auroral zone are larger than the magnitudes presently reported, (2) high-altitude high-energy ion conics are gyrophase bunched near their sources, (3) low-energy conical distributions of molecular ions (*e.g.* O_2^+ and NO^+) exist at ionospheric altitudes in the wakes of moving arcs, and (4) the wave vectors of auroral kilometric radiation lie in planes of constant longitude near the AKR source.

1. INTRODUCTION

The observations of beam-like distributions of precipitating electrons above auroral arcs led researchers to conclude that there are electrostatic-potential drops above auroral arcs that act to accelerate the electrons downward [*e.g. Hoffman and Evans*, 1968; *Heikkila*, 1970; *Albert and Lindstrom*, 1970; *Evans*, 1975; *Borovsky*, 1992]. One form that a potential drop in a plasma can take is a strong double layer, which is a positive-negative charge separation that is maintained by the flow of ions and electrons into the double layer. Strong double layers are robust phenomena that are seen in laboratory plasmas [*Hershkowitz*, 1985] and in computer-simulation plasmas [*Goertz and Borovsky*, 1983]. Electrical energy is converted into beam kinetic energy in a double layer, so a generator is required to supply power to the double layer.

A two-dimensional strong double layer above an east-west-aligned auroral arc is depicted in Figure 1. Owing to the observed conjugacy of auroral arcs in the Northern and Southern Hemispheres, the electrostatic contours of the double-layer structure are thought to enclose the equatorial region of the auroral-arc magnetic-field line and close above an arc in the Southern Hemisphere. By examining the orbits of electrons and ions in such a magnetized-double-layer structure, many observed auroral phenomena can be explained. Note, however, that the double-layer model by itself is not complete: a generator mechanism is needed to supply power to the auroral-arc magnetic-field lines.

Auroral Plasma Dynamics
Geophysical Monograph 80
Copyright 1993 by the American Geophysical Union.

In this paper, a summary of the double-layer model for auroral arcs is presented. Magnetized double layers will be reviewed, auroral phenomena described by the model will be overviewed, predictions of the model will be listed, and evidence against the model will be discussed. The model will not be compared with other auroral-arc models; for a limited comparison between models, see *Borovsky* [1993].

2. PROPERTIES OF MAGNETIZED DOUBLE LAYERS

In a magnetized plasma, when the electric field of a double layer is not parallel to the magnetic field the double layer is said to be oblique. As is depicted in Figure 1, only a very small portion of the auroral double-layer structure is magnetic-field aligned, most of it is oblique. When the potential drop $\Delta\phi$ of a double layer is larger than the temperature of the plasma $k_B T/e$ in which it resides, the double layer is said to be strong. Typical auroral-arc-electron energies are 5 - 15 keV, and typical plasma-sheet electron temperatures are 0.5 keV. Hence, the potential drop $\Delta\phi$ of the double layer is large, and the double layer is strong.

Examining a segment of the potential structure of Figure 1, the auroral double layer is seen to be a thin, planar structure, much like a parallel-plate capacitor. The thickness of a strong oblique double layer is about 20 $\lambda_{De_{\text{mag}}}$, where $\lambda_{De_{\text{mag}}}$ is the Debye length of the magnetospheric electrons. This thickness is independent of the strength and orientation of the plasma magnetic field [*Borovsky*, 1983].

Two beams of particles emanate from a magnetized double layer, each having particle energies slightly greater than $e\Delta\phi$. Electrons flowing into the double layer from the low-potential plasma are accelerated by the electric field to form an electron beam in the high- potential plasma and ions flow-

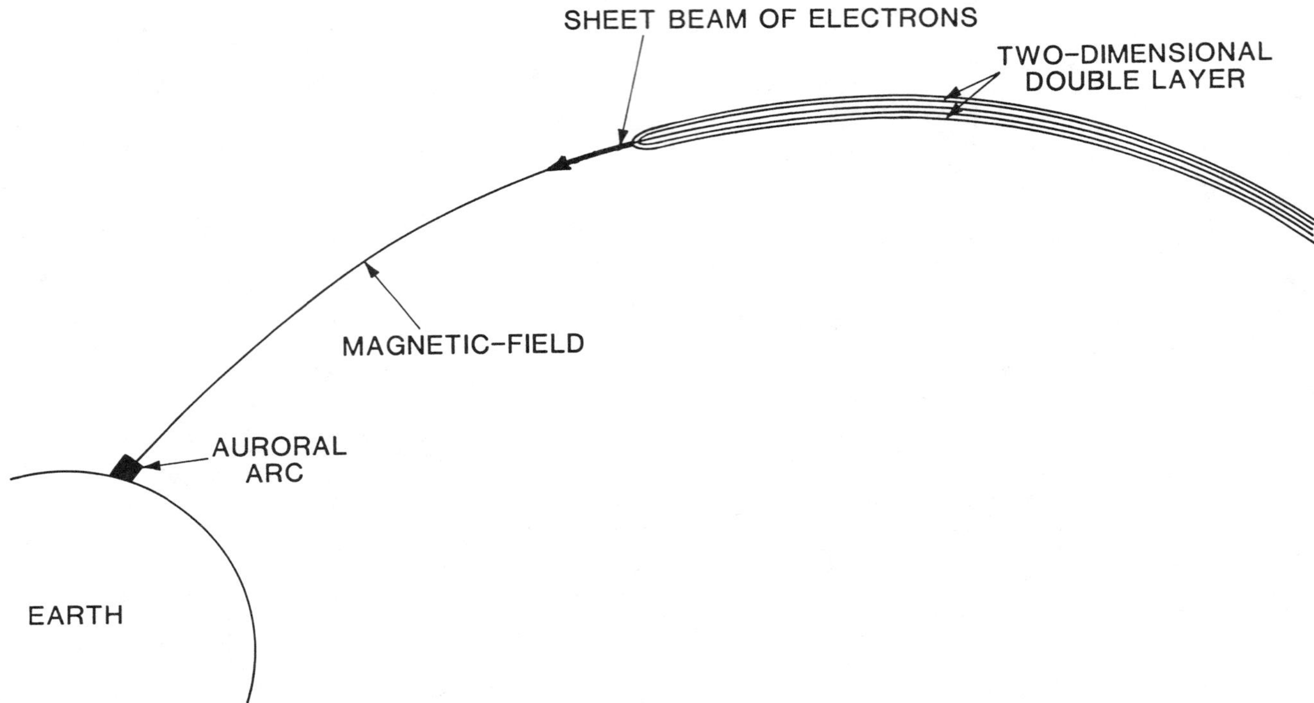

Fig. 1. A sketch (not to scale) depicting the equipotential contours of a double-layer structure above an auroral arc.

ing in from the high-potential plasma are accelerated to form an ion beam in the low-potential plasma. The properties of these beams can vary, owing to variations in the temperature of the plasma sources, the angle between the magnetic field and electric field, and the strength of the magnetic field. For strongly magnetized beam particles [*Borovsky*, 1988a] the beam particles have small pitch angles and are gyrophase bunched. For weakly magnetized beam particles [*Borovsky*, 1984] the beams are conical and strongly gyrophase bunched.

Magnetized double layers can be unstable or they can be stable. When they are unstable, oblique double layers flicker with a period of 4 - 8 ion-plasma periods [*Borovsky*, 1988a], independent of the strength and orientation of the magnetic field and independent of the magnitude of the double-layer potential drop. And when double layers are unstable, the particle beams that they emit flicker in intensity.

Consistent with the flow of particles through the double layer (electrons from the low-potential side and ions from the high-potential side), a current flows through the double layer. The current density vector points in the same direction as the double-layer electric-field vector, indicating that Joule energy is dissipated in the double layer. The magnitude of the current through the double layer is $j_{\mathrm{dl}} = n e v_{\mathrm{Te}}$, where n is the plasma density, e is the electronic charge, and v_{Te} is the electron thermal velocity in the low-potential plasma (magnetospheric side). The current density through

a magnetized double layer changes slightly when the angle between the double-layer electric field and the ambient magnetic field is varied [*Borovsky and Joyce*, 1983a].

A final property of oblique double layers that is of note is that they are surrounded by electrostatic plasma waves, which they generate. These waves can act to disturb particle orbits, drastically changing the flow of thermal particles into the double layer, which can change the thickness of the double layer and can change the amount of current needed to support the double layer [*Borovsky*, 1986a].

3. THE STRONG-DOUBLE-LAYER MODEL OF AURORAL ARCS

The model describes a quiet auroral arc, one that persists for many Alfven-wave transit times between the ionosphere and the equatorial magnetosphere. The arc may slowly drift. In the model the electrostatic potential drop that accelerates auroral-arc electrons downward is contained in a strong double layer above the arc. (see Figure 1). The bottom edge of the double-layer structure resides about 1 - 2.5 R_E above the earth's surface, where the ionospheric and magnetospheric plasmas have densities that are approximately equal [see Appendix of *Borovsky*, 1988b]. The generator mechanism that supplies current and power is not specified.

In the following 11 subsections, the connection of the model to observed auroral phenomena is discussed.

A. Strong Isolated Electric Field

One of the dominant characteristics of a strong double layer is the strong electric field that is isolated within it. As a rule of thumb, the electric-field strength of a magnetized double layer is $E \approx \Delta\phi/(20\,\lambda_{De\mathrm{mag}})$, where $\Delta\phi$ is the potential drop across the double layer. The electric-field strength is independent of the strength and orientation of the magnetic field [*Borovsky*, 1983; *Borovsky and Joyce*, 1983a], providing that the magnetic field and the electric field are not *exactly* perpendicular. (When they are exactly perpendicular, arguments have been made that a wave-produced diffusive spreading of the charge density should act to weaken the electric field [*Smith*, 1986, 1987].) For a potential drop of 5 kV [*Evans*, 1968; *Lin and Hoffman*, 1979], a plasma density of 0.1 cm^{-3} [*Persoon et al.*, 1988; *Louran et al.*, 1990], and an electron temperature of 500 eV [*Louran et al.*; *Lu et al.*, 1991], the predicted electric-field strength is 480 mV/m and it extends over 10 km. This is in the range of electric-field values observed in the auroral-acceleration region [*Mozer et al.*, 1977; *Kletzing et al.*, 1983]. Note, however, that the double-layer model will predict electric fields that are significantly stronger if the potential drop is larger or the Debye length is smaller, as in the predictions of *Borovsky* [1986b].

As can be inferred from Figure 1, since the double-layer structure is tens of kilometers wide (in the north-south direction) and tens of thousands of kilometers long (in the magnetic-field-aligned direction), the electric field is predominantly nearly perpendicular to the magnetic field.

B. Strong Field-Aligned Currents

Owing to the strong flow of electrons and ions through a magnetized double layer, there is a current flowing along the terrestrial magnetic-field lines that contain the auroral arc. The current density at the double layer's position is approximately $j_{\mathrm{dl}} = nev_{\mathrm{Te}}$, where n and v_{Te} pertain to the magnetospheric plasma. For $T_e \approx 500$ eV and n varying from 0.1 to 1 cm^{-3}, this yields $j_{\mathrm{dl}} \sim 10^{-7}$ - 10^{-6} Amp/m^2. For double-layer altitudes of 1 - 2.5 R_E, the ambient magnetic induction is in the range $B_{\mathrm{dl}} \sim 1$ - 8×10^{-2} gauss. According to current continuity, the field-aligned current density of the auroral-arc field line in the ionosphere will be $j_{\mathrm{ionos}} = j_{\mathrm{dl}}\,B_{\mathrm{ionos}}/B_{\mathrm{dl}}$, yielding $j_{\mathrm{ionos}} \sim 5\times10^{-7}$ - 1×10^{-5} A/m^2, which is in the range of typical auroral arcs.

C. Electron Sheet Beams

Sheet beams of electrons emanate downward from the double-layer structures. The electrons of the beams are nearly mono-energetic, although their energies may be spread by electrostatic plasma waves as the beam propagates down from the double layer [*Menietti and Borovsky*, 1989]. The electrons of the beam are magnetic-field aligned near the double layer, but this field alignment will be reduced away from the double layer owing to mirroring effects and perhaps to plasma-wave scattering. Scanning in the north-south direction below a double-layer structure (see Figure 1), the energy of the electrons in the sheet beam is maximum near the center of the double-layer structure and falls off near the edges of the structure [*Borovsky and Joyce*, 1983b]. Near the double layer, the electron sheet beam is characterized by a large value of $\partial f/\partial v_{\parallel}$ on the low-energy edge of the beam [*Borovsky*, 1988a].

The width of the electron sheet beam is 40 $\lambda_{De\mathrm{mag}}$ or more near the bottom of the double-layer structure: this typically yields thicknesses of 20 km [*Borovsky*, 1993], which map down to thicknesses of a few km in the ionosphere. Below the double layer, the electron beam carries the current density of the auroral arc.

D. Upflowing Ion Beams

Sheet beams of energetic, field-aligned, upflowing ions emanate from the double-layer structures: magnetic-field aligned hydrogen and oxygen ions emanate upward from the central portions of the double-layer structures and magnetic-field-aligned hydrogen ions emanate upward from the edges of the structures. The ions are locally mono-energetic, with the energy highest near the center of the structure and the energy falling off near the edges [*Borovsky and Joyce*, 1983b]. Since upward pre-acceleration is expected near 100 km altitudes for ionospheric oxygen ions, oxygen ions obtain greater final kinetic energies than do hydrogen ions [*Borovsky*, 1984; *Borovsky and Joyce*, 1986].

Ion beams are observed in regions of strong electric field in the auroral magnetosphere [*e.g. Redsun et al.*, 1985]. Since the double-layer structures are latitudinally narrow, the sheet beams of upflowing ions that they produce are also narrow, with minimum widths of 20 km or so at altitudes of 1 R_E. This is much too narrow for the present generation of satellites to fully resolve.

E. High-Altitude Hight-Energy Ion Conics

The oblique portions of the double-layer structures can accelerate ions to produce high-energy ion conics [*Lennartsson*, 1980; *Greenspan*, 1984; *Borovsky*, 1984; *Borovsky and Joyce*, 1986], these upward-propagating conical beams originating at the double-layer altitudes (1 - 2.5 R_E). Ion conics are commonly observed on magnetic-field lines that are associated with strong electric fields [*e.g. Redsun et al.*, 1985]. To form a conic, the ions must be weakly magnetized: specifically the ion gyroperiod must be greater than twice the double-layer crossing time [*Borovsky and Joyce*, 1983b]. This condition is easily met for oxygen ions, but is difficult to meet for protons. Hence, the strong-double-layer model predicts a predominance of oxygen conics.

The high-energy ion conics are sheets that are latitudinally narrow. As is the case for field-aligned ion beams, the conics have widths of 10's of km at altitudes of 1 R_E or so, making them impossible to fully resolve with present space-flight instrumentation.

The model also predicts that the ion conics are strongly gyrophase bunched [*Borovsky*, 1984; *Greenspan*, 1984]. At any point in space that is above and near to an oblique double layer, the energetic oxygen ions will not be gyrotropic but rather will form a beam directed with a given pitch angle and a given gyrophase angle. The spatial gradients of the gyrophase angles in the gyrophase-bunched conics is on the order of a km, which is too small to be resolved with present-generation ion detectors.

F. Low-Altitude Low-Energy Ion and Molecular Conics

A standard mechanism for the production of low-altitude low-energy ion conics is the transverse heating of ionospheric ions by electrostatic plasma waves [e.g. *Ungstrup et al.*, 1979; *Chang and Coppi*, 1981]. Another possible source for these conics is the motion of electrostatic potential structures associated with the motion of arcs. When the backscattering of electrons off of the atmosphere is accounted for, the potential drop between the distant magnetosphere and the ionosphere can occur in two steps: the largest part of the potential drop occurs in a double layer at high altitudes and the rest of the drop occurs in a sheath-like structure in the ionosphere. Between the double layer and the sheath, a narrow region with strong perpendicular electric fields forms [*Borovsky and Joyce*, 1983b]; this region shrouds the electron sheet beam. These electric fields reside above an auroral arc, and as the arc drifts the fields move. As ionospheric plasma is passed over by the arc and its fields, the ions are kicked in the transverse direction, gaining perpendicular energy [*Borovsky*, 1984; *Borovsky and Joyce*, 1986]. Typically, 10's of eV of energy can be gained by ions owing to arc motions, with molecular ions such as O_2^+ and NO^+ gaining the largest energies. Owing to the rapid motion in the east-west direction of perturbations in the arcs, kinetic energies of 100's of eV can occasionally be gained. More probable than producing conics, the moving electric fields associated with the drift of auroral arcs will act to heat ionospheric ions and molecular ions, leaving a warmer plasma on the trailing side of the arc.

G. Auroral Kilometric Radiation

The electron beam that emanates from an oblique double layer is gyrophase bunched [*Borovsky*, 1988a]. This means that the electrical current density just below an auroral double layer is not exactly parallel to the ambient magnetic field, but instead it has a spiral pattern characterized by parallel planes of equal gyrophase angle. Owing to plasma-wave fluctuations and to fluctuations of the double layer, the flux of electrons through a double layer is not steady. Because of the electron-flux variation, the spiral current below the double layer has a time variation and the region below the double layer therefore radiates as an antenna [*Borovsky*, 1988b]. The Poynting vector of the electromagnetic radiation is normal to the electron sheet beam emanating from the double-layer structure, so it will be in the north-south direction. This antenna can be very efficient, with $\sim 1\%$ of the energy dissipated in an auroral arc going into radiation.

The frequency of the electromagnetic radiation is near the local electron cyclotron frequency, even though the temporal modulation of the electron flow is at a much-lower frequency. If the electron plasma frequency at the double layer is less than the local electron gyrofrequency, then the radiation from the electron-beam antenna is free to escape the vicinity of the double layer. The electromagnetic radiation is elliptically polarized, containing extraordinary-mode and ordinary-mode components.

The radiation from each segment of each auroral arc is centered at the local electron gyrofrequency and has a frequency spread of about 200 Hz. Therefore the broadband radiation from the entire auroral zone will consist of many fine structures of discrete frequencies [e.g. *Gurnett and Anderson*, 1981]. The frequencies of the fine structures drift with time as the double layer drifts along the magnetic-field line (typical double layer drift velocities are on the order of the ion-acoustic speed [*Singh and Schunk*, 1982; *Borovsky and Joyce*, 1983a]).

H. Whistler-Mode Hiss

The electron beam that emanates from a magnetized double layer into the high-potential plasma is typically unstable to the two-stream instability and so some of the beam kinetic energy is transferred to electrostatic plasma waves. For an oblique double layer, the electrostatic waves driven are magnetized Langmuir waves [*Menietti and Borovsky*, 1989], also known as whistler-mode hiss. The waves driven by the electron beam have frequencies below the local electron plasma frequency.

I. Ion-Acoustic Solitary Waves

In the low-potential plasma adjacent to a magnetized double layer, large-amplitude ion-acoustic-like solitary waves grow [*Joyce and Hubbard*, 1978; *Borovsky and Joyce*, 1983a]. For an auroral double layer, these pulses reside in the current channel above the double layer in the region of the upflowing ion beams and the energetic ion conics. Each pulse consists of a region of positive charge surrounded by a region of negative charge: hence they are positive-potential pulses. The size of a pulse is on the order of 20 $\lambda_{De_{\mathrm{mag}}}$. They propagate downward toward the double layer with velocities that are on the order of the ion-acoustic speed. Presumably they are driven by the relative drift between the thermal electrons and thermal ions above the double layer.

J. Electrostatuc Ion-Cyclotron Waves

Because they are strongly gyrophase bunched, the high-energy ion conic emanating from an oblique double layer drives electrostatic ion-cyclotron waves in the low-potential plasma [*Borovsky*, 1984]. As these waves grow, the perpendicular kinetic energy of the high-energy ions decreases, turning the conical distribution into a more beam-like distribution as it propagates. Unlike other mechanisms for ion-cyclotron-wave generation [e.g. *Kindel and Kennel*, 1971; *Bergmann*, 1984], the ion-conic-driven mechanism leads to the prediction that ion-cyclotron waves would be seen close to strong-electric-field regions that are producing ion conics.

K. Flickering Auroral Arcs

Magnetized double layers can be temporally unstable: when they are they go through a periodic cycle of disruption and reformation. The period of this cycle is 4 - 8 ion-plasma periods [*Borovsky*, 1988a]. For a hydrogen plasma with a density of 0.1 cm^{-3}, this yields a disruption frequency of 8 - 17 Hz. Owing to this cyclic disruption of the double layer, the beam flux is cyclically modulated with the same frequency [*Borovsky*, 1988a]. This beam modulation leads to a periodic variation in the brightness of an auroral arc produced by the double layer. The predicted frequencies are in the range of the the frequencies observed for flickering auroral arcs [*Beach et al.*, 1968; *Spiger and Anderson*, 1985].

4. STANDING PREDICTIONS OF THE MODEL

There are four outstanding predictions of the strong-double-layer model of auroral arcs: the first concerns electric fields in the auroral zone, the second concerns energetic ion conics, the third concerns molecular conics at low altitudes, and the fourth concerns AKR.

The first prediction is that electric fields in the high-altitude auroral magnetosphere exist that are stronger than those that have been reported. The electric field of a double layer is approximately $E = \Delta\phi/(20\lambda_{De_{mag}})$ [*Borovsky*, 1983; *Borovsky and Joyce*, 1983a]: taking the electron temperature to be 500 eV [*Louran et al.*, 1990; *Lu et al.*, 1991], then $E = 0.3 \Delta\phi \, n^{1/2}$ mV/m. If the plasma density n at the sight of the double layer is considerably greater than 0.1 cm^{-3} or if $\Delta\phi$ exceeds 5000 V, then the predicted electric-field strength exceeds 500 mV/m, which exceeds the reported sizes of auroral electric fields. A related prediction of the auroral-arc model is that strong parallel electric fields exist in the high-altitude auroral zone. Strong parallel electric fields are occasionally observed [*e.g. Mozer et al.*, 1980], but strong fields with $E_\parallel > E_\perp$ have not been observed [*F. S. Mozer, private communication*, 1992]. The region of field-aligned electric field on the bottom tip of a double layer structure (see Figure 1) is about 10-km in extent, and the bottom of the structure resides in the altitude range of 1 - 2.5 R$_E$. This means that the chance of encountering an $E_\parallel > E_\perp$ region with a satellite pass in the high-altitude auroral zone is about 1/1000 of the probability of encountering an an $E_\perp > E_\parallel$ region, making encounters rare but not impossible.

The second standing prediction of the strong-double-layer model is that energetic ion conics are gyrophase bunched near their source [*Borovsky*, 1984]. The gradient scale length for gyrophase-angle variations in these upflowing ions is the smaller of the double-layer thickness or the energetic-ion gyroradius. In general, the gradient scale length of the gyrophase-angle dependence will be of the order of 10 km, requiring three-dimensional ion measurements with time resolutions of a second or so before it is possible to detect gyrophase-bunched ion distributions.

The third standing prediction of the model is that conical distributions of molecular ions will exist in the ionosphere [*Borovsky*, 1984]. These distributions should be found in the wakes of moving auroral arcs, since they are produced by low-altitude electric-field structures passing over ionospheric plasma. Typical transverse kinetic energies of O_2^+ and NO^+ are predicted to be 10's of eV.

The fourth standing prediction of the model is that the Poynting vectors of auroral kilometric radiation will lie in constant-longitude planes near the source of the radiation [*Borovsky*, 1988b]. In the double-layer model, AKR is coherently broadcast from the electron sheet beam and the sheet beam is typically aligned in the east-west direction. The radiated power peaks perpendicular to the sheet antenna, so the radiation is launched with wave vectors in the north-south direction. The standard model for the production of AKR is the cyclotron-maser instability, where electromagnetic radiation grows in amplitude as it propagates through an unstable auroral plasma — since plasmas associated with auroral arcs are extended in the east-west direction, one could imagine that a wave propagating in the east-west direction spends a longer time in an unstable auroral plasma than does a wave propagating in the north-south direction. Hence the cyclotron-maser model would make the different prediction that the Poynting vectors of the AKR would lie perpendicular to constant-longitude planes near the source region. AKR having wave vectors that lie in constant-longitude planes is consistent with direction-finding analyses of AKR [*e.g. Calvert*, 1985] and with the observed hollowness of the AKR frequency pattern [*Calvert*, 1987].

5. EVIDENCE AGAINST THE MODEL

There is a major problem with the model when it is compared with satellite data: the electric fields predicted by the model can be stronger than those that are observed. The electric fields predicted by the model can exceed 1000 mV/m for strong auroral arcs, whereas electric fields with strengths much greater 500 mV/m have not been reported (however, electric-field detectors have saturated in the auroral zone [*e.g. Temerin et al.*, 1981], so fields stronger than 500 mV/m have been encountered).

The discrepancy between the strong-double-layer model and observations could be owed to a failure of the model, or it could be owed to a failure of the electric-field instrumentation. In response to this discrepancy between the model and the observations, a simple analysis of the double-floating-probe electric-field-measuring technique was performed by *Borovsky* [1986b] (see also *Laakso et al.* [1992]). That analysis found two drawbacks to the double-probe method of measuring strong electric fields: (1) the instrument response time can be longer than the measurement-sampling time (yielding a false value of the electric field) and (2) high fluxes of beam electrons hitting the probes can cause the double-probe technique to measure gradients in beam flux rather than gradients in the electrostatic potential (yielding a value of the electric field that is too small). For both instrumental drawbacks, cylindrical probes and spherical probes obtain different values of the electric field. In response to such criticisms of double probes, it has been pointed out [*Mozer*, 1991; *F. S. Mozer, private communication*, 1992] that capacitive coupling of double probes to the auroral plasma may overcome the first drawback to yield correct electric-field measurements and it has been pointed out [*F. S. Mozer, private communication*, 1992] that the fluxes of electrons in the auroral zone are seldom strong enough to cause double probes to misbehave, invalidating the second drawback. Hopefully, these theory/instrumentation conflicts will be resolved in the literature.

If the discrepancy between the strong-double-layer model and observations is owed to a failure of the model then two possibilities come to mind. The first possibility is that the estimates used for the plasma density n at the double layer are wrong, *e.g.* when a plasma density of 1 cm^{-3} is used (as in *Borovsky* [1988a,b]), then large potential drops lead to overly strong electric fields. However, if the plasma density near the double layers is 0.1 cm^{-3} or less (which is the mid-to-low range of values of *Persoon et al.* [1988]), then the electric-field values predicted by the model are in agreement with observed electric-field values. A scenario that was explored by *Borovsky* [1986a] is that double layers act to lower the plasma density in their immediate vicinity, *i.e.* they act in concert with electrostatic plasma waves to drive and maintain a density cavity in which they reside. With this self-lowered plasma density, the electric field of the double layer is weaker than a simple theoretical estimate would predict, putting the values of the electric fields predicted by the model into the range of observed electric-field values, even for auroral plasma densities of 1 cm^{-3}. The second possibility that comes to mind is that the strong-double-layer model is wrong. This would imply that the theoretical approach used to obtain the structure of strong electric fields in plasmas is wrong. The approach is based on self-consistently solving Poisson's equation for a plasma in which particles do not suffer from collisions or from very strong wave-particle scattering (so called Poisson-Vlasov solutions or BGK solutions [*e.g. Montgomery and Joyce*, 1969; *Knorr and Goertz*, 1974]). The approach is also supported by particle-in-cell computer simulations, which are less stringent in their assumptions, but which are still collisionless and which probably have reduced wave-particle scattering owing to the limited spatial sizes of the plasmas simulated. (Note that strong double layers appear only in computer simulations with boundary conditions that allow potential drops to form between two plasma sources [e.g. *Goertz and Joyce*, 1975; *Joyce and Hubbard*, 1978; *Singh*, 1980], in analogy to the potential drop between the magnetospheric plasma and the ionospheric plasma on auroral-arc magnetic-field lines [*Mizera et al.*, 1981; *Temerin et al.*, 1981; *Borovsky*, 1992]; strong double layers will not, and should not, form in computer simulations employing periodic boundary conditions [e.g. *Sato and Okuda*, 1981; *Barnes et al.*, 1985; *Gray et al.*, 1992].) The results of the strong-double-layer theory

are also confirmed in laboratory plasmas [*e.g. Coakley et al.*, 1979; *Hollenstein et al.*, 1980; *Carpenter and Torven*, 1987], however there are many laboratory experiments that obtain strong-electric-field structures that are thicker (have weaker electric fields) than the strong-double-layer model would predict [*e.g. Jovanovic et al.*, 1982; *Alport et al.*, 1986]. If the strong-double-layer model is wrong, then there is no self-consistent theory for strong electric fields in the auroral plasma.

A less severe problem with the strong-double-layer model is that the model predicts auroral-arc thicknesses that are on the order of a few km [*Borovsky*, 1993]. This thickness is of the order of the bright portion of an auroral arc, but it is an order of magnitude or more wider than the auroral-arc fine-structure thicknesses measured by ground-based optical imagers [*Maggs and Davis*, 1968; *Borovsky et al.*, 1991; *Borovsky and Suszcynsky*, 1993]. Accordingly, the double-layer model may, however, be valid for describing the larger-scale features of auroral arcs that contain the fine-scale structures [see *Borovsky and Suszcynsky* 1993].

6. THE ASSESSMENT

The positive aspect of the strong-double-layer model of auroral arcs is that it can explain many observed phenomena simultaneously and in a simple manner. The drawback to the model is that it sometimes predicts electric fields that are stronger than those observed. Since the motivation for the model is to explain these strong electric fields that are observed in the auroral zone, this drawback is serious.

To extend the model, work is needed (1) to connect auroral-double-layer structures to generator mechanisms in the plasma sheet, (2) to investigate the electrodynamic properties of magnetized double layers, (3) to further understand the mechanism behind the disruption-reformation cycle of magnetized double layers, and (4) to further understand the interaction between magnetized double layers and long plasmas. Better estimates of the auroral plasma density are crucial if the model is to be compared with observed electric fields and a full analysis of the double-probe method for measuring strong electric fields in the auroral zone is needed.

Acknowledgments. The author wishes to acknowledge Chris Goertz and Glenn Joyce for their help through the years and to acknowledge Forrest Mozer for his comments on this manuscript. The work has been supported by NASA, the NSF, and the Department of Energy.

I wish to dedicate this paper to Chris Goertz, whose inspiration lives on.

REFERENCES

Albert, R. D., and P. J. Lindstrom, auroral-particle precipitation and trapping caused by electrostatic double layers in the ionosphere, *Science, 170*, 1398, 1970.

Alport, M. J., S. L. Cartier, and R. L. Merlino, Laboratory observations of ion cyclotron waves associated with a double layer in an inhomogeneous magnetic field, *J. Geophys. Res., 91*, 1599, 1986.

Barnes, C., M. K. Hudson, and W. Lotko, Weak double layers in ion-acoustic turbulence, *Phys. Fluids, 28*, 1055, 1985.

Beach, R., G. R. Cresswell, T. N. Davis, T. J. Hallinan, and L. R. Sweet, Flickering, a 10-cps fluctuation with bright auroras, *Planet. Space Sci., 16*, 1525, 1968.

Bergmann, R., Electrostatic ion (hydrogen) cyclotron and ion acoustic wave instabilities in regions of upward field-aligned current and upward ion beams, *J. Geophys. Res., 89*, 953, 1984.

Borovsky, J. E., The scaling of oblique plasma double layers, *Phys. Fluids, 26*, 3273, 1983.

Borovsky, J. E., The production of ion conics by oblique double layers, *J. Geophys. Res., 89*, 2251, 1984.

Borovsky, J. E., Parallel electric fields in extragalactic jets: double layers and anomalous resistivity in symbiotic relationships, *Astrophys. J., 306*, 451, 1986a.

Borovsky, J. E., The theory of Langmuir probes in strong electrostatic potential structures, *Phys. Fluids, 29*, 718, 1986b.

Borovsky, J. E., Properties and dynamics of the electron beams emanating from magnetized plasma double layers, *J. Geophys. Res., 93*, 5713, 1988a.

Borovsky, J. E., Production of auroral kilometric radiation by gyrophase-bunched double-layer-emitted electrons, *J. Geophys. Res., 93*, 5727, 1988b.

Borovsky, J. E., Double layers do accelerate particles in the auroral zone, *Phys. Rev. Lett., 69*, 1054, 1992.

Borovsky, J. E., Auroral-arc thicknesses as predicted by various theories, *J. Geophys. Res., 98*, 6101, 1993.

Borovsky, J. E., and G. Joyce, The simulation of plasma double-layer structures in two dimensions, *J. Plasma Phys., 29*, 45, 1983a.

Borovsky, J. E., and G. Joyce, Numerically simulated two-dimensional auroral double layers, *J. Geophys. Res., 88*, 3116, 1983b.

Borovsky, J. E., and G. Joyce, The direct production of ion conics by plasma double layers in *Ion Acceleration in the Magnetosphere and Ionosphere*, edited by T. Chang, American Geophysical Union, Washington, D.C., 1986, pg. 317.

Borovsky, J. E., D. M. Suszcynsky, Optical measurements of the fine-scale structure of auroral arcs, this monograph, 1993.

Borovsky, J. E., D. M. Suszcynsky, M. I. Buchwald, and H. V. De-Haven, Measuring the thicknesses of auroral curtains, *Arctic, 44*, 231, 1991.

Calvert, W., DE-1 measurements of AKR wave directions, *Geophys. Res. Lett., 12*, 381, 1985.

Calvert, W., Hollowness of the observed auroral kilometric radiation pattern, *J. Geophys. Res., 92*, 1267, 1987.

Carpenter, R. T., and S. Torven, The current-voltage characteristics and potential oscillations of a double layer in a triple-plasma device, *IEEE Trans. Plasma Sci., PS-15*, 434, 1987.

Chang, T., and B. Coppi, Lower hybrid acceleration and ion evolution in the supraauroral region, *Geophys. Res. Lett., 8*, 1253, 1981.

Coakley, P., L. Johnson, and N. Hershkowitz, Strong laboratory double layers in the presence of a magnetic field, *Phys. Lett., 70A*, 425, 1979.

Evans, D. S., The observations of a near monoenergetic flux of auroral electrons, *J. Geophys. Res., 73*, 2315, 1968.

Evans, D. S., Evidence for the low altitude acceleration of auroral particles, in *Physics of Hot Plasma in the Magnetosphere*, edited by B. Hultqvist and L. Stenflo, Plenum, New York, 1975.

Goertz, C. K., and J. E. Borovsky, Numerical simulations of plasma double layers in *High-Latitude Space Plasma Physics*, edited by B. Hultqvist and T. Hagfors, Plenum, New York, 1983, pg. 469.

Goertz, C. K., and G. Joyce, Numerical simulation of the plasma double layer, *Astrophys. Space Sci., 32*, 165, 1975.

Gray, P. C., M. K. Hudson, and W. Lotko, Acoustic double layers in multispecies plasma, *IEEE Trans. Plasma Sci., 20*, 745, 1992.

Greenspan, M. E., Effects of oblique double layers on upgoing ion pitch angle and gyrophase, *J. Geophys. Res., 89*, 2842, 1984.

Gurnett, D. A., and R. R. Anderson, The kilometric radio emission spectrum: Relationship to auroral acceleration processes, in *Physics of Auroral Arc Formation*, edited by S.-I. Akasofu and J. R. Kan, pg. 341, American Geophysical Union, Washington, D.C., 1981.

Heikkila, W. J., Satellite observations of soft particle fluxes in the auroral zone, *Nature, 225*, 369, 1970.

Hershkowitz, N., Review of recent laboratory double layer experiments, *Space Sci. Rev., 41*, 351, 1985.

Hoffman, R. A., and D. S. Evans, Field-aligned electron bursts at high latitudes observed by OGO 4, *J. Geophys. Res., 73*, 6201, 1968.

Hollenstein, C., M. Guyot, and E. S. Weibel, Stationary potential jumps in a plasma, *Phys. Rev. Lett., 45*, 2110, 1980.

Jovanovic, D., J. P. Lynov, P. Michelsen, H. L. Pecseli, J. J. Rasmussen, and K. Thomsen, Three dimensional double layers in magnetized plasmas, *Geophys. Res. Lett., 9*, 1049, 1982.

Joyce, G., and R. F. Hubbard, Numerical simulation of plasma double layers, *J. Plasma Phys., 20*, 391, 1978.

Kindel, J. M., and C. F. Kennel, Topside current instabilities, *J. Geophys. Res., 76*, 3055, 1971.

Kletzing, C., C. Cattell, F. S. Mozer, S.-I. Akasofu, and K. Makita, Evidence for electrostatic shocks as the source of discrete auroral arcs, *J. Geophys. Res., 88*, 4105, 1983.

Knorr, G., and C. K. Goertz, Existence and stability of strong potential double layers, *Astrophys. Space Sci., 31*, 209, 1974.

Laakso, H., T. Aggson, and R. Pfaff, Spurious electric fields induced by the electron density and temperature gradients, *EOS Trans. Amer. Geophys. Union, 73* (43), 473, 1992.

Lennartsson, W., On the consequences of the interaction between the auroral plasma and the geomagnetic field, *Planet. Space Sci., 28*, 135, 1980.

Lin, C. S., and R. A. Hoffman, Fluctuations of inverted V electron fluxes, *J. Geophys. Res., 84*, 6547, 1979.

Louarn, P., A. Roux, H. de Feraudy, D. Le Queau, M. Andre, and L. Matson, Trapped electrons as a free energy source for the auroral kilometric radiation, *J. Geophys. Res., 95*, 5983, 1990.

Maggs, J. E., and T. N. Davis, Measurements of the thicknesses of auroral structures, *Planet. Space Sci., 16*, 205, 1968.

Menietti, J. D., and J. E. Borovsky, Numerical simulations of the heating of electrons by beam-driven electrostatic waves, *J. Geophys. Res., 94*, 492, 1989.

Mizera, P. F., J. F. Fennell, D. R. Croley, and D. J. Gorney, Charged particle distributions and electric field measurements from S3-3, *J. Geophys. Res., 86*, 7566, 1981.

Montgomery, D., and G. Joyce, Shock-like solutions to the electrostatic Vlasov equation, *J. Plasma Phys., 3*, 1, 1969.

Mozer, F. S., C. W. Carlson, M. K. Hudson, R. B. Torbert, B. Parady, J. Yatteau, and M. C. Kelley, Observations of paired electrostatic shocks in the polar magnetosphere, *Phys. Rev. Lett., 38*, 292, 1977.

Mozer, F. S., Spherical double probe measurements of DC and AC electric fields, *EOS Trans. Amer. Geophys. Union, 72* (44), 409, 1991.

Mozer, F. S., C. A. Cattell, M. K. Hudson, R. L. Lysak, M. Temerin, and R. B. Torbert, Satellite measurements and theories of low altitude particle acceleration, *Space Sci. Rev., 27*, 155, 1980.

Persoon, A. M., D. A. Gurnett, W. K. Peterson, J. H. Waite, J. L. Burch, and J. L. Green, Electron density depletions in the nightside auroral zone, *J. Geophys. Res., 93*, 1871, 1988.

Redsun, M. S., M. Temerin, and F. S. Mozer, Classification of auroral electrostatic shocks by their ion and electron associations, *J. Geophys. Res., 90*, 9615, 1985.

Sato, T., and H. Okuda, Numerical simulations on ion acoustic double layers, *J. Geophys. Res., 86*, 3357, 1981.

Singh, N., Computer experiments on the formation and dynamics of electric double layers, *Plasma Phys., 22*, 1, 1980.

Singh, N., and R. W. Schunk, Dynamical features of moving double layers, *J. Geophys. Res., 87*, 3561, 1982.

Smith, R. A., Effects of anomalous transport on the potentials of discrete auroral arcs, *Geophys. Res. Lett., 13*, 889, 1986.

Smith, R. A., Anomalous transport in discrete arcs and simulation of double layers in a model auroral circuit, *Laser and Part. Beams, 5*, 381, 1987.

Spiger, R. J., and H. R. Anderson, Fluctuations of precipitated electron intensity in flickering auroral arcs, *J. Geophys. Res., 90*, 6647, 1985.

Temerin, M., M. H. Boehm, and F. S. Mozer, Paired electrostatic shocks, *Geophys. Res. Lett., 8*, 799, 1981.

Ungstrup, E., D. M. Klumpar, and W. J. Heikkila, Heating of ions to superthermal energies in the topside ionosphere by electrostatic ion cyclotron waves, *J. Geophys. Res., 84*, 4289, 1979.

J. E. Borovsky, Space and Atmospheric Group, Mail Stop D466, Los Alamos National laboratory, Los Alamos, NM 87545.

Generalized Model of the Ionospheric Alfvén Resonator

ROBERT L. LYSAK

School of Physics and Astronomy
University of Minnesota

Since the Alfvén speed rises dramatically above ionospheric altitudes, Alfvén waves can be trapped in an effective resonant cavity, which has been termed the ionospheric Alfvén resonator. Previous models of this resonator have considered the increase of the Alfvén speed up to about 2 R_E (geocentric); however, the subsequent decrease of the Alfvén speed above this altitude has not been considered in these models. In addition, although there has been discussion of the role of the resonator in accelerating auroral electrons, the parallel electric fields which are responsible for this acceleration have not been explicitly included in the model. This paper presents first results of a model of the ionospheric Alfvén resonator which includes an arbitrary magnetospheric density profile, dipole geometry in the background magnetic fields, and the effect of parallel electric fields due to electron inertia. Results from this model indicate that enhancements of the background electric spectra at the eigenfrequencies of the resonator (in the neighborhood of 1 Hz) should be observable in the range around the Alfvén speed peak. These results are compared with spectra obtained from the Viking satellite. In addition, the structure of the parallel electric field can be obtained. These results indicate that Alfvén wave structures with short perpendicular wavelength can accelerate electrons by a Fermi type process.

INTRODUCTION

The structure of currents and electric fields in the auroral zone is controlled by the interactions and reflections of Alfvén waves which propagate along auroral field lines. It is well known that Alfvén waves can be reflected from the conducting ionosphere [Scholer, 1970; Maltsev et al., 1977] as well as from gradients in the Alfvén speed [Mallinckrodt and Carlson, 1978]. Above the auroral ionosphere, the plasma density decreases exponentially while the magnetic field decreases more slowly; thus, the Alfvén speed increases dramatically above the ionosphere. This sharp gradient in the Alfvén speed forms an effective resonant cavity, termed the ionospheric Alfvén resonator by Polyakov and Rapoport [1981] and studied extensively by Trakhtengertz and Feldstein [1981, 1984, 1987, 1991] and Lysak [1986, 1988, 1991]. In all of these works, with the exception of some numerical calculations by Lysak [1986], the Alfvén speed profile was simplified to the following:

$$V_A^2(z) = \frac{V_{AI}^2}{\varepsilon^2 + e^{-z/h}} \qquad (1)$$

where V_{AI} is the Alfvén speed at the ionosphere, h is the scale

Auroral Plasma Dynamics
Geophysical Monograph 80
Copyright 1993 by the American Geophysical Union.

height of the ionospheric density, and $\varepsilon \ll 1$ is the ratio of the Alfvén speed in the ionosphere to that in the outer magnetosphere. This profile increases monotonically with increasing altitude, becoming constant for $z \gg 2h \ln(1/\varepsilon)$. This model has the advantage that the solutions to the wave equation can be expressed in terms of Bessel functions of imaginary order and argument.

Although this profile allows for analytic results, it does not allow for the fact that the Alfvén speed in fact decreases above its peak value, which typically occurs in the range of 1.5-3 R_E (geocentric). This decrease occurs because the plasma density is either constant or slowly falling at higher altitudes while the dipole magnetic field continues to decrease. Although this Alfvén speed gradient has longer scale lengths than that below the Alfvén speed peak and so does not reflect waves as strongly, the structure of the electric fields and currents seen above the peak can be modified by the presence of these additional reflections. Thus, a more realistic Alfvén speed profile than that given by equation (1) is necessary to make comparisons with satellite measurements of low frequency electric fields, such as those recently obtained by the Viking satellite [Marklund et al., 1990; Block and Fälthammar, 1990; Erlandson et al., 1990].

In addition to these considerations, Alfvén waves with short perpendicular wavelengths carry a parallel electric field com-

"

ponent, due to electron pressure when the plasma $\beta > m_e/m_i$ [Hasegawa, 1976] and due to electron inertia when $\beta < m_e/m_i$ [Goertz and Boswell, 1979]. The latter limit is appropriate at the lower ends of the field line where the resonator effects are the most important. The resulting inertial parallel electric field can accelerate electrons in a time-dependent fashion, which has implications for the development of the fast feedback instability [Lysak, 1991], and possibly plays a role in the formation of electron conics [Menietti and Burch, 1985; Lundin et al., 1987; Temerin and Cravens, 1990; André and Eliasson, 1992]. In addition, the electron inertial effect modifies the phase speed of the Alfvén wave and changes its reflection characteristics. Thus, short perpendicular scale waves may have a quite different structure than waves with longer perpendicular wavelengths.

It is the purpose of this paper to describe a generalization in the ionospheric Alfvén resonator model which includes more realistic Alfvén speed profiles as well as inertial parallel electric fields. A numerical integration of the wave equation will be performed in order to find the structure of Alfvén waves as a function of altitude, frequency, and perpendicular wavelength. By assuming a power law incident wave spectrum, the spectrum of electric field perturbations at varying altitudes can be constructed and compared with observations from the Viking spacecraft.

Model Equations

It is convenient in the discussion of shear Alfvén waves in a strong background magnetic field to describe the wave in terms of a scalar potential Φ and a vector potential with only a parallel component $\mathbf{A} = A_{\parallel}\hat{\mathbf{z}}$. Use of these potentials has the advantage that they remain constant along field lines, even in a nonuniform magnetic field, under conditions of no parallel electric fields and static field-aligned currents. These potentials can be related by a gauge condition [Lysak, 1988, 1991]:

$$\frac{\partial \Phi}{\partial t} + \frac{V_A^2}{c}\nabla\cdot\mathbf{A} = 0 \tag{2}$$

It should be remembered that the square of the Alfvén speed in Equation (2) must be replaced by $V_A^2/(1+V_A^2/c^2)$ when the Alfvén speed becomes comparable to the speed of light. Note that the perpendicular components of the vector potential vanish for a pure shear mode wave. Thus, the perpendicular magnetic field perturbations can be written $\mathbf{b} = \nabla\times(A_{\parallel}\hat{\mathbf{z}})$ where $\hat{\mathbf{z}}$ is in the direction of the background magnetic field and the perpendicular electric field is $\mathbf{E}_\perp = -\nabla_\perp\Phi$. The field-aligned current in this notation is simply $j_{\parallel} = -(c/4\pi)\nabla_\perp^2 A_{\parallel}$. The parallel electric field has both inductive and potential contributions:

$$E_{\parallel} = -\frac{1}{c}\frac{\partial A_{\parallel}}{\partial t} - \frac{\partial \Phi}{\partial z} \tag{3}$$

In ideal MHD, the parallel electric field vanishes and so the right hand side can be set to zero [as in Lysak, 1988, 1991]; however, when electron inertia is important the parallel electric field is related to changes in the current:

$$E_{\parallel} = \frac{m_e}{ne^2}\frac{\partial j_{\parallel}}{\partial t} \tag{4}$$

Equating the right hand sides of equation (3) and (4) and using the definition of the field-aligned current in terms of the vector potential, one finds:

$$(1-\lambda^2\nabla_\perp^2)\frac{\partial A_{\parallel}}{\partial t} = -c\frac{\partial \Phi}{\partial z} \tag{5}$$

where $\lambda = c/\omega_{pe}$ is the electron inertial length. Equations (2) and (5) are simply the linearized Strauss [1976] equations, modified to include the electron inertia [Seyler, 1988].

First, note a few of the properties of this system in a uniform geometry. If Fourier transforms are made of these equations, it is well known that the dispersion relation of the Alfvén waves becomes [Goertz and Boswell, 1979]:

$$\omega = \frac{k_{\parallel}V_A}{\sqrt{1+k_\perp^2\lambda^2}} \tag{6}$$

Thus, the parallel phase velocity of the wave is decreased by the inertial effect. The inertial terms also affect the reflection of the wave, which depends on the proportionality between the scalar and vector potentials:

$$\Phi = \pm\frac{V_A}{c}\sqrt{1+k_\perp^2\lambda^2}\,A_{\parallel} \tag{7}$$

with the square root now in the numerator. In Equation (7), the sign indicates propagation in the $\pm\hat{\mathbf{z}}$ direction. Thus, when inertial effects are important, the scalar potential (and thus the electric field) is enhanced relative to the vector potential (magnetic field). Alfvén waves are reflected when the gradient scale length of the factor in Equation (7) becomes comparable to the parallel wavelength of the wave and the WKB approximation breaks down.

We wish to model equations (2) and (5) in an isotropic flux tube in which the background magnetic field decreases as $1/r^3$. Note that in a truly dipolar flux tube, the longitudinal and latitudinal coordinates map somewhat differently. This effect introduces additional terms in equations (2) and (5) which are proportional to gradients of the scale factors, which have gradient scale lengths the order of $L\,R_E$. Such factors must be included when considering waves of a global scale, such as in the work of Lee and Lysak [1989]. Since the emphasis here is on perpendicular wavelengths of tens of km and parallel wavelengths less than 1 R_E, these terms are small in the present work. Thus, the assumption of an isotropic flux tube simplifies the problem without obscuring the essential physics.

To determine the mode structure of this system, equations (2) and (5) are Fourier transformed in time. If we further assume that the transverse gradients of the background density and magnetic field are small over the perpendicular wavelength, we can Fourier transform in the perpendicular coordinates, but not in the parallel coordinate. Thus, for a given frequency and perpendicular wave number, equations (2) and (5) become:

$$\frac{\partial A_{\parallel}}{\partial z} = \frac{i\omega c}{V_A^2}\Phi \tag{8}$$

and

$$\frac{\partial \Phi}{\partial z} = \frac{i\omega}{c}(1+k_\perp^2\lambda^2)A_\parallel \qquad (9)$$

Note that because the perpendicular coordinates must expand as $B^{-1/2}$ to maintain constant flux along the flux tube, the perpendicular wavenumber must vary as $B^{1/2} \sim r^{-3/2}$; thus, the quantity $k_\perp^2\lambda^2$ varies as B/n_e along the field line while V_A^2 varies with B^2/ρ, where ρ is the mass density.

These considerations indicate that the structure of the cavity is dependent on the density profile of each species. For each ion species considered, the density profile is assumed to have the form:

$$n_i(z) = n_{0i}e^{-z/h_i} + n_{1i}r^{-\gamma_i} \qquad (10)$$

where z is the altitude above the ionospheric height and $r = r_I + z$ is the geocentric radial distance. Here, $r_I = 1.03\,R_E$ is the radial distance of the ionosphere. Thus, the total mass density is the sum of the ion densities weighted by their mass while the electron density can be determined from quasi-neutrality as the sum of the ion densities weighted by their charge.

Finally, we need to adopt a boundary condition for the ionospheric end of the flux tube. Assuming constant conductivity, the ionospheric Ohm's Law and current continuity give the relation [Lysak, 1991]:

$$A_\parallel + \frac{4\pi\Sigma_P}{c}\Phi = 0 \qquad (11)$$

Thus, the calculation procedure is as follows: an arbitrary unit amplitude is assumed for the value of Φ at the ionosphere. Equation (11) is then utilized to determine the ionospheric value of $A_\parallel$. Equations (8) and (9) can then be integrated using a Runge-Kutta scheme, assuming the dipolar profile for B and the density profile for each species given by Equation (10). Thus, the parameters n_{0i}, n_{1i}, h_i, γ_i for each ion species, together with Σ_P define the model, and $k_\perp$ (defined at the ionosphere) and ω are parameters of the wave modes. Note that the simplified model described by equation (1) may be recovered by taking a single ion species with $\gamma=0$, $\varepsilon^2 = n_1/n_0$ and keeping the background magnetic field constant. The calculation is carried out to an arbitrary altitude above the Alfvén speed peak. At this point, the wave can be defined in terms of Elsässer variables $\Phi^\pm = [\Phi \pm (V_A/c)\sqrt{1+k_\perp^2\lambda^2}\,A_\parallel]/2$, which give the relative amplitudes of up and downgoing waves, respectively. As noted in Lysak [1991], the reflection coefficient $R = \Phi^+/\Phi^-$ evaluated at the upper boundary of the system exhibits a resonant structure at the frequencies corresponding to the resonator eigenfrequencies. For the purpose of displaying the results, the various quantities will be normalized to the amplitude of the downgoing incident wave Φ^- at the top boundary.

RESULTS

A. Eigenmode Structure

Before presenting results from the numerical calculations in a dipole geometry, it is worth while to summarize the results from the straight field line model presented in Lysak [1991]. The main result of that model was that the eigenfrequencies of the resonator were given by $\omega_n = \xi_n V_{AI}/2h$, where V_{AI} and h are defined above and ξ_n are a set of constants which, in the limit of high ionospheric conductivity, become the zeroes of the zero order Bessel function (2.40, 5.52, 8.65,...). In the results presented below, we will adopt a model with an exponential O^+ density profile with density of $n_0 = 5\times10^5\,cm^{-3}$ at the ionosphere and a scale height of $h = 0.025\,R_E$ and a power law H^+ profile with $n_1 = 10\,cm^{-3}$ and $\gamma = 1$. For an ionospheric magnetic field of 0.545 G (corresponding to $L=10$), this gives an Alfvén speed of $V_{AI} = 420$ km/s. For the scale height $h = 0.025\,R_E$ this then yields $V_{AI}/2h = 1.32\,s^{-1}$, or a frequency of .21 Hz. Thus, these parameters will yield mode frequencies in the 0.1-1.0 Hz range observed by Viking [Marklund et al., 1990; Block and Fälthammar, 1990].

The resulting mass density, electron density and Alfvén speed profile if shown in Figure 1. This mass and electron densities decrease exponentially until about 1.4 R_E, at which point the power law takes over with densities between 1 and 10 cm^{-3} throughout the model. As can be seen, the electron density is a factor of 16 less than the mass density in the oxygen dominated exponential region, but they become the same in the hydrogen dominated power law region. As the solid line in Figure 1 shows, the Alfvén speed in this model increases rapidly to a peak value of about 1.3×10^5 km/s at 1.4 R_E, and decreases with increasing altitude thereafter. We assume an ionospheric conductivity of 10 mhos, and a long perpendicular wavelength so that parallel electric fields are not important.

Figure 2 shows the calculated reflection coefficient for this model. Decreases in the reflection coefficient at frequencies of 0.44, 1.00, 1.56, 2.13 and 2.62 Hz indicate the presence of resonator eigenmodes at these frequencies. These frequencies are somewhat lower than would be predicted by the straight

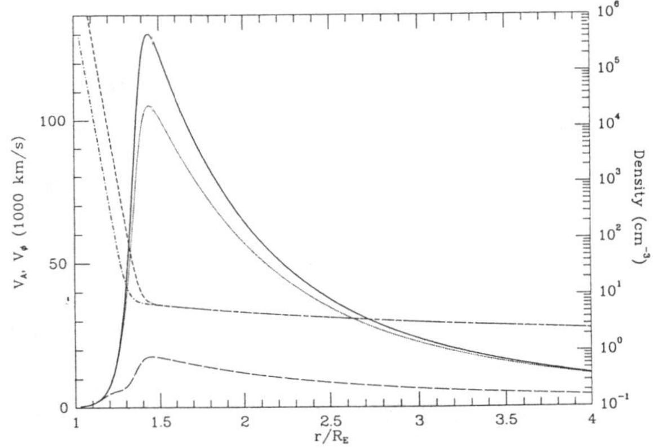

Fig. 1. Short dashed curve gives the mass density profile, in units of the proton mass, and the dot-dash curve gives the electron density profile for a density model including oxygen and hydrogen ions as described in the text. These densities are referred to the right hand scale. Also shown for this model are the Alfvén speed (solid curve), and the parallel wave phase velocity for perpendicular wavelengths at ionospheric altitude of 10 km (dotted curve) and 1 km (long dashes), referred to the left hand scale.

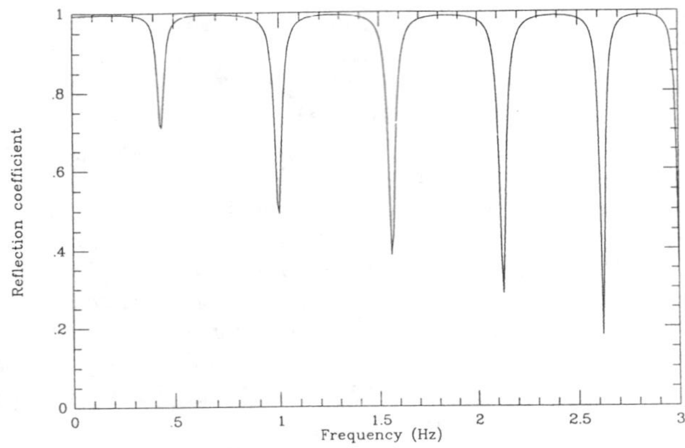

Fig. 2. Reflection coefficient $|\Phi^+|/|\Phi^-|$ for the model of Figure 1. Minima at 0.44, 1.00, 1.56, 2.13 and 2.62 Hz indicate the presence of the resonator eigenmodes.

field line model, having values of ξ_n = 2.1, 4.8, 7.4, 10.1 and 12.5. This reduction in the ξ_n values can be understood when it is realized that in the dipole geometry, the Alfvén speed decreases less rapidly than it would if the magnetic field stayed constant. It is interesting to note that other density profiles with the same values of n_0 and h, but values of n_1 ranging from 0.01 cm^{-3} to 1000 cm^{-3} and values of γ ranging from 0 to 6 yield the same frequencies to within 5%, reinforcing the conclusion that the exponential fall off of the density just above the ionosphere is the critical factor in determining the mode frequencies.

Figure 3 shows the structure of the scalar potential of these modes normalized to the incoming wave at the top of the model. This figure clearly shows the existence of the eigenmodes in the resonator at the frequencies indicated above. The solid contours in this figure represent values of $|\Phi| = 1, 2, 3$ and 4, normalized to the incident wave, while the dotted con-

tours show lower values (0.2, 0.4, 0.6, 0.8) needed to illustrate the resonator structure in the low altitude regime, where Φ is decreased because of the low value of the Alfvén speed. It can also be seen that the higher amplitudes penetrate to lower altitude at the eigenmode frequencies. The highest values of the potential occur just above the Alfvén speed peak, since the waves reflecting from this high altitude gradient in the Alfvén speed reflect in a sense to enhance the electric fields. At very high altitudes, a structure of nodes and antinodes associated with the interference of the downgoing and upgoing waves can be seen. This structure depends on the high altitude density model, with the wavelength increasing for lower densities.

In order to relate this calculation to observed wave spectra, we must assume a form for the input spectrum. To this end, a spectrum $|\Phi^-|^2 \sim f^{-5/3}$ was assumed. This particular spectral index has no real physical significance, but serves as an example of the response of the resonator to a broad band source. The results for spectra calculated at r = 1.5, 2, 2.5, and 3 R$_E$ are shown in Figure 4. (Note that each successive spectra has been multiplied by a factor of 10 for clarity.) Within the resonator at r =1.5 R$_E$ (solid curve), the spectrum exhibits peaks at the resonant frequencies. The peak to valley ratios above the fundamental peaks are as large as two orders of magnitudes. At higher altitudes, the spectral peaks broaden and the spectrum is more accurately characterized as having minima at the resonances. At high altitudes and higher frequencies, the interference between up and downgoing waves noted in Figure 3 leads to a deep minimum at about 2 Hz, a point of destructive interference between the waves.

It is interesting to compare results such as those shown in Figure 4 to the Viking data. First of all, note that the eigenfrequencies of the model shown are in close agreement with the spectral peaks observed in Figure 10 of Block and Fälthammar [1990], which have similar peak to valley ratios as the bottom curve in Figure 4. However, this observation was made at about 3 R$_E$ (geocentric), corresponding to the top curve in Figure 4, which shows dips in the spectrum at these frequencies.

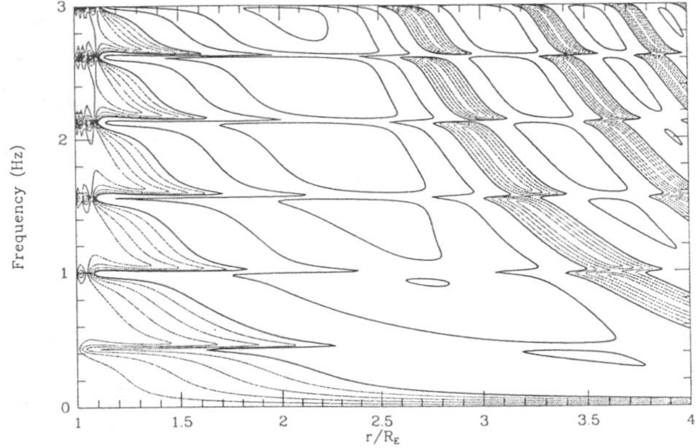

Fig. 3. Magnitude of the scalar potential, normalized to the incident wave, for the model of Figure 1. Heavy solid contours give values of 1, 2, 3, and 4, and the dotted contours give values of 0.2, 0.4, 0.6, 0.8. The structure of the eigenmodes at the frequencies corresponding to the minima in Figure 2 is evident.

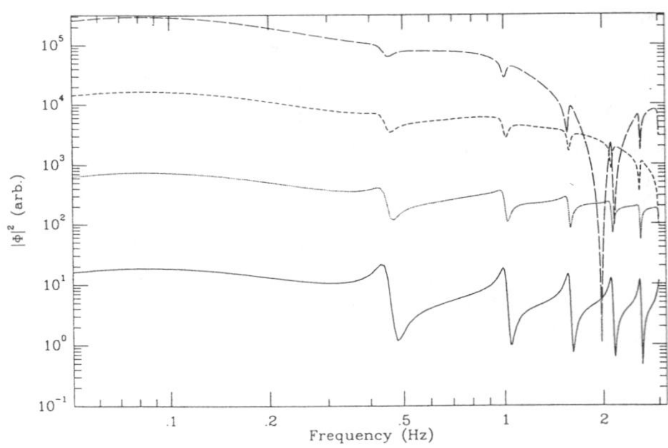

Fig. 4. Simulated spectra for the model of Figure 1 for an input spectrum given by $f^{-5/3}$ at altitudes of 1.5, 2, 2.5 and 3 R$_E$, proceeding from bottom to top. Each successive spectrum has been multiplied by a factor of 10.

A number of possible modifications could possibly reconcile these results. One would be to increase the scale height, so that the Alfvén speed peak occurred at altitudes closer to that of the spacecraft. The difficulty with this is that to maintain the same set of resonant frequencies, the Alfvén speed at the ionosphere would also have to increase, implying a low ionospheric density. One could also assume that the density in the outer magnetosphere was very low (e.g., less than 0.1 cm^{-3}). This extends the region of the Alfvén speed peak to larger altitudes, and does produce peaks in the spectrum at 3 R$_E$. Such low densities are difficult to measure; however, Persoon et al. [1988] have reported densities below 0.1 cm^{-3}, although at somewhat higher altitudes.

On the other hand, not all of the Viking electric field spectra show such nicely structured spectral peaks. Marklund et al. [1990] show examples of more broad-banded spectra. These examples might be more suitable candidates to be explained by the present model. In particular, the top example in their Figure 11 shows a case in which the fundamental frequency is more like 0.15 Hz, which would be consistent with larger scale heights which would allow the resonator waves to be seen at higher altitudes. Marklund et al. [1990] show other examples in which it appears that a monochromatic wave is superimposed on the broadband spectrum. It is likely that these narrowly peaked spectra are associated with another process. Erlandson et al. [1990] suggest that these waves may be due to an instability of electromagnetic ion cyclotron (EMIC) waves. It is also possible that these waves are enhanced by ionospheric feedback [Lysak, 1991], which would emphasize the eigenmodes of the cavity.

In any case, it is encouraging to note that this simple model of a broadband incident spectrum passing through a passive (i.e., constant ionospheric conductivity) ionospheric Alfvén resonator seems to be consistent with at least some of the observations from Viking. It would be interesting to correlate the observed wave spectrum with the local plasma density as well as simultaneous ionospheric parameters (measured from sounding rockets or ground-based radar), in order to quantitatively check the resonator model.

B. Parallel Electric Fields

As noted in Equation (3), Alfvén waves are accompanied by parallel electric fields due to electron inertia. In terms of the Fourier transformed variables, this parallel electric field can be written:

$$E_\| = -\frac{i\omega}{c} k_\perp^2 \lambda^2 A_\| \qquad (12)$$

This equation illustrates that parallel electric fields will be enhanced for higher frequency modes, and in locations in which $k_\perp^2 \lambda^2$ is large. As noted above, this quantity scales as B/n_e, which, for the density profile used above, maximizes at altitudes just above the Alfvén speed peak.

It is convenient to express the perpendicular wavelength of the waves in terms of the wavelength mapped to the ionosphere. For example, for a 10 km wavelength at the ionosphere

and the density profile in Figure 1, $k_\perp^2 \lambda^2$ reaches a maximum of 0.55 at $r = 1.4$ R$_E$. This maximum value is dependent on the form of the high altitude density distribution; e.g., for $n_1 = 1$ cm^{-3} and $\gamma = 0$, the maximum value of $k_\perp^2 \lambda^2$ increases to 3.6. The effect of inertia on the wave dispersion can be seen in the dotted line of Figure 1, which shows the wave parallel phase velocity $V_A/\sqrt{1 + k_\perp^2 \lambda^2}$. In this case, there is only a slight decrease in the phase velocity. Consequently, the wave eigenmode structure is not strongly affected by the presence of the parallel electric field, and Figures 2, 3, and 4 not changed appreciably.

On the other hand, the presence of the parallel electric field due to this wave can have important effects on the acceleration of auroral particles. Figure 5 shows the structure of the amplitude of the parallel electric field, in a similar format to Figure 3. In this Figure, the maximum parallel electric field is 0.41 mV/m, normalized to an incident amplitude in the perpendicular potential of 1 kV. Note that this potential corresponds to an perpendicular electric field of 78.5 mV/m at 4 R$_E$ for a wavelength of 10 km at the ionosphere. Comparison of Figures 3 and 5 shows that the largest parallel electric fields occur at the nodes of the perpendicular potential, as would be expected from Equations (9) and (12). The parallel electric field does not extend to altitudes below about 1.3 R$_E$ since the rapidly increasing density implies that the inertial length becomes very small. However, the resonator modes can be clearly seen in Figure 5 as frequencies in which the node-antinode structure of the parallel electric field is reduced. Since these nodes and antinodes imply parallel electric fields which oscillate in sign, the largest parallel potential drops occur at the eigenfrequencies of the resonator, having magnitudes of 330, 333, 543, and 775 volts for the first four modes, again normalized to an incident perpendicular potential of 1 kV. These potentials are dependent on the location chosen for the top boundary of the calculation, and more importantly are sensitive to the form of the high altitude density profile. For example, for the case $n_1 = 1$ cm^{-3} and

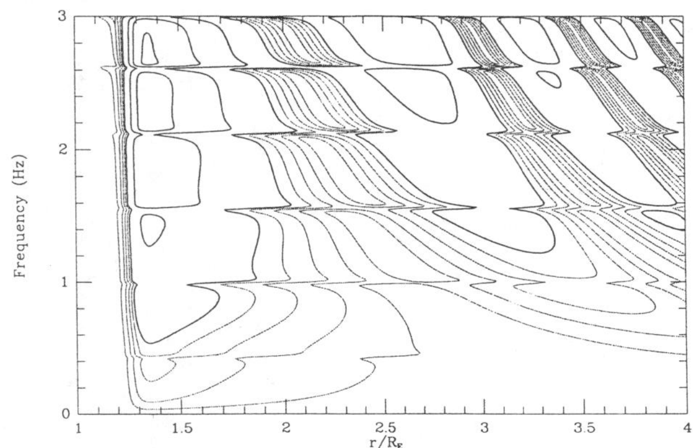

Fig. 5. Contours of parallel electric field magnitude for the model of Figure 1 with a perpendicular wavelength of 10 km at the ionosphere. Solid contours give levels of 0.1, 0.2, 0.3 and 0.4 mV/m, normalized to an incident potential of 1 kV. Dotted contours give levels of 0.02, 0.04, 0.06, and 0.08 mV/m.

$\gamma=0$, the peak parallel electric increases to 1.1 mV/m and the parallel potential drops become the order of 1 kV.

Of course, these parallel electric fields are seen as static potential drops only by particles whose transit time through the structure is much less than the wave period. An electron with an energy of 2 keV will pass through the 3 R_E region shown in 1 second, and so for electrons of lower energy, the time structure of the parallel electric field will become important. Figure 6 shows a representation of the time history of the first and fourth modes. In this Figure, contours of equal parallel electric field strength are plotted as a function of the altitude and the phase of the wave. Solid contours indicate positive (upward) $E_\parallel$, dotted contours denote negative values and the heavy contour is 0. It can be seen from these plots that contours of constant parallel electric field show a downward motion during one cycle of the wave, particularly for the higher frequency mode. This downward motion is related to the net downward energy flow of these waves.

This characteristic motion during the wave cycle implies that these waves may provide a Fermi acceleration to suprathermal electrons with energies of the order of 100 eV or somewhat less, which have bounce times between the acceleration region and the ionosphere comparable to the wave periods. Since the parallel potentials in these modes are the order of a few hundred volts, these electrons will be reflected from the electrostatic mirror and will pick up additional velocity from the phase velocity of the wave. The similarity between the wave periods and the bounce times of electrons between the acceleration region indicates that electrons may be able to make multiple bounces between the parallel electric field and the magnetic mirror, gaining energy on each bounce. Eventually, such electrons may gain enough energy to overcome the parallel electric field, or their bounce times will decrease to the point where they lose resonance with the wave, at which point they may be observed at higher altitudes as upgoing electron conics [Menietti and Burch, 1985; Lundin et al., 1987]. Electrons of thermal energies will also be accelerated somewhat by these waves, which provides the field-aligned current which supports the wave. Hui and Seyler [1992] have shown that in the nonlinear regime, this acceleration may be responsible for a wave breaking phenomenon.

As a final point, at even smaller perpendicular wavelengths, e.g., 1 km at ionospheric heights, the Alfvén wave dynamics can become dominated by the electron inertial effect. This implies that the phase velocity becomes very small, as can be seen in the long dashed curve in Figure 1, which shows the phase velocity for the 1 km case. Thus, for a given frequency, short perpendicular wavelengths lead to short parallel wavelengths and a large ratio between Φ and $A_\parallel$ (see equation 7), indicating a strong reflection at the Alfvén speed peak. These considerations indicate that waves incident from higher altitudes are not able to penetrate very effectively to the low altitude region. In this regime, the electric field becomes nearly electrostatic and the parallel potential drop becomes equal to the perpendicular potential. Thus, the low phase velocity and electrostatic character of these waves may be describable as an electrostatic shock.

DISCUSSION

The results presented above indicate that the effects of inhomogeneity can introduce structure at the ionospheric Alfvén resonator eigenfrequencies into an otherwise smooth input spectrum, even without any ionospheric feedback. These results compare reasonably well with observations of electric field spectra from Viking taken primarily in the dayside region. While these results do not rule out the presence of other mechanisms for waves in this frequency range, notably the generation of EMIC waves by particles injected into the cusp, they do point to the importance of considering the Alfvén wave inhomogeneities in interpreting low frequency wave spectra observed by polar orbiting satellites. Observations taken by high altitude satellites such as Dynamics Explorer or Viking, which primarily take place above the Alfvén speed peak, may show quite different characteristics than those that would be expected from lower altitude satellites, such as the recently launched Freja mission and the upcoming Fast mission. These lower altitude satellites will sample the primary region of the resonator itself, below the Alfvén speed peak, and so should exhibit more sharply peaked spectra than in the case of the typi-

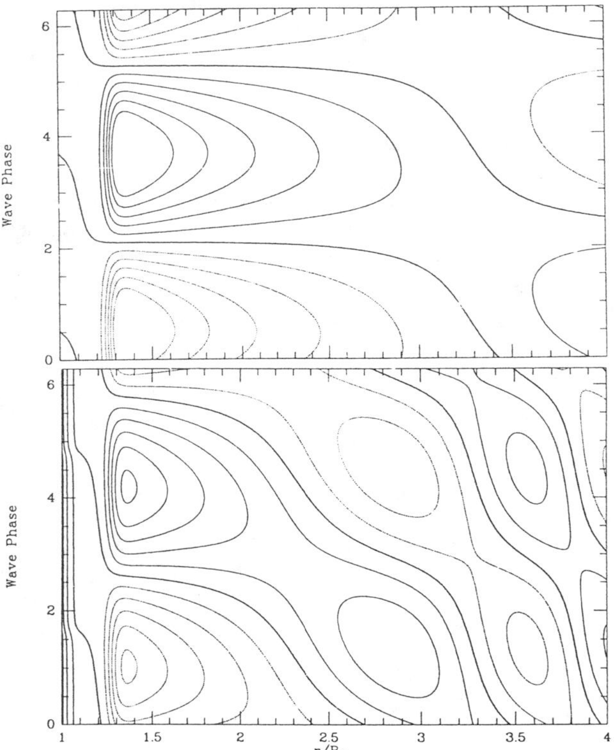

Fig. 6. Contours of parallel electric field strength during the progression of a wave cycle for the eigenmodes at 0.44 Hz (top panel) and 2.13 Hz (bottom panel). Heavy contour gives the zero level, solid contours give positive values and dotted contours negative values in increments of 0.01 mV/m (top) and 0.05 mV/m (bottom), normalized to an incident potential of 1 kV.

cal Viking measurements. Indeed, it might be expected that the spectra observed by these satellites may be similar to that seen in Figure 10 of Block and Fälthammar [1990], which may represent a relatively rare case when the Alfvén speed peak has moved to higher altitudes.

A number of cautions should be noted here. First of all, the results presented here have all been for a relatively high ionospheric conductivity of 10 mhos. This conductivity is reasonable on the dayside, where most of the published Viking wave spectra have been observed. For lower values of conductivity, the Alfvén waves are absorbed more at the ionosphere, and the spectral peaks at the eigenfrequencies are not so well defined. In such cases the resonator structure can still be distinguished below the Alfvén speed peak, but is not apparent above the peak. Thus, nightside observations would not be expected to show such a clear spectrum. However, this conclusion only holds for the passive ionospheric model presented here; feedback interactions may enhance the resonator frequencies in nightside regions, particularly during substorms when small scale enhancements in the ionospheric conductivity have been observed [Kirkwood et al., 1988].

Another point that should be emphasized is that the theoretical spectra shown here are taken for one perpendicular wavelength in the plasma frame. In a spacecraft frame, any spectral peaks should be broadened by the Doppler shift. For a spacecraft velocity of 3 km/s, a spectral peak with a width of 0.3 Hz as is seen in the published spectra implies a wavelength of greater than 10 km at the spacecraft, which maps to about 2 km at the ionosphere. Although spatial irregularities on scales of less than 1 km have been deduced from Viking interferometric measurements [Holmgren and Kintner, 1990], the sharply peaked spectra observed on Viking must have longer wavelengths.

The second major result presented here is that the electron inertial effect due to the Alfvén waves in the resonator may produce significant acceleration of suprathermal electrons. At moderate perpendicular wavelengths (10 km at the ionosphere), parallel potentials of the order of 100 volts can be associated with the cavity modes; at smaller scales the potentials can become larger. It is worth noting here that these parallel electric fields are scale as B/n (Equation 12), and can be greatly enhanced if regions of low density can penetrate to lower altitudes. This could lead to an interesting possible feedback interaction since the low density auroral plasma cavities are thought to be formed by a parallel potential drop which excludes the ionospheric electrons, and in turn the low densities enhance the parallel electric fields. These interactions need to be studied with more detailed simulations.

A possible application of these fluctuating parallel electric fields may lie in the production of the so-called electron conic distributions [Menietti and Burch, 1985; Lundin et al., 1987; Hultqvist et al., 1988] which have been variously attributed to perpendicular electron heating [Menietti and Burch, 1985; Wong et al., 1988], parallel electron heating [Temerin and Craven, 1990], and fluctuating electric fields [Lundin et al., 1987; Hultqvist, 1988; André and Eliasson, 1992]. The model

presented here would fall into this last class; however, while these previous authors arbitrarily assumed a fluctuating parallel electric field over a certain altitude range, the present model specifies the frequencies and the spatial structure of these waves. An additional feature introduced by this model is the downward propagation of the parallel electric field region, which allows for an additional Fermi acceleration of the electrons. Temerin et al. [1986] have considered electron acceleration due to higher frequency EMIC waves which propagate down field lines in a similar manner, and have found significant electron acceleration due in part to this type of Fermi process. While this model can account for the rapid time fluctuations associated with flickering aurora, the present model can deal with the lower frequency component of the precipitation. Again it is clear that more work should be done to examine the effects of these waves on the particles through test particle simulations.

A final comment concerns the relation of this model to the production of narrow (<1 km) scale auroral arcs. The strong electron inertial effect at such short spatial scales leads to a strong reflection of Alfvén waves from the region where the Alfvén speed and the inertial effect maximize. Thus, the regions above and below the Alfvén speed peak may be effectively decoupled electromagnetically at these scales. On the other hand, Hui and Seyler [1992] have recently pointed out that electrons can be effectively accelerated when the electron inertial effects are important. These accelerated electrons are not affected by the Alfvén speed peak and can propagate through to the ionosphere, where they may be responsible for enhancements in the ionospheric conductivity. These electrons may thus become the seed for the development of the fast ionospheric feedback instability [Lysak, 1991] which can lead to the structuring of auroral currents in the region below the Alfvén speed peak. These considerations suggest the complexity of magnetosphere-ionosphere interactions on these small spatial scales, and indicate that this approach may lead to a better understanding of these narrow auroral arcs.

CONCLUSIONS

The work presented here has extended the theory of the ionospheric Alfvén resonator by considering arbitrary density profiles and parallel electric fields due to the electron inertial effects. The main conclusions can be summarized as follows:

1. The model of the ionospheric Alfvén resonator carries over into more generalized geometries, and the fundamental scaling of the resonator frequencies with $V_{AI}/2h$ persists independent of the structure of the density above the Alfvén speed peak.

2. Given a power law input spectrum of Alfvén wave power, the resonator model produces peaks in the spectrum at resonator frequencies at low altitudes, and minima in the spectrum at high altitudes. The computed wave spectra are consistent with the observations from Viking, although the model does not rule out the possibility that the more sharply peaked spectra observed are associated with EMIC instabilities at high altitudes.

3. The electron inertial effect can produce a significant parallel electric field for perpendicular wavelengths the order of 10

km or less at ionospheric heights, which can have an effect on the acceleration of low energy electrons and may play a role in the formation of electron conics.

Acknowledgments. This work as well as my previous work on Alfvén waves has been greatly influenced over the past years by many discussions with Chris Goertz, whose paper on magnetosphere-ionosphere coupling [Goertz and Boswell, 1979] introduced me to the importance of Alfvén waves to auroral acceleration. This work has also been significantly enhanced by discussions with Mats André on the Viking data and on observations of electron conics, and with Mike Temerin on the Alfvén speed profile in the auroral zone. The work was supported in part by NASA grant NAGW-2653 and NSF grant ATM-9111791.

REFERENCES

André, M., and L. Eliasson, Electron acceleration by low frequency electric field fluctuations: electron conics, *Geophys. Res. Lett.*, *19*, 1073, 1992.

Block, L. P., and C.-G. Fälthammar, The role of magnetic-field-aligned electric fields in auroral acceleration, *J. Geophys. Res.*, *95*, 5877, 1990.

Erlandson, R. E., L. J. Zanetti, T. A. Potemra, L. P. Block, and G. Holmgren, Viking magnetic and electric field observations of Pc1 waves at high latitudes, *J. Geophys. Res.*, *95*, 5941, 1990.

Goertz, C. K. and R. W. Boswell, Magnetosphere-ionosphere coupling, *J. Geophys. Res.*, *84*, 7239, 1979.

Hasegawa, A., Particle acceleration by MHD surface wave and formation of aurora, *J. Geophys. Res.*, *81*, 5083, 1976.

Holmgren, G., and P. M. Kintner, Experimental evidence of widespread regions of small-scale plasma irregularities in the magnetosphere, *J. Geophys. Res.*, *95*, 6015, 1990.

Hui, C.-H., and C. E. Seyler, Electron acceleration by Alfvén waves in the magnetosphere, *J. Geophys. Res.*, *97*, 3953, 1992.

Hultqvist, B., On the acceleration of electrons and positive ions in the same direction along magnetic field lines by parallel electric fields, *J. Geophys. Res.*, *93*, 9777, 1988.

Hultqvist, B., R. Lundin, K. Stasiewicz, L. Block, P.-A. Lindqvist, G. Gustafsson, H. Koskinen, A. Bahnsen, T. A. Potemra, and L. J. Zanetti, Simultaneous observation of upward moving field-aligned energetic electrons and ions on auroral zone field lines, *J. Geophys. Res.*, *93*, 9765, 1988.

Kirkwood, S., H. Opgenoorth, and J. S. Murphree, Ionospheric conductivities, electric fields and currents associated with auroral substorms measured by the EISCAT radar, *Planet. Space Sci.*, *36*, 1359, 1988.

Lee, D.-H., and R. L. Lysak, Magnetospheric ULF wave coupling in the dipole model: the impulsive excitation, *J. Geophys. Res.*, *94*, 17097, 1989.

Lundin, R., L. Eliasson, B. Hultqvist, and K. Stasiewicz, Plasma energization on auroral field lines as observed by the Viking spacecraft, *Geophys. Res. Lett.*, *14*, 443, 1987.

Lysak, R. L., Coupling of the dynamic ionosphere to auroral flux tubes, *J. Geophys. Res.*, *91*, 7047, 1986.

Lysak, R. L., Theory of auroral zone PiB pulsation spectra, *J. Geophys. Res.*, *93*, 5942, 1988.

Lysak, R. L., Feedback instability of the ionospheric resonant cavity, *J. Geophys. Res.*, *96*, 1553, 1991.

Mallinckrodt, A. J., and C. W. Carlson, Relations between transverse electric fields and field-aligned currents, *J. Geophys. Res.*, *83*, 1426, 1978.

Maltsev, Y. P., W. B. Lyatsky, and A. M. Lyatskaya, Currents over the auroral arc, *Planet. Space Sci.*, *25*, 53, 1977.

Marklund, G. T., L. G. Blomberg, C.-G. Fälthammar, R. E. Eriandson, and T. A. Potemra, Signatures of the high-altitude polar cusp and dayside aurora regions as seen by the Viking electric field experiment, *J. Geophys. Res.*, *95*, 5767, 1990.

Menietti, J. D., and J. L. Burch, "Electron conic" signatures observed in the nightside auroral zone and over the polar cap, *J. Geophys. Res.*, *90*, 5345, 1985.

Persoon, A. M., D. A. Gurnett, W. K. Peterson, J. H. Waite, J. L. Burch, and J. L. Green, Electron density depletions in the nightside auroral zone, *J. Geophys. Res.*, *93*, 1871, 1988.

Polyakov, S. V., and V. O. Rapoport, Ionospheric Alfvén resonator, *Geomag. Aeronomy*, *21*, 816, 1981.

Scholer, M., On the motion of artificial ion clouds in the magnetosphere, *Planet. Space Sci.*, *18*, 977, 1970.

Seyler, C. E., Nonlinear 3-d evolution of bounded kinetic Alfvén waves due to shear flow and collisionless tearing instability, *Geophys. Res. Lett.*, *15*, 756, 1988.

Strauss, H. R., Nonlinear, three-dimensional magnetohydrodynamics of noncircular tokamaks, *Phys. Fluids*, *19*, 134, 1976.

Temerin, M. A., and D. Cravens, Production of electron conics by stochastic acceleration parallel to the magnetic field, *J. Geophys. Res.*, *95*, 4285, 1990.

Temerin, M., J. McFadden, M. Boehm, C. W. Carlson, and W. Lotko, Production of flickering aurora and field-aligned electron flux by electromagnetic ion cyclotron waves, *J. Geophys. Res.*, *91*, 5769, 1986.

Trakhtengertz, V. Yu., and A. Ya. Feldstein, Effect of the nonuniform Alfvén velocity profile on stratification of magnetospheric convection, *Geomag. Aeronomy*, *21*, 711, 1981.

Trakhtengertz, V. Yu., and A. Ya. Feldstein, Quiet auroral arcs: ionospheric effect of magnetospheric convection stratification, *Planet. Space Sci.*, *32*, 127, 1984.

Trakhtengertz, V. Yu., and A. Ya. Feldstein, Turbulent regime of magnetospheric convection, *Geomag. Aeronomy*, *27*, 221, 1987.

Trakhtengertz, V. Yu., and A. Ya. Feldstein, Turbulent Alfven boundary layer in the polar ionosphere, I, Excitation conditions and energetics, *J. Geophys. Res.*, *96* 19,363, 1991.

Wong, H. K., J. D. Menietti, C. S. Lin, and J. L. Burch, Generation of electron conical distributions by upper hybrid waves in the Earth's polar region, *J. Geophys. Res.*, *93*, 10,025, 1988.

Robert L. Lysak, School of Physics and Astronomy, University of Minnesota, 116 Church Street SE, Minneapolis, MN 55455

Laboratory Work on Transient Currents and its Application to Auroral Arc Currents

J. M. URRUTIA AND R. L. STENZEL

Department of Physics, University of California, Los Angeles

The propagation of switched currents between charged electrodes is studied in a large magnetized laboratory plasma. Transient currents with risetime $1/\omega_{ce} \ll t_{rise} \ll 1/\omega_{ci}$ are carried by electron whistler waves. The field topology of the current density $\mathbf{J}(\mathbf{r}, t) = \nabla \times \mathbf{B}(\mathbf{r}, t)/\mu_0$ and the perturbed magnetic field $\mathbf{B}(\mathbf{r}, t)$ resemble flux ropes since electrons support both field-aligned currents and Hall currents. Induced coaxial return currents establish the closure to the source during the propagation of the current front. The results may be relevant to rapidly fluctuating auroral arcs.

INTRODUCTION

Field-aligned currents play an important role in the physics of auroral arcs. Rapid temporal variations and small spatial scales have been observed in auroral arcs [*Suszcynsky and Borovsky*, 1991] and interpreted as inertial Alfvén wave phenomena [*Seyler*, 1990]. It is possible that, with further increased spatial and temporal resolution, currents in the electron MHD regime [*Kingsep et al.*, 1990] may be discovered. Such currents are supported by low-frequency electron whistler waves ($\omega_{ci} \ll \omega \ll \omega_{ce}$). Being decoupled from the ions, whistlers transport currents much faster than Alfvén waves. Such phenomena have been studied in the present laboratory experiment [*Urrutia and Stenzel*, 1989; *Stenzel and Urrutia*, 1990a]. Transient (switched) currents are established between electrodes immersed into a large magnetoplasma. The time scales can be chosen such that the currents close within the plasma free of boundary effects as in space. Propagation, field topology (from $\mathbf{J}(\mathbf{r}, t) = \nabla \times \mathbf{B}(\mathbf{r}, t)/\mu_0$), return currents and beam-current systems have been investigated, the results of which will be summarized here.

EXPERIMENTAL ARRANGEMENT

The laboratory plasma is produced by a pulsed discharge (Figure 1) in low pressure argon ($p = 3 \times 10^{-4}$ Torr). The cylindrical column of the cathode's diameter (1 m) is confined by a uniform axial dc magnetic field ($\mathbf{B}_0 \simeq 20$ G). The experiments are performed in the afterglow plasma ($n_e \geq$ $10^{11} \mathrm{cm}^{-3}$, $kT_e \leq 2\mathrm{eV}$) which is quiescent, uniform, current-free, and slowly decaying (ms) compared with the time scale of the transient currents (μs). By inserting disk electrodes connected via insulated wires to an external pulse generator, transient plasma currents of various waveforms (steps, pulses) are produced. A vector magnetic probe (B_x, B_y, B_z) is used to measure the magnetic field $\mathbf{B}(\mathbf{r}, t)$ associated with the plasma currents. The spatial dependence is obtained from a single probe, movable in three dimensions, by repeating the experiment under identical conditions. From the comprehensive data set of $\mathbf{B}(x, y, z, t)$ at up to 5000 positions the current density, $\mathbf{J}(\mathbf{r}, t) = \nabla \times \mathbf{B}(\mathbf{r}, t)/\mu_0$, is calculated without assuming field symmetries. $\mathbf{J}(\mathbf{r}, t)$ includes the displacement current density, hence is divergence-free, i.e., currents are always closed. In addition to obtaining the magnetic field, electric measurements are made with emissive probes and plasma parameters are obtained from Langmuir probes.

EXPERIMENTAL RESULTS

Figure 2 shows contours of the field-aligned current density $J_\parallel(y, z, t)$ at different times t after applying a current step function ($t_{rise} \simeq 30$ ns) to a disk electrode. The current front is observed to propagate along $\mathbf{B}_0$ at a speed $v_\parallel \simeq 10^8$ cm/s. The speed does not depend on polarity or current amplitude but increases with B_0 and decreases with n_e. By applying a periodic current waveform and performing interferometry, it has been shown [*Urrutia and Stenzel*, 1989] that the propagating signal corresponds to a low-frequency whistler wave, $\omega \simeq \omega_{ce}(kc/\omega_{pe})^2$ [*Helliwell*, 1965]. Hence, transient currents are carried by whistler wave packets and not by drifting particles as for steady-state currents.

Auroral Plasma Dynamics
Geophysical Monograph 80
Copyright 1993 by the American Geophysical Union.

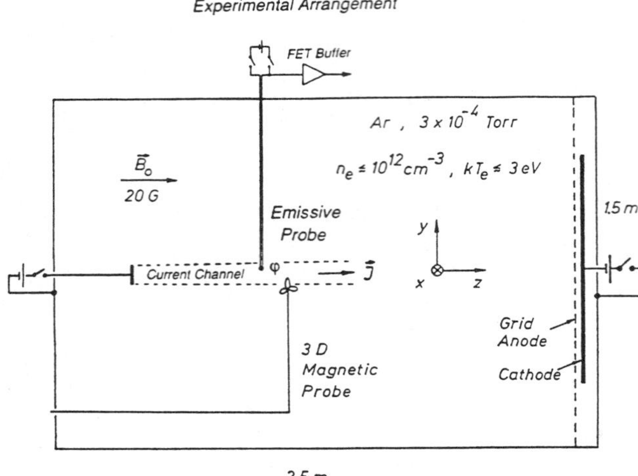

Fig 1. Experimental setup and major diagnostics.

The dashed contours in Figure 2 refer to negative $J_\parallel$, i.e., return currents. Thus, before the current front reaches the far end of the device, the closure occurs via coaxial return currents to the closest end of the device where the electrode is energized. The current closure path is the same whether the return electrode consists of a coaxial ring or the entire chamber wall. However, the return current path depends on the arrangement of the insulated wire connecting the electrodes to the pulse generator: A field-aligned wire induces anti-parallel return currents along $\mathbf{B}_0$, a cross-field wire induces Hall currents which provide current closure across $\mathbf{B}_0$. The former return currents are diffusive, the latter are radiative in the whistler mode [Stenzel et al., 1992].

Figure 3 addresses the field topologies of the perturbed magnetic field $\mathbf{B}(\mathbf{r}, t) = \mathbf{B}_{total} - \mathbf{B}_0$ and the current density $\mathbf{J} = \nabla \times \mathbf{B}/\mu_0$. Contour maps of orthogonal components B_θ ($\simeq B_x$, for $x = 0$) and B_z as well as J_θ and J_z are shown for a current channel of high cylindrical symmetry. The azimuthal magnetic field B_θ is created by field-aligned currents J_z. The axial magnetic field B_z is associated with an azimuthal current $J_\theta = -ne E_r/B_0$ which is an electron Hall current around the positively charged current channel. The resultant field topologies for both $\mathbf{B}$ and $\mathbf{J}$ are as in flux ropes, i.e., nested helixes with radially decreasing pitch lengths. The pitch changes sign off-axis due to the return currents. All fields and currents are well outside the propagating current channel. It should be pointed out that the flux rope topology applies only to the wave magnetic field which is usually much smaller than the uniform background magnetic field $\mathbf{B}_0$. The total magnetic force $\mathbf{J} \times (\mathbf{B}(\mathbf{r}, t) + \mathbf{B}_0)$ is nearly canceled by the electric force $ne\mathbf{E}$ since $\mathbf{J} \times \mathbf{B}(\mathbf{r}, t) \simeq 0$ and $\mathbf{J}_{Hall} \times \mathbf{B}_0 \simeq ne\mathbf{E}$. Thus, the electromagnetic field exerts no significant force on the electrons and even less on the ions such that the plasma can carry high currents ($I \simeq 150$ A, $B(\mathbf{r}, t) \approx B_0$) without

losing stability by pinching, kinking, or tearing [Stenzel and Urrutia, 1990b].

Auroral arcs are characterized by downward precipitating energetic electrons and upward field-aligned currents. It is thought that both the ballistic motion of the super-Alfvénic tail electrons as well as the shear Alfvén waves supported by the background plasma can carry the current [Nakamura et al, 1989]. The corresponding problem in the whistler wave regime has been investigated in the present experiment. A pulsed electron beam (100 eV, 0.1 A, $t_{rise} \simeq 30$ ns, 1 cm diam. $\simeq c/\omega_{pe}$) is injected along $\mathbf{B}_0$ into a uniform magnetoplasma. The return electrode is a disk collecting background electrons from a different flux tube than that of the beam so that one can compare the propagation speeds in both current channels [Stenzel and Urrutia, 1986].

Figure 4 shows contours of the two dominant components of the current density, i.e., field-aligned currents

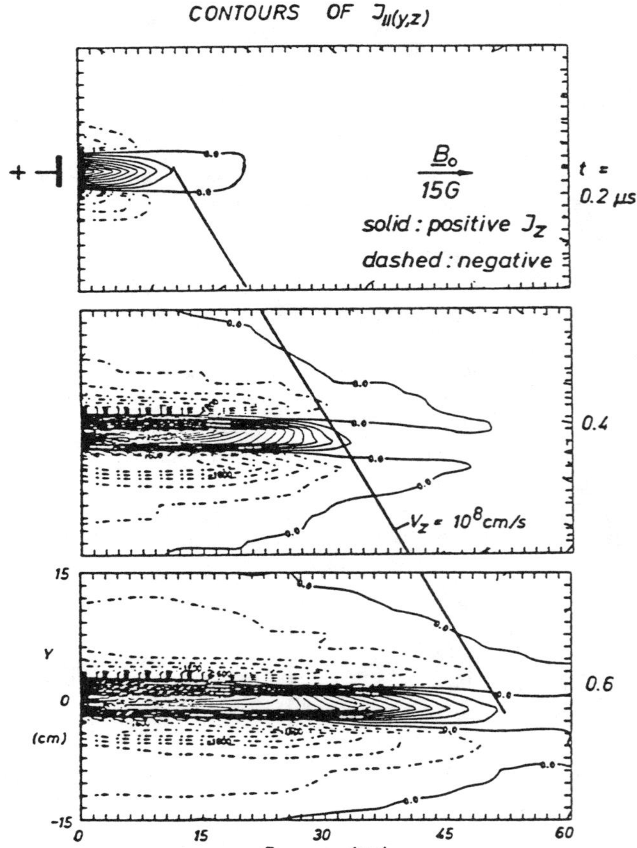

Fig. 2. Contour plots for the field-aligned current density component $J_z(y, z)$ at different times t after applying a current step to the positively biased disk electrode. The propagation of the current front is controlled by an electromagnetic wave, the electron whistler mode. Induced return currents ($J_z \leq 0$, dashed contours, with magnitude 1/10 of that of solid contours) establish the current closure to the source at $z \leq 0$ while front is propagating as into an unbounded plasma.

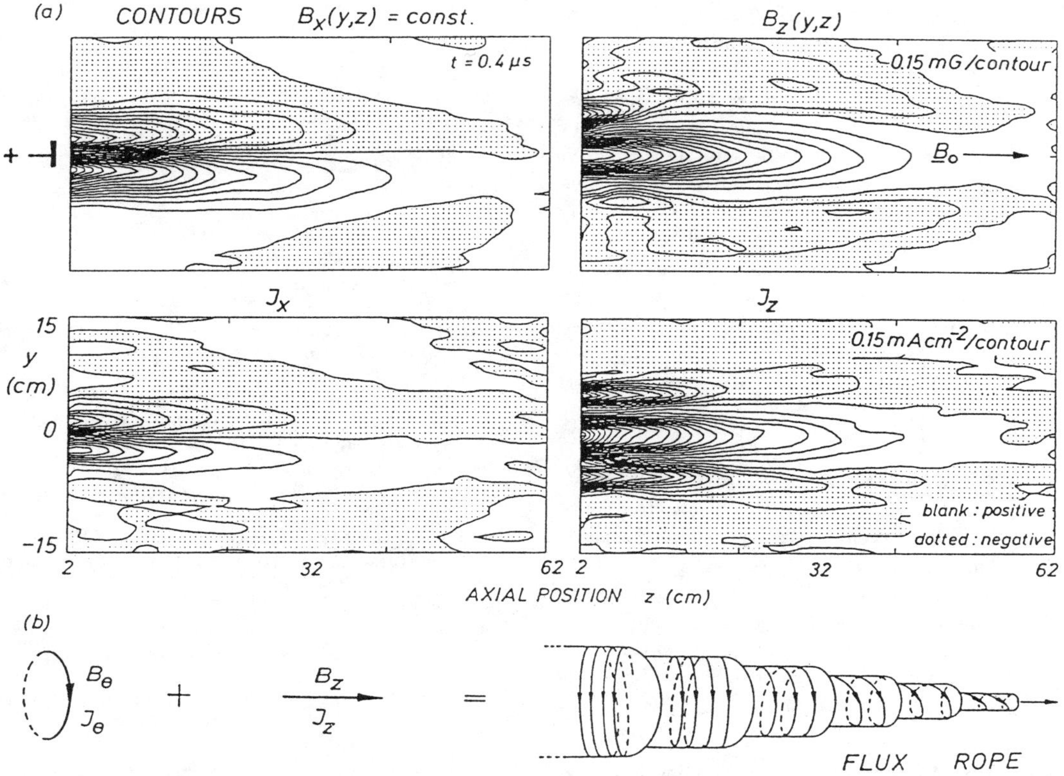

Fig. 3. Field topology of wave magnetic field $\mathbf{B}(\mathbf{r},t)$ and current density $\mathbf{J}(\mathbf{r},t)$. (a) Contour plots of azimuthal components $B_\theta \simeq B_x$, $J_\theta \simeq J_x$ and axial components B_z, J_z in the y-z plane at a fixed time during propagation. (b) Both $\mathbf{B}(\mathbf{r},t)$ and $\mathbf{J}(\mathbf{r},t)$ exhibit a flux-rope like field topology as a result of field-aligned and Hall currents.

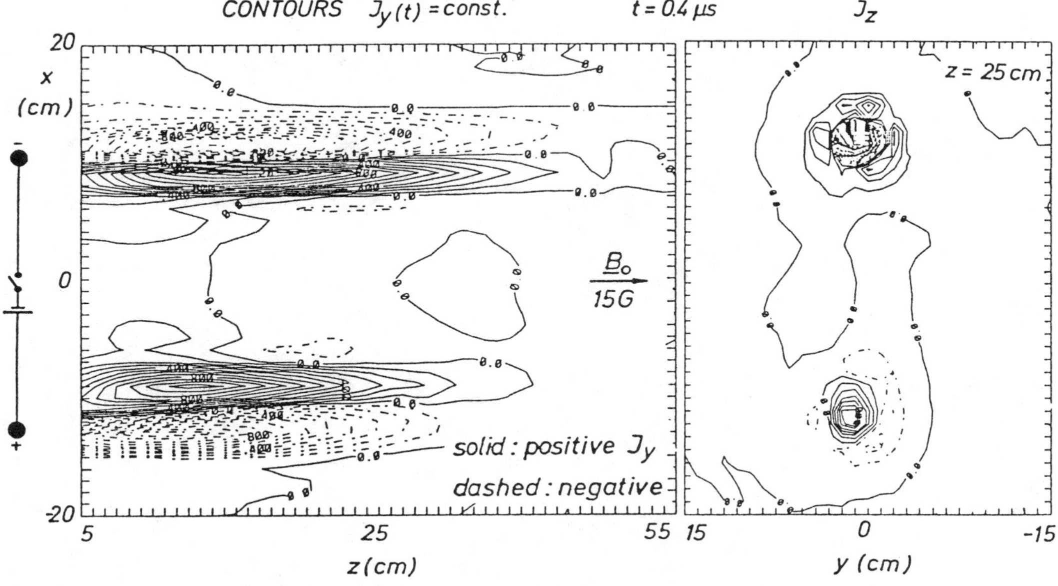

Fig. 4. Comparison of transient currents established with (i) a 100 eV electron beam emitted from the upper negative electrode, and (ii) a disk electrode collecting 1 eV background electrons. Contour plots of $J_\theta(x,z)$ ($\simeq J_y$, for $y = 0$) in the left-hand panel show comparable propagation speeds in both current channels, implying that the current is *not* carried by free-streaming beam electrons ($v_b \simeq 6 \times 10^8$ cm/s $\gg$ $v_{current} \simeq 10^8$ cm/s) but by the collective whistler mode supported by the bulk of background electrons.

$J_z(x,y)$ and Hall currents $J_\theta(x,z)$ $(\simeq J_y$, for $y=0)$. The propagation speed in both channels is comparable $(v_\parallel \simeq 40 \text{ cm}/0.4\mu s = 10^8 \text{ cm/s})$, implying that the current propagates much slower than the speed of the beam electrons $(\simeq 6 \times 10^8 \text{ cm/s})$. Thus, transient currents are *not* carried by the free-streaming beam electrons but by the collective motion of the bulk electrons supporting the whistler wave. The background electrons neutralize the current and space charge of the injected beam electrons beyond the current front where both Hall currents and field-aligned currents are negligibly small. The presence of the free-streaming beam electrons has been confirmed by time and space resolved light emission measurements.

Conclusions

It has been established through laboratory experiments that transient currents can be carried by low-frequency whistler waves. Such currents propagate much faster $\left[\approx (m_i/m_e)^{1/2}\right]$ than those carried by Alfvén waves. They can assume small spatial scales $(\approx c/\omega_{pe})$, rapid fluctuations $(\omega_{ci} \ll \omega \leq \omega_{ce})$, obliqueness to $\mathbf{B}_0$, and large amplitudes $(B(\mathbf{r},t) \approx B_0)$. Although presently not yet resolved, it is possible that such currents are embedded in larger scale auroral arcs and contribute to their dynamics.

Acknowledgements. The authors gratefully acknowledge support for this work by grants from NSF PHY, ATM, and NASA.

References

Helliwell, R. A., *Whistlers and Related Ionospheric Phenomena*, pp. 23-82, Stanford Univ. Press, Stanford, CA. 1965.

Kingsep, A. S., K. V. Chukbar, and V. V. Yankov, Electron magnetohydrodynamics, in *Reviews of Plasma Physics*, ed. by B. B. Kadomtsev, vol. 16, pp. 243-291, Consultants Bureau, N.Y., 1991.

Nakamura, T., J. R. Kan, and T. Tamao, Closure of field-aligned currents carried by super-Alfvénic auroral electrons, *J. Geophys. Res.*, **94**, 5485-5489, 1989.

Seyler, C., A mathematical model of the structure and evolution of small-scale discrete auroral arcs, *J. Geophys. Res.*, **95**, 17199, 1990.

Stenzel, R. L., and J. M. Urrutia, Laboratory model of a tethered balloon-electron beam current system, *Geophys. Res. Lett.*, **13**, 797-800 , 1986.

Stenzel, R. L., and J. M. Urrutia, Currents between tethered electrodes in a magnetized laboratory plasma, *J. Geophys. Res.*, **95** , 6209-6226, 1990a.

Stenzel, R. L., and J. M. Urrutia, Force-free electromagnetic pulses in a laboratory plasma, *Phys. Rev. Lett.*, **65**, 2011-2014, 1990b.

Stenzel, R. L., J. M. Urrutia, and C. L. Rousculp, Pulsed currents carried by whistlers. Part I: Excitation by magnetic antennas, *Phys. Fluids, B*, **5**, 325-338, 1992.

Suszcynsky, D. M., and J. E. Borovsky, The thickness of auroral arcs, *AGU Chapman Conference on Auroral Plasma Dynamics*, Session II, Minneapolis, Minnesota, Oct.21-25, 1991.

Urrutia, J. M., and R. L. Stenzel, Transport of current by whistler waves, *Phys. Rev. Lett.*, **62**, 272-275, 1989.

J. M. Urrutia and R. L. Stenzel, Department of Physics, University of California, Los Angeles, CA 90024-1547.

From Balloons To Chemical Releases—What Do Charged Particles Tell Us About The Auroral Potential Region?

R. A. HOFFMAN

Laboratory for Extraterrestrial Physics, Goddard Space Flight Center, Greenbelt, MD

The effects of an electric potential along the magnetic field on charged particles are reviewed and compared with observations to show that the basic characteristics expected in the particle distributions from a potential are observed. However, a closer analysis of the particle distributions measured in such regions shows five types of discrepancies between the measurements and the properties which cannot be explained by an auroral potential. These include the peak temperature a function of the peak energy, trapped electrons at energies below the peak energy, double peaks, ion energy different from the electron energy, and ion energy dependent upon mass. Each discrepancy is analyzed to determine the type of mechanism which would have to be added to the potential region to explain the discrepancy. The results of chemical release experiments to probe the potential region are summarized and their failure to unambiguously show the existence of a large and extended electric field region is described. Three directions for future work on this subject are suggested, which include continued careful, systematic analyses of the large body of data which already exist, the acquisition of new high resolution data, and a new approach involving the use of barium chemical releases in conjunction with in-situ measurements, accompanied by modeling of the effects of electric fields on injected ion distributions.

INTRODUCTION

On July 1, 1957, at 0330 Universal Time, the first day of the International Geophysical Year, John Winckler made the first discovery of the Geophysical Year, the observation of X-rays at balloon altitudes associated with auroral arcs [Winckler and Peterson, 1957]. On the eve of that day a balloon-born payload was launched from the University of Minnesota airport north of Minneapolis, Minnesota. Just after reaching ceiling altitude, while Winckler was watching the strip chart showing the telemetered counting rates from the detectors, an ion chamber and a Geiger counter, he was astonished by a sudden increase in the rates from both instruments (Figure 1). He went outside and noticed auroral arcs overhead. Subsequent increases in the counting rates were also associated with increased brilliance near the zenith. A clue to the type of radiation causing the increases was obtained from the ratio of counting rates, which jumped by a factor of 7. The radiation was identified as X-rays with energies in the range 50 to 70 keV [Winckler et al., 1958].

While the electrons impinging on the upper atmosphere to produce the X-rays and the visible aurora were thought to have come from an active region on the sun, their transit energy from the sun to the Earth would have been only about 30 eV. The observed time correlation of the X-rays with the overhead aurora caused Winckler et al. [1958] to argue then that some of these low energy solar electrons must have encountered a local acceleration mechanism in the Earth's magnetic field.

It is this local acceleration mechanism postulated by Winckler, now known as the auroral potential, which will be reviewed here, but only from the perspective of charged particles; thus the title, "what do charged particles tell us about the auroral potential region."

BASIC PARTICLE MEASUREMENTS

Measurements made by electron detectors flown on both sounding rockets and satellites over auroral arcs and discrete auroras, in contrast to large scale diffuse auroras, have shown that the electrons encountered have certain general characteristics. Four important characteristics are illustrated in Figure 2, which shows electron spectrograms at the top, one for precipitating electrons and one for upmoving electrons. Note the sequence of bursts of electrons, in most cases with a maximum in the flux in the several hundred eV to the keV energy range, and with this energy usually rising and then falling. These were named inverted-V electron precipitation events by Frank and Ackerson [1971]. Differential electron spectra for electrons moving in the two directions are shown below for one of the events.

Auroral Plasma Dynamics
Geophysical Monograph 80

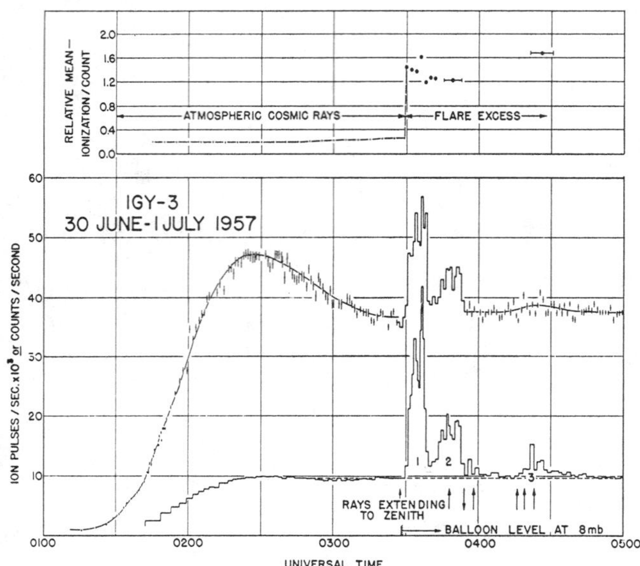

Fig. 1. Record of auroral X-rays obtained on July 1, 1957. In the lower section is plotted the counting rate of a single Geiger counter (upper curve) and pulsing rate of an integrating ion chamber (lower curve) from launch of the instruments on a balloon. The upper curve is the ratio of the counting rates. Between 0130 and 0330 the two instruments showed the normal transition curves as the balloon rose to altitude, with a ratio of about 0.2. After 0330, several enhancements in the counting rates were observed, with ratios 7 times higher (from Winckler and Peterson, 1957).

The four characteristics are as follows:

1. The energy spectrum of the down-moving electrons contains a rather sharp peak, in this case having an energy of about 1.5 keV.
2. The spectrum above the peak falls off very sharply, over two orders of magnitude within a few times the peak energy.
3. The spectrum also has a low energy tail which can be fitted by a power law having a slope of about 2.
4. The energy spectrum of the upgoing electrons has fluxes at low energies comparable to the downgoing fluxes, but there is little sign of a peak in the spectrum.

Measurements of this type, especially the pioneering work of Evans [1968; 1974], led to rapid and wide acceptance by the magnetospheric community during the late 1960s and early 70s of one particular theory for the production of this narrow distribution of auroral electrons: a quasi-static electric field having a component parallel to the magnetic vector. Such an electric field distribution would produce a potential along the magnetic field, and accelerate electrons down the magnetic field into the atmosphere to produce auroras and auroral X-rays, hence the term "auroral potential region".

Let us look at the effects of such a potential region on charged particles, and then compare the predicted effects with observations. We consider six classes of particles (Figure 3):

1. Magnetospheric electrons would be accelerated through the potential region and precipitate into the atmosphere, all receiving the same increase in energy.
2. Magnetospheric ions would be reflected by the potential region, and are, therefore, of no consequence.
3. Similarly, thermal ionospheric electrons would be reflected back into the ionosphere, and would also be of no consequence.
4. Ionospheric ions would be accelerated through the potential region and emerge into the equatorial region of the magnetosphere.

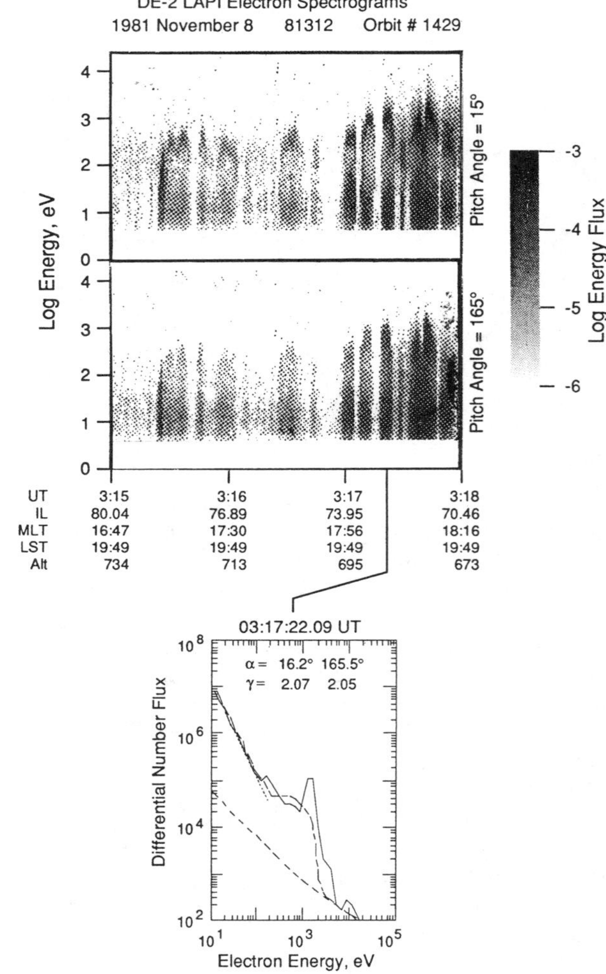

Fig. 2. Electron spectrograms of data obtained from the Low Altitude Plasma Instrument on Dynamics Explorer 2 showing examples of inverted-V electron precipitation events at two pitch angles, 15° and 165°. The lower panel contains differential energy spectra during one of the events, with the lowest curve the instrument sensitivity threshold. The dotted line is a power law fit to the secondary electrons from 20 to 100 eV with a slope of 2.07 (from Fung and Hoffman, 1991).

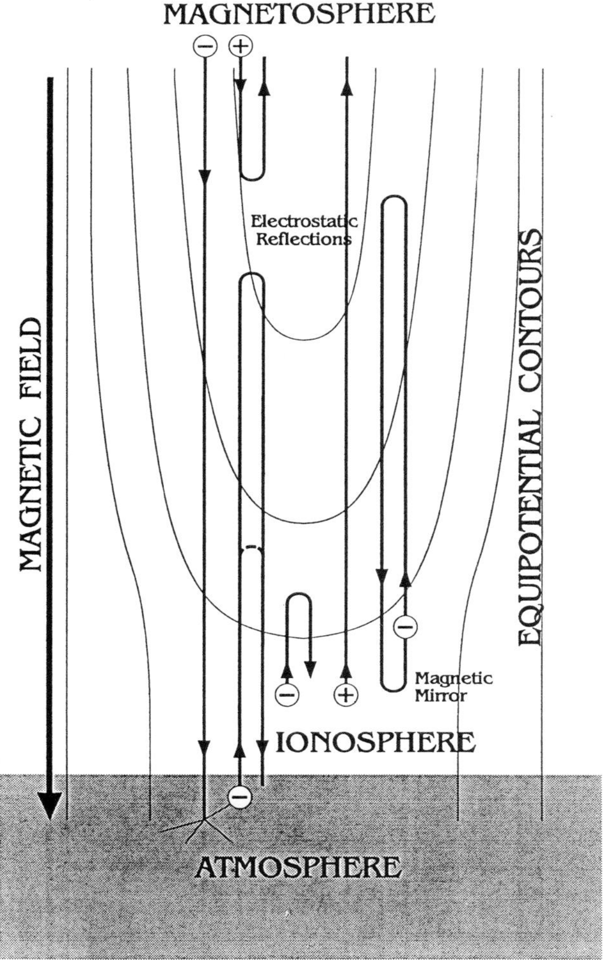

Fig. 3. Six classes of particles involved in a field-aligned potential with charged particle sources in the magnetosphere, ionosphere and atmosphere.

5. Accelerated electrons backscattered out of the atmosphere or secondary electrons produced in the atmosphere would be reflected in the potential region.

6. Electrons of suprathermal energies which magnetically mirror in the ionosphere but have less energy than the potential would be electrically reflected and find themselves trapped. There is no source for these electrons within the context of the auroral potential (more about these later).

We also consider one other aspect of an electrostatic acceleration along the magnetic field: the effect on the pitch angles of the electrons being accelerated. From the first invariant, $E_\perp/B = E \sin^2 \alpha/B = $ constant. From this invariance, the maximum pitch angle which the accelerated particles can have below the electric field region is $\sin^2 \alpha = (B_f /B_i)(E_i /E_f)$, the ratios of the field strengths and particle energies, where

$E_f = E_i + e\Phi$, the final particle energy being the source energy plus the potential energy. If the source is close to the measurement (the B ratio = 1) the electrons would become field-aligned, the degree depending upon the amount of energization. If the source temperature is known and the potential and cut-off pitch angle are measured, the extent along the field line of the potential region can be estimated from the B ratio. One can see, however, if the source magnetic field strength is much smaller than the field in the ionosphere, the monoenergetic peak will be isotropic for the down-moving electrons.

Let us next look at whether these types of particles are observed by instruments flown below or through the potential region (except for no. 6) by first referring to our previous Figure 2:

1. The sharp peak in the spectrum is due to the energization of a thermal distribution of magnetospheric electrons by a constant amount. These we term the primary electrons.

2. The sharp cutoff at energies above the peak is due to the narrow thermal spread of the electron distribution (but more about this later).

3. The low energy tail is due to secondary electrons moving up and out of the atmosphere, which are then reflected back towards the atmosphere. The upgoing and downgoing fluxes should be comparable and have a spectral index near 2 [Fung and Hoffman, 1988].

4. The upgoing electrons having energies nearly those of the primaries are due to degraded primaries backscattered from the atmosphere.

Next we look at the angular distribution of the energized electrons (Figure 4). Here we see the energy of the peak in the spectrum as two detectors, 67° different in phase, spin through pitch angles. They show isotropy in the peak energy through 110°, which includes particles mirroring below the satellite. Thus the potential must extend well above the satellite, like 2 or 3 Earth radii. Additional, excellent examples of electron distributions which agree with the auroral potential model are shown by Lundin and Eliasson [1991], based on Viking satellite data when the satellite was in or close to the potential region.

What is left now are the ions. Figure 5 shows data from a pass of the S3-3 spacecraft through a potential region at an altitude of about 7,250 km [Cladis and Sharp, 1979]. The top portion of the figure contains the potential below the satellite derived from two methods based on measurements of the distributions of upcoming electrons, as well as the total potential, which shows that most of the potential was below the satellite. The bottom three panels contain upflowing ion beams at three points across the potential region. One notes, like the potential below the satellite, the ion beam peak energy increases from the edges of the potential region to the center.

These are the basic charged particle data pertaining to the auroral potential region, and if all data were so consistent with the expected properties of such an acceleration region, we could be satisfied that the problem has been solved. Unfortunately, nature is not so simple, and furthermore, she likes to confuse the issue.

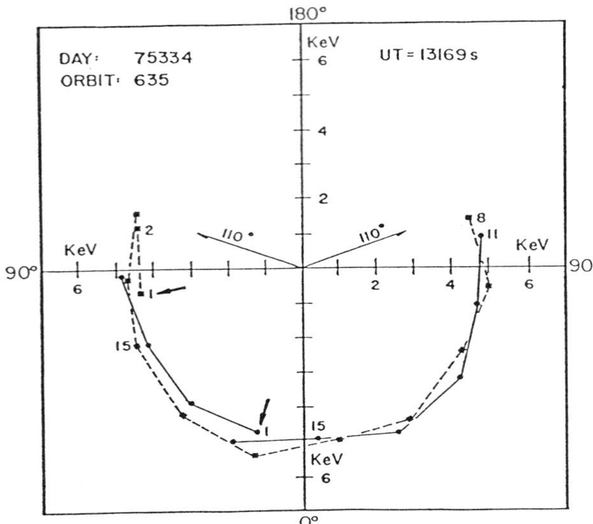

Fig. 4. Peak energy variation as a function of pitch angle for an inverted-V event from two detectors phased 67° apart on the Atmosphere Explorer C satellite when it was spinning. Commonly numbered points were acquired simultaneously to show the stability of the event.

Before we continue, one fact must be made clear: there is NOT a one-to-one relationship between inverted-V electron precipitation events and auroral arcs. Auroral arcs are usually much narrower than typical inverted-Vs [Lin and Hoffman, 1979a]. Furthermore, most inverted-V events have insufficient energy flux to produce visible auroras. However, microstructures within the inverted-V precipitation profiles seem to have sufficient energy flux to produce the visible arcs [Lin and Hoffman, 1979b]. In addition, the auroral potential is not necessarily the only mechanism for energizing and/or precipitating auroral electrons, so care must be taken in selecting events for analysis of this mechanism.

DISCREPANCIES

While their voices have not been very loud, and have generally been ignored by the scientific community, there have been, through the years, investigators who have raised a number of objections to the auroral potential model, based on apparent discrepancies between observations and predictions of the model [Whalen and Daly, 1979; Bryant, 1990, which contains a review of these discrepancies]. We use the term "apparent" for three reasons: (1) Some data sets have been analyzed in terms of the potential model, yet either no clear evidence is presented that the data pertain to such a situation, or the data seem to be contaminated with particles not directly associated with a potential region, such as events in the central plasma sheet (CPS). Examples appear in Arnoldy et al. [1974]; Whalen and Daly [1979]; and Sharp et al. [1980]. (2) The data

set is incomplete, in the sense that either only a small subset of energy-pitch angle space is sampled, or considerable aliasing of the data with time or space occurred due to the sequential sampling of energy-pitch angle space [e.g., Whalen and McDiarmid, 1972; Reasoner and Chappell, 1973; Whalen and Daly, 1979]. (3) Considerable sounding rocket data have been acquired when the rocket was in the secondary electron production region (i.e., below 350 km altitude, and for some effects, below even 800 km) [Arnoldy and Choy, 1973; Maehlum and Moestue, 1973; Bryant et al., 1978]. For such

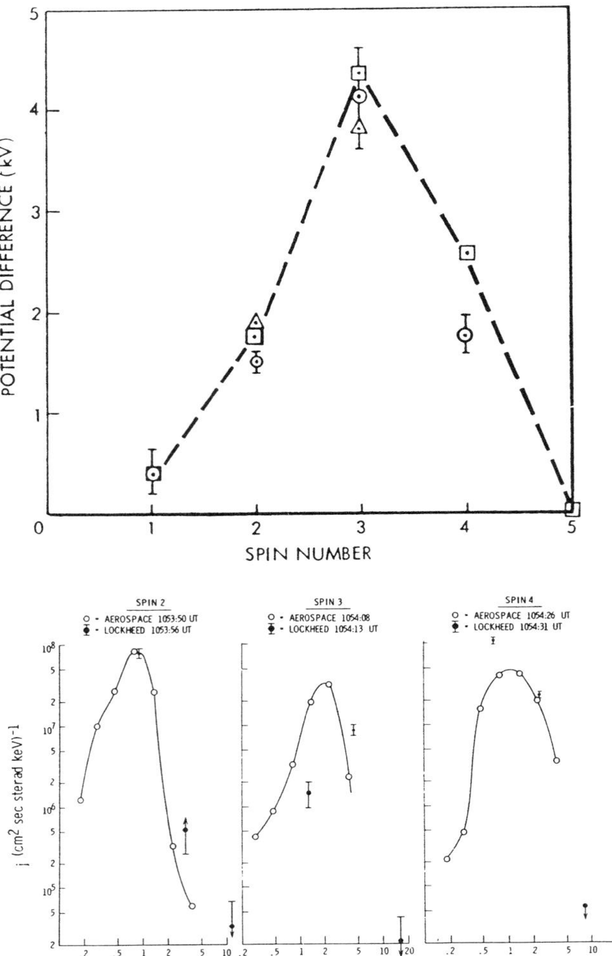

Fig. 5. Electrostatic potential differences below the S3-3 satellite inferred from measurements of electron distributions on a pass of the satellite through a potential region and the total potential (squares) (top). Energy spectra of ions during three points across the potential region (bottom). (Adapted from Cladis and Sharp, 1979.)

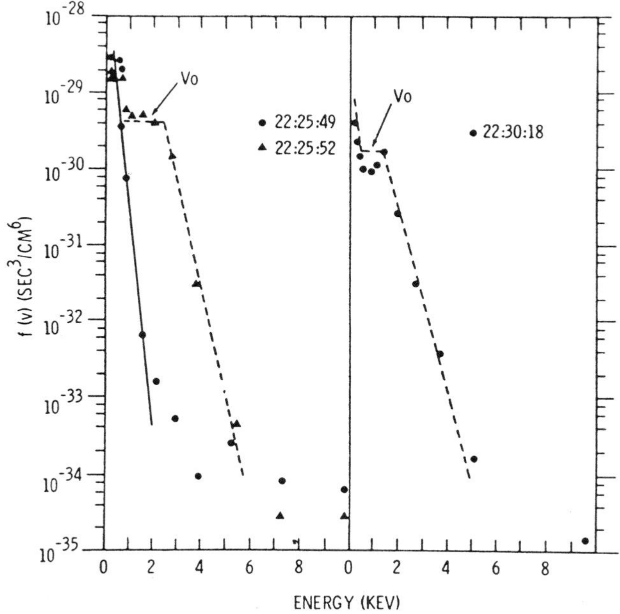

Fig. 6. Examples of electron distribution functions in inverted-V electron precipitation events. Vo is the displacement in energy of the primary beam from the low-energy continuum and is a measure of the electron acceleration. The slope of the lines above about 1 keV gives a measure of the temperature of the accelerated electrons. (From Burch et al., 1976.)

data, the low energy portion of the spectrum must be very carefully analyzed, and may not be isotropic because of transport and loss processes.

What appear to be real discrepancies pertain to the following topics:

1. Peak temperature as a function of peak energy.
2. Trapped electrons at energies below the peak energy.
3. Double peaks.
4. Ion energy different from electron energy.
5. Ion energy dependent upon mass.

We will now look at each of these discrepancies and what they mean to the concept of the auroral potential region.

Peak Temperatures

The temperature of the monoenergetic beam is difficult to measure and can only be obtained from the slope of the spectrum at energies above the peak energy, because of possible contamination to the beam from backscattered and mirroring particles at energies below the peak. The temperature is determined by using the fact that this part of the monoenergetic component can be fitted by a shifted Maxwellian distribution function.

We show in Figure 6 such fits with the data plotted as phase space density, which is proportional to the flux divided by the energy, resulting in temperature being a straight line on the

semi-log plot. This was first investigated by Burch et al. [1976], who found a systematic increase in temperature with increases in the accelerating potential. Three examples of this relationship are shown in Figure 7. But acceleration across a potential drop should produce no change in the temperature of the source. Thus, assuming that the creation of a potential is independent of the magnetospheric source temperature, it is necessary to introduce a heating process into the acceleration region to broaden the electron distribution.

In spite of the need to add a heating mechanism to the potential region, the measurement of these very low temperatures (in the 100 to 1000 eV range) by fitting data over several decades of phase space density at energies above Vo (Figure 6) is an important argument for the auroral potential concept. We know of no other mechanism that can energize electrons to the typical inverted-V energies of one to a few keV [Lin and Hoffman, 1979a] (note: >10 keV inverted-Vs are rare) and maintain such low temperatures of the accelerated distribution, especially any mechanism involving a stochastic process. (See Bryant et al. 1990; Bryant, 1990, for electron distributions from resonant acceleration by lower hybrid waves.)

Trapped Electrons at Energies Below the Peak Energy

Figure 8 shows the variation in percent of the 85 eV flux, which is a typical secondary electron energy, as a function of pitch angle, averaged over 106 inverted-V events. Note that the flux at 95°, i.e., mirroring electrons, has the same intensity as all other pitch angles for these secondary electrons. Above the "top of the atmosphere", these electrons shouldn't exist at mirroring pitch angles, since there is no source. Thus the addition of a scattering mechanism into the

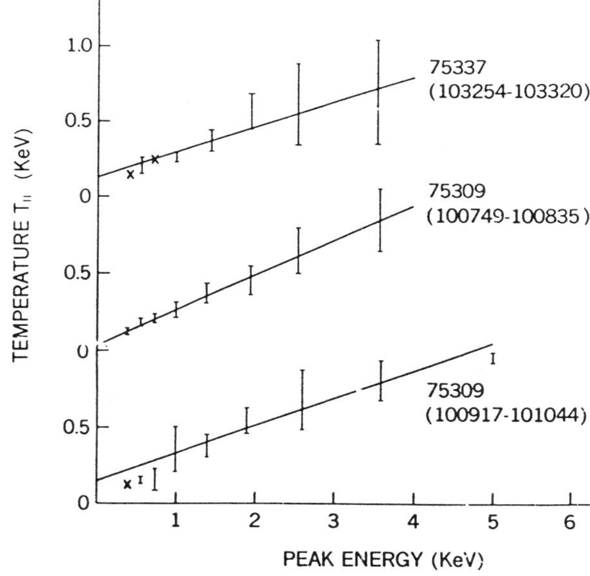

Fig. 7. Electron temperature versus peak energy for three inverted-V events (from Lin and Hoffman, 1979a).

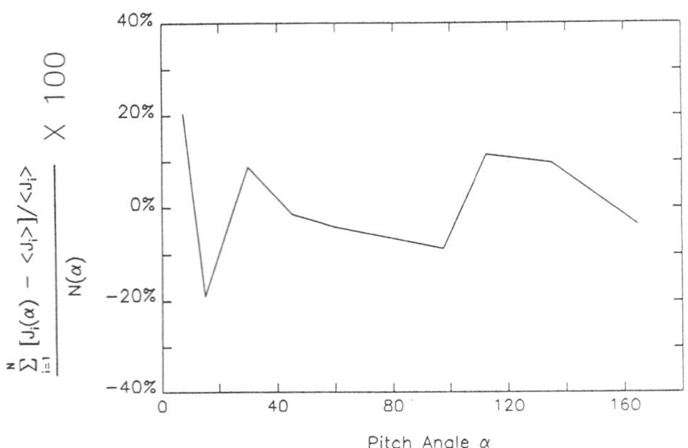

Fig. 8. Variation in percent of the flux at 85 eV secondary electrons as a function of pitch angle, averaged over 106 inverted-V events (from Fung and Hoffman, 1988).

potential region is required to populate this forbidden pitch angle domain.

Double Peaked Spectra

This is probably the most fascinating discrepancy. Arnoldy et al. [1974] first clearly identified the existence of secondary peaks in the spectrum usually with intensities greater than the primary peak (Figure 9). Subsequent observations by our Atmosphere Explorer experiment [Lin and Hoffman, 1979a] and more recently by the Berkeley group with very high resolution angle/energy measurements [McFadden et al., 1990] confirm that these secondary peaks only exist at small pitch angles, and are transient in comparison with the primary peaks. Their maximum energies can be nearly that of the primary peak.

The dual sounding rocket high resolution Berkeley data, shown in Figure 10, and the recent extremely high resolution

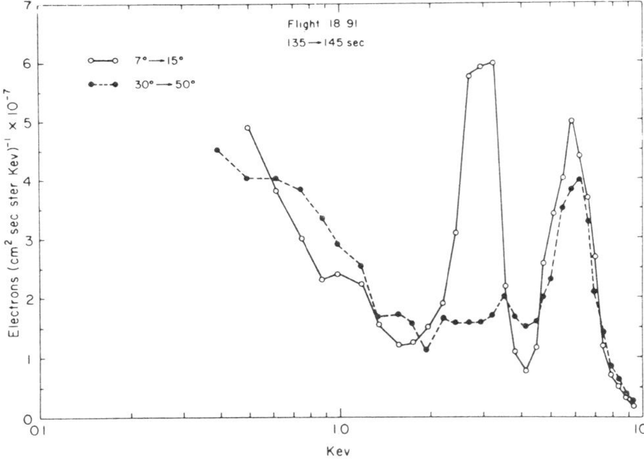

Fig. 9. Example of secondary peak in inverted-V spectra at small pitch angles (from Arnoldy et al., 1974).

ARCS4 sounding rocket of Arnoldy et al. [1990] show extreme structure within the inverted-V, with dispersive and nondispersive field-aligned bursts prominent in the small pitch angle data. But note that the basic inverted-V structure remains relatively steady, seemingly oblivious to the antics of the lower energy electrons.

Since the secondary peak is very field-aligned, it argues for a source close to the spacecraft, simultaneously with a source extending to considerable distances to produce the isotropic primary peak. There is some evidence that this situation exists primarily near the edges of the inverted-V structures. McFadden et al. [1990] argue that an auroral arc potential structure that consists of two separate potential drops, with cold plasma convected between them, is consistent with these measurements.

Ion Energy Different From Electron Energy

The Dynamics Explorer pair of spacecraft placed into coplanar polar orbits provided a planned unique opportunity to search for the presence and extent of the potential region by

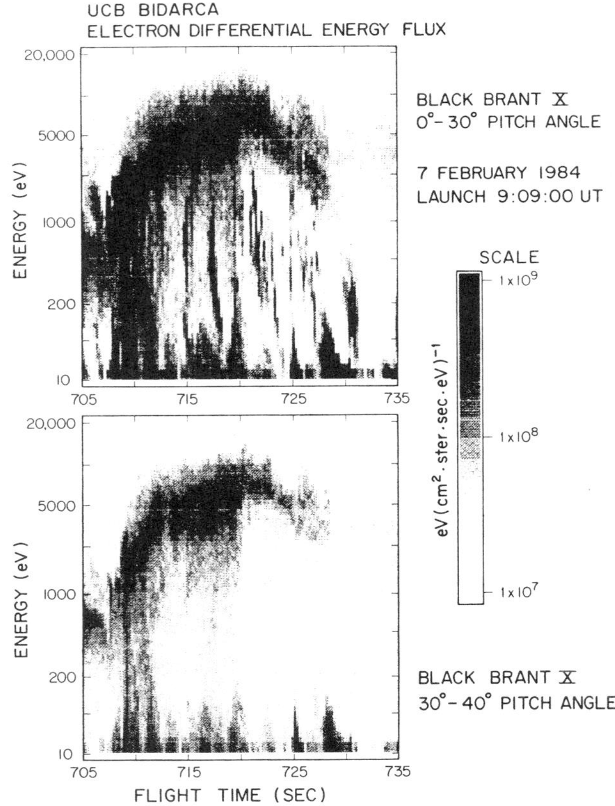

Fig. 10. Very high time resolution electron spectrograms from a sounding rocket flight through an inverted-V event, showing fine structure of dispersive and non-dispersive fluxes at small pitch angles below the primary electrons (from McFadden et al., 1990).

sampling, nearly simultaneously, the velocity-space distributions of ions and electrons at two altitudes during near magnetic conjunctions. Reiff led an extensive analysis of charged particle data from the two spacecraft, using the auroral images to assure the temporal stability and general form of the auroral features crossed [Reiff et al., 1988]. She used several techniques to obtain the total potential difference. The total potential was obtained from the peak in the electron spectrum at DE 2. No potentials have been observed below the DE-2 altitude range from 300 to 1000 km [Fung and Hoffman, 1991]. The potential below the DE-1 satellite was obtained by the enlargement of the electron loss cone for upcoming electrons, due to the fact that they are loosing parallel energy to the potential, causing the pitch angles to spread. Just as for DE 2, the potential difference above DE 1 could be obtained from a peak in the downcoming electron distribution. Finally, the upcoming ions displayed a peak in their spectrum from their acceleration, but as will be discussed, the ion average energy was also determined.

For the conjunction passes, the total potential differences should be the same, independent of technique. First it should be pointed out that the loss cone method is quite imprecise because of the difficulty in defining its edge from the particle distribution measurements. Uncertainties approaching 50% in the measurement of the potential are typical.

A conjunction case for which the potential above DE 1 was near zero is shown in Figure 11. The inverted-V potential from DE 2 displays a single region with the classical increase and decrease in potential. The loss cone method from DE 1 data shows a major region narrower than seen by DE 2, with a second narrow region within the single DE 2 region. The ion peak energy follows the loss cone profile, but at considerably lower magnitudes. Thus the basic data show apparent discrepancies with the auroral potential model in what one would hope to be the definitive experiment.

But these measurements need some interpretation. First, Reiff looked at the ion situation and found that the ions had unusually high temperatures for having been accelerated out of the ionosphere (Figure 12). We see that the velocity space distributions have a break at very nearly the same energy (E peak) independent of mass, as they should, and the high energy tails can be represented by a temperature. This is the distribution one expects if an ionospheric source of particles is heated within or above the acceleration region. These thermal energies varied considerably, but were typically 25% of the electron beam energy.

When the ion average energy was calculated and compared with the energies calculated by the loss cone method, instead of using the ion peak energies, the agreement was remarkably good, both in magnitude and structure for three of the four cases studied (Figure 13). Thus it appears that we need to add an ion heating mechanism also to the auroral potential region.

The second interpretative comment is that the two data sets were acquired about 4 minutes apart. Earlier the temporal stability of inverted-V events was studied using data from the two satellites when the DE 1 perigee was near the southern

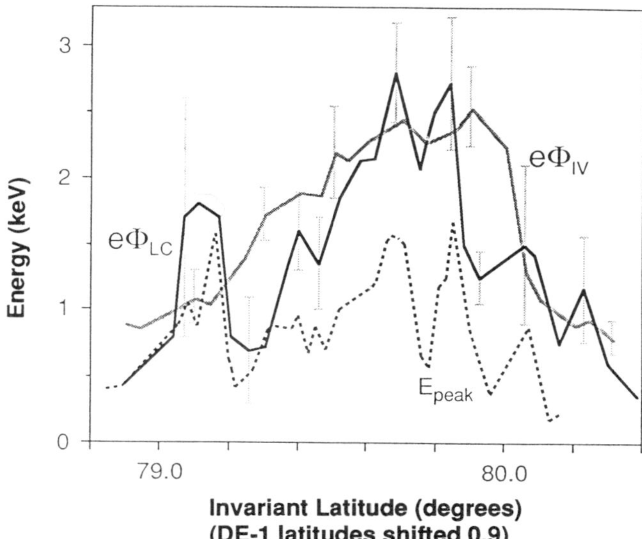

Fig. 11. Potentials (in keV) across an inverted-V measured during a conjunction pass between the DE 1 and DE 2 by three methods (from Reiff et al., 1988). Φ_{IV} is the potential above DE 2 from the peak in the spectrum, Φ_{LC} the potential below DE 1 from the loss cone method, and Φ_i the potential below DE 1 obtained from the upgoing ions. The shift in latitude between the data from the two spacecraft is required to better align the data, and is attributed to the slightly different local times of the two spacecraft orbits which were not traveling transverse to the arc.

auroral oval, so that the two satellites had similar altitudes [Thieman and Hoffman, 1985]. Passages through commonly located inverted-V events provided a measure of the temporal stability of the existence of inverted-Vs, as well as stability of the peak potential (Figure 14). For such pairs of observations, most of the time the potential remained constant, but could increase or decrease. What isn't shown is the fact that in the time interval between the satellite passages through the same location, an inverted-V could either change characteristics sufficiently so as not to be identified as the same event by the other satellite, or could appear or disappear. Thus temporal variations in the inverted-V potential can explain the discrepancies between the DE 1 and DE 2 potential profiles.

Ion Energy Dependent Upon Mass

A final objection to the potential model is the apparent difference in the mean energies of ions with different masses after they have been energized. An electrostatic potential would energize all singly charged particles by the same amount. Ghielmetti et al. [1987] had shown that the mean energies of oxygen and hydrogen measured above an apparent potential region, which was detected by the widening loss cone method for electrons, were quite different (Figure 15). Reiff and coworkers found the same property, which would be compatible

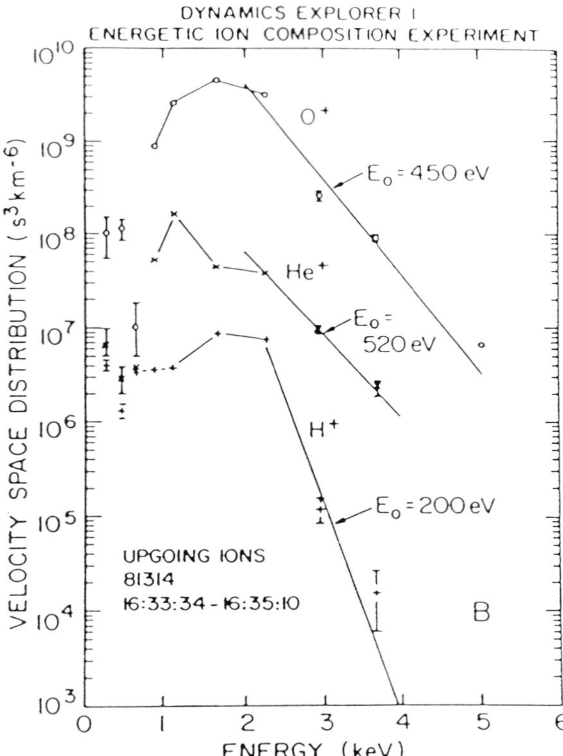

Fig. 12. Upgoing ion distribution functions measured during a conjunction pass between DE 1 and DE 2 for three ion species, with the temperatures of the different ions labeled in eV (from Reiff et al., 1988).

with a two-stream instability, where the faster moving hydrogen would give energy to the heavier oxygen [Kaufmann and Ludlow, 1986]. Thus we add still another mechanism to the auroral potential region.

CHEMICAL RELEASES

A number of investigators have released vaporized barium from sounding rockets and, in two cases, satellites for the purpose of probing the auroral potential region. The barium cloud ionizes in sunlight, and continues to radiate by resonance scattering of sunlight. The dominant emission lines from both neutrals and ions are in visual wavelengths, making possible ground-based optical observations of the barium motion. The barium ions spiral up the magnetic field due to their initial parallel velocity and the magnetic mirror force. The initial velocity distribution is measured immediately following the release, and the motion of the ions in the magnetic and gravitational fields can be calculated very accurately. Thus any deviations from this motion, especially of the leading edge of the jet, can be attributed to electric fields.

Unfortunately, none of these experiments have unambiguously shown the existence of a large and extended electric field parallel to the magnetic field. In each case when a

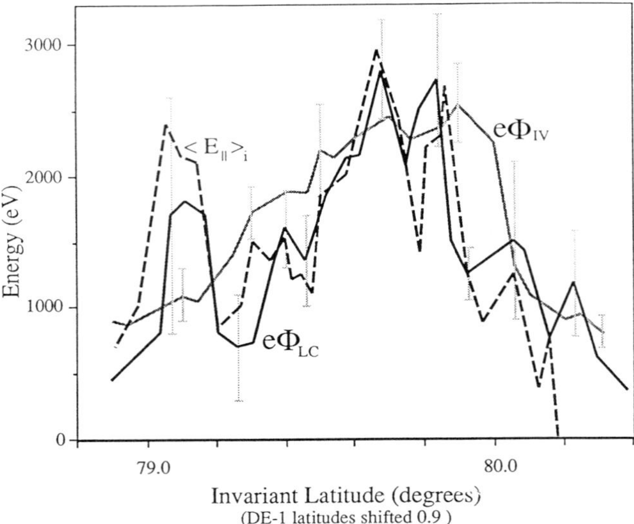

Fig. 13. Potentials (in keV) across the same inverted-V from which the data in Figure 11 are shown, but with the ion average energy $<E>_i = E(peak) + 2 E_0$ plotted instead of just E(peak) (from Reiff et al., 1988).

large potential was reported, the interpretation of the ground-based optical images depended upon a major assumption, or was based on a single data point. The most promising release was the shaped charge experiment named Limerick of the University of Alaska [Stenbaek-Nielsen et al., 1984]. In this case the high altitude tip of the barium jet moved up along the field line as expected for its initial velocity of 11 km/sec with gravity as the only force. But after 14 minutes at an altitude of 8100 km the observable barium tip altitude ceased increasing, but didn't brighten (which would indicate a pile-up of the ions), suggesting a sudden removal of barium from the tip (Figure 16). The only plausible explanation is a removal of ions by a sudden acceleration upwards from an upward dc electric field,

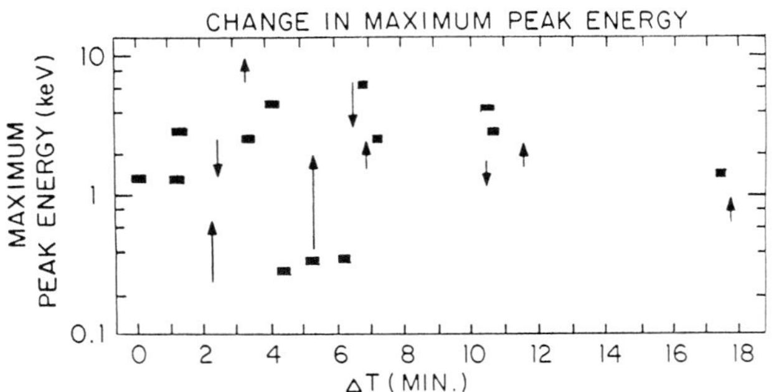

Fig. 14. The maximum peak energy (short bars) or the change in this energy (arrows) of inverted-V events measured by both the DE 1 and DE 2 spacecraft at about the same altitude, but at slightly different times ΔT (in minutes) (from Thieman and Hoffman, 1985).

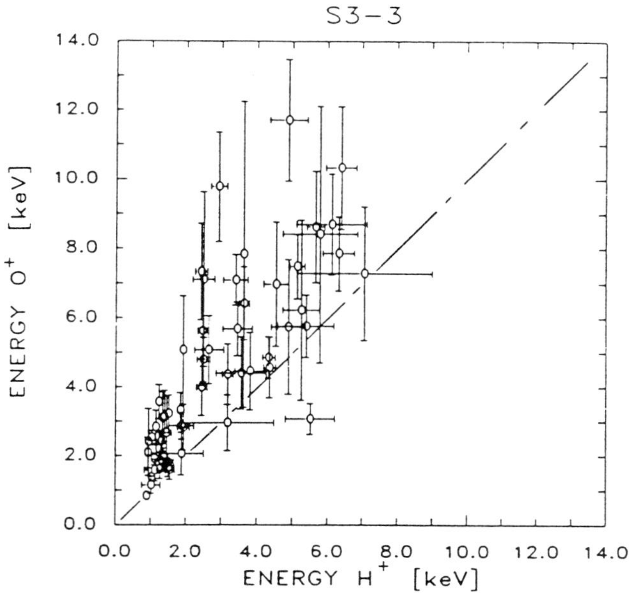

Fig. 15. Scatter plot of average energy of H^+ and O^+ upward flowing ions within individual inverted-V events. Dashed line indicates equal energy condition (from Ghielmetti et al., 1987).

producing a large decrease in the density of the ions. The potential required to produce the tip density reduction over a distance of about 200 km is in excess of 1 keV, giving a field magnitude of at least 5 mV/m.

Quite surprisingly, none of these analyses utilized modeling of the barium distribution along the portion of the field line containing the purported electric potentials to determine whether the optical emissions were consistent with the observations of unusual barium behavior. In all cases, also, the barium was released at altitudes well below where we now think is the expected lower boundary of the potential region, and it was often difficult to determine whether the barium ions, which had moved up the field lines, were on lines conjugate to an aurora. This powerful and very sensitive technique shows much promise for the determination of the electric field distribution along the magnetic field, something that in-situ charged particle measurements can only determine very crudely. However, the releases must be performed much closer to the lower boundary of the potential region and the barium emission distributions must be simulated to prove that they are truly consistent with the existence of a parallel electric field acting on the barium ions.

Conclusions

In summary, then, what do charged particles tell us about the auroral potential region? When charged particle data are properly selected and carefully analyzed, they reveal the following:

1. There is an auroral potential region. There doesn't seem to be any fundamental argument for abandoning this concept.

2. The primary potential region is large scale extending along magnetic field lines from roughly altitudes of one Earth radius to two to three Earth radii.

3. Superimposed on the primary region may be:
   - low altitude transient potentials;
   - ion and electron heating mechanisms;
   - waves to pitch angle scatter;
   - energy exchange mechanisms to transfer energy between ion species.

Perhaps the requirement to superimpose other physical mechanisms onto the primary potential region is not surprising, and in fact, should rather be expected, since the resulting particle distributions arising from a parallel acceleration should be unstable to various wave growths. Lundin and Eliasson [1991] also point out effects of possible fluctuating field-aligned potentials and their probable importance for creating some of the other unusual signatures seen in the particle distributions. These effects have not been included in this review.

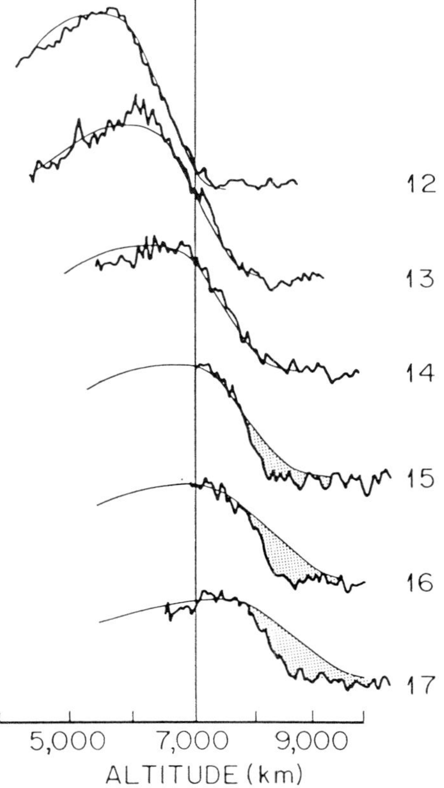

Fig. 16. Relative intensity traces of a barium ion jet near its tip between 12 and 17 minutes after the release. The heavy line gives the intensity traces derived from ground-based TV data, while the smooth line is the intensity calculated based on the initial velocity distribution and motion along the magnetic field with only gravitational force. The "missing" barium is indicated by the shading (from Stenbaek- Nielsen, 1984).

Future work on this subject requires three directions of investigation:

1. Continued careful, systematic analyses of the large body of excellent data which already exist, in order to sort out the true characteristics of inverted-V electron precipitation events and distinguish them from the effects of other processes.

2. More very high resolution data within and below the events, which should be acquired by the upcoming FAST SMEX satellite [Temerin et al., 1990].

3. A new approach to probing the potential region for details of the electric field distribution, such as with barium releases into the lower portion of the expected potential region based on signatures of the existence of a potential region from real-time telemetered data of charged particle distributions, and accompanied by modeling of the effects of the fields on the resulting injected ion distributions.

REFERENCES

Arnoldy, R. L., and L. W. Choy, Auroral electrons of energy less than 1 keV observed at rocket altitudes, *J. Geophys. Res., 78*, 2187, 1973.

Arnoldy, R. L., P. B. Lewis, and P. O. Isaacson, Field-aligned auroral electron fluxes, *J. Geophys. Res., 79*, 4208, 1974.

Arnoldy, R, L., K. Lynch, M. Popecki, P. M. Kintner, L. J. Cahill, Jr., T. E. Moore and C. J. Pollock, ARCS 4 sounding rocket high time resolution auroral electron distribution function measurements, abstract, *EOS, 71*, 1555, 1990.

Bryant, D A., Auroral electron acceleration, *Physica Scripta, T30*, 215, 1990.

Bryant, D. A., D. S. Hall and D. R. Lepine, Electron acceleration in an array of auroral arcs, *Planet. Space Sci., 26*, 81, 1978.

Bryant, D. A., D. S. Hall and R. Bingham, Auroral electron acceleration: a case for the stochastic alternative, *Auroral Physics*, ed. C.- I. Meng, M. J. Rycroft and L. A. Frank, Cambridge University Press, p. 119, 1990.

Burch, J. L., S. A. Fields, W. B. Hanson, R. A. Heelis, R. A. Hoffman and R. W. Janetzke, Characteristics of auroral electron acceleration regions observed by Atmosphere Explorer C, *J. Geophys. Res., 81*, 2223, 1976.

Cladis, J. B., and R. D. Sharp, Scale of electric field along magnetic field in an inverted-V event, *J. Geophys. Res., 84*, 6564, 1979.

Evans, D. S., The observations of a near monoenergetic flux of auroral electrons, *J. Geophys. Res., 73*, 2315, 1968.

Evans, D. S., Precipitating electron fluxes formed by a magnetic field aligned potential difference, *J. Geophys. Res., 79*, 2853, 1974.

Frank, L. A, and K. L. Ackerson, Observations of charged particle precipitation into the auroral zone, *J. Geophys. Res., 76*, 3612, 1971.

Fung, S. F., and R. A. Hoffman, On the spectrum of the secondary auroral electrons, *J. Geophys. Res., 93*, 2715, 1988.

Fung, S. F., and R. A. Hoffman, A search for parallel electric fields by observing secondary electrons and photoelectrons in the low-altitude auroral zone, *J. Geophys. Res., 96*, 3533, 1991.

Ghielmetti, A. G., E. G. Shelley and D. M. Klumpar, Correlation between number flux and energy of upward flowing ion beams, *Physica Scripta, 36*, 362, 1987.

Kaufmann, R. L. and G. R. Ludlow, Interaction of H+ and O+ beams: observations at 2 and 3 Re, *Ion Acceleration in the Magnetosphere and Ionosphere,* ed. Tom Chang, Geophys. Monograph 38, American Geophysical Union, p. 92, 1986.

Lin, C. S., and R. A. Hoffman, Characteristics of inverted-V events, *J. Geophys. Res., 84*, 1514, 1979a.

Lin, C. S., and R. A. Hoffman, Fluctuations of inverted-V electron fluxes, *J. Geophys. Res., 84*, 6547, 1979b.

Lundin, R. and L. Eliasson, Auroral energization processes, *Ann. Geophysicae 9*, 202, 1991.

Maehlum, B. N., and H. Moestue, High temporal and spatial resolution observations of low energy electrons by a mother-daughter rocket in the vicinity of two quiescent auroral arcs, *Planet. Space Sci., 21*, 1957, 1973.

McFadden, J. P., C. W. Carlson and M. H. Boehm, Structure of an energetic discrete arc, *J. Geophys. Res., 95*, 6533, 1990.

Reasoner, D. L., and C. R. Chappell, Twin payload observations of incident and back-scattered auroral electrons, J. Geophys. Res., 78, 2176, 1973.

Reiff, P. H., H. L. Collin, J. D. Craven, J. L. Burch, J. D. Winningham, E. G. Shelley, L. A. Frank, and M. A. Friedman, Determination of auroral electrostatic potentials using high- and low-altitude particle distributions, *J. Geophys. Res., 93*, 7441, 1988.

Sharp, R. D., E. G. Shelley, R. G. Johnson and A. G. Ghielmetti, Counterstreaming electron beams at altitudes of ~1 Re over the auroral zone, *J. Geophys. Res., 85*, 92, 1980.

Stenbaek-Nielsen, H. C., T. J. Hallinan and E. M. Wescott, Acceleration of barium ions near 8000 km above an aurora, *J. Geophys. Res., 89*, 10,788, 1984.

Temerin, M. A., C. W. Carlson, C. A. Cattell, R. Ergun, J. P. McFadden, F. S. Mozer, D. M. Klumpar, W. K. Peterson, E. G. Shelley and R. C. Elphic, Wave-particle interactions on the FAST satellite, in *Physics of Space Plasmas*, ed. T. Chang, G. B. Crew and J. R. Jasperse, Scientific Publishers, Inc., p. 343, 1990.

Thieman, J. R., and R. A. Hoffman, Determination of inverted-V stability from Dynamics Explorer satellite data, *J. Geophys. Res., 90*, 3511, 1985.

Whalen, B. A., and P. W. Daly, Do field-aligned auroral particle distributions imply acceleration by quasi-static parallel electric fields? J. Geophys. Res., 84, 4175, 1979.

Whalen, B. A., and I. B. McDiarmid, Observations of magnetic-field-aligned auroral electron precipitation, *J. Geophys. Res., 77*, 191, 1972.

Winckler, J. R., and L. Peterson, Large auroral effect on cosmic-ray detectors observed at 8 g/cm atmospheric depth, *Phys. Rev., 108*, 903, 1957.

Winckler, J. R., L. Peterson, R. Arnoldy, and R. Hoffman, X rays from visible aurorae at Minneapolis, *Phys. Rev., 110*, 1221, 1958.

R.A. Hoffman, Laboratory for Exterrestrial Physics, Goddard Space Flight Center, Greenbelt, MD 20771.

On the High- and Low-Altitude Limits of the Auroral Electric Field Region

P. H. Reiff,[1] G. Lu,[2] J. L. Burch,[3] J. D. Winningham,[3] L. A. Frank,[4] J. D. Craven,[5]
W. K. Peterson,[6] and R. A. Heelis [7]

Using measurements from the High Altitude Plasma Instrument (HAPI) on the Dynamics-Explorer 1 (DE-1) spacecraft and the Low Altitude Plasma Instrument (LAPI) on Dynamics Explorer 2 (DE 2), we investigate both the high altitude and low altitude extents of the auroral acceleration region. To infer the high altitude limit, we searched the HAPI data base for evidence of upward-directed auroral electric fields located above the spacecraft when the HAPI spacecraft is above 9000 km altitude. We find that such acceleration is common when DE-1 flies through the auroral oval at an altitude of 9,000-11,000 km. At altitudes above 11,000 km, the fraction of the orbits with evidence of at least a 1000 V potential drop above the spacecraft falls, becoming essentially zero above an altitude of 15,000 km. Above that altitude, small (100 V) potential drops are frequently observed, but only rarely are ~1 kV potentials observed, typically associated with polar cap or "theta" arcs or westward traveling surges. To investigate the low-altitude limit of the auroral acceleration region, we use conjunctions of DE 1 and DE 2 along auroral field lines and match the upgoing fluxes of ionospheric ions observed by DE 2 with the flux of accelerated upgoing ions observed at DE 1. Calculating the ionospheric scale height from the ion and electron temperatures and assuming that the parallel flow velocity is independent of height above 800 km, we calculate the altitude at which the upwelling ionospheric ions are effectively completely lost to upward acceleration. The initial lowest-altitude acceleration process could be either a perpendicular acceleration or a parallel electric field, but it must be sufficient to give the entire distribution escape energy. We find that in the two cases studied, near the region of peak auroral potential drop the altitude of this acceleration was around 1700 km (near the O/H neutral crossover altitude), but was significantly higher (~2000 km) near the edges of the arc, where the potential was lower. The composition of the upgoing ion beam was consistent with these heights, being predominately H^+ near the edges and O^+ near the peak.

INTRODUCTION

The S3-3 spacecraft was the first to routinely fly through the altitude range of the auroral electric field and infer auroral electric fields both from the electric field and particle instruments. From the particle distributions, Mizera and Fennell [1977] showed that, for a case at 7300 km, roughly

1/3 of the 3 kV total potential drop was located above the spacecraft. The potential drop above the spacecraft "Φ_{HI}" was inferred from precipitating electrons and the potential below the spacecraft "Φ_{MID}" was inferred by upgoing ions to yield the total potential drop "Φ_{TOT}" (terms introduced in Reiff et al. [1986]). Mozer et al. [1980] showed that electrostatic shocks, which they suggest can account for the entire auroral parallel electric field, can be observed as low as 1000 km and increase in occurrence frequency with altitude, up to the 8000 apogee of S3-3. They showed that frequent observations of these shocks start at an altitude of about 1/2 R_E, consistent with contemporary models of the auroral acceleration region [Kindel and Kennel, 1971]. From ISEE-1 data, Mozer [1981] extended that study to show that electrostatic shocks can be observed up to 7 R_E altitude, but with decreasing frequency and intensity. The purpose of this paper is to use the Dynamics Explorer 1 spacecraft to probe that upper altitude limit. In addition, we use magnetic conjunctions of DE 1 and DE 2 to investigate the low altitude limit. For the high altitude limit, we performed a statistical analysis; for the low altitude limit, we examine in detail two auroral conjunctions. This first portion of the study is contained in more detail in Reiff et al.

[1]Department of Space Physics and Astronomy, Rice University, Houston, Texas
[2]HAO, NCAR, Boulder, Colorado
[3]Southwest Research Institute, San Antonio, Texas
[4]Department of Physics and Astronomy, University of Iowa, Iowa City, Iowa
[5]Geophysical Institute and Department of Physics, University of Alaska, Fairbanks, Alaska
[6]W. K. Peterson, Lockheed, Palo Alto, California
[7]R. A. Heelis, Center for Space Science, University of Texas at Dallas, Richardson, Texas

Auroral Plasma Dynamics
Geophysical Monograph 80

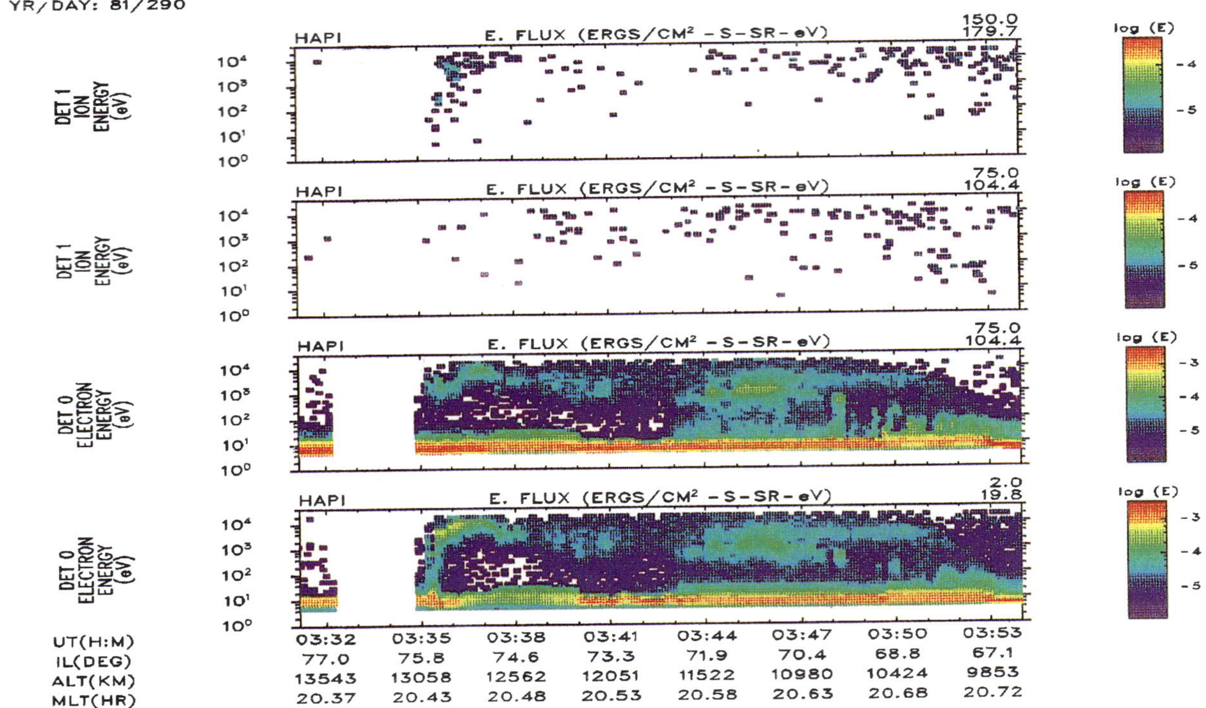

DYNAMICS EXPLORER
JOINT DATA PRESENTATION
YR/DAY: 81/290
HAPI   E. FLUX (ERGS/CM² − S − SR − eV)   150.0 179.7
log (E)
DET 1 ION ENERGY (eV)
HAPI   E. FLUX (ERGS/CM² − S − SR − eV)   75.0 104.4
DET 1 ION ENERGY (eV)
HAPI   E. FLUX (ERGS/CM² − S − SR − eV)   75.0 104.4
DET 0 ELECTRON ENERGY (eV)
HAPI   E. FLUX (ERGS/CM² − S − SR − eV)   2.0 19.8
DET 0 ELECTRON ENERGY (eV)
UT(H:M)   03:32   03:35   03:38   03:41   03:44   03:47   03:50   03:53
IL(DEG)   77.0   75.8   74.6   73.3   71.9   70.4   68.8   67.1
ALT(KM)   13543   13058   12562   12051   11522   10980   10424   9853
MLT(HR)   20.37   20.43   20.48   20.53   20.58   20.63   20.68   20.72

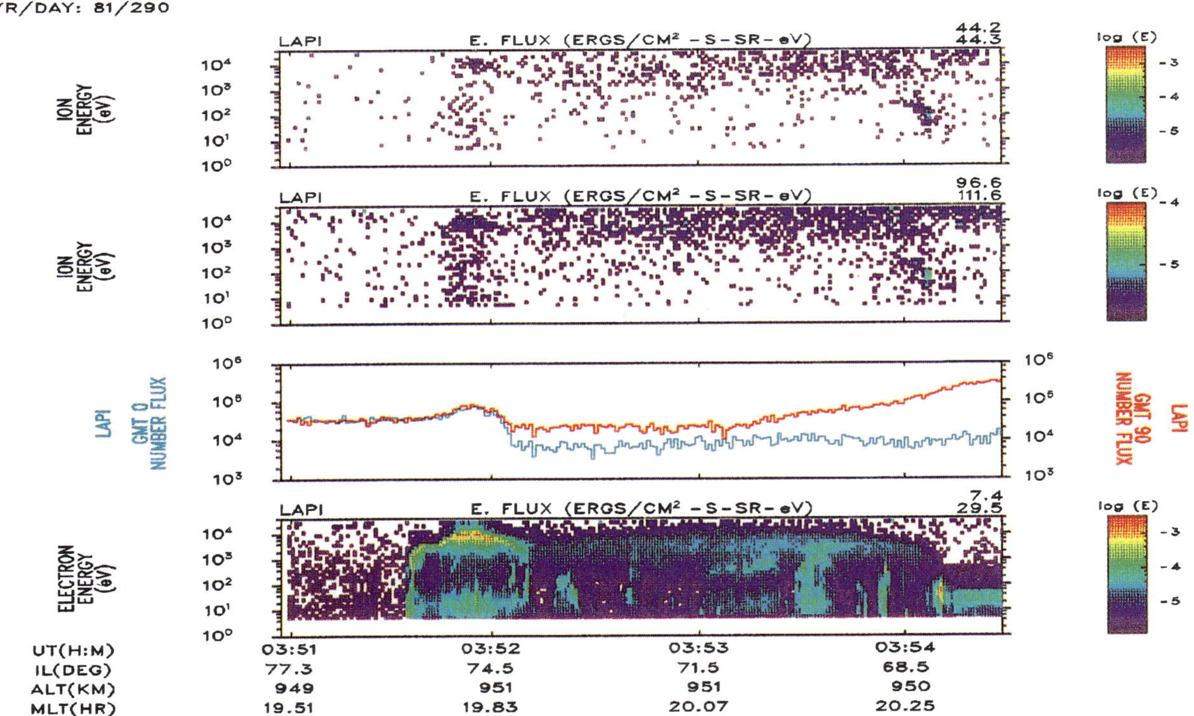

DYNAMICS EXPLORER
JOINT DATA PRESENTATION
YR/DAY: 81/290
LAPI   E. FLUX (ERGS/CM² − S − SR − eV)   44.2 44.3
log (E)
ION ENERGY (eV)
LAPI   E. FLUX (ERGS/CM² − S − SR − eV)   96.6 111.6
ION ENERGY (eV)
LAPI   GMT 0 NUMBER FLUX
LAPI   GMT 90 NUMBER FLUX
LAPI   E. FLUX (ERGS/CM² − S − SR − eV)   7.4 29.5
ELECTRON ENERGY (eV)
UT(H:M)   03:51   03:52   03:53   03:54
IL(DEG)   77.3   74.5   71.5   68.5
ALT(KM)   949   951   951   950
MLT(HR)   19.51   19.83   20.07   20.25

[1993]; the second portion of the study is contained in Lu et al. [1992]; thus only a relatively brief summary will be presented here.

APPROACH

The Dynamics Explorer 1 spacecraft, with a 23,000 km apogee is ideally suited to study the high altitude limit of the auroral electric field. The HAPI instrument [Burch et al., 1981] is an ion-electron spectrometer using 5 electron and 5 ion coplanar detectors which covers ~2.8π steradian of look directions in a six-second spin. As a practical matter, the highest and lowest pitch angles are typically observed in the detectors mounted on the spacecraft equatorial plane. The highest pitch angle resolution is obtained by using data from two consecutive spins. The Low-Altitude Plasma Instrument (LAPI) on the non-spinning DE 2 spacecraft [Winningham et al., 1981] is an essentially similar instrument; pitch angle information is obtained from an array of detectors mechanically slewed to fixed angles with respect to the magnetic field direction.

For this study, we first examined the DE 1 microfiche database for all orbits that reached an altitude of at least 9000 km in the auroral zone. For this study we operationally defined the auroral zone as being from 60-75 degrees invariant latitude on both the morning and evening sides (the orbital plane was roughly 1000-2200 MLT ($\pm$ ~2 hours) during the lifetime of the HAPI instrument). Orbits were classified as "yes" (the electron energy flux spectrogram showed evidence of downward electron acceleration of at least 1 keV), "no," or "maybe." "Maybe" orbits were examined in detail using the distribution function line plots at 0-15° pitch angle, to determine whether acceleration over 1 keV existed, using the technique of Reiff et al. [1988]. Data from the part of the crossing above 75° invariant were also examined, if available. A 1 keV threshold was chosen so that auroral acceleration could be readily judged from the microfiche summary plots, and could be confirmed by optical measurements from the imager.

Orbits were classified as "no" only if data were obtained during the entire auroral oval crossing (60-75° invariant) without evidence of $\Phi_{HI} > 1$ kV. The altitude listed for the "no" orbits was the altitude of the spacecraft where it crossed 70° invariant latitude on the evening side. The altitude used for the "yes" orbits was the altitude of the maximum Φ_{HI} (which could have occurred anywhere in the 60-85° range, but was typically near 70°).

We emphasize that the fraction of the potential that occurs at high altitudes is not a constant for a particular orbital pass.

Plate 1 (top) shows a HAPI energy flux spectrogram for day 290 (October 17, 1981), with upgoing (150-180° pitch angle) ions on top and down coming (0-20° pitch angle) electrons on bottom. The middle two panels show mirroring particles, with the upper being ions and lower being electrons. The energies $e\Phi_{HI}$ (shown dotted) and $e\Phi_{TOT}$ (e times the total potential drop) derived for this pass (shown solid) are given in Figure 1a (left side). Here $e\Phi_{HI}$ is estimated from examining the peak in the zero-pitch angle distribution function from line plots of the precipitating electrons. Similarly $\Phi_{TOT} = \Phi_{HI} + \Phi_{MID}$, where Φ_{MID}, the potential drop between the spacecraft and the ionosphere, is estimated from the energy of the upgoing ion beam (also from line plots of the zero-pitch-angle distribution functions). The fraction of the potential that occurs above DE-1 (Φ_{HI}/Φ_{TOT}) varies considerably through the pass, and perhaps reflects the fact that the total parallel potential drop is not spread uniformly in altitude but is concentrated in two or more structures separated in altitude [Gurgiolo and Burch, 1988]. If these structures are "double layers," then they can propagate in altitude in a relatively short time [e.g., Block and Fälthammar, 1991 and references therein]. The bottom portion of Plate 1 gives the nearly-simultaneous ion and electron spectrogram from the low altitude spacecraft DE 2, and Figure 1b (right side) shows the total potential drop Φ_{TOT} estimated from the energy of the peak distribution function of LAPI precipitating electrons, again from line plots at zero pitch angle. For both Φ_{TOT} and Φ_{MID}, the uncertainty of the estimated potential is roughly one half energy step or about 16% (if two neighboring steps have approximately the same value of the distribution function, the point was placed halfway between).

RESULTS

In the four months of HAPI data, 60 orbits were found with potentials $\Phi_{HI} > 1$ kV at altitudes > 9000 km. Figure 2a (left) shows the integral number of "no" (light shaded) and "yes" (dark shaded) orbits above a given altitude, with the upper envelope giving the sum. Thus, above ~9000 km, the total "no"(64) and "yes"(60) orbits above that height were nearly the same. As the altitude increases, the fraction of orbits with evidence of at least 1 keV energy of downward acceleration, which we will call HAA for "high altitude acceleration," decreases precipitously, compared to the numbers of "no" orbits above the same height. Above 16,000 km only three orbits were observed with HAA, one of which occurred at an altitude near 20,000 km, one near 22,500 km and one near

---

Plate 1. (top) Energy-time spectrogram from the High Altitude Plasma Instrument (HAPI) on DE 1 for day 81290 (October 17, 1981). The colors indicate differential energy flux (note that the scale differs between the electrons and the ions). The top panel shows upgoing ions (pitch angles pa=165-180°); the second shows mirroring (pa~90°) ions; the third, mirroring electrons; and the bottom, downcoming electrons. A significant downward acceleration is observed in the downcoming electrons at the same time that a strong flux of upgoing ions is observed (0335-0338 UT) indicating a passage of the spacecraft through the center of an auroral acceleration region. (bottom) Energy-time spectrogram for the Low Altitude Plasma Instrument (LAPI) on DE 2 at nearly the same time. Here the top panel is precipitating ions (pa ~45°); then, mirroring ions, next, the Geiger-Mueller tube for mirroring (red) and precipitating (blue) energetic electrons, and on bottom, precipitating electrons.

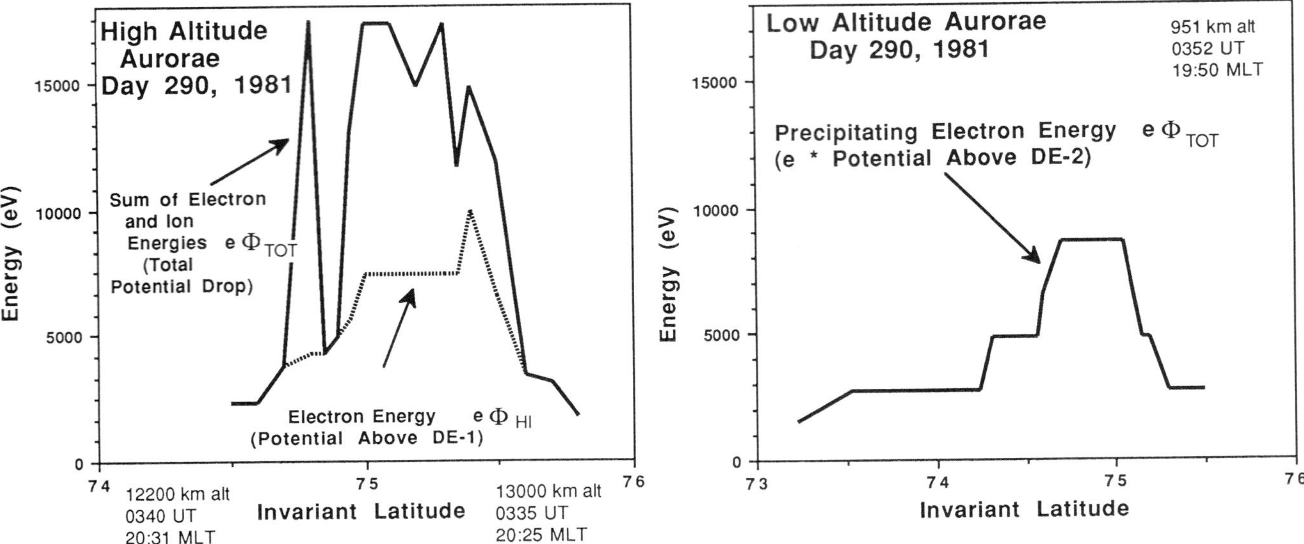

Fig. 1. (left) Field-aligned potential drop inferred from DE 1 HAPI data. The potential above DE 1 (Φ_{HI}, shown dotted) is determined from the energy of the precipitating electrons. The total potential (Φ_{TOT}, shown solid) is the sum of $\Phi_{HI} + \Phi_{MID}$, where Φ_{MID} is calculated from the energy of the upflowing ions. (right) Low-altitude potential drop inferred from DE 2 LAPI data. The total potential is substantially lower than that inferred at high altitudes - however, note the 15 minutes in UT and 30 minutes in MLT separating the passes.

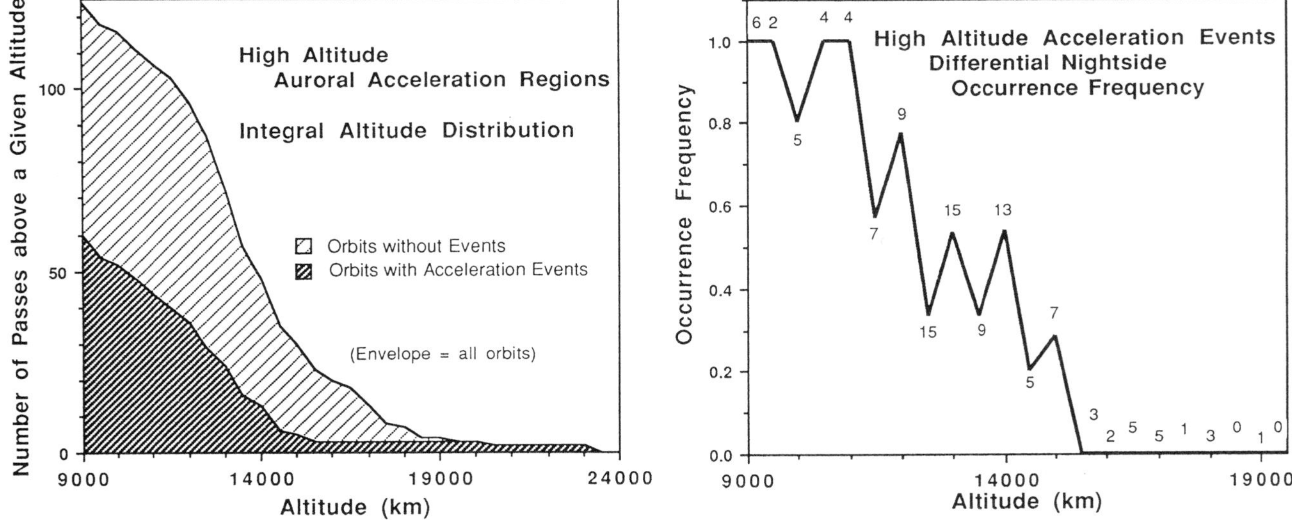

Fig. 2. (left) Integral altitude distribution of orbits with high altitude acceleration (HAA) events (heavy shading) and orbits without high altitude acceleration (light shading). The envelope gives the total number of orbits above a given altitude. The altitude listed for a "no" orbit was the altitude at which the spacecraft passed 70° invariant. In all, 124 orbits passed our criteria, of which 60 had high altitude acceleration and 64 did not. Only three orbits above 16,000 km showed evidence of HAA, and these were all observed on the dayside. (right) Differential occurrence frequency of HAA, in 500-km altitude bins. Only eveningside orbits were considered in these statistics (57 "yes," 64 "no"). The number of total ("yes" plus "no") orbits in each bin is given near the trace.

23,000 km. All but those three events were observed on the evening side. These three morning side passes with $\Phi_{HI} > 1$ kV all occurred in the polar cap during times of strongly northward IMF and appear to be associated with transpolar arcs [Frank et al., 1986]. Since only three of the "yes" orbits occurred on the dayside, the remaining statistics will pertain to evening crossings only. The eveningside "no" orbits totaled 64, meaning that nearly half of the orbits in the DE 1 lifetime (fall of 1981) showed evidence of upward directed electric fields of at least 1 kV total potential drop above DE 1.

To look at differential occurrence frequency (Figure 2b, right) we separated the eveningside orbits into 500 km altitude bins. Thus if the location of the peak potential drop was between 12,750 and 13,250 km (as in Plate 1), the orbit was placed as a "yes" in the 13,000 km altitude bin. The "no" orbits, as above, were placed at the altitude at which the spacecraft crossed 70 degrees invariant. The occurrence frequency of HAA for each altitude bin, calculated as the number of "yes" orbits divided by total number of orbits at that altitude (the sum of "yes" + "no" orbits), is shown as the solid curve. To give the distribution of total orbital coverage, and thus the statistical significance, the number of total passes in that altitude bin ("yes" + "no") is written at the top. (The zero observed distribution above 15,500 km would be meaningless if in fact no "no's" had been seen in that altitude range.) We see that over 95% (20 of 21) of the orbits in the altitude range between 9000 - 11,000 km show at least 1 kV of downward acceleration at least somewhere during the crossing. This occurrence frequency falls rapidly as the altitude rises, to near 50% in the 12,000 to 14,000 km altitude range. No eveningside acceleration events were observed above 16,000 km, even though 20 "no" passes were recorded in this altitude range; thus 16,000 km appears to be the practical high altitude limit for typical auroral electric fields of potential drop > 1 kV. We note, however, that the DE 1 orbit reached its apogee above the dayside auroral oval, and fell in altitude throughout the eveningside auroral oval. Thus we did not sample significantly the evening side above 20,000 km. Had we taken 75°, rather than 70°, to determine the altitude for the "no" events, the average altitude for the "no" cases would have risen somewhat. This orbital effect of falling in altitude through the auroral oval implies that our very highest altitude events are predisposed to occur at the highest latitudes as well. This is borne out in the case of day 290, where the peak potential drop occurs at 75° invariant.

CASE STUDIES OF HIGH ALTITUDE ACCELERATION

The sixteen cases of HAA observed at altitudes $\geq$ 13,000 km became immediately more interesting, and all of them were examined in more detail. Two high altitude events occurred near times of DE 1/DE 2 conjunctions, and these were of particular interest. One of them is day 81290 (October 26, 1981), shown in Plate 1a (top), showing DE 1 particle data, and in Plate 1b (bottom) showing DE 2 data. A second DE 1 / DE 2 conjunction event with high altitude acceleration occurred on day 81325 (November 21, shown in Plate 2).

One observational difficulty in determining the total potential drop at such high altitudes is that any upgoing ionospheric ions have the appearance of a beam (having their perpendicular energies adiabatically changed into parallel energy by the mirror force), even if they were only initially accelerated perpendicularly in a "conic" [Klumpar, 1979]. Thus to check that possibility (that Φ_{MID} was in fact zero), we examined the total potential drop observed at DE 2 [see technique in Reiff et al., 1988]. The total potential at low altitudes for day 290 is shown in Fig 1b (right). One observes that the maximum energy electron observed at DE 2 is only ~9 kV, much less than the total potential inferred at DE 1 (but about the value of either Φ_{HI} or Φ_{MID}, taken separately). This suggests that the upgoing ion beam might have originated as a conic, and therefore, is not due to an electrostatic acceleration. We had previously shown that the potentials observed by DE 1 and 2 at near-conjunctions were nearly the same [Reiff et al., 1988]. We then examined the EICS ion spectrometer data from DE 1 [Shelley et al., 1981], (Plate 3, for day 325), and found that, although the ions do show some conical pitch angles (near 1235 UT), at the peak of the acceleration (from 1236 - 1237 UT), the energy of the upflowing Oxygen (middle panel) is the nearly the same as the upflowing Hydrogen (top panel). The Oxygen energy is slightly larger than the ion energy, which could be attributed to a small amount of perpendicular acceleration or minor heating from the two-stream instability. Examination of the detailed Hydrogen and Oxygen distribution functions shows only a modest difference in energy [W. Peterson, private communication, 1992]. Thus it is much more likely that the bulk of the acceleration is caused by a parallel electric field and not a perpendicular acceleration. (The data from day 290 is similar, but the EICS instrument was not in as favorable an analysis mode.)

Why then is the potential measured by DE 1 so different from that at DE 2? We note that the conjunction was not perfect, that is, at least a half-hour of magnetic local time and nearly fifteen minutes of Universal Time separated the two passes. We then examined the images from the Scanning Auroral Imager (SAI) [Frank et al., 1981]. We immediately observed the reason for the difference in total potential drop: the footprint of the DE 1 spacecraft passed through an auroral enhancement at 0336 UT, possibly associated with a westward traveling surge (Plate 4, left), whereas the DE 2 footprint passed through a less intense aurora at 0352 UT (Plate 4, right). Note that the time of crossings of the (magnetically mapped) footprint of DE 1 and DE 2 through the peak auroral intensity exactly corresponds to the times of peak particle energizations in Plate 1. A more dramatic example occurred for day 81325 (Plate 5): the DE 1 footprint (right) passed exactly through the surge (crossing the peak auroral intensity at 1235 UT) whereas the DE 2 footprint passed just to the west of it, crossing the peak auroral intensity at 1230:30 UT (left). Again the times of peak luminosity from the magnetic mapping agree exactly with the times of the peak particle acceleration in Plate 2. The potentials measured on this day (Figure 3) are similar to that on day 81290: the total potential drop measured by DE 1 is larger (and covers more distance) that of DE 2 at the peak of the inverted V precipitation, although DE 2 did see an extremely high (~19 keV) potential for just one second, at 12:30:36. This extreme potential yielded fluxes which were also observed in the Geiger-Mueller tube (third panel of Plate 2b), as well as false counts in the ion channels (top two panels). In addition, Figure 3 shows the jump in

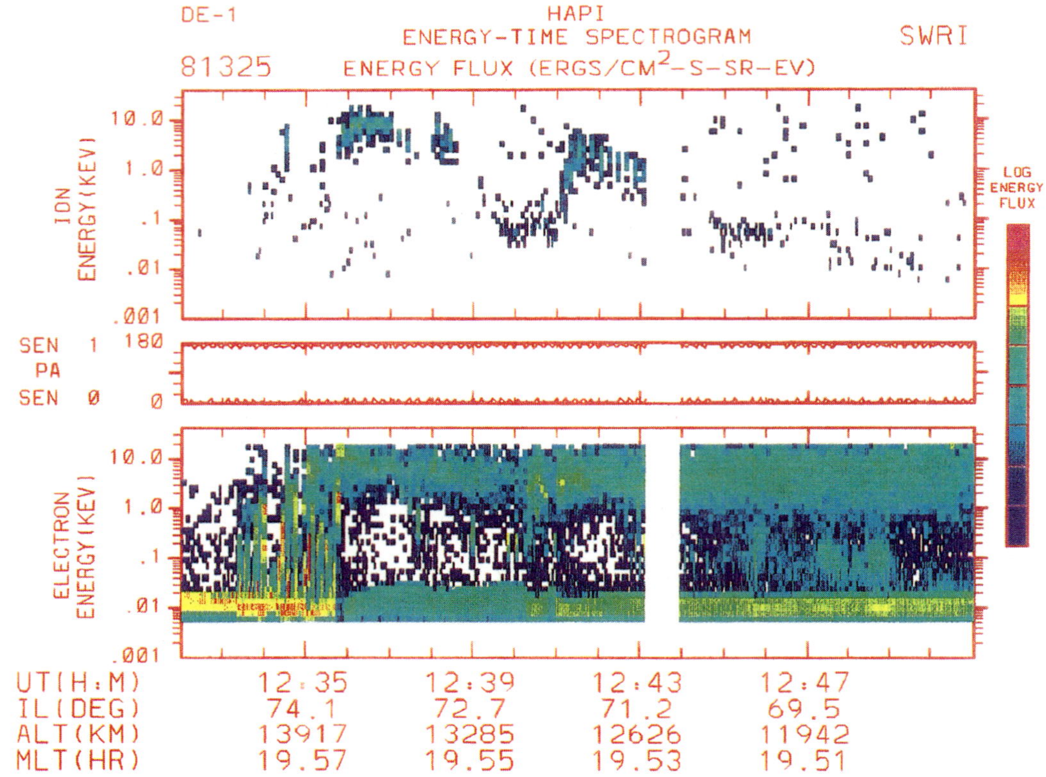

DYNAMICS EXPLORER
JOINT DATA PRESENTATION

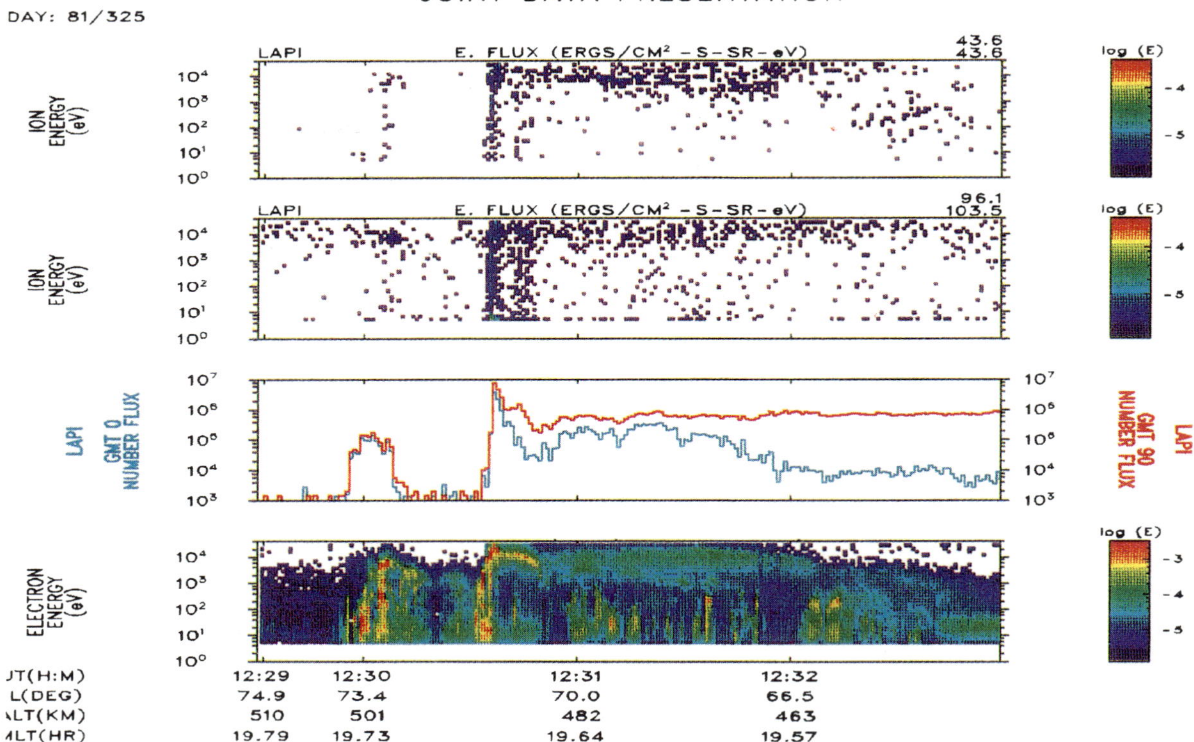

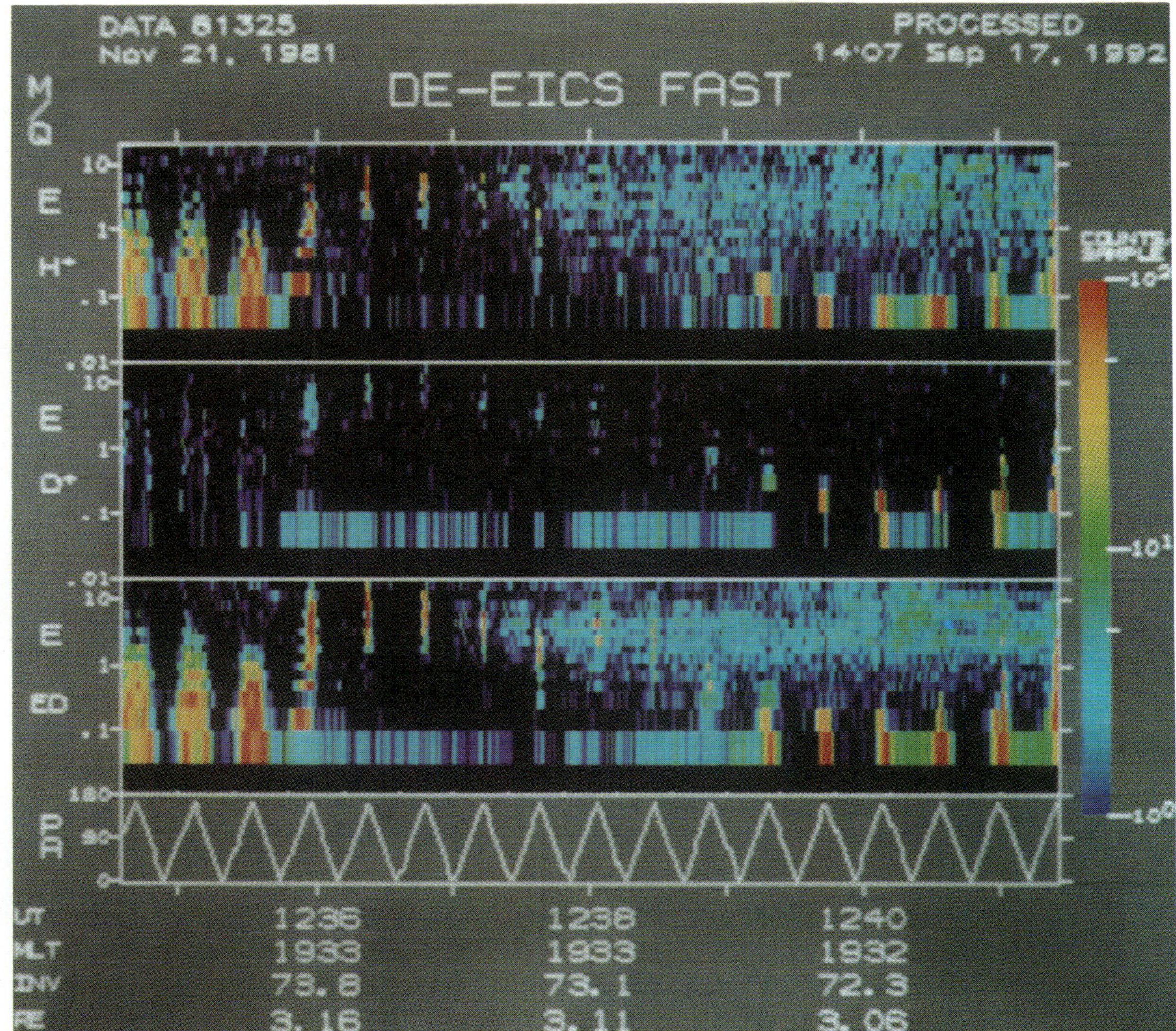

Plate 3. Energy-time spectrogram for the Energetic Ion Spectrometer (EICS) on DE 1 for Nov. 21, 1981. Hydrogen is shown on top, singly-charged Oxygen in the center. The lowest panel shows the pitch angle: thus the upgoing keV ions from 1236-1239 are principally Hydrogen, but with an Oxygen component of only slightly higher energy.

latitude of the precipitation with the passage of the bulge: from near 71 degrees at DE 2 to 73.5 degrees at DE 1, also visible in the images (Plate 5).

Thus it is likely that occurrences of very high altitude acceleration are most common during substorms, and more specifically during westward traveling surges. With our data sets alone we cannot tell whether this additional acceleration is relatively nearby or whether it occurs well out along the field line. However, the study of Gurgiola and Burch [1988] indicate that such significant separation is not uncommon. It has been noted that a tail x-line should have a parallel electric field along it [Birn and Hesse, 1991]; thus it is possible that the highest-altitude ~10 kV of auroral potential occurs out in the distant magnetotail; the DE 1 orbit merely crossed that

Plate 2. (top) HAPI energy-time spectrogram for day 81325 (Nov. 21, 1981). Upgoing ions are on top, and downcoming electrons on bottom. (bottom) LAPI spectrogram for day 81325, in the same format at in Plate 1 (bottom). HAPI flew through the peak of the acceleration region at 1235-1237 UT; LAPI at 1230:40. The high Geiger fluxes indicate a very large total potential drop at LAPI; the ions observed at all energies at the same time are most likely crosstalk from the extremely high flux of energetic electrons.

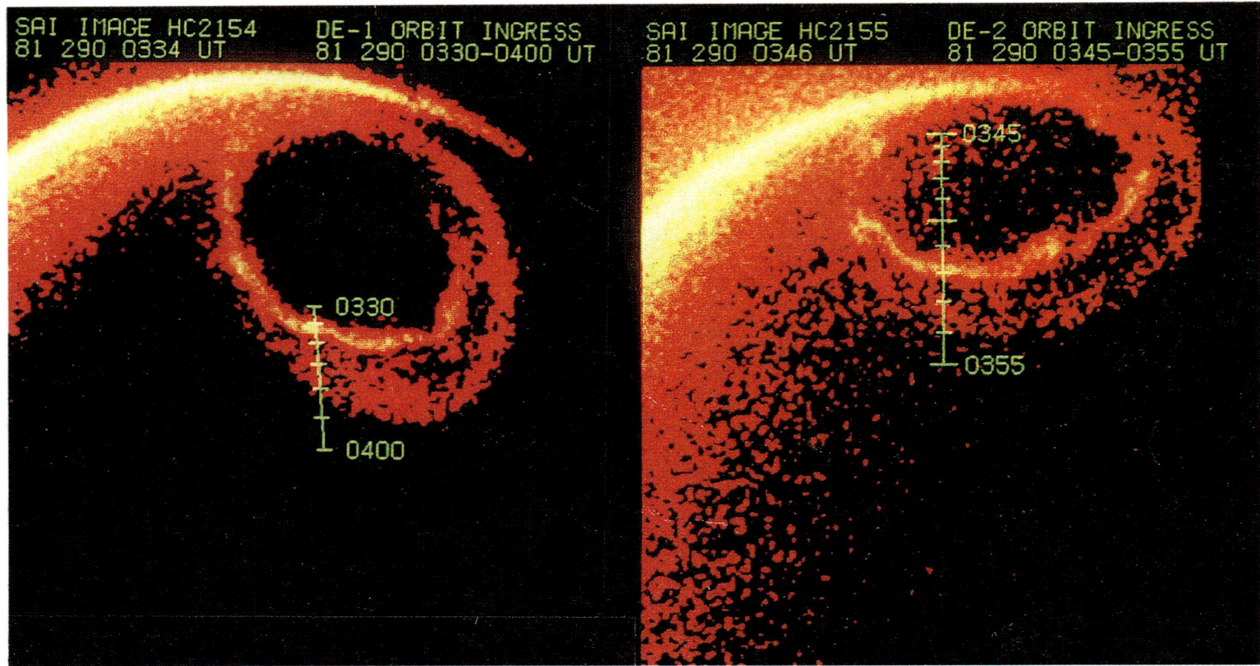

Plate 4. DE 1 ultraviolet auroral images from the Scanning Auroral Imager (SAI) on October 17, 1981. (left) image taken from 0334-0346 UT with DE 1 orbit footprint magnetically mapped to 100 km altitude. Midnight is to the right, noon to the left. (right) image taken from 0346-0358 UT, with the DE 2 orbit footprint similarly mapped.

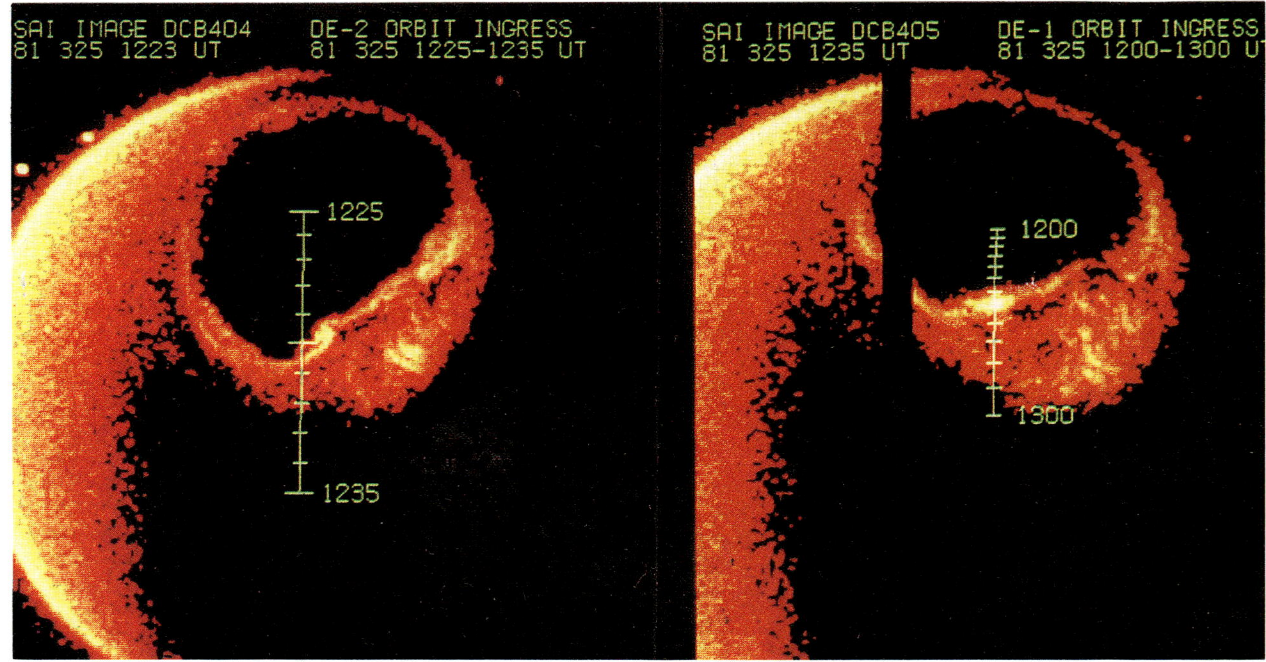

Plate 5. SAI ultraviolet images from November 21, 1981. (left) image from 1223-1235 UT with the DE 2 footprint magnetically mapped. (right) image from 1235-1247 UT with the DE 1 footprint shown. DE 1 passed just through the head of a westward traveling surge, whereas DE 2 passed just westward of it.

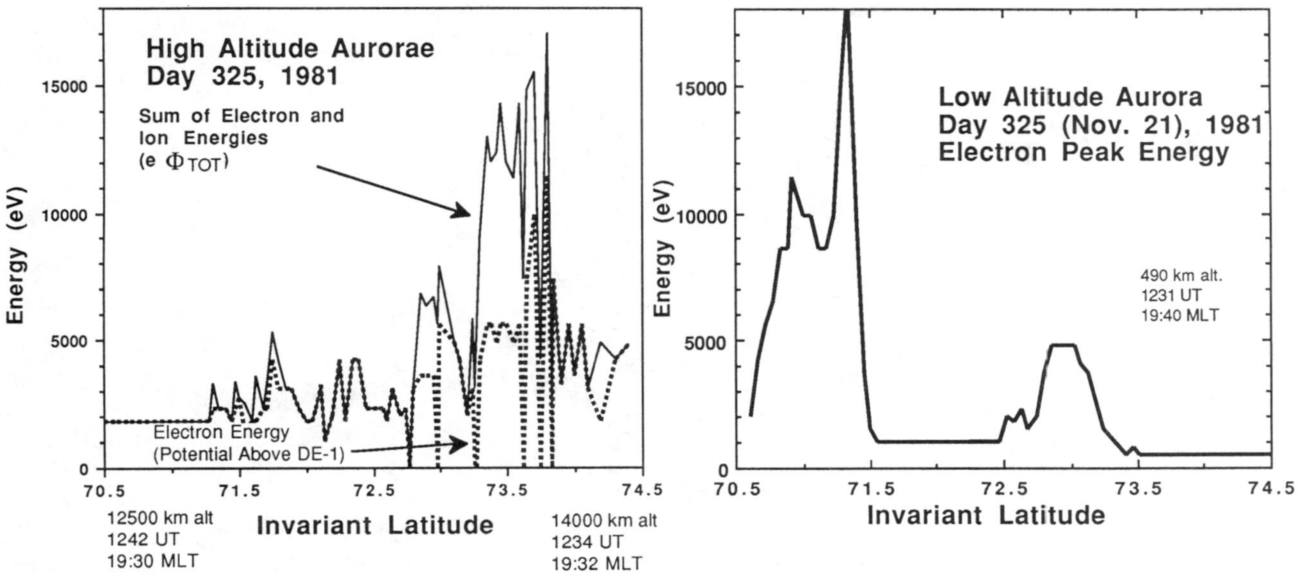

Fig. 3. Similar to Figure 1, but for day 81325 (November 21, 1981). LAPI only sees the very edge of the bulge, at 1230:36 (invariant latitude 71.35), whereas HAPI crossed the main portion of the surge at 1235-1237 (invariant latitude near 73.5).

field line at a 14,000 km altitude. Alternatively, the surge field line may simply have a much larger demand for parallel current; thus the field must extend to higher altitudes to satisfy the demand for current by increasing the mirror factor (ratio of magnetic field strength between low altitudes and the top of the potential drop). For a discussion of the effect of the mirror ratio on currents, see Knight, [1973]; Fridman and Lemaire, [1980].

THE LOW ALTITUDE ACCELERATION LIMIT

Previous estimates of the low-altitude limit of the auroral electric field have been based on the presence of electrostatic shock structures [Bennett et al., 1983], or the photoelectron spectrum [Fung and Hoffman, 1991; see also references therein]. In the former study, they found shocks down to an altitude range of 1000-2000 km, but almost none below 1000 km altitude. In the latter study, the authors compared measured low-energy electron spectra with an upgoing photoelectron spectrum having been retarded by a parallel electric field of various values of the total potential drop. That procedure, very sensitive to potentials of a few volts, indicated a typical potential drop of ~2 V between 400 and 800 km altitude, not enough to give O^+ ions escape energy, but certainly enough to help in their escape. Direct measurement of transversely (locally) accelerated ions have been presented by Garbe et al. [1992] and Arnoldy et al. [1992]. Arnoldy et al. showed that the acceleration at 1000 km altitude was not uniform over space, but was concentrated into regions of a few tens of meters wide, with a filling factor of perhaps 5 - 10%.

In order to estimate the altitude of the bottom of the potential drop, we again examine a DE 1/DE 2 conjunction, but now a case where essentially all of the potential drop occurs between the altitudes of the two spacecraft [Reiff et al., 1988]. In Figure 4, we present three components of the ion drift velocity, the O^+ density, and the O^+ temperature observed by RPA and IDM [Hanson et al., 1981] on the DE 2 satellite for day 81308 (November 4, 1981). In this study we use a left-handed Cartesian coordinate system: the x axis is positive southward, the y axis is positive toward west, and the z axis points from the Earth to the spacecraft. On the bottom of Figure 4, the solid line shows the upward flowing ion flux parallel to the magnetic field line (assuming a dipole magnetic field), and the dashed line shows the ion flux along the z axis. In this case, they are about the same in magnitude. We note a reasonably good correlation between the ion temperature and the perpendicular ion drift velocity in this case, as was first observed by St.-Maurice and Hanson [1982] from Atmosphere Explorer C measurements.

In the region between 68° and 69.6° invariant latitude, the thermal O^+ ions are moving upward at velocities exceeding 100 m/s with number densities between 9×10^4 and 1.3×10^5 cm^{-3}. The average corresponding upward flowing O^+ flux is about 1.2×10^9 cm^{-2} s^{-1}. On the other hand, the local energetic ion plasma in the energy per charge range from 5 eV to 32 keV measured by the low-altitude plasma instrument (LAPI) [Winningham et al., 1981] has a net upward number flux of less than 6×10^6 cm^{-2} s^{-1} (see Figure 3 of Lu et al. [1992]). At high altitudes, though, the ion fluxes are much more energetic, and are mostly O^+ and partially H^+ (Figure 5), with a total upflux of nearly 10^8 cm^{-2} $s^{-1} sr^{-1}$.

We can use the ratio of the upgoing ion fluxes at DE 1 to that at DE 2 to estimate the altitude at which the thermal ions are given escape velocity and are thus lost. We make two

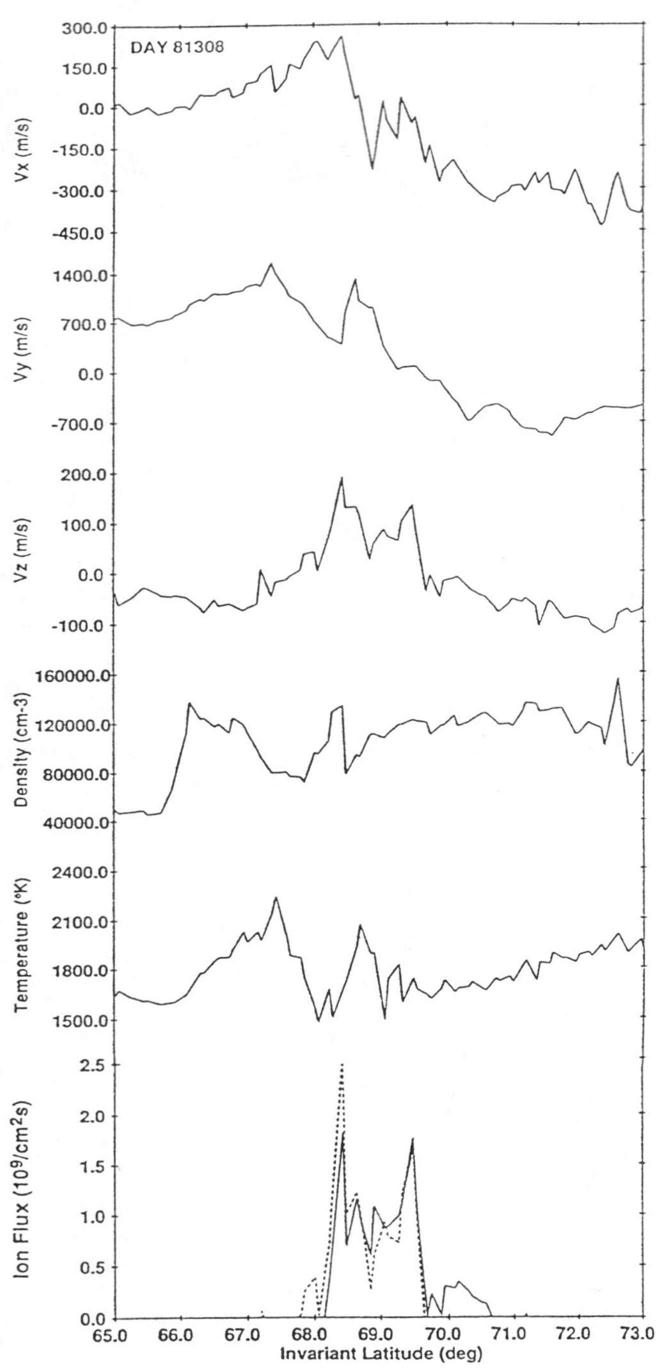

Fig. 4. Three components of the ion drift velocity, the O^+ density, and the O^+ temperature observed by RPA and IDM on the DE 2 satellite for day 81308 (November 4, 1981). The bottom panel shows the outgoing ion flux, calculated from the density and the three components, of drift to yield the true vertical component (dashed) and (separately) the upward component parallel to the magnetic field (solid). The ionospheric outflows observed here were matched to the high altitude outflows of Figure 5 to yield an effective source altitude [From *Lu et al.*, 1992].

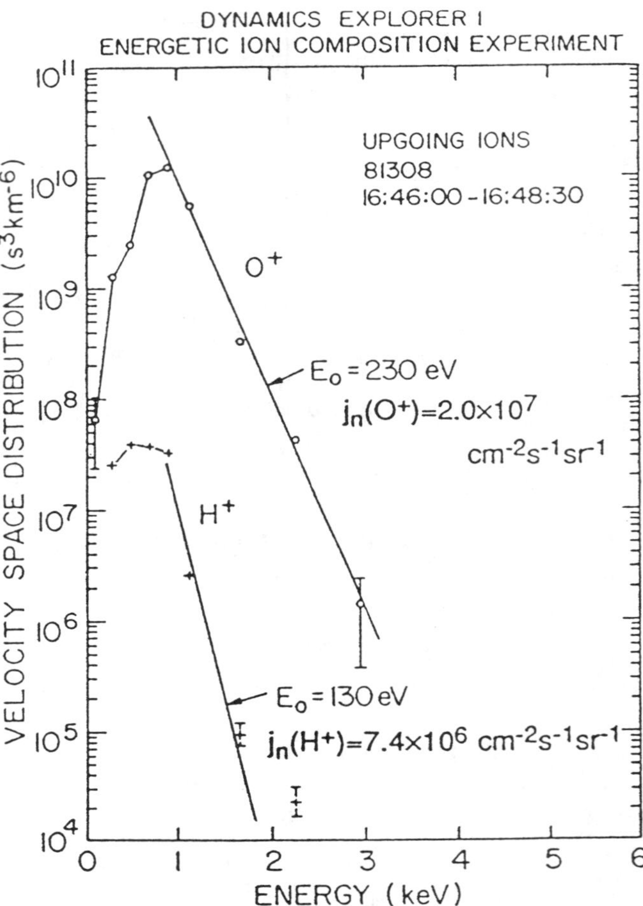

Fig. 5. Upflowing singly-charged Oxygen and Hydrogen measured on November 4, 1981. The total outflux was nearly 10^8 ions/cm²-s-sr [From *Lu et al.*, 1992].

assumptions: (1) the density of the thermospheric ions decreases in altitude with an exponential scale height corresponding to the plasma temperature inferred from the local measurements of T_e and T_i, and (2) their upward parallel velocity is constant with height. With these two assumptions, the magnitude of the upgoing thermal ion flux should fall exponentially with altitude, until the ions reach the acceleration altitude. At that point the ions are given escape velocity and the integrated upflux from then on should remain constant (however, the pitch angle distribution changes because of the accelerating electric field and diverging magnetic field geometry). For this case, the altitude of the acceleration region is determined to be around 1700 km, consistent with the predominance of O^+ in the ion beam. A second case presented in Lu et al. yielded a similar result. At the edge of the arc, however, where the potential drop is less, the upflowing ions are predominantly H^+ and are in the range of the RIMS (Retarding Ion Magnetic Spectrometer). Again by matching fluxes, we find [Lu et al., 1992], that the low altitude limit of the acceleration is higher: 2000 km and 2500 km for

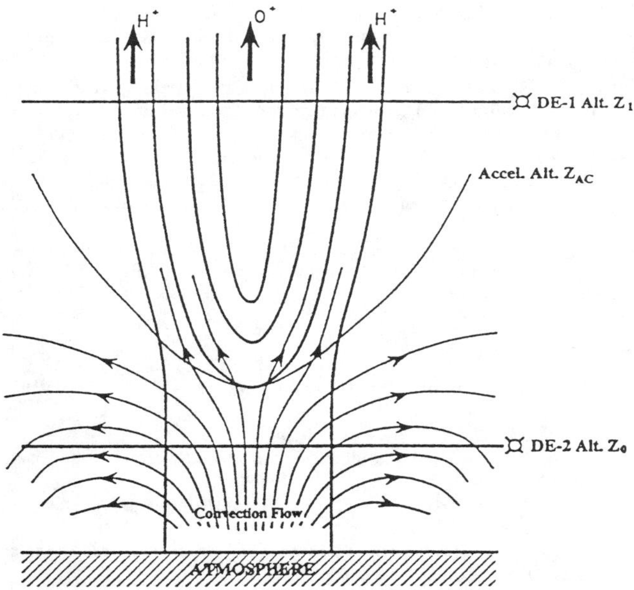

Fig. 6. Schematic of the auroral accelerating potential contours and their effects on the ion outflow. The outflow is principally O^+ in the center and H^+ at the edges [From *Lu et al.*, 1992].

the two cases studied in that paper. Near the maximum potential, the acceleration altitude is near the neutral H/O crossover altitude and thus is consistent with the preponderance of O^+ in the ion beam. Near the edges, however, the acceleration altitude is well above the H/O crossover altitude and thus the flux is predominantly H^+ there. We stress that the acceleration altitude may not necessarily correspond to the low altitude limit of the auroral electric field; this acceleration may well be a two-step one, with perpendicular acceleration at lowest altitudes followed by parallel electric field acceleration higher up. However, this sets a lower limit for the base of the parallel electric field, and sets an "effective acceleration altitude" where the bulk of the ionosphere is effectively lost. Note that this altitude is compatible with the results of Arnoldy et al.: we assume an acceleration of the entire distribution, and calculate an altitude of ~1700 km. If one, on the other hand, has an acceleration mechanism that is patchy or sporadic and operates only over 5-10% of the distribution, as shown by Arnoldy, the altitude of the acceleration mechanism must be correspondingly less. Figure 6 summarizes the conclusions: the auroral acceleration penetrates to lower altitudes near the center of the arc, whereas at the edges, the acceleration is higher in altitude.

CONCLUSIONS

We have shown that, under typical conditions, the auroral electric field region is confined between ~1400 km and 14,000 km altitude. The highest altitude acceleration events (>14,000 km) are rare and are apparently associated with westward traveling surges or polar cap arcs. In those cases, the highest acceleration might even occur in the magnetotail. The low altitude effective limit of the auroral acceleration region, determined by matching high and low altitude upgoing ion fluxes, occurs near 1700 km in regions of high potential drop (near the H/O neutral crossover altitude), and significantly higher (2000-2500 km) in regions of low parallel potential drop (the actual acceleration altitude may be lower if the process is intermittent in time or space). This difference in acceleration altitude is reflected in the composition of the upgoing ion beam, which is substantially O^+ in the region of peak potential drop but is essentially completely H^+ at the edges where the potential drop is smaller.

Acknowledgments. The work at Rice University was supported in part by NASA under grants NAG5-775 and NAGW-1655 and by the National Science Foundation under grant ATM91-03440. The work at Southwest Research Institute was supported by NASA under grant NAS5-33030. The work at the University of Iowa was supported by NASA under grant NAG5-483, the work at Lockheed was supported by NASA under grant NAS-5-33032, and the work at the University of Alaska Fairbanks was supported by NASA under grants NAGW-2735 and NAG5-1915. The authors thank Reneé Beveridge James for her help in the data analysis.

REFERENCES

Arnoldy, R. L., K. A. Lynch, P. M. Kintner, J. Vago, S. Chesney, T. E. Moore and C. J. Pollock, Bursts of transverse ion acceleration at rocket altitudes, *Geophys. Res. Lett., 19,* 413-416, 1992.

Bennett, E. L., M. Temerin, and F. S. Mozer, The distribution of auroral electrostatic shocks below 8000-km altitude, *J. Geophys. Res., 88,* 7107-7120, 1983.

Birn, J. and M. Hesse, The substorm current wedge and field-aligned currents in MHD simulations of magnetotail reconnection, *J. Geophys. Res., 96,* 1611-1618, 1991.

Block, L. P. and C.-G. Fälthammar, Characteristics of magnetic-field aligned electric fields in the auroral acceleration region, in *Auroral Physics*, edited by M. J. R. C.-I. Meng and L. A. Frank, pp. 109-118, Cambridge Univ. Press, Cambridge, England, 1991.

Burch, J. L., J. D. Winningham, V. A. Blevins, N. Eaker, W. C. Gibson, and R. A. Hoffman, High-altitude plasma instrument for Dynamics Explorer-A, *Space Sci. Instrum., 5,* 455, 1981.

Frank, L. A., J. D. Craven, K. L. Ackerson, M. R. English, R. H. Eather, and R. L. Carovillano, Global auroral imaging instrumentation for the Dynamics Explorer Mission, *Sp. Sci. Instr., 5,* 369, 1981.

Frank, L. A., J. D. Craven, D. A. Gurnett, S. D. Shawhan, D. R. Weimer, J. L. Burch, J. D. Winningham, C. A. Chappell, J. H. Waite, R. A. Heelis, N. C. Maynard, M. Sugiura, W. K. Peterson and E. G. Shelley, The theta aurora, *J. Geophys. Res., 91,* 3177, 1986.

Fridman, M. and J. Lemaire, Relationship between auroral electron fluxes and field-aligned electric potential difference, *J. Geophys. Res., 85,* 664, 1980.

Fung, S. F. and R. A. Hoffman, A search for parallel electric fields by observing secondary electrons and photoelectrons in the low-altitude auroral zone, *J. Geophys. Res., 96,* 3533-3548, 1991.

Garbe, G. P., R. L. Arnoldy, T. E. Moore, P. M. Kintner, and J. L. Vago, Observations of transverse ion acceleration in the topside auroral ionosphere, *J. Geophys. Res., 97,* 1257-1269, 1992.

Gurgiolo, C. and J. L. Burch, Simulation of electron distributions within auroral acceleration regions, *J. Geophys. Res., 93,* 3989-4003, 1988.

Hanson, W. B., R. A. Heelis, R. A. Power, C. R. Lippincott, D. R. Zuccaro, B. J. Holt, L. H. Harmon, and S. Sanatani, The retarding potential analyzer for Dynamics Explorer-B, *Space Sci. Instrum., 5,* 503, 1981.

Heelis, R. A., W. B. Hanson, C. R. Lippincott, D. R. Zuccaro, L. H.

Harmon, B. J. Holt, J. E. Doherty, and R. A. Power, The ion drift meter for Dynamics Explorer-B, *Space Sci. Instrum.*, *5*, 511, 1981

Kindel, J. M. and C. F. Kennel, Topside current instabilities, *J. Geophys. Res.*, *76*, 3055-3078, 1971.

Klumpar, D. M., Transversely accelerated ions: An ionospheric source of hot magnetospheric ions, *J. Geophys. Res.*, *84*, 4229, 1979.

Knight, S., Parallel electric fields, *Planet. Space Sci.*, *21*, 741-750, 1973.

Lu, G., P. H. Reiff, T. E. Moore and R. A. Heelis, Upflowing Ionospheric Ions in the Auroral Region, *J. Geophys. Res.*, in press, 1992.

Mizera, P. F. and J. F. Fennell, Signatures of electric fields from high and low altitude particles distributions, *Geophys. Res. Lett.*, *4*, 311-314, 1977.

Mozer, F. S., ISEE-1 Observations of electrostatic shocks on auroral zone field lines between 2.5 and 7 earth radii, *Geophys. Res. Lett.*, *8*, 823, 1981.

Mozer, F. S., C. A. Cattell, M. K. Hudson, R. L. Lysak, M. Temerin and R. B. Torbert, Satellite measurements and theories of auroral acceleration mechanisms, *Space Sci. Rev.*, *27*, 155, 1980.

Reiff, P. H., H. L. Collin, E. G. Shelley, J. L. Burch, and J. D. Winningham, Heating of upflowing ionospheric ions on auroral field lines, in *Ion Acceleration in the Magnetosphere and Ionosphere*, *Geophys. Monogr. Ser.*, vol. 38, edited by T. S. Chang, p. 83, AGU, Washington, D. C., 1986.

Reiff, P. H., H. L. Collin, J. D. Craven, J. L. Burch, J. D. Winningham, E. G. Shelley, L. A. Frank, and M. A. Friedman, Determination of auroral electrostatic potentials using high- and low-altitude particle distributions, *J. Geophys. Res.*, *93*, 7441, 1988.

Reiff, P. H., J. L. Burch, J. D. Winningham, W. K. Peterson, L. A. Frank and J. D. Craven, On the high-altitude limit of the auroral acceleration region, *J. Geophys. Res.*, (submitted), 1993.

Shelley, E. G., D. A. Simpson, T. C. Sanders, E. Hertzberg, H. Balsiger, and A. Ghielmetti, The energetic ion composition spectrometer (EICS) for Dynamics Explorer-A, *Space Sci. Instrum.*, *5*, 443, 1981.

St.-Maurice, J.-P. and W. B. Hanson, Ion frictional heating at high latitudes and its possible use for an in situ determination of neutral thermospheric winds and temperatures, *J. Geophys. Res.*, *87*, 7580, 1982.

Winningham, J. D., J. L. Burch, N. Eaker, V. A. Blevins, and R. A. Hoffman, The low altitude plasma instrument (LAPI), *Space Sci. Instrum.*, *5*, 465, 1981.

P. H. Reiff, Department of Space Physics and Astronomy, Rice University, Houston, TX 77005

G. Lu, HAO, NCAR, Boulder, CO 80307

J. L. Burch and J. D. Winningham, Southwest Research Institute, San Antonio, TX 78284

L. A. Frank, Department of Physics and Astronomy, University of Iowa, Iowa City, IA 52242

J. D. Craven, Geophysical Institute and Department of Physics, University of Alaska, Fairbanks, AK 99775

W. K. Peterson, Lockheed, Palo Alto, CA 94304

R. A. Heelis, Center for Space Science, University of Texas at Dallas, Richardson, TX 75083

The Acceleration of Electrons by Electromagnetic Ion Cyclotron Waves

M. TEMERIN, C. CARLSON, AND J. P. McFADDEN

Space Sciences Laboratory, University of California, Berkeley

Electromagnetic ion cyclotron waves provide an important mechanism, in addition to parallel electric fields, for accelerating electrons in the aurora. Such waves may be responsible for flickering aurora and for the intense field-aligned electron component of the aurora. Electrons are accelerated by the small parallel electric field component of the wave at the Landau resonance. In an inhomogeneous magnetic field the phase velocity of the wave can increase and electrons that are near resonance can be accelerated to large energies. Test particle simulations using these waves show that electron conics, intense field-aligned electron fluxes, and the trapped electron distribution can all be explained using a single wave together with a parallel electric field. Specifically, it is shown that ~100 Hz electromagnetic ion cyclotron waves can produce the observed ~100 Hz oscillations in the 30 keV electron flux in intense aurora. These results show the importance of both waves and parallel electric fields in accelerating auroral particles.

INTRODUCTION

Electron and ion acceleration in the aurora remains a topic of great interest. Current research focuses on two primary mechanisms of electron acceleration: quasi-static parallel electric fields and inertial Alfvèn waves [*Mallinckrodt and Carlson*, 1978; *Goertz and Boswell*, 1979; *Knudsen et al.*, 1990; *Boehm et al.*, 1990; *Hui and Seyler*, 1992; and papers presented during this meeting]. In the auroral acceleration region the inertial Alfvèn wave has a finite parallel electric field which can interact with the electron distribution to accelerate electrons. 'Inertial' Alfvèn waves are sometimes called 'kinetic' Alfvèn waves, but the latter term is best reserved for waves in regions where the Alfvèn velocity is less than the electron thermal velocity and where, as a consequence, the thermal terms are the dominant correction to the Alfvèn dispersion relation. For the inertial Alfvèn wave, it is the inertia of the electrons, a term that is automatically taken into account in the normal cold plasma dispersion relation, that is the dominant 'correction'. Since the inertial effects are important only when the perpendicular wavelength is on the order of or less than the electron skin depth (a few kilometers in the auroral acceleration region) the inertial Alfvèn waves are suspected of contributing to the electron acceleration in small-scale auroral structures. However, it is often hard, because of the difficulty of distinguishing spatial from temporal variations, to find convincing evidence that Alfvèn waves are in fact involved, or to ascertain from the available data, such as the measured electron distribution, the acceleration mechanism. A striking exception is the flickering aurora. In this case, ground optical observations [*Kunitake and Oguti*, 1984] have established the approximate temporal and spatial variations and rocket measure-

ments [*McFadden et al.*, 1987] have established the electron distributions responsible for the observed optical variations. The periodic nature of the variations provide convincing evidence that a wave mode is involved and a model of the wave has been shown to be able to reproduce the data [*Temerin et al.*, 1986]. Thus, flickering aurora provides convincing evidence of an acceleration process that may be more generally applicable.

In the flickering aurora model of *Temerin et al.* [1986], the modulation of the electron flux is caused by an electromagnetic oxygen cyclotron wave propagating from the auroral parallel-electric-field acceleration region at about 6000 km. At the altitude of the acceleration region, the frequency of the wave is approximately the local oxygen cyclotron frequency and the wave has a fairly small parallel phase velocity which enables it to be resonant with the tail of the secondary electron distribution (those electrons produced by the collisions of the primary electron beam with the atmosphere at lower altitudes). As the wave propagates down the field line its phase velocity increases and those electrons that were initially in phase can be accelerated to large energies. This model is schematically illustrated in Figure 1. That such a wave could accelerate ionospheric electrons was shown using a test particle model by *Temerin et al.* [1986]. This model, however, did not produce ionospheric backscattered and secondary electrons self-consistently from the incident electron flux and used only field-aligned electrons so the relative flux of the field-aligned component to the primary incident electrons could not be calculated.

The electromagnetic oxygen cyclotron wave propagating obliquely to the magnetic field in the auroral acceleration region is basically an inertial Alfvèn wave with corrections for the finite wave-to-ion-gyrofrequency ratio. In a multi-ion species plasma there are similar but distinct modes associated with each ion species. The obliquely propagating hydrogen and helium cyclotron modes usually do not propagate all the way to the ionosphere because they reflect near the two-ion hybrid frequency. The hydrogen cyclotron [*Gurnett and Frank*, 1972; *Temerin and Lysak*, 1984;

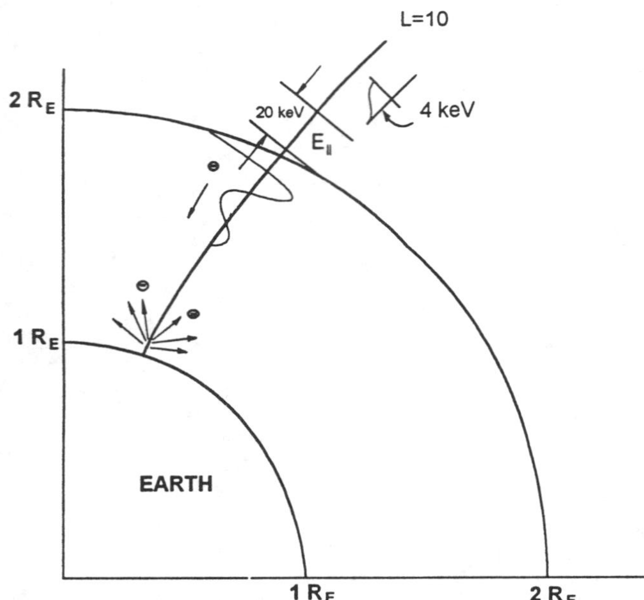

Fig. 1. A schematic illustration showing the properties of the model which include parallel electric fields, waves, and backscattered and secondary electrons produced from the primary beam of electrons that is modelled as a half-Maxwellian above the potential drop.

Saito et al., 1987; *Gustafsson et al.*, 1990] and the helium cyclotron [*Gustafsson et al.*, 1990] modes have, however, been observed in satellite data.

Because of the existence of these additional modes, *Temerin et al.* [1986] suggested that electron flux modulations should also be seen at frequencies corresponding to the hydrogen cyclotron frequency, which is typically between 80 and 120 Hz in the auroral acceleration region. This modulation signature was subsequently identified by electron measurements made on the UC Berkeley/NASA Alaska 88 sounding rocket flight.

The instruments for the Alaska 88 experiment extended the high frequency range of the particle measurements but were otherwise similar to those on the BIDARCA rocket flight that identified the electrons responsible for flickering aurora [*McFadden et al.*, 1987]. Two separate electron instruments collected complementary data. The Fast Electron Spectrograph (FES) measured a 16-channel energy spectrum of magnetic field-aligned electrons at energies between 300 eV and 60 keV every 4 ms. A pitch-angle imaging 'top hat' electrostatic analyzer measured the pitch-angle distribution of 30 keV electrons with 10° angle resolution and 2 ms time resolution. This combination of rapid angle and energy distribution measurements was able to resolve important features of the high frequency electron flux modulations.

The energy-time spectrogram in Figure 2 provides an overview of the electron precipitation during this flight. The rocket traveled northward during the flight, crossed through a series of auroral arcs and finally entered the polar cap at about 1100 s. The most prominent feature of the electron precipitation was the characteristic inverted V electron signature that extended to peak energies of 40 keV. In addition, there were sporadic bursts of electron precipitation that extended from the energy of the inverted V peak down to the lowest energy channel at 300 eV. These events are most evident at flight times between 200 and 500 s, although they can be identified throughout the flight. This feature is the signature of the flickering aurora electrons that have modulation frequencies of a few Hz.

Although not apparent in the E-T spectrogram, electrons with energies above the inverted peak were modulated at frequencies of 80 to 120 Hz during this same general time interval when the flickering aurora electrons were seen. These modulations are easily seen in Figure 3, which presents time series plots of differential

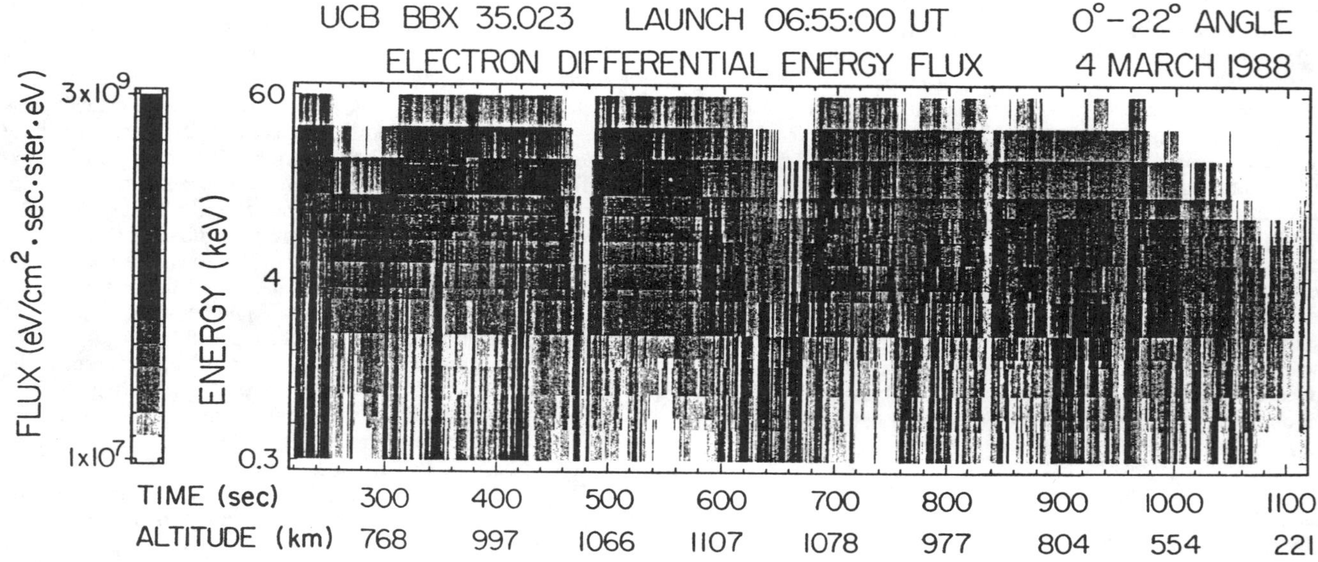

Fig. 2. An overview of the Alaska 88 electron rocket data showing an energy-time spectrogram of the electron flux.

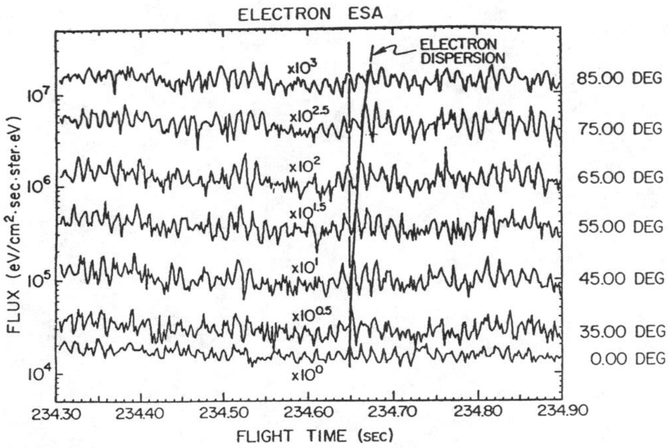

Fig. 3. A high time resolution view of the 30 keV electron flux at different pitch angles during an interval when there were 80 Hz oscillations in the Alaska 88 rocket data.

fluxes in 7 pitch-angle intervals between 0 and 90 degrees. The individual traces are offset by factors of $10^{1/2}$ to avoid overlap. The modulation frequency during this interval was about 80 Hz, and the modulation amplitude ranged between 30% and 100%. The energy of these electrons lies just above the inverted V energy peak, where the energy spectrum is steeply falling. Consequently the 100% amplitude modulation represents only a few keV energy modulation of the distribution function.

An important property of these oscillations is the phase shift that increases systematically with increasing pitch angle. Two lines are drawn in Figure 3 to illustrate this phase shift. The vertical line is a fixed phase reference line and the slightly curved line connects the fixed phase maximum of one of the oscillation features at each pitch angle. The total phase delay between the 0 and 85 degree electrons is about 25 ms.

A simple model was constructed to fit these data. This model assumes that a stationary inverted V electron distribution is created at high altitude and that the modulation takes place in a lower, narrow altitude range. The modulation varies the energy of the electrons in the high energy tail of the distribution function. Below this localized modulation source the electrons undergo adiabatic motion in the Earth's magnetic field. The phase delay for larger pitch angle particles simply results from the increased path length of their spiral trajectories. A comparison of this model with several of the measured events is illustrated in Figure 4. The delay times were measured by performing cross-correlation fits between the 0° channel and each of the other pitch angle channels. Events with a range of oscillation periods between 83 and 122 Hz were selected to see if there was a relation between source altitude and frequency. The apparent modulation source altitude falls in the range of 2400 km to 3200 km for the five selected events. The simple model gives a good fit over the entire pitch-angle range, suggesting that the approximation of a narrow altitude modulation source is reasonable.

TEST-PARTICLE SIMULATION

In this section we describe a test particle model that simulates the ~100 Hz oscillations in the 30 keV electron flux. In the model 10,000 test particle electrons drawn from a 4 keV temperature half

Maxwellian electron distribution represent the incident plasmasheet electrons above a 20 keV potential drop. The 20 keV potential drop is represented by a 20 mV/m parallel-to-the-magnetic-field electric field extending for 1000 km above an altitude of 6370 km (~two R_E geocentric). Below the parallel field there is an obliquely propagating electromagnetic hydrogen cyclotron wave with a frequency equal to the gyrofrequency of hydrogen at 6370 km (~100 Hz). In an uniform plasma the dispersion relation for an electromagnetic hydrogen cyclotron wave is given by

$$\omega = ck_{\parallel}\varepsilon_1^{1/2}\left(1 + c^2 k_{\perp}^2 / \omega_p^2\right)^{1/2} \tag{1}$$

where

$$\varepsilon_1 = 1 + \sum_s \frac{\omega_{ps}^2}{\omega_{cs}^2 - \omega_{ps}^2} \tag{2}$$

and ω, ω_{ps}, and ω_{cs} are the wave, plasma, and gyro frequencies of the electrons and the ion species, ω_p is the plasma frequency, $k_{\parallel}$ and $k_{\perp}$ are the parallel and perpendicular wave numbers. In a nonuniform plasma with changing magnetic field and density we find the parallel electric field as a function of altitude using the method of *Temerin et al.* [1986]. We assume that the magnetic field varies as

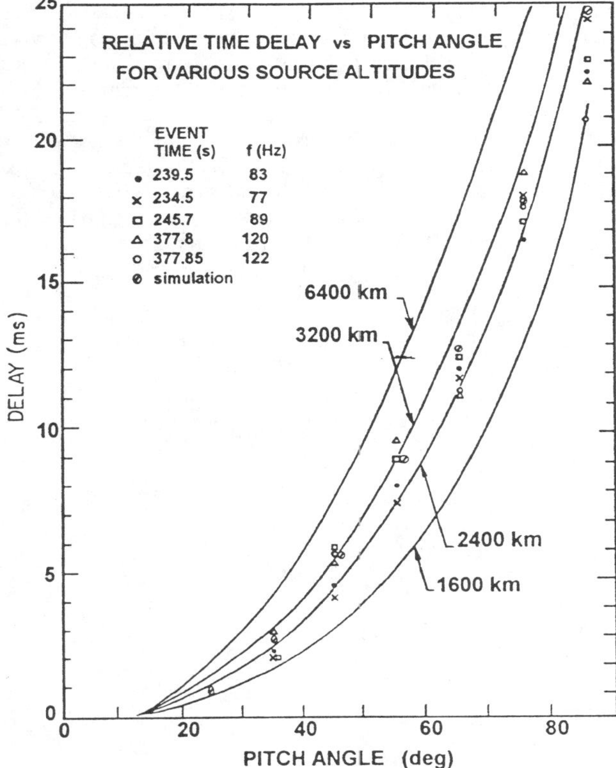

Fig. 4. A plot of the relative time delay of the electron flux at different pitch angles for several different intervals during the Alaska 88 rocket plotted along with a model calculation that assumes that the time delay is due to the increased effective travel distance of electrons with greater pitch angle.

r^3, where r is the geocentric distance, and the density varies as r^2 with a total ion density of four ions per cm^3 at 6370 km altitude and that three quarters of the ions are H^+ and the rest He^+. The perpendicular wavelength is assumed to be 15.7 km ($k_\perp = .4$ km^{-1}) at 6370 km altitude varying with geocentric distance as $r^{3/2}$ and the wave frequency is assumed to be the H^+ cyclotron frequency at 6370 km altitude or about 100 Hz. Such a perpendicular wavelength when projected to the ionosphere gives a scale size comparable to that measured optically for the flickering aurora [*Kunitake and Oguti*, 1984]. The parallel component of the wave vector is determined from the dispersion relation. We assume the wave propagates down without damping or growth. Reflection near the two-ion hybrid frequency then produces a standing wave pattern in the electric field. Given these assumptions, the parallel electric field of the standing wave pattern can be found. The envelope of the parallel electric field component of this standing wave pattern is shown in Figure 5. The resulting wave is similar to the model for the electromagnetic hydrogen cyclotron wave used in *Temerin et al.* [1986], except that previously total reflection at the two-ion hybrid resonance was not imposed.

The parallel phase velocity of the wave, based on the dispersion relation using the local magnetic field and density is shown in Figure 6. As the wave approaches the two-ion hybrid resonance, its parallel phase velocity approaches infinity. Thus we expect that 30 keV electrons, whose velocity is about 10^{-5} km/s, will be resonant with the wave at an altitude of about 3000 km.

The test particle electrons are injected above the potential drop at random times with respect to wave phase during one wave period. They are accelerated by the parallel potential drop and interact with the wave. The mirror force of the magnetic field is taken into account and so some of the electrons mirror out of the system after one pass. Other electrons interact with the atmosphere to produce backscattered and secondary electrons. Backscattered electrons are calculated according to the method of *Strickland et al.* [1976] and secondary electrons are calculated according to the method of *Evans* [1974], adjusted slightly for better agreement with the measured secondary electron data. The backscattered and

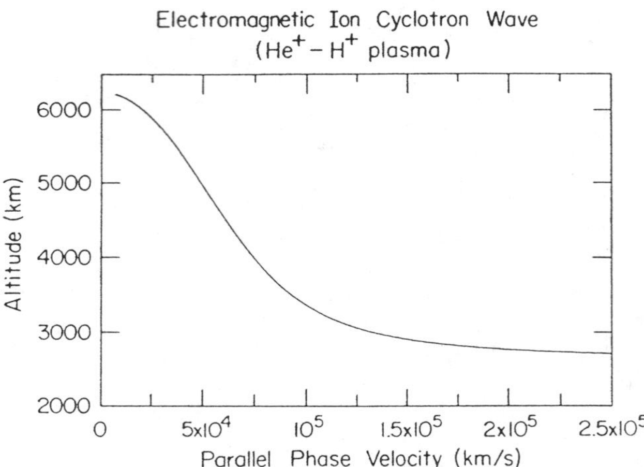

Fig. 6.   The parallel phase velocity of the electromagnetic ion cyclotron wave that is used in the simulation described in the text.

secondary electrons are followed until they escape from the top of the potential drop or are again lost to the atmosphere where they can make further secondary and backscattered electrons. Some of the primary electrons interact with the wave and become trapped between their mirror points and the potential drop. Such electrons can execute many bounces before further interactions with wave lead to loss either to the atmosphere or out of the top of the system. The electron flux as a function of energy, pitch angle, and wave phase can then be determined at any altitude by counting all the electrons crossing at that altitude as a function of wave phase. Electrons are injected above the potential during only one wave period. However, they are counted at various altitudes over many subsequent periods. Following the electrons that start during one wave period until they and all the secondary electrons they produce are finally lost from the system and counting them at a given altitude as a function of wave phase determines the flux as a function of wave phase. Except for a greater statistical variation, this is an equivalent way of determining the flux as a function of wave phase as counting during one wave period all the electrons that cross a given altitude from all electrons injected into the system during all previous times. In the presence of a periodic wave the electron flux is also periodic and electron flux as a function of wave period can be found by following all the electrons injected during one wave period.

Figure 7 shows the perpendicular and parallel electron velocities of the electrons that pass through 630 km altitude (a typical rocket altitude) during one quarter of a wave period. Each electron in the figure has been plotted twice with both a positive and a negative perpendicular velocity for ease of viewing. The circle indicates the energy corresponding to the potential drop. In the absence of the wave and of backscattered and secondary electrons there would be no electrons within the circle. Figure 8 shows a similar simulation with backscattered and secondary electrons included but without the wave. There are three notable differences between the simulation with and without the wave. The most notable difference is that the primary down going electrons, that is, those near the circle, are modulated in the presence of the wave. There are what appear to be several waves of electrons coming down. These successive 'waves' are electrons that were modulated by successive crests and troughs of the wave. Another difference is that the pie shaped

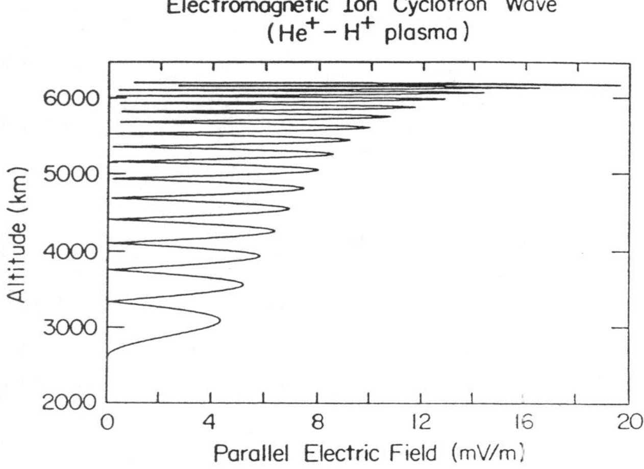

Fig. 5.   The envelope of the standing wave pattern produced by an electromagnetic ion cyclotron wave propagating from the parallel electric field region. The wave reflects near the two-ion hybrid frequency to produce a standing wave pattern.

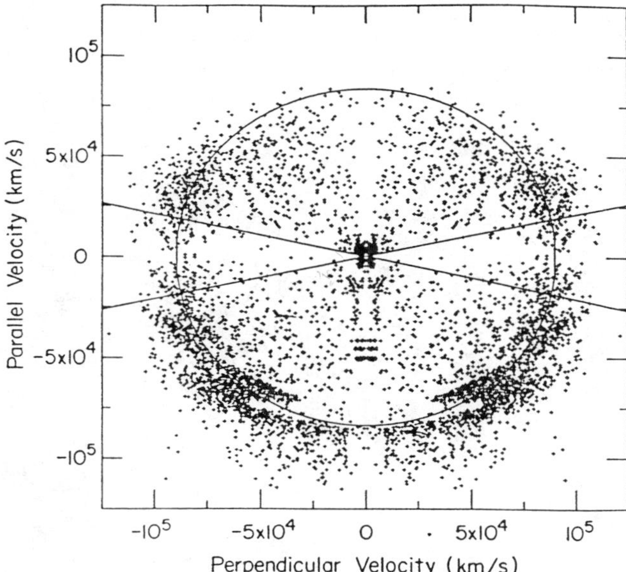

Fig. 7. The perpendicular and parallel electron velocities of the electrons crossing through 630 km altitude in the simulation during one quarter of a wave period. Each perpendicular velocity is plotted as both a negative and a positive velocity for symmetry.

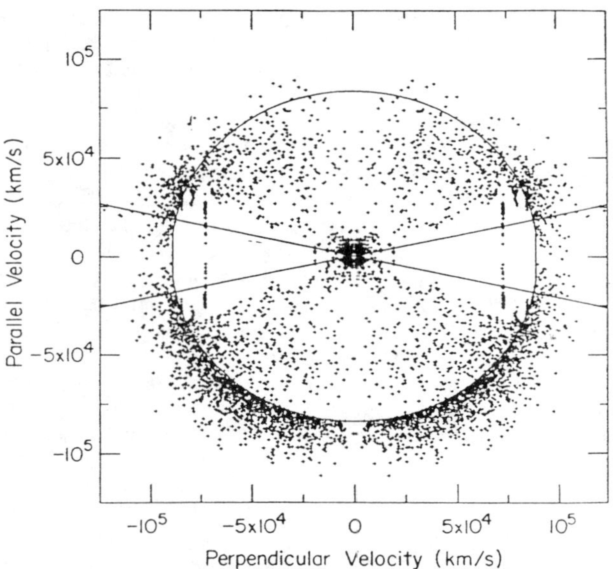

Fig. 8. This figure is similar to Figure 7, but the simulation had no wave. The purpose of this figure is to offer an contrast with the Figure 7 so that the reader can judge the effect of the wave on the electron distribution.

regions near 90° are filled in. This is the forbidden region of electrons trapped between their mirror points above the atmosphere and the parallel field. At 630 km this region is fairly small but becomes much larger at higher altitudes. However, an electron entering this forbidden region by a chance scattering at a higher than normal altitude will bounce many times before further scattering removes it from the trapped region. One such electron is seen at about ±73000 km/s perpendicular velocity in Figure 8. This can be considered a statistical fluctuation due to the use of a limited number of electrons. The third difference is that there is a field-aligned beam at about 45000 km/s (~6 keV). These are secondary electrons that have come up the field line, been reflected by the parallel electric field, caught up by the wave and accelerated as in the previous analysis of the flickering aurora [*Temerin et al.*, 1986]. In looking at Figure 7 and 8 one should note that the density of dots is not indicative of the phase space density of electrons, but rather of the flux of electrons through a given altitude. Because of this, the density of dots near $v_{\parallel} = 0$ and $v_{\perp} = 0$ is less relative to the actual phase space density there. This is because the flux through a given altitude is proportional to the parallel velocity and because the phase space volume as function of perpendicular velocity is proportional to the perpendicular velocity.

Figure 9 shows the relative flux for different pitch angles as a function of time. The 0-10 degree flux arrives first as expected because it has a shorter travel time to the observing altitude. The time delays for the simulations have been plotted in Figure 4 along with those derived from the data. We see that the time delay indicates dispersion from an altitude of about 3000 km, consistent with the resonance of the electrons with the wave at that altitude.

DISCUSSION

The simulation, because it is in good agreement with the data, provides evidence that electromagnetic ion cyclotron waves pro-

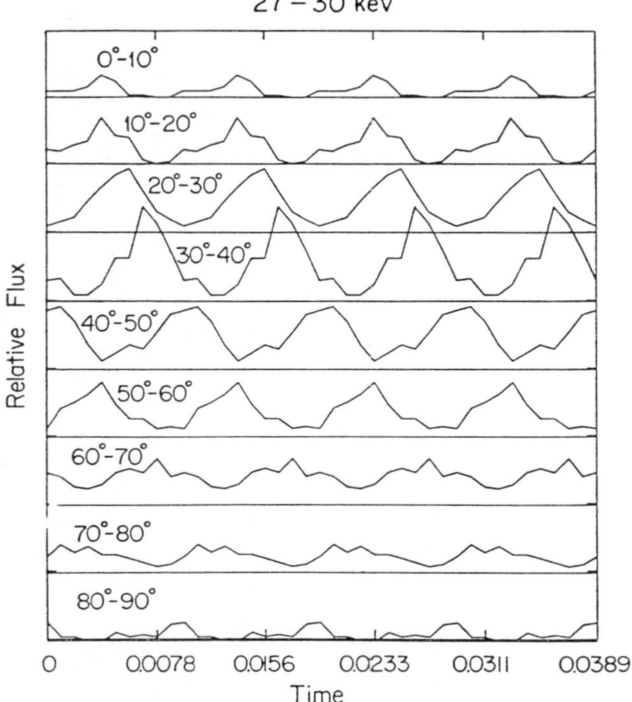

Fig. 9. The relative electron flux in the energy range 27-30 keV for different pitch angles as a function of time illustrating the dispersion of the electrons in pitch angle because of the different pitch-angle-related travel times.

duced the ~100 Hz modulation in the electron flux in this specific instance and modulations at other frequencies in other similar cases. Modulations are expected and observed for a range of frequencies corresponding to the gyrofrequencies of the H^+, O^+, and He^+ ions at the different altitudes at which electrons are accelerated by parallel potential drops. In addition, the wave model also reproduces other features of the electron distribution that have long puzzled auroral researchers. These features include the acceleration of secondary field-aligned electron beams, the filling in of the forbidden electron region of trapped electron flux, and the production of electron conics. Figure 10 shows the flux spectrum at various pitch angles for the simulation, (note the secondary peak in the field-aligned flux at 6 keV energy. Other simulations produce broader field-aligned electron fluxes). Figures 11 and 12 illustrate the results from another simulation using smaller potential drops (10 keV) and wave fields (approximately half that of Figure 7) and lower initial electron temperatures (1 keV). In this case, instead of a narrow field-aligned peak, there is a much broader enhancement in the field-aligned electron flux at energies below the energy of the 'monoenergetic' peak. Both types of distributions are often seen in auroral electron data.

The amplitudes required of the wave field in these simulations are substantial. In the simulation described in Figure 7 the parallel electric field in the region where the wave is in resonance with 20-30 keV electrons is about 0-5 mV/m but the corresponding perpendicular electric field is on the order of the square root of the mass ratio of the hydrogen to the electron $(M_H/m_e)^{1/2}$ larger or about 300 mV/m. Arguments for the existence of such large amplitudes

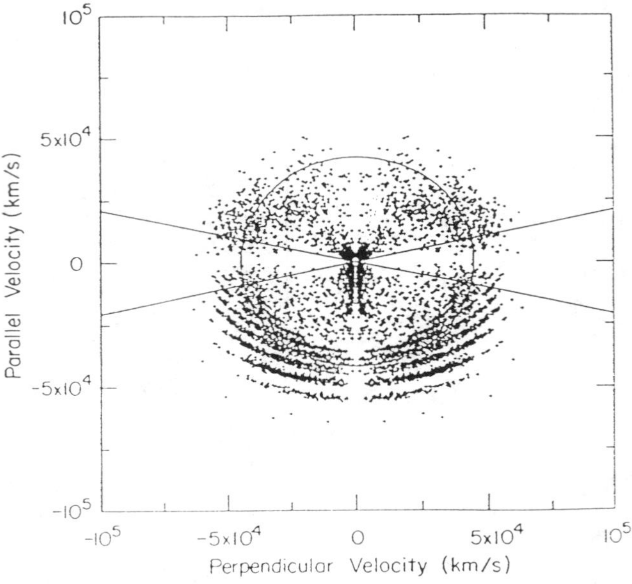

Fig. 11. Similar to Figure 7 but from a different simulation using lower energy initial electrons, smaller potential drops, and a slower lower-amplitude wave.

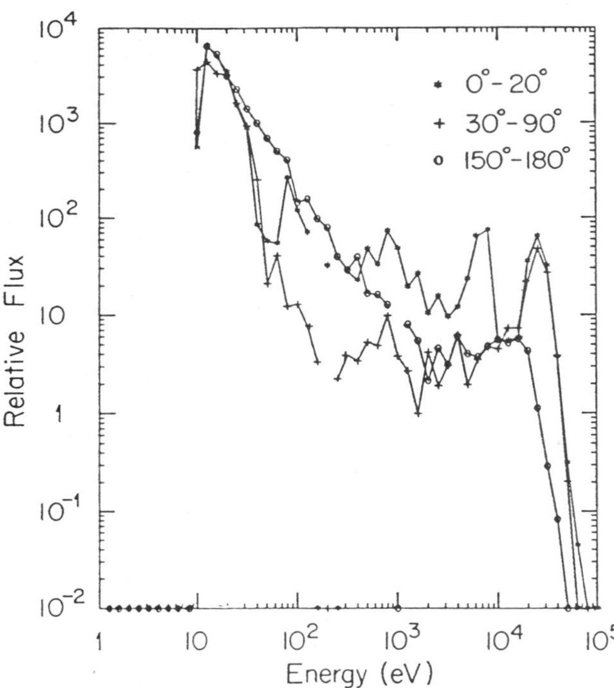

Fig. 10. The relative electron flux for different pitch angles as a function of energy. The purpose of this figure is to illustrate the simulation results in a format often used to plot auroral electron data. Note the secondary peak in the field-aligned electron distribution.

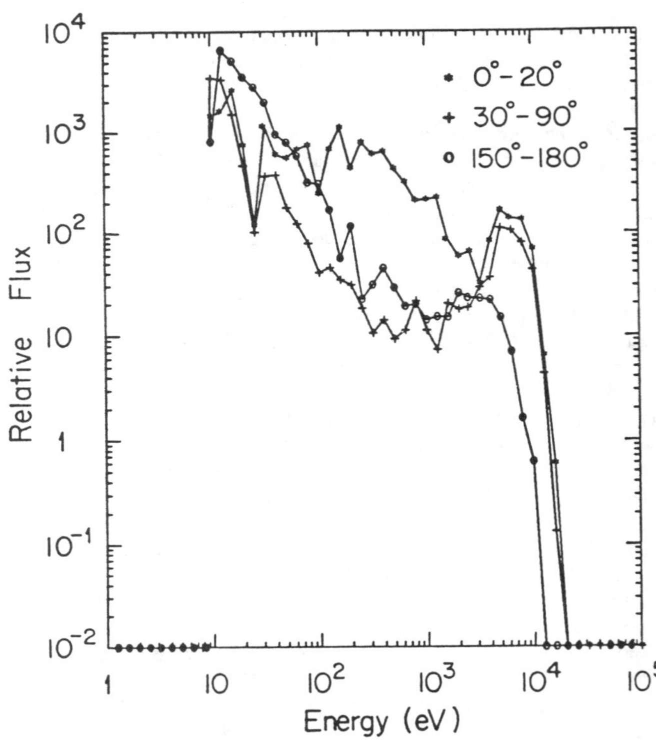

Fig. 12. Similar to Figure 10 but for the simulation illustrated in Figure 11. The purpose is to illustrate that broader enhancements in the field-aligned electron flux, in addition to the secondary peaks in the electron distribution as in Figure 10, can result from the interaction of electrons with electromagnetic ion cyclotron waves.

are stated in *Temerin et al.* [1986]. However, it should be borne in mind that this was a very energetic aurora. In the simulation shown in Figure 11 the wave amplitude was half as large and the parallel potential drop and initial electron temperature was a quarter of that in Figure 7. In addition, the plasma in this simulation had the same density, but consisted of hydrogen and oxygen which because of the oxygen's larger mass resulted in a smaller parallel wave phase velocity. The result was a much more distinct modulation of the electron flux. It takes large amplitude waves to significantly modulate energetic electrons, but less energetic fluxes can be modulated and accelerated by smaller amplitude waves. It is easy to estimate the required parallel electric field amplitude required. As an example, to modulate significantly (factor of ~2), an electron distribution with a 4 keV temperature would take about a 3 keV potential drop or an average wave parallel electric field of 3 mV/m over the 1000 km altitude range that the wave is approximately in resonance with the electrons.

In our simulation, we have used a single coherent wave. In fact, the wave spectrum may be expected to have a finite bandwidth. Experimental data, however, shows [*Temerin and Lysak*, 1984] that these waves can be sufficiently narrowbanded so that a single wave can be a good model. An estimate of when the wave can be considered coherent for the purpose of interacting with the electron distribution is given by $\Delta\omega\Delta t < 1$ where $\Delta\omega$ is the bandwidth of the wave and Δt is the time it takes for an electron to pass through the interaction region or about 0.03 s. This implies that $\Delta\omega < 300$ s^{-1} which is more than consistent with the ~100 s^{-1} bandwidth that is observed.

The data and the simulations show that electromagnetic ion cyclotron waves play an important role in accelerating and modulating the auroral electron flux. In addition, such waves can also accelerate ions to produce ion conics [*Temerin and Mozer,* 1984; *Temerin and Roth*, 1986]. Recently *Temerin and Roth* [1992] have applied these results to show that electromagnetic ion cyclotron waves can accelerate ^{3}He to MeV energies in ^{3}He-rich solar flares. In the simulation it is also possible to look at the electron distribution at any altitude. At altitudes above the monoenergetic peak the electron distribution shows an electron conic as previously suggested by *Temerin and Cravens* [1990]. For other recent models of electron conic generation see *André and Eliasson* [1992] and *Menietti et al.* [1992]. Within the potential drop region the simulation shows the trapped electron region completely and fairly uniformly filled in as is seen in the data. For other models of the filling in of the trapped electron region see *Eliasson et al.* [1979] and *Louarn et al.* [1990], who discuss the filling in of the region by a temporal increase in the potential drop. A thorough discussion of these results is contemplated for a future publication.

Acknowledgments. This research was supported by NASA grants NAGW-1626 and NAG6-10 and by National Science Foundation grant ATM-8918774. The computations were performed at the San Diego Supercomputer Center.

References

André, M., and L. Eliasson, Electron acceleration by low frequency electric field fluctuations: electron conics, *Geophys. Res. Lett.*, *19*, 1073, 1992.

Boehm, M. H., C. W. Carlson, J. P. McFadden, J. H. Clemmons, and F. S. Mozer, High-resolution sounding rocket observations of large-amplitude Alfvèn waves, *J. Geophys. Res.*, *95*, 12157, 1990.

Eliasson, L., L.-A. Holmgren, and K. Ronnmark, Pitchangle and energy distribution of auroral electrons measured by the Esro 4 satellite, *Planet. Space Sci.*, *27*, 87, 1979.

Evans, D. S., Precipitating electron fluxes formed by a magnetic field aligned potential difference, *J. Geophys. Res.*, *79*, 2853, 1974.

Goertz, C. K., and R. W. Boswell, Magnetosphere-ionosphere coupling, *J. Geophys. Res.*, *84*, 7239, 1979.

Gurnett, D. A., and L. A. Frank, ELF noise bands associated with auroral electron precipitation, *J. Geophys. Res.*, *77*, 4311, 1972.

Gustafsson, G., M. André, L. Matson, and H. Koskinen, On waves below the local proton gyrofrequency in auroral acceleration regions, *J. Geophys. Res.*, *95*, 5889, 1990.

Hui, C.-H., and C. E. Seyler, Electron acceleration by Alfvèn waves in the magnetosphere, *J. Geophys. Res.*, *97*, 3953, 1992.

Knudsen, D. J., M. C. Kelley, G. D. Earle, J. F. Victory, and M. Boehm, Distinguishing Alfvèn waves from quasi-static field structures associated with the discrete aurora: sounding rocket and HILAT satellite measurements, *Geophys. Res. Lett.*, *17*, 921, 1990.

Kunitake, M., and T. Oguti, Spatial-temporal characteristics of flickering spots in flickering auroras, *J. Geomagn. Geoelectr.*, *36*, 121, 1984.

Louarn, P., A. Roux, H. de Féraudy, D. Le Quéau, M. André, and L. Matson, Trapped electrons as a free energy source for the auroral kilometric radiation, *J. Geophys. Res.*, *95*, 5983, 1990.

Mallinckrodt, A. J., and C. W. Carlson, Relations between transverse electric fields and field-aligned currents, *J. Geophys. Res.*, *83*, 1426, 1978.

McFadden, J. P., C. W. Carlson, M. H. Boehm and T. J. Hallinan, Field-aligned electron flux oscillations that produce flickering aurora, *J. Geophys. Res.*, *92*, 11133, 1987.

Menietti, J. D., C. S. Lin, H. K. Wong, A. Bahsen, and D. A. Gurnett, Association of electron conical distributions with upper hybrid waves, *J. Geophys. Res.*, *97*, 1353, 1992.

Saito, H., T. Yoshino, and N. Sato, Narrow-banded ELF emission over the southern polar regions, *Planet. Space Science*, *35*, 745, 1987.

Strickland, D. J., D. L. Book, T. P. Coffey, and J. A. Fedder, Transport equation techniques for the deposition of auroral electrons, *J. Geophys. Res.*, *81*, 2755, 1976.

Temerin, M., and R. L. Lysak, Electromagnetic ion cyclotron-mode (ELF) waves generated by auroral electron precipitation, *J. Geophys. Res.*, *89*, 3945, 1984.

Temerin, M., and F. S. Mozer, Observations of the electric fields that accelerate auroral particles, *Proc. Indian Acad. Sci. Earth and Planet. Sci.*, *93*, 227, 1984.

Temerin, M., and I. Roth, Ion heating by waves with frequencies below the ion gyrofrequency, *Geophys. Res. Lett.*, *13*, 1109, 1986.

Temerin, M., and I. Roth, The production of ^{3}He and heavy ion enrichments in ^{3}He-rich flares by electromagnetic ion cyclotron waves, *Astrophys. J.*, *391*, L105, 1992.

Temerin, M., J. McFadden, M. Boehm, C. W. Carlson, and W. Lotko, Production of flickering aurora and field-aligned electron fluxes by electromagnetic ion cyclotron waves, *J. Geophys. Res.*, *91*, 5769, 1986.

Temerin M., and D. Cravens, Production of electron conics by stochastic acceleration parallel to the magnetic field, *J.Geophys. Res.*, *95*, 4285, 1990.

M. Temerin, C. Carlson, and J. P. McFadden, Space Sciences Laboratory, University of California, Berkeley, CA 94720

Statistical Distributions of the Auroral Electron Albedo in the Magnetosphere

D. M. KLUMPAR

Lockheed Palo Alto Research Labs., Space Sciences Laboratory, Palo Alto, CA

We present occurrence distributions of counterstreaming electrons on closed field lines in the earth's near-equatorial magnetosphere. Counterstreaming electrons measured by the AMPTE Charge Composition Explorer (CCE) are observed near the magnetospheric midplane at radial distances beyond 6 R_e. The probability of occurrence increases with increasing radial distance out to the 8.8 R_e radial distance limit sampled by the CCE satellite. The greatest probabilities of occurrence exceed 50% in the near midnight magnetosphere between 8.5 and 8.8 R_e. The occurrence rate decreases away from midnight and reaches a minimum in the 0700-1100 local time sector. We compare and contrast the local time distribution of these occurrence probabilities to the distribution of precipitating keV auroral electrons. These occurrence distributions are consistent with the previously published hypothesis that these near-equatorial counterstreaming electrons result from the precipitation of electrons at the ionospheric ends of the flux tubes. These equatorial electrons provide natural tracers for mapping auroral oval flux tubes into the magnetosphere.

INTRODUCTION

Counterstreaming field-aligned electrons are a common occurrence on closed field lines near the equatorial plane in the earth's magnetosphere. They are routinely observed from the AMPTE/CCE satellite [Klumpar et al., 1988; Klumpar, 1990] at radial distances from about 6 R_e out to the satellite's apogee at 8.8 R_e. Highly field-aligned electrons within or near the edge of the bounce loss cone are difficult to detect in the near-equatorial magnetosphere at and beyond geosynchronous orbit. In order to observe the bounce loss cone beyond L≈6 a particle detector must look within about 2° of the magnetic field. Despite this difficulty there have been previous scattered reports of field-aligned or counterstreaming electrons in the more distant near-equatorial magnetosphere [McIlwain, 1975; Parks et al., 1977; Borg et al., 1978; Lin et al., 1979; Richardson et al., 1981; Hada et al., 1981; Moore and Arnoldy, 1982; Arnoldy, 1986; Kremser et al., 1988; Klumpar et al., 1988; Klumpar, 1990]. Most of the above observations were made in geosynchronous orbit. The first reported detection of highly field-aligned electrons in the near-equatorial magnetosphere was made by McIlwain [1975] who identified them as auroral in origin. McIlwain reported intense beams of electrons travelling parallel to the local magnetic field with energies rising to several kilovolts. The observations by Hada

Auroral Plasma Dynamics
Geophysical Monograph 80
Copyright 1993 by the American Geophysical Union.

et al. [1981] are particularly interesting because they were observed (with IMP-6) in the outer magnetosphere between 10 and 30 R_e. Out of a 200-hour observing interval, about 10% of the pitch angle distributions had bi-directional anisotropy (i.e., peaked in both directions along the field line). Particularly inside of about 12 R_e they reported the existence of very sharply peaked distributions with a half-width of about 10° centered on **B**.

The detection of upstreaming or counterstreaming electrons near auroral field lines at low and intermediate altitudes (out to ~20,000 km) have in the past received considerable attention [Sharp et al., 1980; Klumpar and Heikkila, 1982; Collin et al., 1982; Lin et al., 1982; Hultqvist et al., 1988; Hultqvist, 1988; Lundin and Eliasson, 1991]. At these high latitudes where the field-line mapping is relatively well behaved and where there is other direct evidence of simultaneous auroral activity there is little doubt that these upstreaming and counterstreaming electrons are related to auroral processes.

It is highly probable that many, if not most, of these low to mid-altitude observations of upstreaming or counterstreaming electrons as well as the high altitude near-equatorial counterstreaming electrons are related to a common cause or class of mechanisms. We have suggested [Klumpar et al., 1988; Klumpar, 1990] that substantially all are associated with auroral precipitation. Specifically we have advocated that strongly peaked fluxes of electrons at the edge of or within the atmospheric loss cone observed in the distant near-equatorial magnetosphere result from backscattering of auroral primaries and injection of atmospheric secondaries at pitch angles

"

greater that 90°. The strong magnetic gradient between the ionosphere and the equatorial plane forces the upstreaming electrons into a closely field-aligned pitch angle distribution as they cross the geomagnetic equator. Active auroral processes such as plasma wave heating, electric field trapping, and low-altitude oscillating parallel electric fields will also contribute to the equatorial flux at the edge of the loss cone. An important consequence is that the flux tubes connected to the aurora are painted with a very distinctive electron distribution that provides a unique and natural tracer of the extension of auroral flux tubes into space.

In this paper we report occurrence distributions of counterstreaming electrons observed earthward of 8.8 R_e in the earth's magnetosphere. We utilize data from the Hot Plasma Composition Experiment (HPCE) on the AMPTE Charge Composition Explorer (CCE) satellite [Shelley et al., 1985]. The AMPTE/CCE was launched in August 1984 into a 4.8° inclination orbit with apogee of 8.8 R_e and perigee of 1108 km. During its 15.7 hour orbital period it spends most of its time beyond 6 R_e. The satellite was spin stabilized at 10 rpm with its spin axis pointed approximately toward the sun. The HPCE instrument incorporated a set of eight fixed-energy electron spectrometers which operated simultaneously and nearly continuously between August 1984 and January 1989. Each magnetic spectrometer had an energy resolution of 50% and together the eight spectrometers covered the energy range from 50 eV to 25 keV (Table 1). The spectrometers were co-aligned with the centers of their fields-of-view perpendicular to the spacecraft spin axis. Each was collimated with a 5° full width conical field of view. Measurements were made in unison by all eight spectrometers every 155 msec. With the 6-sec satellite spin period an 8-point energy spectrum was thus obtained every 9.5° of satellite rotation.

DATA SET

The data base used for this investigation was extracted from the high resolution data by computing averages of the differential electron flux at each of the eight energies listed in Table 1 over 6.4 minute intervals. The data were binned into pitch angle bins 10 degrees wide and collapsed into nine pitch bins relative to the magnetic field direction. Thus, for example, electrons in pitch bin 1 all had pitch angles within 10° of ±**B**, and those in bin 9 had pitch angles of 90±10°. In this process the information on the direction of travel of the electrons was thus lost (electrons with pitch angles between 0° and 10° were averaged together with those between 170° and 180°). As will be explained below this method of compression had no adverse effect on the study. It is important to note that pitch angles referred to in this paper are true pitch angles derived from simultaneous on-board measurement of the magnetic field vector. It would not have been possible to conduct this study using pitch angles derived from a model of the magnetic field. The instantaneous magnetic field direction and magnitude in the near equatorial magnetosphere can have very large departures from a model field owing to deviations related to substorm variations. Magnetometer data used to compute pitch angles were supplied by the Johns Hopkins University/Applied Physics Lab magnetometer experiment [Potemra et al., 1985].

Data used in this study were taken from all available orbits between August 1984 and December 1986. At the time this survey was conducted, processed data were not available from July 1985 through April 1986. During the more than 530 days of observation time (~12,500 hours) the detectors were not always able to view the field-aligned direction owing to the tilt of the geomagnetic field. Figure 1 shows the distribution in radial distance versus local time of those 6.4 minute samples in which the centers of the detector view cones came within 10° of **B**. This condition was met about 38% of the time (~4750 hours). Each datum on the figure represents the satellite location at the beginning of a 6.4 minute interval. For the period of study included here the satellite covered essentially all local times. Some coverage is absent at large radial distances in the early afternoon sector. In the pre-midnight to midnight local time sector the coverage becomes more sparse in part due to the relatively poorer seeing conditions. In this sector the magnetic field direction often lies outside the fields of view as the geomagnetic field tends to depart more often from near dipolar in this sector in response to substorm growth phase activity.

An anisotropy "index" was formed for each energy channel for each 6.4 minute interval throughout the orbit by calculating the ratio between the flux in the field-aligned bin (0°-10°) to that in the trapped bin (80°-90°). The eight flux ratios thus formed for each averaging interval constitute the measure which is analyzed in this paper. In this paper we use index values exceeding 1.5 as indicators of counterstreaming electrons.

To verify and "calibrate" this measure the raw high resolution (155 msec) data were examined for hundreds of averaging intervals. Flux versus time plots like the example shown in Figure 2a,b were produced and compared to the indices produced from the averages (in the example, only the third and fourth energy channels are plotted for clarity). The anisotropy indices corresponding to this interval were 8.1 at .342 keV

TABLE 1. AMPTE/CCE Electron Spectrometer Characteristics

Chan. No.	Center Energy (keV)	Lower Energy Limit (keV)	Upper Energy Limit (keV)	Minimum Flux[*] (1 count/sample)
1	0.067	0.050	0.838	8.56×10^6
2	0.151	0.113	0.189	2.20×10^6
3	0.340	0.255	0.425	1.20×10^6
4	0.768	0.576	0.960	5.75×10^5
5	1.735	1.30	2.17	2.59×10^5
6	3.92	2.94	4.90	1.20×10^5
7	8.85	6.64	11.1	6.38×10^4
8	20.0	15.0	25.0	3.58×10^4

[*] electrons/cm^{2}sec sr keV

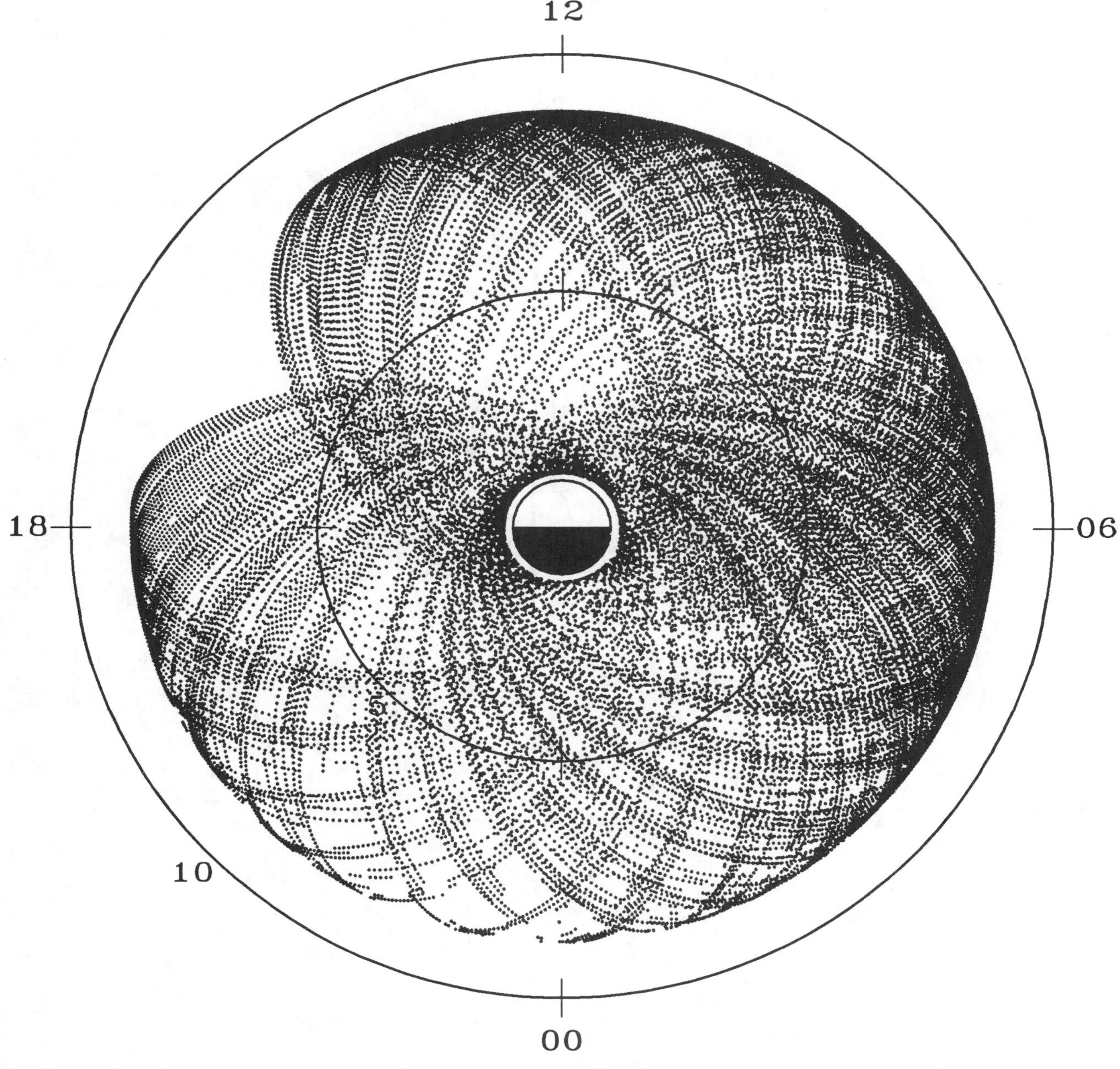

Fig. 1. Locations of the CCE satellite plotted in local time versus radial distance for each 6.4-minute interval used in this study. Only those intervals are shown for which the electron spectrometers sampled within 10° of the field line. The period covered in this survey was August 1984 through June 1985 and May - December, 1986.

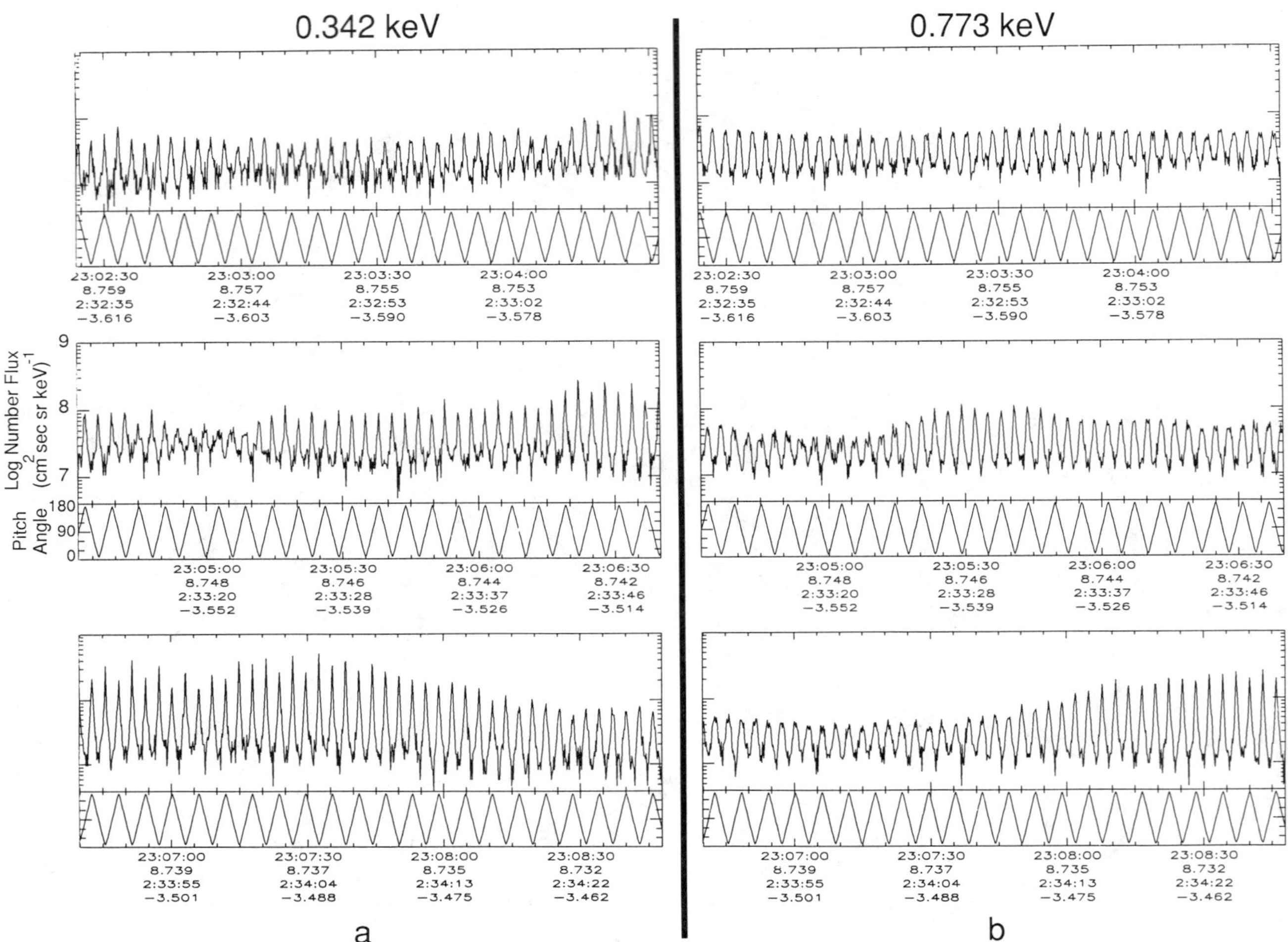

Fig. 2. Example of high resolution data for a 6.4 minute interval. The indicated anisotropy indices from the computer search were 8.1 at .342 keV (Fig 2a) and 4.6 at 0.773 keV (Fig. 2b).

(Figure 2a) and 4.6 at 0.773 keV (Figure 2b). Visual examination of Figures 2 a and b confirms that strong flux anisotropies exist throughout most of this 6.4 minute interval. In the latter half of the interval shown the field-aligned flux exceeds the trapped flux by more than an order of magnitude at both energies shown. These are not unusually large anisotropies. By examining large numbers of such examples it was found that anisotropy indices >1.5 were a reliable indicator of the presence of counterstreaming electrons within the 6.4 minute average. In this paper we refer to these periods of >1.5 anisotropy as counterstreaming electrons. It should be clear that our averaging technique, which combines 0-10° and 170-180° together, actually precludes us from distinguishing between unidirectional streaming and counterstreaming electrons. In fact the verification stage confirmed to us that

anisotropy index values exceeding 1.5 were almost never due to unidirectional streaming.

This verification stage also revealed that anisotropy measures exceeding 50 were most often indicative of telemetry noise, as were random single (limited to one energy channel) large anisotropies. In order to obtain a clean data set it was prudent to put additional constraints or selection criteria on the index data prior to performing the analysis presented in this paper. To lessen the effects of poor statistics we eliminated any 6.4 minute interval that had less than 6.2 minutes of data (such occurrences can arise from telemetry dropouts). It has been found that true counterstreaming events are almost never monoenergetic [Klumpar et al., 1988]. When they occur, field-aligned electron anisotropies are spread over a range of energies and therefore almost always involve several

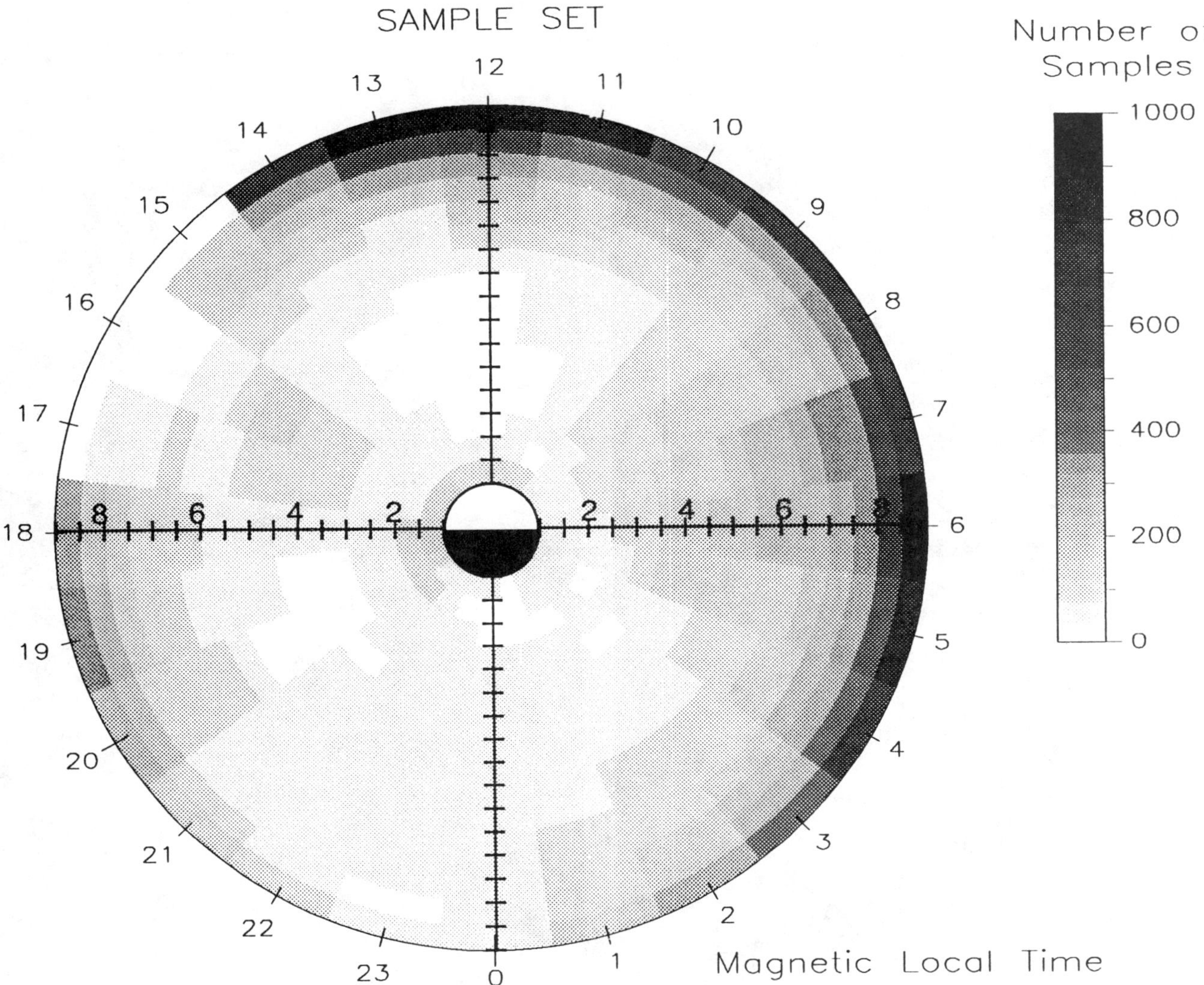

Fig. 3.   Data from Figure 1 binned into 0.5 R_e by one hour magnetic local time bins.   The number of 6.4-minute observing periods for each bin is indicated by the level of grey shading.

STATISTICAL RESULTS

instrument energy channels.   In contrast, telemetry noise spikes are often limited to a single channel (while the data from the 8 spectrometers are acquired in parallel, they are telemetered out serially).   In order to greatly diminish the possibility that telemetry noise can contribute false positive anisotropies we required that the anisotropy threshold be exceeded on two or more channels before the indices were considered for further analysis.

Figure 3 shows the sample set from Figure 1 organized into 1-hour Magnetic Local Time by 0.5 R_e radial distance bins. Represented by grey-scale shading are the number of 6.4-minute-average samples in each bin during which pitch angles were sampled within 10° of **B**.   This sample set then constitutes the ensemble of measurements from which counterstreaming electrons could have been observed if present.   This sample set forms the basis for computing the probability distributions shown in subsequent figures.   The least populated bins at large radial distances near 1600 MLT contain no samples.   The most heavily populated bins at large radial distances from early morning thru dawn and through noon to about 1400 MLT contain between 500 and 880 samples, each of 6.4-minutes duration.   In the relatively sparsely sampled area between 2100 and midnight and beyond 6 R_e there were as few as 38 samples (8.25 R_e at 2300 MLT), but most bins in this region had more than 50 samples or more than 5 hours of observing time each.

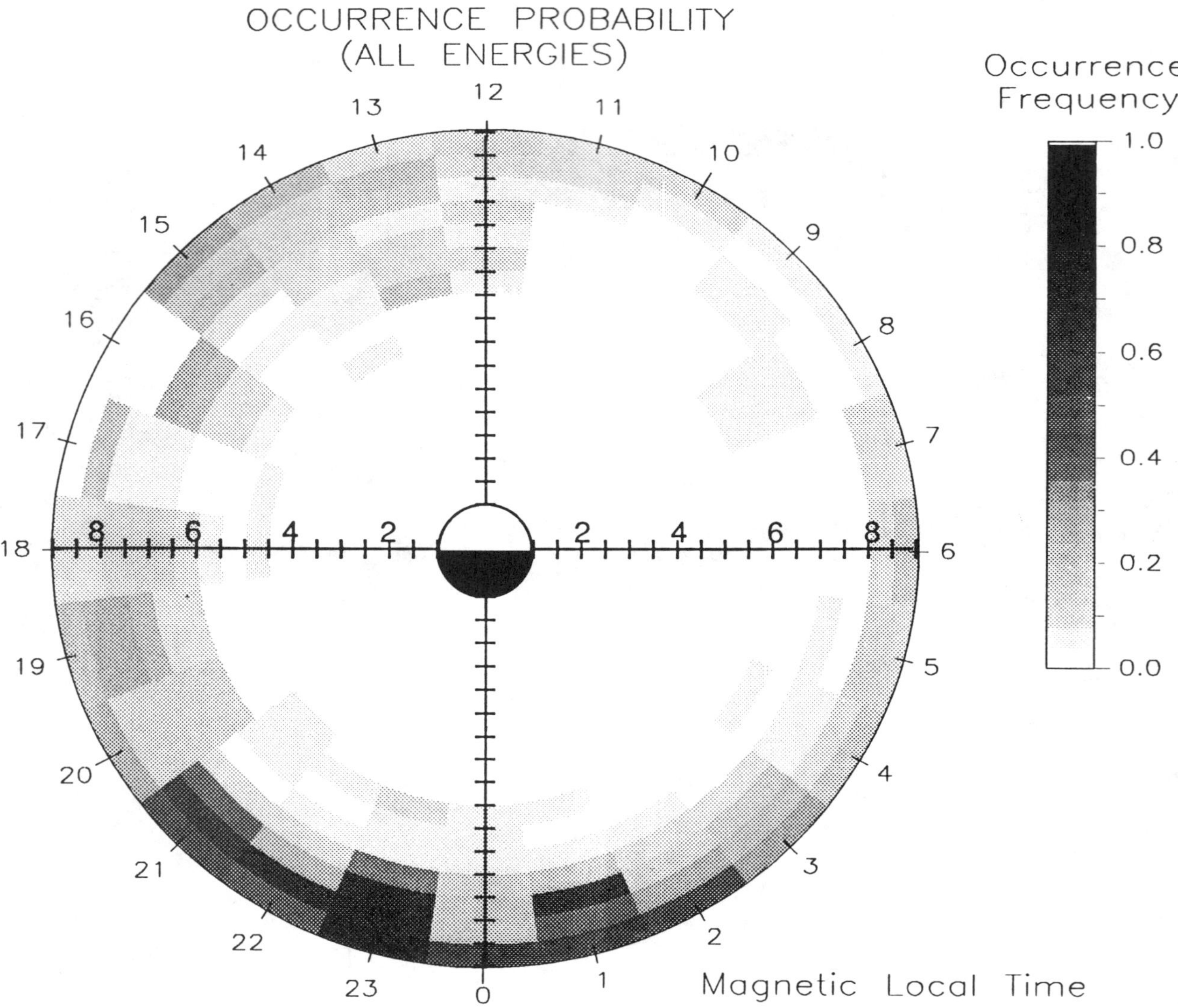

Fig. 4. The probability of occurrence of anisotropy index exceeding 1.5 in any two energy channels.

A more important consideration is that some of these bins are dominated by two or three orbits. Thus the particular geophysical conditions extant on these orbits can have a strong influence on the resulting occurrence probabilities. We recognize this shortcoming and are in the process of expanding the study to include all available data for the >4-year lifetime of the satellite.

Figure 4 shows the probability that the 6.4-minute anisotropy index exceeds 1.5 in any two of our eight energy channels between 0.067 and 20.0 keV. The figure shows that, at 8.75 ±.25 R_e, the greatest probability for detection of field-aligned counterstreaming electrons occurs near midnight and the lowest probability for detection between 0700 and 1100 MLT. Occurrence probabilities ≥40% were found in this radial distance range from 2100 MLT to 0200 MLT. Between 0700 and 1100 MLT occurrence probabilities were less than 10% at 8.75 ±.25 R_e. There appears to be a very real and striking pre-noon post-noon asymmetry in the occurrence distribution about noon with a clear preponderance of counterstreaming electrons in the afternoon sector in comparison with the morning sector. The radial distribution is also striking. We find little or no probability for occurrence of counterstreaming electrons inside of 6 R_e at any local time. The occurrence probabilities generally increase beyond 6 R_e with increasing radial distance out to the apogee of the CCE satellite.

In order to get some sense of the energy distribution of the occurrence of counterstreaming electrons we show in Figure 5 the probability of occurrence of index >1.5 in one of the four highest energy channels (E>1.3 keV). This figure shows that the highest probability for detection of counterstreaming

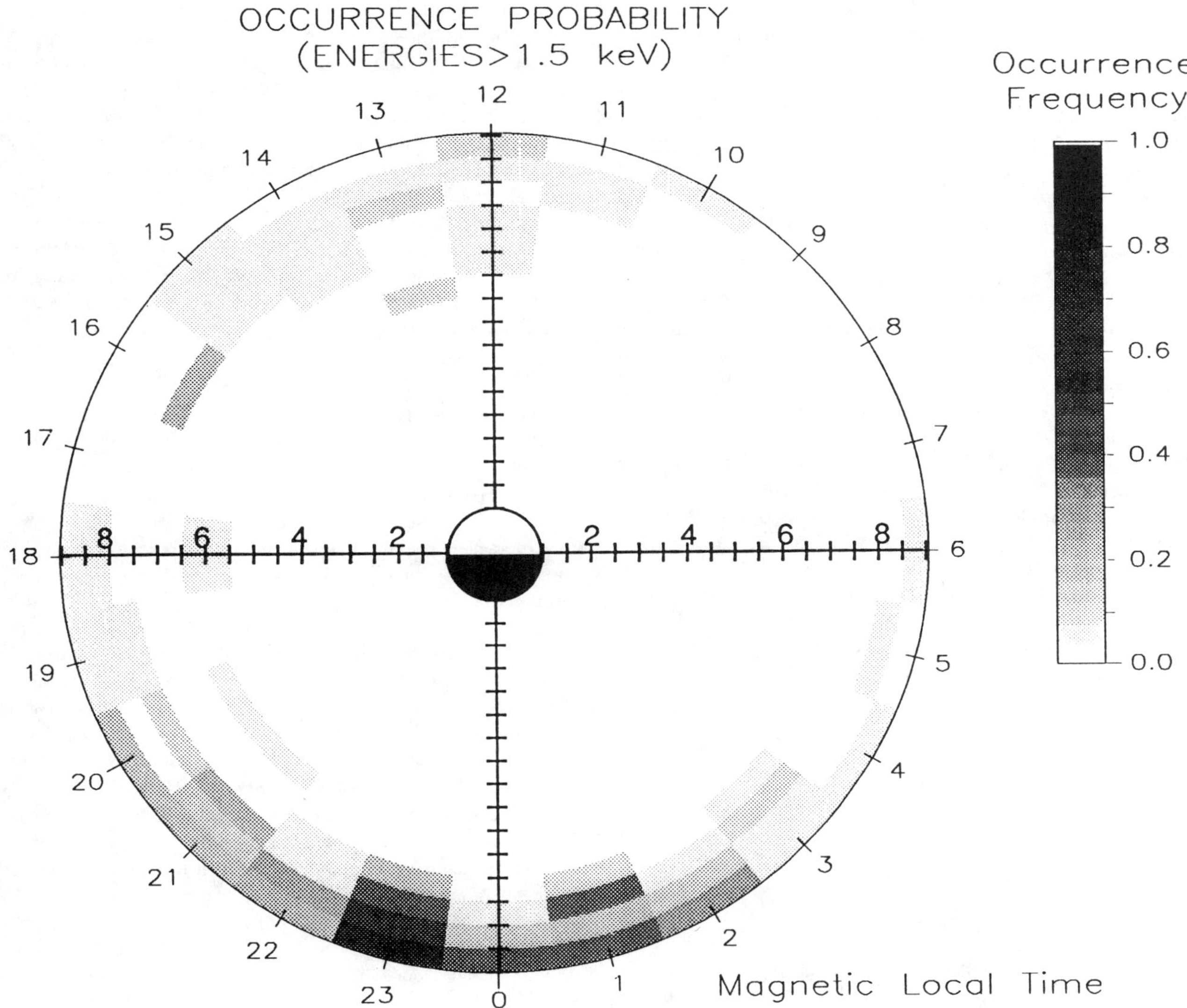

Fig. 5. The probability of occurrence of an anisotropy index exceeding 1.5 in two energy channels with one of them above 1.5 keV.

kilovolt electrons occurs near midnight. At 8.75 ±.25 R_e the occurrence rate exceeds 30% between 2300 and 0200 MLT, but then drops precipitously around the dawn sector with probabilities below 5% between 0600 and noon. No significant detection probability exists earthward of 7 R_e for counterstreaming keV electrons.

Comparing figures 4 and 5 illustrates that the more energetic events are likely to be found at larger radial distances and at local times throughout the dusk and midnight sectors.

DISCUSSION

We have shown that for electrons between 50 eV and 25 keV highly anisotropic pitch angle distributions are common in the magnetosphere at radial distances beyond 6 R_e. Association of these distributions with auroral mechanisms has previously been suggested [McIlwain, 1975; Klumpar et al., 1988; Klumpar, 1990], owing to the extreme field-alignment of the electrons and to the general location where such distributions are seen. The measurements are made on L-shells that map to the statistical location of the auroral oval. While we offer no mapping of our counterstreaming electrons in this paper, Klumpar [1990] demonstrated that for several specific events observed from CCE the field-line mapping using the Tsyganenko field model showed the ionospheric projections lie within the statistical auroral oval. The pitch angle distributions, peaked within 10° of the local magnetic field, indicate that the electrons have mirror points at altitudes below 12000 km (assuming a dipole field and an equatorial pitch angle of 10° at L=7.5).

The present statistical studies of the occurrence distribution of near-equatorial counterstreaming electrons shows a striking similarity to the general occurrence morphology of the aurora. Figure 6 shows average intensity contours of 1.3 keV electrons measured at 1400 km above the auroral zone by ISIS-2 [after McDiarmid et al., 1975; fig. 6]. This plot, in invariant latitude magnetic local time coordinates, indicates weakest precipitation intensity of 1.3 keV electrons in the morning sector of the oval between 0600 and 1200 MLT. The strongest precipitation intensities at 1.3 keV were found by McDiarmid et al [1975] between 1500 and 1800, between 2100-2200 and between 0100 and 0500 MLT where intensities greater than 10^7 electrons/cm^{2}sec sr keV were observed. These local time distributions bear a striking resemblance to our counterstreaming electron occurrence distributions in Figures 4 and 5.

The behavior of the occurrence distributions restricted to high energies shown in Figure 5 are consistent with what one might expect for an auroral source. The highest probabilities were found near midnight near the outer limits of our observing region (near 8.8 R_e). Similarly the most intense and dynamic aurora are often observed near midnight. If backscattered primary electrons (distinguishing here from atmospheric secondaries) play a role in the counterstreaming electron distributions one would expect the highest energy counterstreaming electrons to locate on field lines that map to the most intense and most dynamic aurora.

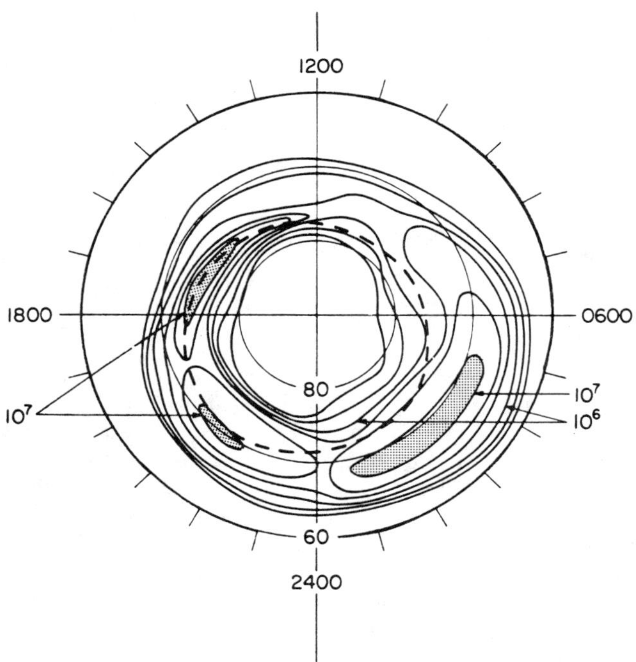

Fig. 6. Average intensity contours for 1.3 keV electrons measured at 1400 km by ISIS-2. Units are cm^{-2}s^{-1}sr^{-1}keV^{-1}, and adjacent contours differ in intensity by a factor of about 2.2. The dashed line is the 35-keV Alouette 2 background boundary; $K_p \le 3_0$, and $\theta < 45°$. Coordinates are invariant latitude and MLT. From McDiarmid et al., 1975.

In our view there are several possible mechanisms related to the aurora and/or electron precipitation in the ionosphere that can be responsible for populating closed magnetospheric flux tubes with counterstreaming electrons. It is well known that active processes are at play on auroral field lines. Upward accelerated electrons have been observed in the auroral region as pointed out earlier, as have counterstreaming electrons on auroral field lines near the auroral acceleration region. Those low altitude electrons are constrained to follow the field lines and would be expected to fill the flux tubes near the loss cone, along its entire length. Thus the presence of counterstreaming electrons crossing the magnetic equator on field lines that map to the auroral zone should not be unexpected.

SUMMARY

We have analyzed the occurrence distribution of highly field-aligned counterstreaming electrons in the terrestrial magnetosphere out to 8.8 R_e. We have found that counterstreaming electrons within 10° of **B** at energies between 50 eV and 25 keV have high probabilities (>50%) of being observed near midnight in the outermost portion of the region sampled. We have previously associated the occurrence of counterstreaming electrons with auroral electron precipitation at the ionospheric ends of the flux tubes [Klumpar et al., 1988; Klumpar, 1990] and find the statistical distributions presented here to be consistent with that hypothesis. We suggest that there are a number of auroral processes, both active and passive, which contribute to the flooding of closed auroral flux tubes throughout the magnetosphere with low to medium energy counterstreaming electrons. This heretofore largely neglected population of particles is important in two respects. As an operational tool these electrons provide an opportunity to further quantify the mapping between low and high altitudes in the magnetosphere. The proper mapping of magnetic field lines from the ionosphere to the magnetosphere is especially controversial particularly at latitudes which thread the auroral zone. The presence of a very distinctive electron population such as that discussed here can serve as a highly efficient natural tracer of the equatorial crossing point of closed auroral flux tubes. Studies are now in progress with this data set to utilize simultaneous auroral imaging near the ionospheric footpoints of CCE field lines in an attempt to relate specific auroral features to specific streaming events.

A second possibly important implication is that the highly anisotropic distributions contained in these counterstreaming electrons may be unstable to the generation of plasma waves. Instability analyses need to be performed to determine whether these electrons may be responsible for the generation of waves which may have a destabilizing influence on trapped populations.

Acknowledgements. This work was supported by NASA under contract number NAS5-30565.

REFERENCES

Arnoldy, R. L., Fine Structure and Pitch Angle Dependence of Synchronous Orbit Electron Injections, *J. Geophys. Res., 91,* 13411, 1986.

Borg, H., L. -A. Holmgren, B. Hultqvist, F. Cambou, H. Reme, A. Bahnsen, and G. Kremser, Some Early Results of the keV Plasma Experiment on GEOS-1, *Space Science Reviews, 22*, 511, 1978.

Collin, H. L., R. D. Sharp, and E. G. Shelley, The Occurrence and Characteristics of Electron Beams over the Polar Regions, *J. Geophys. Res., 87*, 7504, 1982.

Hada, T., A. Nishida, and T. Terasawa, Bi-Directional Electron Pitch Angle Anisotropy in the Plasma Sheet, *J. Geophys. Res., 86*, 11211, 1981.

Hultqvist, B., R. Lundin, K. Stasiewicz, L. Block, P.-A. Lindqvist, G. Gustafsson, H. Koskinen, A. Bahnsen, T. A. Potemra, and L. J. Zanetti, Simultaneous Observation of Upward Moving Field-Aligned Energetic Electrons and Ions on Auroral Zone Field Lines, *J. Geophys. Res., 93*, 9765, 1988.

Hultqvist, Bengt, On the Acceleration of Electrons and Positive Ions in the Same Direction along Magnetic Field Lines by Parallel Electric Fields, *J. Geophys. Res., 93*, 9777, 1988.

Klumpar, D. M., Near Equatorial Signatures of Dynamic Auroral Processes, in *Physics of Space Plasmas (1989)*, SPI Conference Proceedings and Reprint Series, Number 9, pg. 265, Scientific Publishers, Inc, Cambridge MA, 1990.

Klumpar, D. M., J. M. Quinn, and E. G. Shelley, Counter-Streaming Electrons at the Geomagnetic Equator near 9 R_e, *Geophys. Res. Lett., 15*, 1295, 1988.

Klumpar, D. M. and W. J. Heikkila, Electrons in the Ionospheric Source Cone: Evidence for Runaway Electrons as Carriers of Downward Birkeland Currents, *Geophys. Res. Lett., 9*, 873, 1982.

Kremser, G., A. Korth, S. L. Ullaland, S. Perrault, A. Roux, A. Pedersen, R. Schmidt, and P. Tanskanen, Field-Aligned Beams of Energetic Electrons (16 keV $\leq$ E $\leq$ 80 keV) Observed at Geosynchronous Orbit at Substorm Onsets, *J. Geophys. Res., 93*, 14453, 1988.

Lin, C. S., B. Mauk, G. K. Parks, S. DeForest and C. E. McIlwain, Temperature Characteristics of Electron Beams and Ambient Particles, *J. Geophys. Res., 84*, 2651, 1979.

Lin, C. S., J. L. Burch, J. D. Winningham, J. D. Menietti, and R. A. Hoffman, DE-1 Observations of Counterstreaming Electrons at High Altitudes, *Geophys. Res. Lett., 9*, 925, 1982.

Lundin, Rickard, and Lars Eliasson, Auroral Energization Processes, *Ann. Geophysicae, 9*, 202, 1991.

McDiarmid, I. B., J. R. Burrows, and E. E. Budzinski, Average Characteristics of Magnetospheric Electrons (150 eV to 200 keV) at 1400 km, *J. Geophys. Res., 80, 73*, 1975.

McIlwain, C. E., Auroral Electron Beams near the Magnetic Equator, in *Physics of Hot Plasma in the Magnetosphere*, Plenum Publishing Corp., pg. 91, 1975.

Moore, T. E. and R. L. Arnoldy, Plasma Pitch Angle Distributions near the Substorm Injection Front, *J. Geophys. Res., 87*, 265, 1982.

Parks, G. K., C. S. Lin, B. Mauk, S. DeForest, and C. E. McIlwain, Characteristics of Magnetospheric Particle Injection Deduced from Events Observed on August 18, 1974, *J. Geophys. Res., 82*, 5208, 1977.

Potemra, T. A., L. J. Zanetti, and M. H. Acuna, The AMPTE/CCE Magnetic Field Experiment, *IEEE Transactions on Geoscience and Remote Sensing, GE-23*, 246, 1985.

Richardson, J. D., J. F. Fennell, and D. R. Croley, Jr., Observations of Field-Aligned Ion and Electron Beams from SCATHA (P78-2), *J. Geophys. Res., 86*, 10105, 1981.

Sharp, R. D., E. G. Shelley, R. G. Johnson, and A. G. Ghielmetti, Counterstreaming Electron Beams at Altitudes of ~1R_e over the Auroral Zone, *J. Geophys. Res., 85*, 92, 1980.

Shelley, E. G., A. Ghielmetti, E. Hertzberg, S. J. Battel, K. Altwegg-VonBurg, and H. Balsiger, The AMPTE/CCE Hot Plasma Composition Experiment (HPCE), *IEEE Transactions on Geoscience and Remote Sensing, GE-23*, 241, 1985.

D. M. Klumpar, Space Sciences Laboratory, Dept. 91-20, Bldg. 255, Lockheed, 3251 Hanover Street, Palo Alto, CA 94304

Diffusion of Echo 7 Electron Beams During Bounce Motion

ROBERT J. NEMZEK[1]

Los Alamos National Laboratory, Los Alamos, NM

The Echo 7 sounding rocket experiment injected electron beams into the magnetosphere and detected them after one or more bounces along field lines near $L = 6.5$. Waves with equatorial amplitudes of a few mV/m diffused the beams so that only $\approx 20\%$ of the initial current returned to the rocket altitude in the northern hemisphere. On successive bounces the electron flux continued to drop at the same rate. These results imply a lifetime of ≈ 1.7 s for 20 keV electrons just outside of the loss cone. Comparison with other Echo flights shows that the beam return is dependent upon geomagnetic conditions: low activity causes there to be less scattering, while high activity can actually prevent detection of the returning beam.

INTRODUCTION

The Electron Echo program at the University of Minnesota comprised seven flights of sounding rockets carrying high-power electron accelerators. The first flight was from Wallops Island, VA; the second from Churchill, Man.; and the last five from Poker Flat, AK. All of the flights had a common experimental method: electron beams were injected up magnetic field lines into the magnetospheric tail region, so that they would mirror or scatter at the southern conjugate point, and then were detected by instruments on the payload when they returned to the northern hemisphere. This "conjugate echo" process is demonstrated schematically in Figure 1. The injection of electron beams into the ionospheric plasma creates a wide variety of plasma physics phenomena (see the review by Szuszczewicz [1985]), but the echo technique allows the beams to be used to measure aspects of the large-scale magnetosphere as well. For instance, determining the bounce time and energy of the echoing electrons makes possible the calculation of the length of magnetospheric magnetic field lines [Nemzek et al., 1992], and finding the net drift of the electrons after their bounce allows one to deduce magnetospheric electric fields [Nemzek and Winckler, 1991]. This paper deals with a third measurement of the echoing electrons: not when they returned or where, but how much of the original beam survived the entire trip through the magnetosphere in nearly unchanged form. Scattering of the original beam occurs most easily near the equatorial region, where pitch angles are smallest and wave-

induced pitch angle diffusion is most effective. The pitch angle diffusion affecting the echoing artificially-injected electrons is the same as the diffusion that causes the natural trapped radiation to precipitate and create the diffuse aurora. Lyons and Thorne [1973] calculated that for outer belt electrons lifetimes determined by pitch angle diffusion can be orders of magnitude smaller than those due to Coulomb collisions. Also, during their initial bounce, the electrons remained well above the atmosphere. Thus collisions were negligible, and we will assume that the measured lifetime for the beam electrons was determined solely by wave-particle interactions. Using measurements of the conjugate echoes, then, we will estimate parameters important in the behavior of the trapped radiation: equatorial wave strengths and electron lifetimes.

THE ECHO 7 EXPERIMENT

Echo 7 [Winckler et al., 1989] consisted of four instrumented subpayloads lofted to 292 km altitude by a Black Brant IX launch vehicle. In flight, three of the subpayloads- the Nose, the Plasma Diagnostics Package (PDP), and the Energetic Particles Package (EPP)- were spring-ejected from the Main (or Accelerator) payload. The ejected subpayloads drifted away from the Main at about 1 m/s, measured normal to B. The PDP was directed magnetically SW, and the EPP NW, so the separation between them was primarily in a N-S direction, which will become important in the echo analysis. Nose data will not be used in this study. The subpayloads were actually injected upward at an angle to B, but since the echoing beams moved quite rapidly parallel to the magnetic field, the subpayload separation along B was unimportant. The exact directions and separation velocities of the subpayloads are in general not needed for conjugate echo analysis; however it was vital that the subpayloads carried the instruments away from the Main. Injection of the powerful electron beam created a large disturbed region around the

---

[1]Formerly at: School of Physics and Astronomy, University of Minnesota, Minneapolis, MN.

Auroral Plasma Dynamics
Geophysical Monograph 80
Copyright 1993 by the American Geophysical Union.

accelerator payload which would have seriously disrupted attempts to measure echoes on the beam-emitting payload itself.

The data used in this study came from the PDP and EPP. These were outfitted with identical sets of echo-detecting instruments: four scintillation detectors sensitive to the total electron energy flux > 1 keV, and two electron spectrometers (TED's) capable of measuring the energy spectrum between 2 and 40 keV every 52 ms. These instruments are described fully in Nemzek [1990]. The detectors were evenly spaced in azimuth about the payload spin axis. Since the payloads were tipped with respect to B, the instruments scanned through both azimuth and pitch angle as the payloads rotated.

The Echo 7 electron gun was a diode type, capable of up to 250 mA current at 40 kV acceleration. The actual current emitted into space was not measured directly, just the current leaving the filament, but only a few percent of the filament emission was lost inside the gun [P.R. Malcolm, personal communication, 1991].

The program used to control the sequence of accelerator pulses was 10 s long, and repeated without interruption throughout the flight. There were two distinct types of injections: Discrete and Continuous. The Discrete pulses were quasi-DC in voltage and current, most of them being at 36 keV and 180 mA. These were intended primarily for provoking plasma physics effects and providing "marker pulses" (artificial auroral streaks) to aid in optical detection of echoes.

The Continuous injections were designed so that during each millisecond of a pulse, the injected energy would decrease exponentially from 40 keV to 8 keV. Since the Echo 7 diode had space-charge-limited emission, the emitted current dropped as well, from 210 mA to 20 mA. The net result of the Continuous pulses was an injected beam with a continuous spectrum of energies from 40 keV to 8 keV. This was vital for the successful detection of echoes, as will be explained.

Continuous and Discrete pulse lengths were mostly 50, 100, or 150 ms with similar gaps between. The Continuous injections had a complex coding of on-off times constructed so that any sequence of these injections longer than 400 ms was unique within the entire gun program. This was the basis for identifying bounce times in the conjugate echo data. All of the Continuous and many of the Discrete injections had 110° pitch angles. With this pitch angle, assuming adiabatic motion, the injected electrons mirrored at several-hundred-kilometer altitudes at the southern conjugate point but were within the loss cone upon their return to the north.

Echo 7 was launched from Poker Flat, AK Feb 8, 1988 at 2316:49 local time, or 0816:49 Feb 9 1988 UT. Except for a few minor problems, the flight went as planned and met all of its objectives. The magnetosphere during the flight was moderately disturbed (Kp = 3-), and the rocket's westerly trajectory kept it within a region of intense diffuse auroral precipitation throughout the flight. This precipitation will figure in the interpretation of the echo results.

CONJUGATE ECHOES

In order to understand the results of the Echo 7 pitch angle diffusion measurements, one must have some familiarity with the physics of conjugate echoes. What follows will be a brief summary; more complete versions in various styles may be found in publications such as Winckler [1982], Malcolm [1986], Nemzek [1990], and Nemzek and Winckler [1991]. To begin, we can assume that the injected electrons travel through the magnetosphere adiabatically. This can be inferred both from measurements made on the conjugate echoes themselves [Nemzek, 1990; Nemzek et al., 1992], and from the Larmor spiral of the beam recorded by a low-light video camera carried by the PDP [Winckler et al., 1989; Franz, 1991]. While the electrons execute their single-particle motion through the magnetosphere, they drift across magnetic field lines due to the effect of magnetic gradient-curvature (V_{GC}) drifts and electric field (ExB) drifts (Figure 1). The gradient-curvature drift is energy dependent, and so high energy electrons drift further to the west than low energy ones. In order for an echo to be detected upon its return to the northern ionosphere, these particle drifts must be matched by the payload's own drift normal to B. The situation can be visualized by reference to Figure 2. This depicts the payload and returning Continuous-injection echoes in a coordinate system centered on the payload, moving with the payload velocity. The plane of the figure is normal to B. The initially small Continuous beam has spread out several kilometers in an E-W direction, occupying a curve called the "echo locus." Figure 2 was constructed for constant electric and magnetic fields. The three loci shown correspond to three different times; changes in the payload's N-S magnetic velocity caused the change in position of the echo locus. The locus labelled as 350 s intersects the payload: this is the flight time when an echo would have been detected.

The point of interest in this discussion of the echo locus is that the payload and the echo intersect at one point, and so the returning echo is measured as a monoenergetic burst of electrons, even though the initial injection had a continuous spectrum. This makes it easy to see why the Discrete injections should be very uncommon as echoes: they can't undergo energy dispersion, and so occupy a very small region of space when they return,

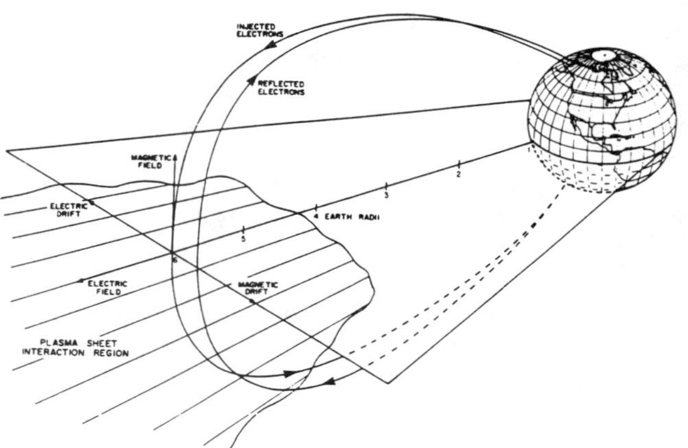

Fig. 1. A schematic depiction of electrons being injected from the northern hemisphere, mirroring, and returning.

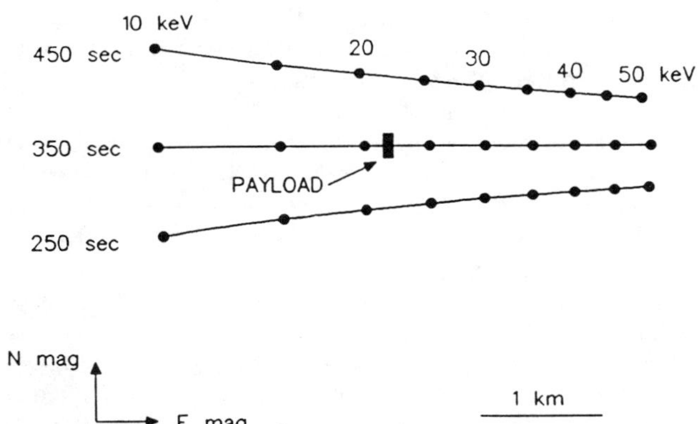

Fig. 2. The "echo locus," the region of space occupied by a returning Continuous energy injection. The coordinate system moves with the injecting payload. Energies are in keV.

requiring a precise match of particle and payload drifts. The Continuous injections have a wide range of electron drifts and so can accommodate a wide range of payload drifts.

Figure 2 was constructed for a constant electric field, but this was not the case during the flight of Echo 7. The electric field actually varied by a large amount, moving the echo locus around the payload in a complex fashion, rather than the simple south-to-north progression shown in Figure 2. This caused echoes to be detected at many different energies at many different flight times [Nemzek and Winckler, 1991].

These ideas about conjugate echoes can be illustrated with examples from the Echo 7 data. Figure 3 shows the best bounce time measurement available in the data set. The measurements come from two of the PDP scintillators. The distinct pulses rising

above the background level are conjugate echoes. The data are plotted on a 5-decade log scale: some of the echoes are actually about 1000x more intense than the background flux. The open rectangles below the data depict the sequence of gun pulses. The scintillator peaks had no relation to concurrent injections, but one region did match pulse-for-pulse with a sequence of injections that began 2.77 s earlier, as shown by the dotted lines. This identifies the bounce time. Figure 4 shows the energy measurement for this same string of conjugate echoes. The nine consecutive scans through the energy range of the spectrometer registered nothing but background levels except at 18 keV. So, we know that during the flight of Echo 7, 18 keV electrons required 2.77 s to travel to the conjugate point and back.

"LOSS" OF ECHOING ELECTRONS

Determining what fraction of the injected beam returned to the payload requires a calculation of the electron current arriving at the payload altitude during an echo pulse. This can then be compared directly to the emitted beam current. The comparison must be done for a specific energy, because the echoes were monoenergetic and the beam current varied with acceleration voltage. The calculation of the current at one energy in a measured echo is a simple process once certain characteristics of the returning beam have been estimated, namely its energy spread, pitch angle distribution, and spatial extent. These values will then be applied to the measurement of the echo flux made by the electron spectrometers.

The spectrometer measurements of the echoes had a full-width-at-half-maximum of about 15%. However, the spectrometers only had a full-width resolution of 8%, so the true energy width of the echoes must have been less than 15%. In addition, the broad pitch angle distribution of the echoes would have increased their

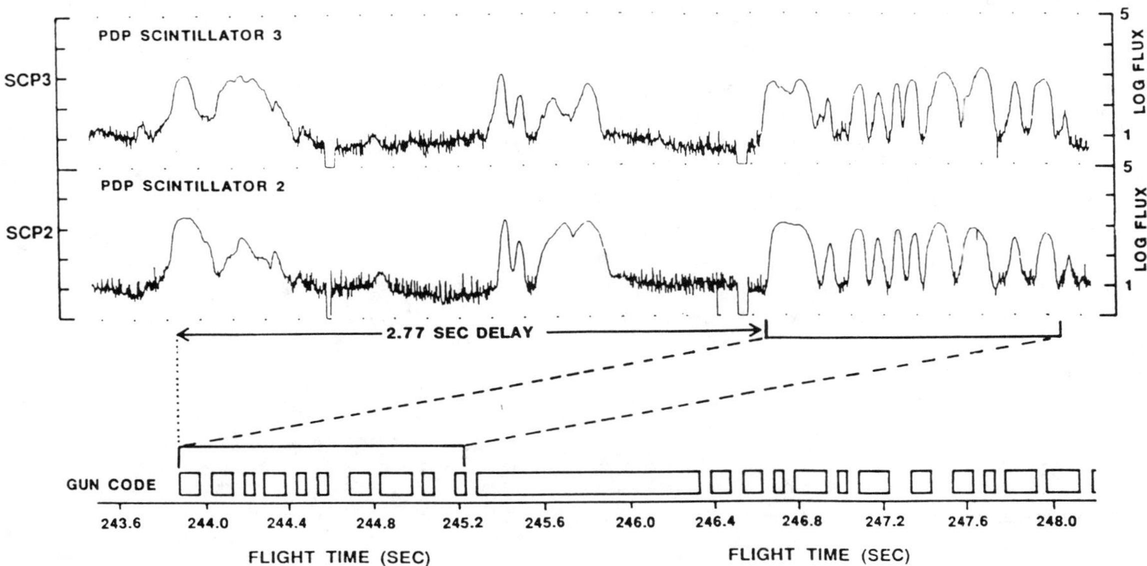

Fig. 3. As shown by the dotted lines, this episode of conjugate echoes originated in a set of injections that occurred 2.77 s earlier.

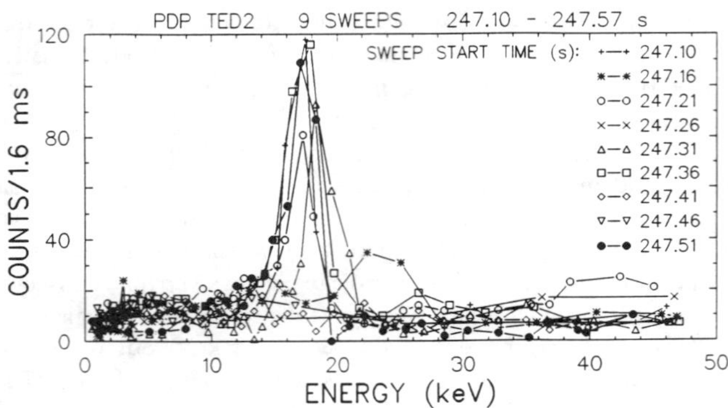

Fig. 4. A set of spectrometer traces corresponding to the echo train of Figure 3, showing the monoenergetic nature of the echoes.

apparent energy spread. For the accuracy required here, we can take the width to be 0, i.e., assume that the echoes were truly monoenergetic at each point of the locus.

As they scanned through pitch angles, the scintillation detectors measured echo electrons with pitch angles down to 63°. There was little modulation of echo intensity with pitch angle. The University of New Hampshire supplied electron detectors that crossed part of the echo energy range; these detected some echoing electrons at pitch angles down to 40°. With these two facts, we will assume that the echoes actually were isotropic over the entire down-coming hemisphere. This is reasonable also in light of the pitch angle diffusion indicated by the diffuse aurora. Strong pitch angle diffusion isotropizes electrons over the equatorial loss cone, which encompasses most of the down-coming hemisphere at rocket altitudes.

An estimate of the physical size of the echo locus at a particular energy can be done in several ways. First of all, echoes as seen at the PDP and EPP can be compared. In general, echoes were detected on either the PDP or EPP, but not on both at the same time. At a few times echoes were observed simultaneously on both; however, such examples never had the same pulse sequence on the two payloads. This seems to point to an unusual, spatially-varying electric field configuration that greatly modified the simple echo locus picture of echo detection [Nemzek, 1990]. The first echoes were observed near 200 s flight time, when the two payloads were separated by about 100 m in a north-south direction. This was during a "normal" time, when echoes appeared only on one subpayload. Thus the N-S width occupied by a single echo energy could not have been much more than 100-200 m.

The two-payload comparison identifies the echo locus as a relatively small structure, at least in one dimension. Confirmation of this can be found by examining the time history of individual echoes as they passed over one of the payloads. Because the echo drift never completely cancels the payload motion, there is some residual velocity that drags the locus over the payload. If the echoes in Figure 3 were displayed on a linear rather than log scale, the echoes would not be rounded blocks but rather sharp triangular forms. This means that the echoes had an intense core

that convected over the payload in a time short compared to the pulse width. The echo full-widths-at-half-maxima were in almost all cases 40-50 ms. The total convection speed was on the order of 1.3 km/s [Nemzek and Winckler, 1991]; this places an upper limit on the echo core of 65 m. The residual drift velocity was undoubtedly much less than the total convection velocity; it could have been as little as a few m/s. The central, most intense part of the echoes must have been very small indeed.

The best determination of the size of the echoes comes from examining the azimuthal anisotropy of the echoes as they passed over the payload. This in turn can be used to discover the guiding center density of the echoes as a function of time. Figure 5 (upper) shows the azimuthal record of an echo as a function of time. Each of the four scintillators is represented by an arrow pointing in the look direction of the detector; the set of arrows rotates with time due to payload spin. The directions are plotted in a magnetic coordinate system as shown in the upper left. Each cross represents an average over 12.8 ms (8 samples). The length of the arrow is scaled according to the log flux registered by the detector. In this example, the first two samples are at background level. In successive samples, the echo begins, builds, and once again dies away to background. The story of this echo is more easily seen in the anisotropy plot, Figure 5 (lower). Here, each

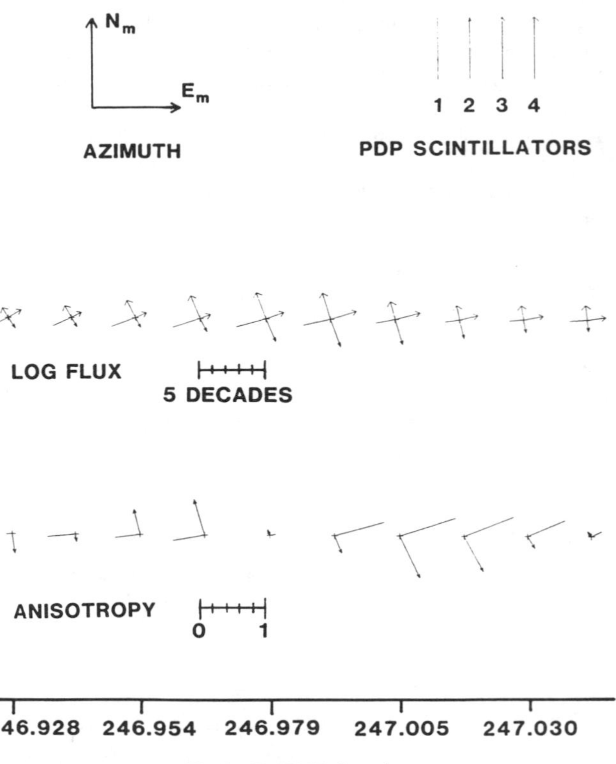

Fig. 5. (upper) The azimuthal character of a conjugate echo is demonstrated by this vector plot. Each arrow represents the log flux measured by a scintillator, and points in the look direction of that scintillator. (lower) The linear anisotropy derived from the azimuthal plot.

diametrically-opposed pair of scintillators has been used to construct an anisotropy, given by the difference between the two signals divided by their sum, using linear values for the scintillator flux. This quantity ranges from 0 (equal fluxes measured on both sides) to ± 1 (all flux measured on one side). The resulting vector points toward the look direction from which most of the flux originated. In this representation, the background flux is mostly isotropic. The flux then becomes highly anisotropic towards the NW as the echo intensity builds. At the center of the echo, the flux becomes almost exactly isotropic (vectors near zero length). Anisotropic fluxes return as the echo moves off of the payload, but now are to the SE. Finally, the background flux level is once again nearly isotropic. In terms of guiding center density, this means that the core of the echo approached from the SW, moved across the payload, and retreated to the NE.

All of the echoes followed this same basic pattern. The flux was initially strongly anisotropic, isotropized at the center of the echo, and become anisotropic again as the echo moved away from the payload. The direction of the echo drift changed from echo to echo, as did the amount of time the echo was isotropic. A typical length of time for the isotropic part of the echo was 30-40 ms. This indicates a very small period of time in which guiding centers were located on both sides of the payload, which in turn implies that the echoes were not much more than 2 gyrodiameters across. Such a geometry gives highly anisotropic edges combined with a small central isotropic region. Since there is no reason to expect that the lateral spreading at some energy should be greater in the EW direction than the NS direction, we will adopt 2x2 gyrodiameters as the dimensions of the echo region at each energy. At the highest energies, this amounts to a spot 50 m across.

With this estimate of the size of the echo area, it finally is possible to determine the current contained in the echo at some energy. We have completed this calculation for six groups of echoes which had measurable bounce times [Nemzek et al., 1991]. The resulting currents ranged from 4.6×10^{14} e/s at 12 keV to 1.9×10^{15} e/s at 23 keV.

Next the current emitted from the gun at the echo energies must be determined. The acceleration energy and emission current were linked through the relation $I[mA] = .833 \cdot V[kV]^{3/2}$. But, since the acceleration voltage decayed from 40 kV to 8 kV each millisecond, a particular energy was emitted for only a short period of time. In effect, a nearly instantaneous burst of current at some energy was spread out over the 1 ms decay until the corresponding part of the next acceleration voltage decay cycle. Thus the current at one energy was reduced by the duty cycle of the accelerator drive. The spectrometers sampled over a 1.6 ms period, so they were sensitive to this average current.

If a calculation of the average current was done using an exact energy, the result would be zero because any particular energy was emitted for a vanishingly small time. The echoes, however, could not be truly monoenergetic, since each energy occupied a finite region of the echo locus. The size of the locus derived above therefore can be used to calculate the energy spread of the detected echo. In a coordinate system that moves with the payload (as used in Figure 2), the distance between the injection

and detection positions of an echo of energy E_2 is just the net drift velocity times the bounce time: $X_2 = \tau_2(V_{GC}(E_2) + V_{ExB} - V_R)$, where V_{GC} is the gradient curvature drift velocity, V_{ExB} the ExB drift, and V_R the payload velocity normal to B. If an echo was detected at an energy E_1, then X_1 is by definition 0, i.e., $V_{GC}(E_1) + V_{ExB} = V_R$. The ExB and Payload velocities are the same for both energies; therefore $V_{GC}(E_1)$ can be substituted for them in the first equation. Finally, V_{GC} is directly proportional to the electron energy (in the non-relativistic regime), and so $\Delta X = 41.3(E_2-E_1)\tau_2$, where ΔX is X_2-X_1, and 41.3 [m/s/keV] is a gradient-curvature drift proportionality constant appropriate for the field lines on which Echo 7 flew (Nemzek and Winckler, 1991]. This now gives the distance between the guiding center locations of two energies on the echo locus. This equation can be inverted to give energy as a function of distance. From the estimate of echo size, we can say that an echo was composed of electrons ranging from a central energy to plus or minus the energy located one gyrodiameter away on each side. Putting in all of the numbers for the six echoes mentioned above gives an energy spread (ΔE) of .2 keV to .4 keV, the larger value for higher central energies. This ΔE is small enough compared to the central energy that it does not disturb the assumption of a monoenergetic echo used to calculate the returning echo current.

The next task is to calculate the time the gun's acceleration voltage sweep spent passing through the echo ΔE. This can be determined by using the exponential decay of the acceleration voltage. It turns out that for all energies the interval spent within the echo energy window was 1.7×10^{-5} s; the increasing duty cycle at low energies (on the "flat" part of the exponential decay) compensated almost exactly for the decreasing ΔE.

The total number of electrons available to an echo from each 1 ms cycle of the gun drive is just the emitted current integrated over the 1.7×10^{-5}-s interval. The current is proportional to $V^{3/2}$, and V is changing with time, but since the energy interval under consideration is so small (about 1% of the central energy) there is very little error incurred in simply multiplying the current appropriate to the central energy by the time interval. Finally, dividing the total number of emitted electrons by the 1 ms cycle time gives the average initial current that went into creating the echo. The resulting numbers were 3.7×10^{15} - 9.6×10^{15} e/s, the larger values corresponding to higher-energy echoes.

Payload charging must also be considered. A positive payload potential builds up very quickly as a result of electron emission. This reduces the energy of the emitted electrons as they travel out into the magnetosphere; they are detected at the reduced energy. This effectively alters the gun's energy-to-current relationship. Payload charging during Continuous pulse injection was about 2 kV [Winckler et al., 1989]. Thus, although the gun emitted 210 mA at 40 keV, for the echo measurement this was in effect 210 mA at 38 keV. At each echo energy, the appropriate current is that corresponding to an energy 2 keV higher. So, the currents in the previous paragraph should be slightly larger for each echo energy.

Figure 6 shows the result of all of these calculations: the percentage of the emitted beam current that returned to the payload altitude in the form of a conjugate echo. The values are

RETURNING BEAM FRACTION
PERCENT OF INITIAL BEAM CURRENT

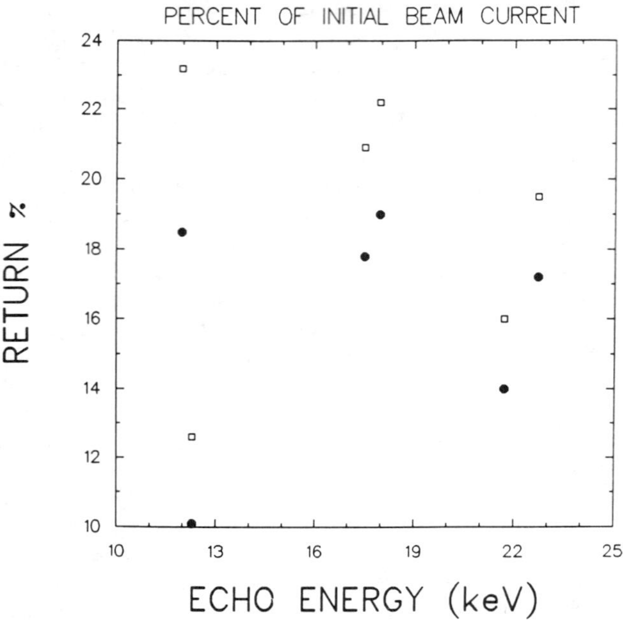

Fig. 6. Calculated fraction of the beam current that returned in various echoes. Open symbols show the direct calculation, filled symbols the correction expected with 2 kV payload charging.

around 10% - 20%; inclusion of payload charging causes them to drop slightly because each emitted gun current is effectively higher. There is a slight trend toward lower returns at higher energies, but it is difficult to put much stock into the trend with so few points. The obvious question is, what happened to the other 80% of the beam? Some of it was truly lost by being scattered into the southern hemisphere loss cone and impacting the conjugate atmosphere, but a large portion was simply diffused to larger equatorial pitch angles, causing it to mirror somewhere above the payload altitude, where it could not be detected.

Swanson [1983] performed a pitch angle diffusion calculation for the Echo 5 beam parameters. The Echo 5 accelerator was quite similar to the Echo 7 version, and so the calculations can be applied directly to the present situation. Swanson considered the effect on the beam of electrostatic turbulence at the equatorial plane. His results show that for a 20% beam return, wave fields of only 2-3 mV/m at the equator are required. This is well within the range of wave amplitudes that have been measured in the equatorial region [Kurth et al., 1979]. The presence of diffuse auroral precipitation on the Echo 7 field lines implies strong pitch angle diffusion. Wave amplitudes of ≈10 mV/m are sufficient for strong pitch angle diffusion near the equatorial region conjugate to Echo 7 [Lyons, 1974].

5. Comparison to Prior Echo Flights

Of the seven Echo flights, four successfully detected echoes:

Echo 1

Echo 1 flew at low latitude, receiving all of its echoes by scattering beams off the conjugate atmosphere. This resulted in a return of .75%, well below the expected value [Hendrickson, 1972].

Echo 2

Echo 2 detected no echoes, apparently because of active conditions and great amounts of pitch angle diffusion [Arnoldy et al., 1974].

Echo 3

Echo 3, the first flight from Poker Flat, detected echoes, but the percentage of beam current returned was never calculated.

Echo 4

Echo 4 had the highest return of any of the Echo flights, 30%. This number was determined from the analysis of echoes measured by a ground-based television camera [Hallinan et al., 1990]. Echoes were also detected by scintillation counters on the payload.

Echo 5

Echo 5 relied primarily on ground-based optical and space-based bremsstrahlung x-ray measurements. No echoes were detected. Swanson et al. [1986] reported an upper limit to the return of 13%, and attributed this to strong pitch angle diffusion by waves with amplitudes of 13 mV/m. In a re-analysis of ground-based data, Hallinan et al. [1990] reduced the upper limit to 2%. This makes the pitch angle diffusion hypothesis less tenable because wave fields on the order of 100 mV/m would be required to sufficiently diffuse the beam.

Echo 6

Echo 6 also measured no echoes. Optical conditions during this flight were so poor that an upper limit of 90% return was all that could be determined [Hallinan et al., 1990]. Malcolm [1986] explained the lack of rocket-detected echoes as the result of unexpectedly tail-like magnetic field lines.

Echo 7

Echo 7 had two return percentages calculated: the space-based measurements as reported here and an upper limit of 8% derived from the lack of optically-observed echoes [Hallinan et al., 1990]. There seems to be a discrepancy between the two numbers. However, the optical upper limit was based on the amount of light that could be expected to be produced by a 36 keV Discrete injection. The 20% return calculated here is for Continuous injections. The power contained in a Discrete pulse is much higher than the power of a Continuous pulse. If the optical upper limit had been calculated for Continuous injections as well, it would have been greater than the 20% derived from the in-situ spectrometer measurements.

The five flights from Poker Flat make for an interesting comparison. Echo 4, which flew during the quietest conditions and the weakest aurora, had the highest return. Echo 7, flown into a region of more intense diffuse auroral precipitation, had a somewhat lower return. Two flights flown across active discrete arcs (5 and 6) had no detectable beam return at all. Thus the amount of the beam which remains in the payload-altitude loss cone after a bounce seems to be dependent upon the level of geomagnetic activity [Winckler et al., 1988]. This must be treated with some caution, though, as it is not clear how large a role beam interactions have in disrupting the beam. The comparison of Echo 7 optical and particle measurements presented above implies that the type of pulse does have some bearing on the returned current. The Discrete injections, with an apparently lower percentage return, were higher power, more dense, and apparently susceptible to some sort of beam interaction [Nemzek, 1989]. The Echo 5 pulses used to determine the 2% upper limit were likewise very high power Discrete injections; they had even more current than the Echo 7 Discrete injections. So, it is an open question how much of the Echo 5 loss was due to natural effects and how much was due to the physics of the beam injection. Continuous pulses seemed to undergo very little inter-action with their surroundings upon injection [Nemzek, 1989]; the 20% beam current return from these likely reflects true magnetospheric effects.

MULTIPLE BOUNCE ECHOES

Because a large fraction of the electron beam was scattered onto trajectories out of the atmospheric loss cone, it was able to make several round trips to and from the conjugate point before becoming so weak that it blended into the background flux. Generally, the magnetospheric electric field was so turbulent that the echo locus intercepted a subpayload only for a short time, making the reception of an echo on successive bounces unlikely. Some of the weaker pulses received probably were second-or-later-bounce echoes, but without an identifiable primary (first bounce) echo, it is impossible to say for sure. The echoes shown in Figure 3, though, came during a time when the electric field was unusually stable [Nemzek and Winckler, 1991], which of course is the reason that this sequence is such a good example for energy and bounce time determination. The electric field turned out to be stable enough that the echoes in Figure 3 could actually be measured through four successive bounces, as shown in Figure 7. This figure is a continuous strip of data from a PDP scintillator, cut up into 2.8-s segments. 2.8 s was the bounce time for the echoes. The progression is: a set of injections (rectangles at top), followed in 2.8 s by a sequence of echoes; 2.8 s later, those echoes created the second set of echoes; after another 2.8 s, those returned as the third set, and so on. Some of the pulses disappeared after one or two bounces, but a number continued throughout all four. After bounce four, the echoes merged with the background, so a fifth bounce was not detected.

Tracing the amplitude changes of these multiple bounce echoes gives another way to look at the change in the echoing electron population. Figure 8 shows the reduction in the amplitudes of 4 individual echoes from one bounce to the next. The intensity (on a log scale) decayed almost exponentially, dropping by 80% after each bounce. This is the same as the value derived for the fraction of the initial beam "lost" after one bounce. It is not clear

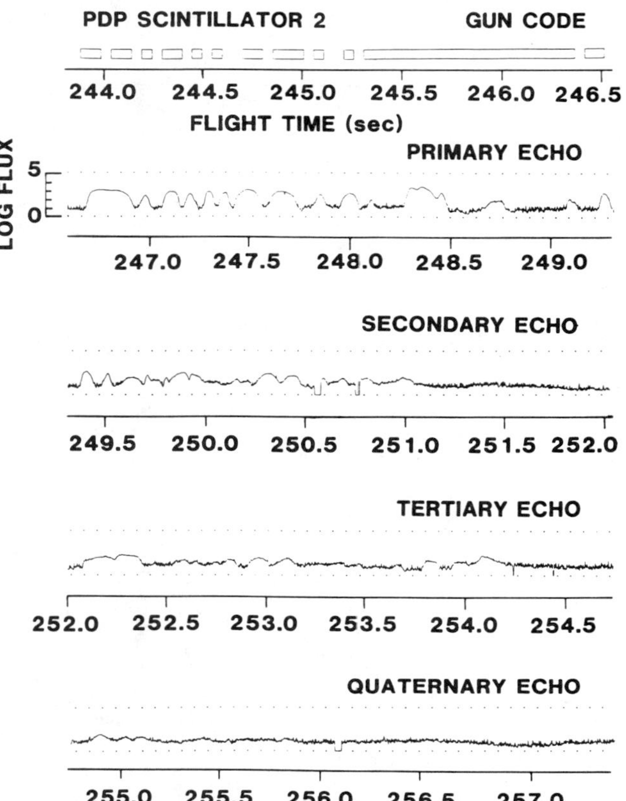

Fig. 7. This continuous string of scintillator data is cut into sections 1-bounce-time long. The pulse pattern repeats from one line to the next, showing that one set of injections created detectable echoes after each of four consecutive bounces.

that these two numbers necessarily should be the same, as the situation of going from the initial well-collimated beam to a diffuse structure is not the same as going from one diffused population to another.

The 80%/bounce loss rate for the echoes means a 1.7 s lifetime for 20 keV electrons. This can be contrasted with the calculated strong pitch angle diffusion lifetime for 20 keV electrons at L = 6.5 of 400-500 s [Kennel, 1969]. There is not necessarily a discrepancy between these two numbers. The standard pitch angle diffusion lifetime is calculated for an equatorial pitch angle distribution that is almost isotropic. The diffusion isotropizes a full 4π distribution across a very small loss cone, about 2.1° at the equator for L = 6.5. The typical electron in this distribution is very far from field-aligned, and so requires many bounces to diffuse over to the loss cone. The echo situation is very different. The 110° injection pitch angle became only 1.7° at the equator. In fact, the entire measured echo pitch angle distribution (which was assumed to be isotropic at 200 km altitude) mapped into a cone just over 2° wide at the equator. The echo pitch angle distribution probably was never much wider than the loss cone. Diffusion from the beam injection angle to the edge of the loss cone required only a few eV change in perpendicular energy for

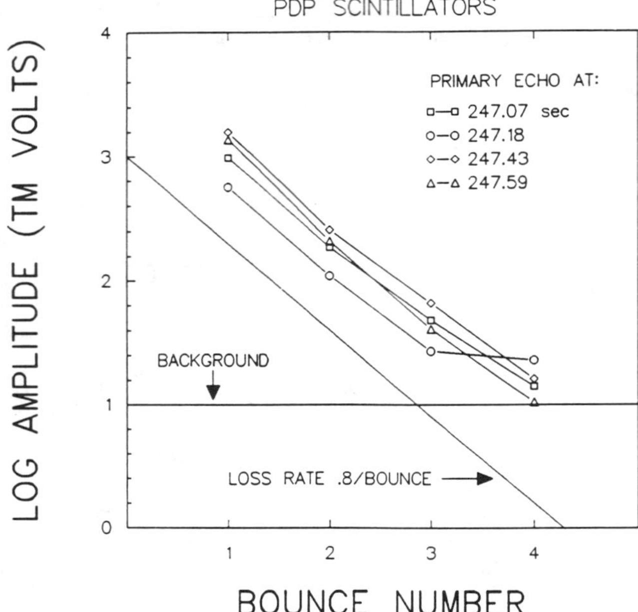

Fig. 8. This plot traces the amplitude of 4 echoes selected from Figure 7 through the four bounces. The amplitude dropped exponentially, at a rate of 80%/bounce.

20 keV electrons. Diffusing out to 10° would have required hundreds of eV. So while it was easy for electrons to get out of the loss cone, diffusion very far from it would have required a very long time. Since the echo electrons remained near the loss cone, their lifetime was drastically reduced from that of typical trapped electrons.

The width of the cone occupied by the echoing electrons at the equator can be estimated by using the 80% loss rate. The loss rate implies that the solid angle of the cone must be such that 80% of it is taken up by the loss cone. There is a slight difference in the northern hemisphere and southern hemisphere loss cones (2.18° vs. 2.03°) due to the asymmetry in the Earth's magnetic field, so I will use an average loss cone size of 2.1° ($\equiv \alpha_{LC}$). A more detailed calculation using the actual sizes of both loss cones yields essentially the same result. For these small angles, the solid angle can be taken as $\sin^2$ of the angle. If the half-angle of the echo cone is α, then the fraction of the flux remaining after each mirroring is $(1-\sin^2\alpha_{LC}/\sin^2\alpha)$. After a full bounce, then, the ratio R of the final echo flux to the initial flux is $(1-\sin^2\alpha_{LC}/\sin^2\alpha)^2$. From the multi-bounce echo measurements, R = .2, implying an α of 2.8°. This is the estimated angular size of the entire echoing electron pitch angle distribution at the equator. As expected, it is not much larger than the loss cone. The size of this cone is actually an upper limit, as the calculation assumed that all of the electrons within the loss cone were actually lost from the system. Inclusion of backscattering which allowed loss cone electrons to make another bounce through the magnetosphere would have meant that the calculated region of

phase space containing the beam would have been even smaller. Likewise, if the beam had been considered to continually undergo radial diffusion the size of the beam cone would have been smaller. Note that this analysis is not valid for the initial bounce, when the beam "loss" was primarily a process of escape from the loss cone, to mirror points above the payload.

SUMMARY

Production and measurement of conjugate echoes provides a unique way of examining the large-scale magnetosphere from a small-scale spacecraft. In the current example, the echoing electrons, some in the loss cone and some trapped, mimicked the motion of the natural trapped radiation. The two populations, while distinct, responded to the same influences. Thus by using the artificial electron beams as tracers it was possible to estimate the magnitude of waves causing equatorial pitch angle diffusion (a few mV/m) and the lifetime of electrons at the edge of the loss cone (1.7 s).

Acknowledgments. The author thanks J.R. Winckler, P.R. Malcolm, and J.C. Ingraham for valuable discussions regarding the interpretations contained within this paper. The Echo project at the University of Minnesota is supported under NASA Grant NSG 5088. R.J. Nemzek is supported at Los Alamos by the Department of Energy.

REFERENCES

Arnoldy, R.L., R.A. Hendrickson, and J.R. Winckler, Measurement of low level electron and x-ray intensities at rocket altitudes during quiet magnetic conditions at Churchill, Manitoba, *Cosmic Phys. Tech. Rep., 165,* Univ. of Minn., Minneapolis, 1974.

Franz, R.C., *A study of the luminosity produced by an electron beam-emitting rocket in the polar ionosphere: Echo 7,* Ph.D. Thesis, 260 pp., Univ. of Minn., Minneapolis, 1991.

Hallinan, T.J., J.R. Winckler, P.R. Malcolm, H.C. Stenbaek-Nielsen, and J. Baldridge, Conjugate echoes of artificially injected electron beams detected optically by means of new image processing, *J. Geophys. Res., 95,* 6519-6532, 1990.

Hendrickson, R.A., *The Electron Echo experiment: observations of the charge neutralization of the rocket and analysis of the echoes from electron beams artificially injected into the magnetosphere,* Ph.D. Thesis, Univ. of Minn., Minneapolis, 214 pp., 1972.

Kennel, C.F., Consequences of a magnetospheric plasma, *Rev. Geophys., 7,* 379-419, 1969.

Kurth, W.S., J.D. Craven, L.A. Frank, and D.A. Gurnett, Intense electrostatic waves near the upper hybrid resonance frequency, *J. Geophys. Res., 84,* 4145-4155, 1979.

Lyons, L.R., Electron diffusion driven by magnetospheric electrostatic waves, *J. Geophys. Res., 79,* 575-580, 1974.

Lyons, L.R. and R.M. Thorne, Equilibrium structure of radiation belt electrons, *J. Geophys. Res., 78,* 2124-2149, 1973.

Malcolm, P.R., *Electron Echo 6 - A study by particle detectors of electrons artificially injected into the magnetosphere,* Ph.D. Thesis, 309 pp., Univ. of Minn., Minneapolis, 1986.

Nemzek, R.J., Immediate and delayed high-energy electrons due to Echo 7 accelerator operation, in *Proc. of the Spacecraft Charging Tech. Conf.,* edited by R.C. Olsen, pp. 404-427, Naval Postgraduate School, Monterey, CA, 1989.

Nemzek, R.J., *Echo 7- magnetospheric properties determined by artificial electron beams,* Ph.D. Thesis, 329 pp., Univ. of Minn., Minneapolis, 1990.

Nemzek, R.J., P.R. Malcolm, and J.R. Winckler, Comparison of Echo 7 field line length measurements to magnetospheric model predictions, *J. Geophys. Res., 97,* 1279-1287, 1992.

Nemzek, R.J., and J.R. Winckler, Parallel electric fields detected via conjugate electron echoes during the Echo 7 sounding rocket flight, *J. Geophys. Res.*, *96*, 11475-11483, 1991.

Swanson, R.L., *Electron Echo 5 - detection of artificially injected electron by optical and x-ray methods*, Ph.D. Thesis, 127 pp., Univ. of Minn., Minneapolis, 1983.

Swanson, R.L., J.E. Steffen, and J.R. Winckler, The effect of strong pitch angle scattering on the use of artificial auroral streaks for the echo detection-Echo 5, *Planet. Space Sci.*, *34*, 411-427, 1986.

Szuszczewicz, E.P., Controlled electron beam experiments in space and supporting laboratory experiments, *J. Atmos. Terr. Phys.*, *47*, 1189-1210, 1985.

Winckler, J.R., The use of artificial electron beams as probes of the distant magnetosphere, in *Artificial Particle Beams in Space Plasma Studies*, edited by B. Grandal, pp. 3-34, Plenum, New York, 1982.

Winckler, J.R., P.R. Malcolm, R.L. Arnoldy, W.J. Burke, K.N. Erickson, J. Ernstmeyer, R.C. Franz, T.J. Hallinan, P.J. Kellogg, S.J. Monson, K.A. Lynch, G. Murphy, and R.J. Nemzek, Echo 7- An electron beam experiment in the magnetosphere, *Eos Trans. AGU*, *70*, 657,666-668, 1989.

Winckler, J.R., R.J. Nemzek, R.L. Arnoldy, and T.J. Hallinan, ECHO electron beam experiments in the diffuse aurora (abstract), *Eos Trans. AGU*, *69*, 1368, 1988.

R. J. Nemzek, Group SST-9, MS-D436, Los Alamos National Laboratory, Los Alamos, NM 87545.

Ion Acceleration in the Low- and Mid-Altitude Auroral Ionosphere

ANDREW W. YAU AND BRIAN A. WHALEN

Herzberg Institute of Astrophysics, National Research Council Canada

Direct observations of the thermal and suprathermal ion distributions from the EXOS D (Akebono) satellite and a number of recent sounding rockets in and near regions of transverse ion energization (TAI) have demonstrated the highly dynamic nature of the low- and mid-altitude (<6000 km altitude) auroral ionospheric plasma, and the importance of the TAI process in the plasma dynamics, both in the dayside and the nightside. We review these observations as well as those from earlier satellites. Sounding rocket observations in the nightside show that during active auroral conditions, the TAI process frequently occurs at 500-to 1000-km altitude. At times, ions are energized in the perpendicular direction up to a few hundred eV in regions of intense lower hybrid waves; the energized (10-500 eV) ions have characteristic temperature, $kT_\perp$, of the order of 10-20 eV while the core (<10 eV) ions have temperatures of the order of 1-2 V. The EXOS D observations show that the TAI region exists on auroral field lines, coincident with regions of large, turbulent electric field, at altitudes which vary between 3000 and 6000 km in the dayside and extend below 2000 km altitude on the nightside. The altitude range over which the energization occurs is narrow, often less than 100 km. In the energization region, all ion species are energized to approximately the same energy perpendicular to the magnetic field. At mid-altitudes, the energized plasma forms a conic distribution that displays significant departure from that expected from conservation of the first adiabatic invariant, and the characteristic temperature, $kT_\perp$, of the energized ions increases from ≤10 eV at 2000 km to ~100 eV at 6000 km. Simultaneous EXOS D and DE 1 ion composition observations in magnetic conjunction directly confirm the occurrence of multi-step ion energization, wherein ions were transversely energized to $kT_\perp = 10$ eV at Akebono altitude, and further energized to 100 eV between Akebono and DE 1.

1. INTRODUCTION

An extensive body of observations currently exists on the transverse energization of ionospheric ions and the formation of conical ion distributions of energized ionospheric ions above the auroral ionosphere; a comprehensive bibliography of the earlier observations and related theoretical work was given in the overview of [Klumpar 1986]. This body of observations was made at a wide range of altitudes and local times under a variety of geophysical conditions, and confirms the importance and ubiquity of the transverse ion energization (TAI) process in the auroral ionosphere. However, important questions remain concerning the fundamental energization mechanism and free energy source for TAI, its connection to other auroral processes, and its precise relationship with conical ions and other upflowing ion distributions at higher altitudes.

To avoid confusion between transversely energized and conical ions, we use the following rules of terminology in this review. The term transversely energized ions (TEI) or transversely

accelerated ions (TAI) will be used to refer to ionospheric ions that appear to have been energized over a narrow altitude range in the predominantly transverse direction to the local magnetic field, **B**. Observationally, ions with pitch angle peaks within ≤5° of 90° and angular width of ≤10° will be identified as TAI and the observation will be deemed to be inside the "TAI source region". Note that it is not possible to distinguish between an ion that is energized near 90° locally from one that is energized precisely at 90° immediately below the point of observation, because of the finite acceptance angle of the particle detector and the finite pitch angle resolution of the ion measurement. The term "ion conics" will be used to refer to ions that have been transversely energized at a lower altitude. Observationally, such ions have pitch angle distributions peaked along the surface of a cone whose apex half-angle is significantly less than 90°. Note that the pitch angle peak of the observed ions is in general energy-dependent, in the case of transverse ion energization over an extended distance along the field line ("extended" ion conics, [Peterson et al., 1992; Andre et al., 1990]); and in the case of additional ion energization in the parallel direction ("bimodal" conics, [Klumpar et al., 1984]).

We will focus on TAI and conical ion observations below the so-called "parallel acceleration", $1\text{-}R_E$ altitude region. We will use

Auroral Plasma Dynamics
Geophysical Monograph 80
Published in 1993 by the American Geophysical Union.

the term "mid-altitude" to refer to the region between the H-O crossover altitude and the 1-R_E altitude region, and the term "low altitude" below it. The reason is two-folded. A number of experimental evidences and theoretical arguments suggest that TAI in the O^+-dominated ionosphere below the H-O crossover altitude has a profound influence on ionospheric plasma transport and energization in the H^+-dominated ionosphere above. At auroral latitudes, the H-O crossover altitude occurs in the 1000-2500 km altitude range, and is moderated by the prevailing solar and atmospheric scale height conditions and the polar wind flow. Existing TAI observations fall naturally into two categories: those made on sounding rockets below ~1000 km, and those on satellites at higher altitudes. Sounding rockets have proven to be much better platforms for observing TAI at low altitudes because of their slower speed (<1 km/s versus ~7 km/s for satellites at these altitudes), faster payload spin rate (~1 Hz versus ~0.1 Hz) and higher telemetry rate and data resolution.

The transverse energization of ionospheric ions at low altitude is believed to play an important role in the ion composition of the magnetosphere [Yau et al., 1985; Chappell et al., 1987]. Our current recognition that ionospheric ions constitute an important source of the magnetospheric plasma, traces back to the early observation of [Shelley et al., 1972], which revealed a large O^+ component in the precipitating (magnetospheric) energetic ion flux during magnetic storms. The fact that the observed O^+/H^+ flux ratio was large compared with the corresponding ratio in the solar wind ruled out the solar wind as the dominant source for the observed magnetospheric O^+ ions, and prompted the search for energetic ions of ionospheric origin on S3-3 [Sharp et al., 1977].

Energetic upflowing ions (UFI) were first observed on S3-3 [Shelley et al., 1976; Sharp et al., 1977], where they appeared as large fluxes of energetic (0.5-17 keV/q) ions moving away from the ionosphere, and were interpreted to originate from the ionosphere below the point of observation. Two types of UFI were identified: those moving along the local magnetic field **B** which were subsequently referred to as "ion beams" in the literature; and those moving at an angle to **B**, which were referred to as "ion conics" in the subsequent literature. The latter were interpreted to result from the transverse energization (to **B**) of the observed ions at a lower altitude and the adiabatic motion of the energized ions under the grad **B** force [Sharp et al., 1977]. On S3-3, they were often dominated by O^+ ions (O^+/H^+ flux ratio ≥ 1, [Collin et al., 1981]) and peaked at 130°-140° pitch angle, corresponding to an inferred source altitude of ~5000 km, assuming adiabatic invariance ($\sin^2\beta(z) = B(z) / B(z_0)$, where $B(z)$ and $\beta(z)$ are the magnetic field strength and ion pitch angle at altitude z, respectively, and z_0 is the source altitude).

Such an inferred source altitude implied the presence of ambient O^+ ions as a dominant species up to ~5000 km altitude. It was also in contrast with the direct evidences of transverse ion energization at lower altitudes (500-2000 km) from observations of lower-energy ions on ISIS-2 [Klumpar 1979; Ungstrup et al., 1979], sounding rockets [Whalen et al., 1978; Yau et al., 1983], and S3-3 [Gorney et al., 1981]. It is now clear from these lower-energy ion observations that the energization of ionospheric ions is often a multi-step process, and that transverse ion energization is an important step at low altitude, where it can impart ~1-1000 eV of energy to the ions.

2.   LOW ALTITUDE ION ACCELERATION

Observations from the ISIS-1 and -2 satellites suggest that the occurrence of TAI at low altitudes is seasonally dependent. On ISIS-2 at 1400 km altitude, Klumpar [1979] found TAI and conical ion events to be a common occurrence in the winter but almost totally absent in the summer: only 8 events were observed in the northern hemispheric summer of 1971, compared with over 500 events in the winter. In the winter, the occurrence probability distribution coincided with the auroral oval; it varied between 0.3 and 0.6 between 65° and 85° invariant in the nightside, and was lower in the dayside. The observed ions were peaked at 90° to ~120° pitch angle, corresponding to an apparent altitude of transverse energization of 1400- to ~700-km. Their differential flux typically followed a power law in energy with an index near 1.8 (i.e. number intensity $j(E) \propto E^{-\alpha}$ where $\alpha = -1.8$) in the few-eV to few-hundred-eV range. In the summer, TAI and conical ion events were generally absent below 2750 km.

A number of observations from sounding rockets below 1000 km altitude [Whalen et al., 1978; LaBelle et al., 1986; Arnoldy et al., 1992] have demonstrated that TAI frequently occurs down to ~500 km altitude in the nightside in conjunction with active aurora. These observations were all made in the winter. The first such observation was made by Whalen et al. [1978] on rocket IVB-33 in the expansive phase of an auroral substorm, in which intense fluxes up to $10^8(cm^2s$ sr $keV)^{-1}$ of 0.1-0.5 keV/q ions confined to 90°-110° pitch angles were observed above 400 km altitude. Figure 1 shows the measured pitch angle distributions of ions in the 0.09-5.3 keV range at 690 km altitude. The ions at 0.09, 0.15, and 0.34 keV appeared as ion beams confined to pitch angles near 100°, with angular widths narrower than the response of the particle spectrometer (~10°) at all three energies. The absence of equivalent beams at 80° pitch angle indicates that these ions were not a mirroring population, but rather were ions originating from lower altitudes, moving away from the ionosphere. This is in contrast with the energetic (5.3 keV) ions, which exhibited the usual isotropic distribution out to the atmospheric loss cone, indicating that they were precipitating particles. The fact that the observed TAI at different energies had the same pitch angle peaks and narrow angular widths suggests that their source region was narrowly confined in altitude (<100 km), and that the source mechanism must accelerate the ions at precisely 90° to within a very small range of angles. This is because if the transverse energization were distributed in altitude, ions from the distant parts of the energization region below the rocket would have appeared at larger pitch angles, resulting in a broad angular distribution.

The observed ions were consistent with ion injection at ~630 km altitude, assuming adiabatic invariance. Figure 2*a* shows the pitch angles of the 0.1-0.5 keV ions at peak intensity as a function of time. Figure 2*b* shows the rocket trajectory, and the inferred source altitudes of the observed ions, on the assumption that the observed ions were injected at 90° pitch angle at a lower altitude and then spiralled up the magnetic field while conserving their magnetic moment. Between 400 and 450 km on both the upleg and the downleg, large enhancements were observed at 90°. At higher altitudes, the ions appeared at larger pitch angles, up to about 110° near the rocket apogee (~730 km). No variation in the

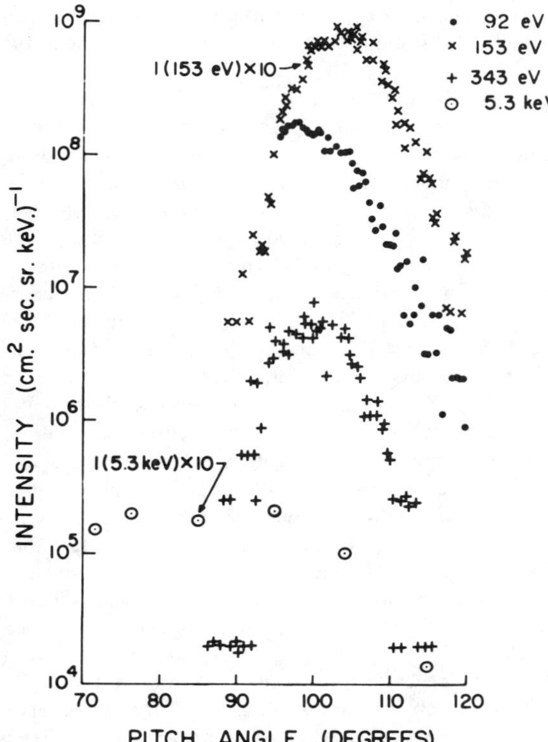

Fig. 1. Ion pitch angle distributions observed at 690 km altitude in IVB-33 rocket. The 92-343 eV ions appeared as beams near 100°, with beam widths less than the angular response (~10°) of the detector. The energetic (5.3 keV) ion precipitation was isotropic out to the loss cone. [Yau et al., 1983].

angular width with altitude was noticeable. For most of the rocket flight, the observed data were consistent with a source region in the 400- to 500-km altitude range, except for the interval from 05:30 to 05:33 when the apparent source altitude moved up about 100 km. Below 400 km, weak enhancements were present near but detectably smaller than 90° pitch angles. The precise origin for these ions is uncertain. TAI with a small downward velocity component have been observed elsewhere [Yau et al., 1987] and Gorney et al. [1985] attributed them to the trapping of the TAI by downward parallel electric fields whose magnitudes are large enough to locally balance the magnetic mirror force on the ions.

Figure 3 shows the velocity phase space distribution of the observed TAI from the MARIE rocket [Yau et al., 1987]. In both the upleg and the downleg, the observed TAI exhibited a characteristic power-law intensity spectrum, i.e., the number intensity $j(E)$ varied with $E^{-\alpha}$ where α = ~2.2. Such power-law intensity spectra were also observed in TAI on ISIS-2 and ion conics on S3-3, and they suggest a heating type mechanism rather than one that accelerates all ions to the same energy. Figure 3 shows that the velocity phase space spectra of the observed TAI fit a Maxwellian in the 40- to 200-eV ion energy range, with a characteristic temperature (energy) of ~20 eV, inside the TAI source region both in the upleg (top left panel) and in the

downleg (bottom panels), as well as above the TAI source region (top right panel). The ambient ion temperature was of the order 0.1-0.2 eV. In other words, the ratio of the characteristic temperature of the TAI to that of the ambient ions was of the order of 100. The ions above 200 eV had a higher characteristic temperature, and the distribution displayed a high-energy tail. In general, only a small fraction (<10%) of the ambient ions are energized to suprathermal energies (≥10 eV) in TAI, and the characteristic temperature of the lowest energy ions (<10 eV, core ions) is typically much lower [c.f. Figure 3 in Yau et al., 1983], resulting in a bi-Maxwellian distribution in the ~eV to ~100 eV ion energy range. For example, in the TAI event on the TOPAZ-2 rocket (Fig. 9 in Garbe et al., [1992]), the characteristic temperatures of the core (<10 eV) and suprathermal (>10 eV) ions were about 1.7 eV and 15 eV, respectively.

The detailed morphological relationship between the auroral substorm and the TAI process provides important clues to the

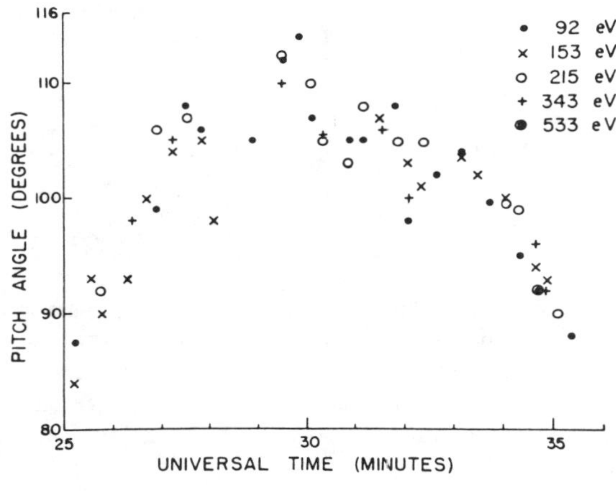

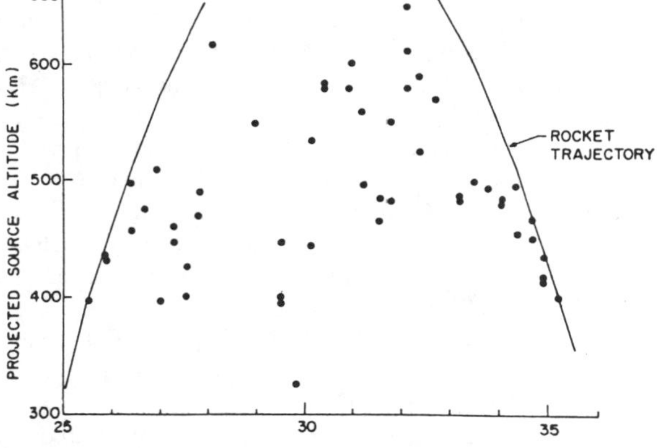

Fig. 2. (a) Angles of the ion pitch angle distribution peaks as a function of time after 05:00 UT in IVB-33 rocket. (b) inferred ion source altitudes using the data in (a) and assuming that the ions were injected at 90° pitch angle and then spiralled up the magnetic field line while conserving their magnetic moment. [Whalen et al., 1978].

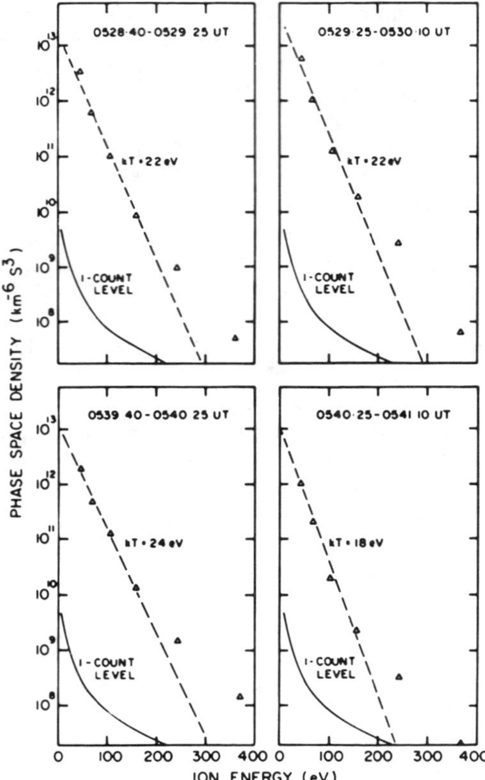

Fig. 3. Velocity phase space density of 90°-120° transversely accelerated ions observed in the MARIE rocket. Top panels: ion acceleration region in the upleg; bottom panels: ion acceleration region in the downleg. Data in the 40-200 eV range fitted a Maxwellian. Data at higher energy displayed a "suprathermal tail". [Yau et al., 1987].

nature of the energization process. Figure 4 shows the auroral intensity keogram (top panel) and ground magnetograms (bottom 3 panels) during the MARIE rocket flight. The keogram shows the measured optical intensity at 5577Å from the ground meridian scanning photometer as a function of time and elevation angle. Its individual elevation scans provide an instantaneous, one-dimensional profile of the changing morphology in the vicinity of the field line traversed by the rocket payload. The three activation (expansion and brightening) phases of the substorm near 05:10, 05:21, and 05:37 UT are clearly evident in the negative excursion of the X-magnetogram. In particular, the third poleward expansion and brightening occurred at 05:36:50 UT, reaching a peak intensity of 200 kR, from a level of <10 kR in the preceding several minutes. Intense hundred-eV H⁺ ions peaking at 105° were observed starting at 05:37:14 UT and 895 km altitude (see Figure 3 and 4 of Yau et al., [1987]). Assuming that these ions were injected at 90° at their source altitude (~600 km), their transit time from the source altitude to the point of observation was about 12 s. Thus, the observed data indicate that the TAI process commenced within seconds of 05:37:02, 10-15 s after the onset

of the optical auroral expansion and brightening at 05:36:50 UT, and that the energization process occurred on a time scale of ~10 s or less.

In the IVB-33 rocket, Whalen et al. [1978] correlated the TAI and precipitating energetic electron intensities, and found the maximum TAI intensities to coincide with sharp decreases in electron precipitation. The correlation suggests the occurrence of the TAI at the edges of auroral precipitation. These authors also noted the presence of intense field aligned precipitating electrons in the region of TAI. In the IVB-36 rocket [Yau et al., 1983], TAI were observed both inside and outside regions of intense energetic electron precipitation. The lack of ground optical morphology data precludes the precise determination of the spatial relationship between the TAI and the energetic electrons in this case. In the MARIE rocket [Yau et al., 1987], TAI were clearly present both inside and near the regions of energetic electron precipitation, and coincided with regions of field-aligned keV electrons in both directions. Similar correlations of TAI with auroral electron precipitation and field-aligned electron bursts were also noted in the TOPAZ-3 rocket [Arnoldy et al., 1992]. The fact that the observed region of TAI generally extends beyond the region of energetic auroral electrons suggests that, if the latter is indeed the source of free energy for the TAI, the energization mechanism of TAI is capable of propagating the free energy across the field line. The significance of field-aligned electrons in TAI regions is probably more subtle. The fact that they were not observed in some of the TAI events does not necessarily mean that they were absent in these events. This is because they are often highly

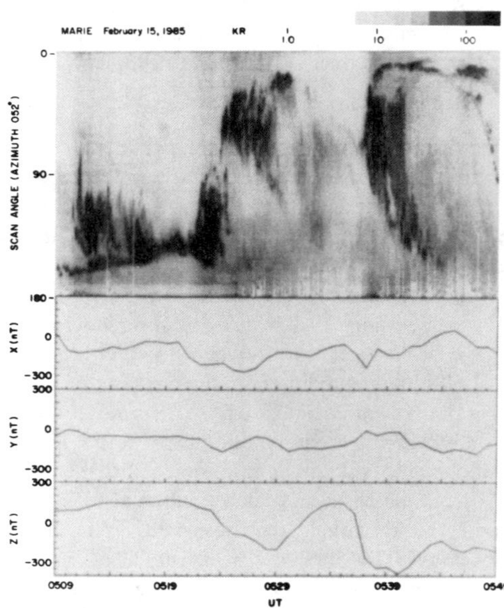

Fig. 4. Keogram of integrated auroral intensity at 5577Å from ground photometer as a function of scan angle and time (top panel) and ground magnetograms (bottom panels) at Churchill, Canada, at the time of the MARIE rocket flight, showing the three activation (expansion and brightening) phases of the substorm near 05:10, 05:21, and 05:37 UT. [Yau et al., 1987].

collimated along the field line and particle detectors on rockets and satellites do not always sample particles along the field line direction (i.e., at 0° and 180° pitch angle).

On the MARIE rocket, magnetic perturbation up to the order of 100 nT was observed in the region of the most intense TAI. The perturbation was consistent with the presence of a strong field-aligned current sheet, whose spatial extent was much more confined than the TAI observed at the time. No perturbation was observed coincident with the other two TAI source regions. Thus, if field-aligned currents were the source of free energy for the observed TAI, the energization mechanism must be capable of propagating the free energy across the field line.

Klumpar [1981] found an apparent relationship between field-aligned current density, ambient plasma density, and ion conics observed at 1400 km on ISIS-2. Figure 5 is a scatter plot of the measured ambient O^+ density versus the field-aligned current density from a number of the satellite orbits: the solid circles denote orbits in which TAI were present; the crosses denote the absence of TAI. It was found that ion conics were observed when the field-aligned current density exceeded a threshold value that was proportional to the ambient ion density. This suggests that the higher the ambient plasma density is, the larger field-aligned

current the TAI energization mechanism requires. This is consistent with the observed seasonal variation of TAI at low altitude: the plasma density at a given altitude is higher in sunlight than in solar darkness; therefore TAI may be expected to occur at a higher altitude for a given current density.

The unique physical process of transverse ion energization and its important role in magnetospheric ion composition have led to many proposed theoretical energization mechanisms. Several of these invoke electrostatic ion cyclotron waves resulting from drifting electrons, which either directly or stochastically energize the ions [Lysak et al., 1980; Okuda and Ashour-Abdalla, 1981]. A number of the other mechanisms invoke energization by electrostatic lower hybrid waves [Chang and Coppi, 1981; Retterer et al., 1986], narrow oblique double layers [Yang and Kan, 1983; Borovsky, 1984], and broadband electromagnetic ion cyclotron waves [Chang et al., 1986]. This last category applies to ion energization over an extended altitude region, such as in central plasma sheet ion conics, and is not relevant to TAI at low altitude. In all cases, the free energy must come from some other source, such as drifting or precipitating electrons.

The definitive identification of TAI energization mechanism using plasma wave data inside regions of TAI has proven difficult, because of the frequent presence of several candidate wave modes, the lack of wavelength information in most measurements, and the Doppler shift of the low frequency waves in satellite measurements. Signatures of electrostatic ion cyclotron waves have been observed in association with TAI in a number of rockets. In the TOPAZ-2 rocket, Kintner et al.[1989] observed large-amplitude ($\leq$ 5 mV/m RMS) electrostatic oxygen ion cyclotron waves (linearly polarized electric field emissions near the oxygen cyclotron frequency) coincident with a region of ion conics at 900 km altitude, where ion conics with a characteristic temperature of about 15 eV were observed up to 700 eV. However, these authors concluded that their coherence was inadequate to effect sufficient stochastic energization to the observed ion energies. As well, they noted the simultaneous presence of lower hybrid and hydrogen Bernstein waves. In both the IVB-33 and IVB-36 rockets, Yau et al. [1983] observed electron density fluctuations ($\delta n/n$) of <10% and a spectral signature consistent with Doppler shifted electrostatic oxygen cyclotron wave, in association with TAI up to 500 eV at 500- to 700-km altitude. In both cases, several of the observed features were consistent with ion cyclotron energization; however, the lack of detailed spectral data precluded an unambiguous identification of the energization mechanism. In the MARIE rocket, the electric field amplitude near the O^+ cyclotron frequency (20-60 Hz) were mostly less than 3 mV/m (RMS) and there was no spectral evidence of electrostatic ion cyclotron waves in association with the regions of observed TAI. On S3-3, Kintner and Gorney [1984] surveyed the S3-3 data and found only one 90° TAI event at 2600 km altitude and this event was accompanied by plasma wave of unidentified modes. The wave amplitude was somewhat smaller than expected. On DE 1, a similar survey of the energetic ion and wave data [Peterson et al., 1988] failed to find any unambiguously correlated 90° TAI and associated low frequency plasma waves. In summary, no definitive example of TAI energization by electrostatic ion cyclotron waves as yet exists.

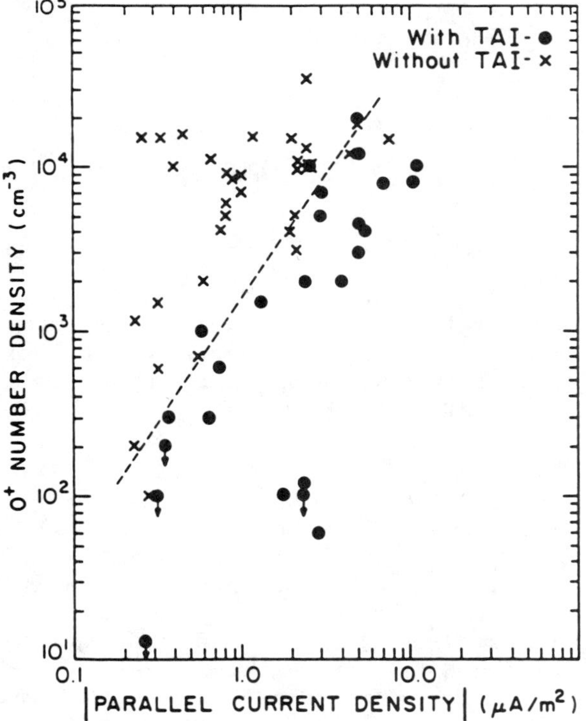

Fig. 5.   Scatter plot of oxygen ion number density and field-aligned current density for auroral crossings on ISIS-2, both with and without transversely accelerated ions (TAI). The absence of TAI in the upper left hand portion of the parameter space might indicate an ionospheric stability to the TAI acceleration mechanism with a current-to-number density ratio threshold. [Klumpar, 1981].

Intense VLF waves were observed in both the TAI regions on MARIE and the ion conic region on TOPAZ-2 [Kintner et al., 1991]. In both cases, spectral structures near and above the lower hybrid frequency were present. These structures were ordered by the hydrogen gyrofrequency, electrostatic, and associated with auroral hiss, and occasionally had short wavelengths. They were called structured auroral hiss (SAH) and were interpreted as wave modes which connect the lower hybrid mode with the hydrogen Bernstein modes. The SAH waves and the auroral hiss correlated strongly with the TAI. Figure 6 is a scatter plot of the energy flux of TAI versus the electric field amplitude in the lower hybrid frequency range (3-6 kHz) on MARIE. A roughly linear correlation exists between the ion energy flux and the electric field amplitude over about three orders of magnitude; the slower-than-linear increase in energy flux at the highest electric field amplitude suggests the presence of a saturation mechanism. Assuming that the observed VLF waves were parallel propagating, the Poynting flux in the TAI source region was about 8 to 80 times the power carried by the TAI. Thus, if they were responsible for energizing the ions, the efficiency of the wave-to-particle energy conversion would appear to be between 1 and 10%

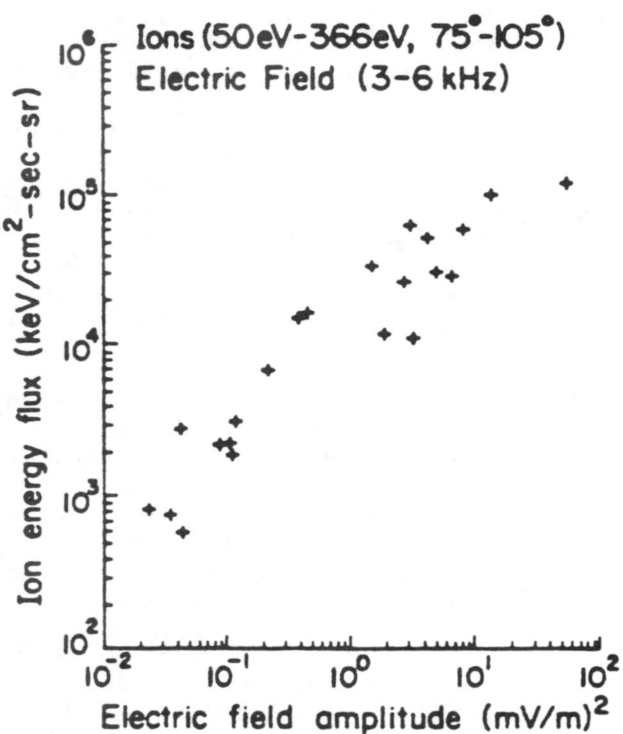

Fig. 6. Scatter plot of transversely accelerated ion energy flux and lower hybrid (3-6 kHz) electric field amplitude observed inside a region of transverse ion energization in the MARIE rocket. The plot indicates a roughly linear relation over three decades; the slower-than-linear increase in ion energy flux at high electric field amplitude suggests the presence of a saturation mechanism. [Kintner et al., 1986].

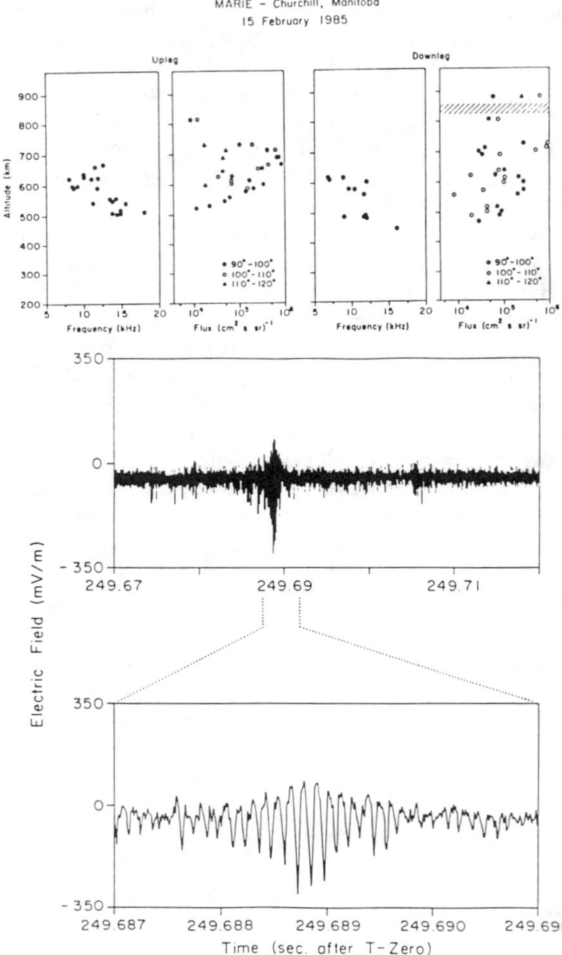

Fig. 7. Top panels: a comparison between the spikelet occurrence and the transversely accelerated ion fluxes in the MARIE rocket. Bottom panels: an example of one of the 30-40 spikelet events observed. [LaBelle et al., 1986].

On TOPAZ-3, Vago et al. [1992] recently found one-to-one correspondence between intense (up to 300 mV/m) lower hybrid waves and the TAI. The observed wave-particle interaction occurred in field-aligned filamentary density cavities at 500- to 1000-km altitude, and was interpreted in terms of lower hybrid wave collapse. On MARIE, LaBelle et al. [1986] reported the presence of bursts of near-monochromatic waves in TAI regions which were indicative of large-amplitude lower hybrid waves. Figure 7 shows an example of such large-amplitude bursts on MARIE (middle and bottom panels), and compares their occurrence and frequency with the observed TAI fluxes (top panels). These "spikelets" were observed in the 450- to 650-km altitude region both in the upleg and the downleg. They had an averaged time scale of 5 ms and an amplitude of 100-150 mV/m, and their frequency varied between 7 and 18 kHz. They are believed to be closely related to the large-amplitude lower hybrid waves observed in Vago et al. [1992]. Their occurrence coincided exactly with the altitude region of the TAI, and exhibited a one-to-one correspondence with discontinuities in the DC electric field of similar time scales (5-10 ms). However, as pointed out by LaBelle et al. [1986], the amplitude of the latter was too weak to sufficiently energize the observed ions.

To summarize this section, transverse ion energization is a frequently occurring phenomenon in the low-altitude (<1500 km) nightside auroral ionosphere, particularly at times of active aurora in the winter. It occurs on a time scale of ~10 sec or less at the expansion onset of the auroral substorm, and results in ion energization predominantly in the transverse direction (to $\mathbf{B}$). The energized ions are characterized by a power-law intensity spectrum ($j(E) \propto E^{-\alpha}$, where $\alpha \approx 1.8\text{-}2.2$) and a bi-Maxwellian velocity distribution (of characteristic temperatures of the order of ~1 eV and ~10-20 eV for <10-eV and >10-eV ions, respectively); also, a high-energy tail is often present at higher energies (>300 eV). The precise source(s) of free energy and energization mechanism(s) of low-altitude TAI are less certain. The best documented events (those on the MARIE, TOPAZ-2 and -3 rockets) strongly suggest wave-to-particle energy transfer from lower hybrid waves connected to the ion Bernstein mode as the primary - if not the only - energization mechanism responsible for the observed TAI. Whether these events constitute definitive experimental proofs remains somewhat debatable [see, for example, Retterer et al., 1992]. In any case, it is conceivable that other mechanism(s) might dominate the energization process under very different geophysical conditions.

3. MID-ALTITUDE ION ENERGIZATION

A number of detailed or statistical studies of ionospheric ion energization at mid-altitude exist in the literature. However, few refer to observations in the transverse energization region itself [Kintner et al., 1986; Peterson et al., 1988] or contain a complete characterization of the ion population (thermal and suprathermal) in and near the source region [Whalen et al., 1991]. In general, the data sets from different experiments and satellites have different ion energy ranges, mass composition resolving capabilities, detection sensitivities, and measurement resolutions. In particular, the earlier data sets on S3-3 were limited to higher ion energies. The importance of lower-energy ions and energization processes at lower altitudes became apparent in the more recent data sets. Also, the different data sets were acquired during different periods of the previous and the current 11-year solar cycles. Therefore, it is necessary to take into account long-term variations in atmospheric and ionospheric composition, and their possible influence on ion energization, when comparing observations of different time periods.

On S3-3 (<8000 km), Gorney et al. [1981] found that the majority of ion conics were below 400 eV. The occurrence distribution of ion conics in the 90- to 400-eV range generally tracked the statistical auroral oval. It had a broad daytime maximum at quiet times and became relatively uniform at active times. At quiet times ($Kp < 3$), the occurrence frequency was 0.2-0.3 on the dayside and 0.1-0.2 on the nightside, and was uniform in altitude above 1000 km. At active times ($Kp > 3$), it was 0.3-0.4 in the auroral oval both on the dayside and the nightside, and

displayed a relative minimum near 2000 km altitude. Also, ion conics that have large cone angles (within a few degrees of 90°) indicative of TAI were observed primarily in the 1500- to 2000-km altitude range. Whereas the ion conics dominated over the ion beams on the dayside in terms of their occurrence, the reverse was true on the nightside. This may be attributed to the generally lower source region of TAI and to parallel energization at mid-altitudes on the nightside. Ion composition measurements above 0.5 keV [Sharp et al., 1983], which did not distinguish between ion conics and beams, showed that the O^+/H^+ flux ratio of the UFI increased with magnetic activity [Collin et al., 1984], and that both the flux ratio and the energy ratio exhibited long-term variations. Near the solar minimum in 1976, the H^+ were generally more intense and the O^+ were more energetic. In comparison, the O^+/H^+ flux ratio increased while the O^+/H^+ energy ratio decreased with the rise in the solar activity level near 1978.

On DE 1, Lockwood et al. [1985] identified subtle differences in the energization characteristics of the ions in the 10- to 100-eV range between the dayside and the nightside. In the nightside, the observed energization was in the perpendicular direction, and displayed very similar detailed characteristics to TAI observed at sounding rocket altitude (see previous section). The transverse heating affects the bulk of the ion distribution, and perpendicular temperatures of the order of 10 eV were produced. In some cases, however, all of the thermal ions appeared to be energized, leaving no cold plasma and resulting in a toroidal distribution with a minimum at zero speed in the plasma frame. Often, the transverse energization region extended over a broad latitudinal region [Moore et al., 1985]. Also, the rate of "folding" of the evolving conical distribution decreased with increasing ion mass, reflecting the larger gravitational effects on the heavier ions in counteracting the magnetic mirror force.

In comparison, the observed ions in the dayside displayed a parallel component of energization in addition to the transverse component. When observed at 2000 to 5000 km, the ions comprised about ~90% of O^+ ions, <10% of H^+, and smaller percentages of other ions [Pollock et al., 1990], and they exhibited transverse heating to ~10 eV, as well as a significant particle and heat flux (>10^8 cm^{-2}s^{-1} and >3×10^4 erg cm^{-2}s^{-1} for O^+). All observed ion species were energized. They were termed "upwelling ions". They were observed exclusively in the morning sector of the auroral oval and the lower latitudes of the polar cap; their low-latitude edge was closely associated with regions of field-aligned currents. Their occurrence was seasonally dependent, and was highest in the summer. The distinction between upwelling ions in the dayside and transversely energized ions in the nightside is not sharp - some overlap exists, and is a matter of degree to which transverse energization affects the thermal core of the ion distribution.

On EXOS D (Akebono), Whalen et al. [1991] recently reported direct observations of TAI inside and near the source region. The TAI events occurred in time spans that were comparable with the measurement resolution on Akebono, and perhaps explain the difficulty of their detection in earlier satellites, in view of the improved sensitivity and capabilities of the Akebono mass spectrometer relative to previous instruments. The ion measurements on Akebono covered both the thermal and the suprathermal energy ranges, from ~0.1 to >10^4 eV, and revealed a number of interesting ion features inside and near the TAI source region.

Figure 8 shows the ion composition data from an auroral pass on Akebono near 10 MLT and between 6000 and 7000 km altitude, on February 20, 1990. In this pass, the Akebono suprathermal mass spectrometer (SMS) was in a fast scan mode in which it measured thermal ions for eight seconds (one spin period) and then switched to energetic ions for eight seconds in each measurement cycle. Each of the six spectrogram panels in the figure is a concatenated series of spin angle versus energy spectrograms. In the thermal mode (panels a-d), SMS stepped rapidly through eight RPA steps between 0 and 16.3 V, relative to the spacecraft ground, successively for each of the four ion mass species (H^+, He^{++}, He^+, and O^+), 32 times (energy/mass sequences) per spin. In the energetic mode (panels f-g), it stepped through four ion energy steps between 150 and 2000 eV/q for each ion mass species. In each panel, the ion counts in each 16-sec interval are color coded, and displayed as a function of spin angle and retarding potential (panels a-d) or ion energy-per-charge (panels f-g). In the spin angle axis, 0° corresponds to ions with minimum pitch angle, ±90° to ions with 90° pitch angle, ±180° to ions with maximum pitch angle; the dashed curve corresponds to ions observed nearest to spacecraft ram.

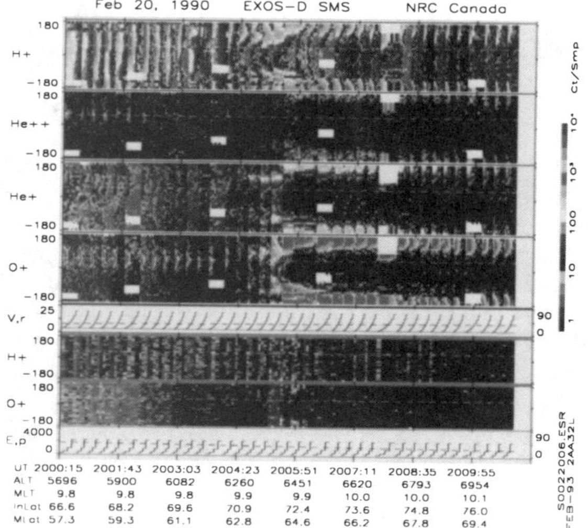

Fig. 8. A summary plot of observed ion composition in a dayside auroral pass on Akebono on February 20, 1990, in concatenated spin angle/energy spectrogram format, showing the occurrence of transverse energization near 6300 km (20:04:30 UT). Panels a-d (top four panels): thermal (0-25 eV) H^+, He^{++}, He^+ and O^+ ion spectrograms, respectively; panel e: RPA energy steps and minimum ram angle (dashed line); panels f and g: suprathermal (>50 eV) H^+ and O^+ ion spectrograms, respectively; panel h (bottom panel): suprathermal energy steps and minimum pitch angle sampled. 0° spin angle corresponds to ions with minimum pitch angle; the dashed curves in the spectrogram panels indicate the spin angle nearest spacecraft ram. [Whalen et al., 1991].

At 5700 km and 67° invariant, the thermal O^+ ion distribution in panel d was strongly peaked in the ram direction, and the count rate dropped to a low value at about 5 volt retarding potential. Such a distribution is consistent with O^+ ions being stationary in the earth's corotating frame and having a temperature of a few thousand degrees or less. The He^+ angular distribution was wider and its RPA cutoff voltage was lower, whereas the cutoff voltage of H^+ was higher. This indicates that both He^+ and H^+ were approximately stationary, and that the H^+ was significantly heated.

At 2002 UT, the H^+ ions developed a strongly upward field-aligned flow. At 2003 UT, the He^+ also developed a field-aligned flow. Soon thereafter, the O^+ flow also approached field-alignment. At 2004:30 UT, the ion distributions of all three masses changed from ones characterized by strong field alignment to ones characterized by intensity peaks at 90° to the magnetic field line, extending up to about 100 volts. Over the next 30 seconds, the peaks rapidly shifted to conic-like distributions so that by 2006 UT only heated energetic ion beams were present. The composition ratio of the thermal-energy ions nearby and inside the ion energization region was very similar to that of the energized ions inside the region. Also, the energized ion distributions had peaks within 10° of perpendicular to the local magnetic field, and the bulk momentum parallel to **B** was conserved by each mass species as it passed through the ion energization region. However, above the source region, the energetic light ions were more closely field-aligned than the heavy ions, suggesting a change in the magnetic moments of some if not all of the ions between the source altitude and the observation point. If one were to make a rough estimate of the source altitude assuming ion energization at 90° and conservation of the first adiabatic invariant, one would find a region of ion energization that was elongated in local time and sloped to lower altitudes at higher latitudes. The observation in this figure is representative of a number of similar events observed on Akebono. The region immediately earthward of the TAI source was characterized by low density, a downward density gradient, and upward flow of all major ions. In the TAI source region, all ions were energized to approximately the same energies transverse to the local magnetic field; $kT_\perp \approx$ 10-20 eV. The TAI source region was at times relatively stable over periods of several hours, and had a typical extent of many hours in local time. It coincided with the region of large and turbulent electric field, and had an altitude thickness of less than 100 km.

Miyake et al. [1992] surveyed the altitude dependence of the apex angle and temperature of ion conics in the dayside (6-18 MLT, 65°-85°Λ) on Akebono. The top panel of Figure 9 is a scatter plot of the apex angle versus altitude. The four curves represent the expected altitude variations due to transverse energization followed by magnetic mirroring and conservation of adiabatic invariant. It clearly shows that none of the observed conics mapped to a mirroring altitude below the topside ionosphere, and that the apex angle decreases with increasing altitude at a much slower rate than expected from conservation of adiabatic invariant. The bottom panel of Figure 9 is a scatter plot of the temperature of the conics versus altitude. The temperature was derived from the measured ion velocity phase space distribution in the 30 eV - 10 keV ion energy range. It increases

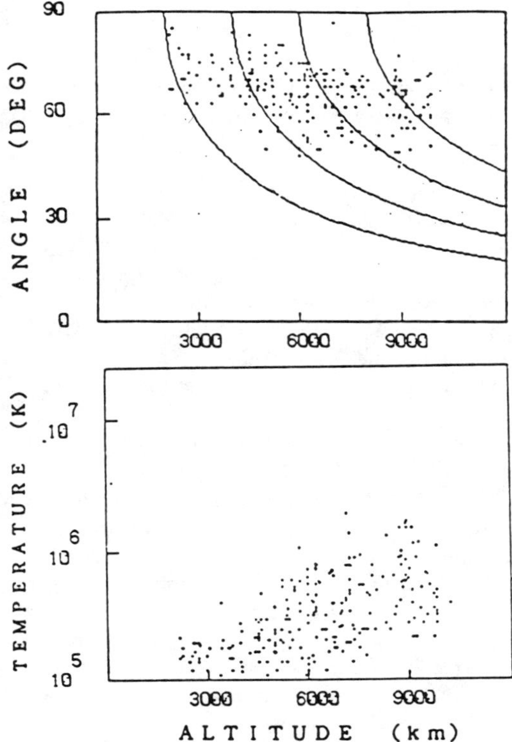

Fig. 9. Scatter plot of apex angle (top) and temperature (bottom) of observed ion conics on Akebono versus altitude. The four curves in the top panel represent the expected altitude variations due to transverse energization followed by magnetic mirroring and conservation of adiabatic invariant. [Miyake et al., 1993].

from the order of ~10 eV near 2000 km altitude, to ≤100 eV (10^6 °K) near 6000 km altitude. The result shows that TAI in the dayside are generally less energetic than their nightside counterparts, and perhaps explains the under-estimation of ion conic occurrence in earlier observations that were restricted to higher ion energies. Also, it suggests that individual ion conics originating in the 2000- to 4000-km altitude range in the dayside experience further transverse energization above 4000 km in the course of their transit to higher altitudes. Miyake et al. [1992] excluded the extended (bimodal) conics in their analysis. Peterson et al. [1992] recently applied image processing techniques to delineate the relative occurrence of restricted versus extended conics at DE 1 altitudes, i.e., those restricted to a narrow angular range versus those having a significant flux of field-aligned ions. Extended conics were found to constitute more than one third of the He^+ and O^+ conics in the 8000- to 24000-km altitude region. Also, the two types of ion conics had different altitude dependences. The result suggests that ion conic formation by localized transverse energization is not the dominant mechanism responsible for producing ion conic distributions above 8000 km.

Statistically, it is evident that the energetic upflowing ions (ion beams and conics) observed at high altitudes are intimately related to ionospheric ions energized at low altitudes, and that the ion energization process is typically a multi-step one. Peterson et al. [1993] recently presented a detailed case study of the multi-step nature of the energization process using simultaneous observations of H^+ and O^+ ions near the cusp at the Akebono and DE 1 altitudes. Figure 10 summarizes the two distinct ion populations that were observed at the two altitudes in their study. At the Akebono altitude, the first population appeared as nearly stationary thermal O^+ ions with intensity comparable to that of H^+ at altitudes of several thousand kilometers on the dayside. As pointed out by Peterson et al. [1993] this population is related to upwelling ions observed in the high-altitude cusp. The second population was observed slightly above a region of TAI at 20:05 UT, and was significantly energized in the transverse direction ($kT_\perp \approx 10$ eV, c.f. Figure 3); both populations had similar ambient temperature. On the basis of the measured poleward ion convection velocity data, the second population was likely the source of the intense stream of upflowing ions observed at the DE 1 altitude at 19:30 UT; the latter were more energetic ($kT_\perp \approx 100$ eV, $kT_/ \approx 80$ eV) and had a conic-like distribution, indicative of significant energization nearby (between the Akebono and DE 1 altitudes). The Peterson et al. [1993] study was the first report of ion conics occurring at widely separated altitudes on the same field line. It illustrates directly the fact that the plasma conditions required for ion conic formation span a wide range, and that it is reasonable to expect more than one plasma process for ion conic generation.

To summarize this section, low-energy ion observations on the ISIS-1, S3-3, DE 1, and Akebono satellites clearly establish the importance of transverse ion energization at mid-altitudes in the dayside auroral ionosphere. In particular, detailed ion observations in both the thermal and the suprathermal ion energy ranges were

**TRANSPORT OF O^+ IN THE POLAR CAP
WITH AND WITHOUT ENERGIZATION.**

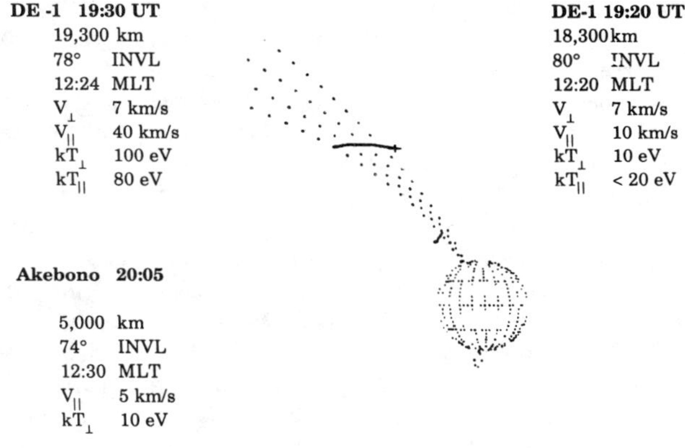

Fig. 10. Schematic summary of simultaneous (magnetically conjugate) ion observations near the cusp at Akebono and DE 1 altitudes, in which O^+ ions were transversely energized to $kT_\perp \approx$ 10 eV at Akebono altitude, and underwent further transverse energization to $kT_\perp \approx 100$ eV above Akebono and below DE 1. [Peterson et al., 1993].

recently available from Akebono, both inside and near the source regions of TAI. These observations show that the characteristics of transverse ion energization in the dayside are very similar to those in the nightside, aside from the lower characteristic temperature and higher altitude of occurrence in the dayside. The energization is predominantly in the perpendicular direction. The region of energization is highly localized, both in altitudinal and latitudinal extents. At the same time, the process of energization is evidently operative over a wide altitude range, and the energization of ionospheric ions at ionospheric altitudes to energies of upflowing ions at high altitudes (hundred-eV or greater) is a multi-step process.

4 DISCUSSIONS

The existing body of observations on transverse ion energization and conical ion formation clearly establishes the importance and fundamental characteristics of the transverse ion energization process in the auroral ionosphere below the 1-R_E ("parallel acceleration") altitude region, both in the nightside and the dayside. At the same time, a number of important gaps in our observational knowledge remain. These include the unambiguous identification of the free energy source and energization mechanism for TAI and ion conic generation; the source (pre-energization) of O^+ ions that undergo transverse energization at mid-altitudes; the detailed evolution of an ion conic along the field line; the dependence of the energization process on the ambient ionospheric and atmospheric environment; the possible coupling between the transverse ion energization process at low- and mid-altitudes; and the additional ion energization and outflow at high altitudes.

Efforts to identify the source of free energy and energization mechanism have proven difficult, particularly for the satellite observations at mid-altitudes, due to the apparent complexity and the presence of multiple wave modes and nonlinear structures in the energization process, and to the lack of high-resolution measurements and comprehensive wave-plasma diagnostics. Observations from the recent TOPAZ 3 rocket [Arnoldy et al., 1992; Vago et al., 1992] clearly showed the filamentary nature of transverse ion energization regions, and the limitations of earlier measurements in resolution; the high-resolution measurements on the Akebono and Freja satellites are expected to significantly improve our observational knowledge in this regard.

The presence of thermal-energy (un-energized) O^+ ions as a dominant species at mid- and high-altitudes points to a pre-energization process that imparts slightly more than sufficient energy on ions at ionospheric altitudes (to enable them to overcome gravitation). On Akebono, Yau et al. [1993] observed the occurrence of molecular upflowing ions (NO^+ and N_2^+) in the ~10-eV range at high altitudes (>7000 km) during periods of sustained auroral activity. Their presence at high altitude indicates the enhanced production of molecular ions in the F-region and topside ionosphere and the immediate energization of these ions, in view of their short dissociative recombination lifetime at ionospheric levels (order of seconds). The process of ion pre-energization may well prove to play a fundamentally important role in the overall ionospheric ion energization. It is interesting to note that pre-energization is important for a number of wave-particle interaction processes, which are more efficient on moderately energized ions.

Acknowledgements We acknowledge support from the Canadian Space Agency on the Akebono Suprathermal Mass Spectrometer (SMS) experiment.

REFERENCES

Andre, M., G. B. Crew, W. K. Peterson, A. M. Persoon, C. J. Pollock, and M. J. Engebretson, Ion heating by broadband low-frequency waves in the cusp/cleft, *J. Geophys. Res.*, *95*, 20809, 1990.

Arnoldy, R. L., K. A. Lynch, P. M. Kintner, J. Vago, S. Chesney, T. E. Moore, and C. J. Pollock, Bursts of transverse ion acceleration at rocket altitudes, *Geophys. Res. Lett.*, *19*, 413, 1992.

Borovsky, J. E., The production of ion conics by oblique double layers, *J. Geophys. Res.*, *89*, 2251, 1984.

Chang, T., and B. Coppi, Lower hybrid acceleration and ion evolution in the suprauroral region, *Geophys. Res. Lett.*, *8*, 1253, 1981.

Chang, T., G. B. Crew, N. Hershkowitz, J. R. Jasperse, J. M. Retterer, and J. D. Winningham, Transverse acceleration of oxygen ions by electromagnetic ion cyclotron resonance with broad band left-hand polarized waves, *Geophys. Res. Lett.*, *13*, 636, 1986.

Chappell, C.R., T. E. Moore, and J. H. Waite, Jr., The ionosphere as a fully adequate source of plasma for the Earth's magnetosphere, *J. Geophys. Res.*, *92*, 5896, 1987.

Collin, H. L., R. D. Sharp, E. G. Shelley, and R. G. Johnson, Some general characteristics of upflowing ion beams over the auroral zone and their relationship to auroral electrons. *J. Geophys. Res.*, *86*, 6820, 1981.

Collin, H. L., R. D. Sharp, and E. G. Shelley, The magnitude and composition of the outflow of energetic ions from the ionosphere, *J. Geophys. Res.*, *89*, 2185, 1984.

Garbe, G. P., R. L. Arnoldy, T. E. Moore, P. M. Kintner, and J. L. Vago, Observations of transverse ion acceleration in the topside auroral ionosphere. *J. Geophys. Res.*, *97*, 1257, 1992.

Gorney, D. J., A. Clarke, D. Croley, J. Fennell, J. Luhmann, and P. Mizera, The distribution of ion beams and conics below 8000 km. *J. Geophys. Res.*, *86*, 83, 1981.

Gorney, D. J., Y. T. Chiu, and D. R. Croley, Jr., Trapping of ion conics by downward parallel electric fields, *J. Geophys. Res.*, *90*, 4205, 1985.

Klumpar, D. M., Transversely accelerated ions: an ionospheric source of hot magnetospheric ions. *J. Geophys. Res.*, *84*, 4229, 1979.

Klumpar, D. M., Transversely accelerated ions in auroral arcs, in *Physics of Auroral Arc Formation, Geophysical Monograph Series, vol. 25*, edited by S. I. Akasofu and J. R. Kan, p. 122, 1981.

Klumpar, D. M., A digest and comprehensive bibliography on transverse auroral ion acceleration, in *Ion Acceleration in the Magnetosphere and Ionosphere, Geophysical Monograph Series, vol. 38*, edited by T. Chang, M. K. Hudson, J. R. Jasperse, R. G. Johnson, P. M. Kintner, M. Schulz, and G. B. Crew, p. 389, 1986.

Klumpar, D. M., W. K. Peterson, and E. G. Shelley, Direct evidence for two-stage (bimodal) acceleration of ionospheric ions, *J. Geophys. Res., 89,* 10779, 1984.

Kintner, P. M., J. LaBelle, W. Scales, A. W. Yau, and B. A. Whalen, Observations of plasma waves within regions of perpendicular ion acceleration. *Geophys. Res. Lett., 13,* 1113, 1986.

Kintner, P. M., W. Scales, J. Vago, R. Arnoldy, G. Garbe, and T. Moore, Simultaneous observations of electrostatic oxygen cyclotron waves and ion conics, *Geophys. Res. Lett., 16,* 739, 1989.

Kintner, P. M., W. Scales, J. Vago, A. Yau, B. Whalen, R. Arnoldy, and T. Moore, Harmonic H^+ gyrofrequency structures in auroral hiss observed by high-altitude auroral sounding rockets, *J. Geophys. Res., 96,* 9627, 1991.

LaBelle, J., P. M. Kintner, A. W. Yau, and B. A. Whalen, Large amplitude wave packets observed in the ionosphere in association with transverse ion acceleration. *J. Geophys. Res., 91,* 7113, 1986.

Lockwood, M., M. O. Chandler, J. L. Horwitz, J. H. Waite, Jr., T. E. Moore, and C. R. Chappell, The cleft ion fountain, *J. Geophys. Res., 90,* 9736, 1985.

Lysak, R. L., M. K. Hudson, and M. Temerin, Ion heating by strong electrostatic ion cyclotron turbulence, *J. Geophys. Res., 85,* 678, 1980.

Miyake, W., T. Mukai, and N. Kaya, On the evolution of ion conics along the field line from EXOS-D observations, *J. Geophys. Res., 98,* 1993.

Moore, T. E., C. R. Chappell, M. Lockwood, and J. H. Waite, Jr., Superthermal ion signatures of auroral acceleration processes, *J. Geophys. Res., 90,* 1611, 1985.

Okuda, H., and M. Ashour-Abdalla, Formation of a conical distribution and intense ion heating in the presence of hydrogen cyclotron waves, *Geophys. Res. Lett., 8,* 811, 1981.

Peterson, W. K., E. G. Shelley, S. A. Boardsen, D. A. Gurnett, G. G. Ledley, M. Suguira, T. E. Moore, and J. H. Waite, Transverse ion energization and low-frequency plasma waves in the mid-altitude auroral zone: a case study, *J. Geophys. Res., 93,* 11405, 1988.

Peterson, W. K., H. L. Collin, M. F. Doherty, and C. M. Bjorklund, O^+ and He^+ restricted and extended (bimodal) ion conic distributions, *Geophys. Res. Lett., 19,* 1992.

Peterson, W. K., A. W. Yau, and B. A. Whalen, Simultaneous observations of H^+ and O^+ ions at two altitudes by the Akebono and Dynamics Explorer-1 satellites, *J. Geophys. Res., 98,* 1993.

Pollock, C. J., M. O. Chandler, T. E. Moore, J. H. Waite, Jr., C. R. Chappell, and D. A. Gurnett, A survey of upwelling ion event characteristics, *J. Geophys. Res., 95,* 18969, 1990.

Retterer, J. M., T. Chang, and J. R. Jasperse, Ion acceleration by lower hybrid waves in the suprauroral region, *J. Geophys. Res., 91,* 1609, 1986.

Retterer, J. M., T. Chang, J. R. Jasperse, The ion conic observed by MARIE, *J. Geophys. Res., 98,* 1993.

Sharp, R. D., R. G. Johnson, and E. G. Shelley, Observation of an ionospheric acceleration mechanism producing energetic (keV) ions primarily normal to the geomagnetic field direction, *J. Geophys. Res., 82,* 3324, 1977.

Sharp, R. D., A. G. Ghielmetti, R. G. Johnson, and E. G. Shelley, Hot plasma composition results from the S3-3 spacecraft, in: *Energetic Ion Composition in the Earth's Magnetosphere,* edited by R. G. Johnson, p. 163, Terra Scientific Publishing Co., Tokyo, 1983.

Shelley, E. G., R. G. Johnson, and R. D. Sharp, Satellite observations of energetic heavy ions during a geomagnetic storm, *J. Geophys. Res., 77,* 6104, 1972.

Shelley, E. G., R. D. Sharp, and R. G. Johnson, Satellite observations of an ionospheric acceleration mechanism, *Geophys. Res. Lett., 3,* 654, 1976.

Ungstrup, E., D. M. Klumpar, and W. J. Heikkila, Heating of ions to superthermal energies in the topside ionosphere by electrostatic ion cyclotron waves, *J. Geophys. Res., 84,* 4289, 1979.

Vago, J. L., P. M. Kintner, S. W. Chesney, R. L. Arnoldy, K. A. Lynch, T. E. Moore, and C. J. Pollock, Transverse ion acceleration by localized hybrid waves in the topside auroral zone, *J. Geophys. Res., 97,* 16935, 1992.

Whalen, B. A., W. Bernstein, P. W. Daly, Low altitude acceleration of ionospheric ions, *Geophys. Res. Lett., 5,* 55, 1978.

Whalen, B. A., S. Watanabe, and A. W. Yau, Observations in the transverse ion energization region, *Geophys. Res. Lett., 18,* 725, 1991.

Yang, W. H., and J. R. Kan, Generation of conic ions by auroral electric fields, *J. Geophys. Res., 88,* 465, 1983.

Yau, A. W., B. A. Whalen, A. G. McNamara, P. J. Kellogg, and W. Bernstein, Particle and wave observations of low-altitude ionospheric ion acceleration events, *J. Geophys. Res., 88,* 341, 1983.

Yau, A. W., E. G. Shelley, W. K. Peterson, and L. Lenchyshyn, Energetic auroral and polar ion outflow at DE-1 altitudes: magnitude, composition, magnetic activity dependence, and long-term variations, *J. Geophys. Res., 90,* 8417, 1985.

Yau, A. W., B. A. Whalen, F. Creutzberg, and P. M. Kintner, Low altitude transverse ion acceleration: auroral morphology and in-situ plasma observations, in *Physics of Space Plasma (1985-7), SPI Conference Proceedings and Reprint Series, no. 6,* edited by T. Chang, J. Belcher, J. R. Jasperse, and G. B. Crew, p. 77, Scientific Publishers, Inc., Cambridge, MA, 1987.

Yau, A. W., B. A. Whalen, C. Goodenough, E. Sagawa, and T. Mukai, EXOS-D (Akebono) observations of molecular NO^+ and N_2^+ upflowing ions in the auroral ionosphere, *J. Geophys. Res., 98,* 1993.

A. W. Yau and B.A. Whalen, Herzberg Institute of Astrophysics, National Research Council of Canada, 100 Sussex Dr., Ottawa, Ontario, Canada K1A 0R6

Transverse Ion Acceleration by Active Experiments

R. L. ARNOLDY

Institute for the Earth, Oceans and Space, University of New Hampshire, Durham, NH

Recent high altitude and high time resolution rocket measurements made over nightside auroral events have shown that the initial transverse energization of ionospheric hydrogen and oxygen ions that enables them to escape to the magnetosphere occurs within solitary structures of dimension transverse to B the order of 100 m containing 100-400 mV/m electrostatic lower hybrid waves (lower hybrid solitary structures--LHSS). The transverse heating of ions appears to be the mechanism that damps the collapse of auroral hiss into these caviton structures of depleted ambient density. The intent of this paper is to review several active sounding rocket experiments, which have also resulted in the transverse acceleration of ions (TAI), to see if the LHSS model briefly outlined above is the origin of the active experiment TAI's. Although the active experiments did not have the resolution to identify caviton structures as seen in the natural aurora, the common feature of all experiments is that transverse ion acceleration is related to the presence of waves near the lower hybrid frequency either directly injected into space or created by electron beams.

INTRODUCTION

Satellite measurements have shown that there is an abundant outflow of heavy terrestrial ions from the ionosphere populating the magnetosphere. This process cannot be explained by a single-step parallel acceleration mechanism. However, an increase of their perpendicular energy will allow them to escape via their adiabatic motion. This increase of perpendicular energy has been readily observed in particles detected above acceleration regions. Due to their adiabatic motion, these ions have a cone shaped distribution function in velocity space when measured above their point of acceleration and have thus been referred to as ion conics. While many studies of transversely accelerated ions (TAI) have been performed at orbiting spacecraft altitudes, a minimal number of measurements have been made at the lower altitudes accessible to rockets where the transversely heated ions commence. A significant number of the rocket observations of transversely accelerated ions have been related to active experiments.

The intent of this paper is not to review all rocket measurements of TAI but only the active experiment TAI and attempt to relate them to the naturally occurring auroral TAI. Before presenting the data from the active experiments, a review of recent rocket data which shows that auroral TAI are accelerated in localized regions of intense lower hybrid wave

activity [Arnoldy et al., 1992; Kintner et al., 1992; and Vago et al., 1992] will be made. The acceleration region dimension across geomagnetic field lines is about 50-100 m corresponding to 5-10 thermal ion gyroradii or one hot ion gyroradius. Within the acceleration region lower hybrid waves reach peak-to-peak amplitudes of 100-400 mV/m and ions are accelerated transversely with characteristic energies the order of 10 eV. These observations are in general agreement with theories of lower hybrid wave collapse forming lower hybrid solitary structures (LHSS) [Musher and Sturman, 1985; Sotnikov et al., 1978; Retterer et al., 1986; Retterer et al., 1992; Shapiro et al., 1992; Vago et al., 1992]. Although the active experiment data is not of sufficient resolution to make a detailed comparison with the LHSS model, one can see if the active experiments could produce conditions where the LHSS might form.

Three different series of active experiments have been performed by a collaboration of scientists from the U. of New Hampshire, Cornell University, the U. of Minnesota, Marshall Space Flight Center, and TRW which have resulted in the transverse acceleration of ions. The ARCS series injected 100 and 200 eV argon plasma, the ECHO series injected 10-40 keV electrons, and the WISP rocket injected electromagnetic waves with frequencies from 40 Hz to 10 kHz.

AURORAL TAI (TOPAZ ROCKET FLIGHTS)

The transverse acceleration of ionospheric ions below 1500 km is very significant in magnetospheric physics because it is the mechanism by which the ionosphere provides a significant percentage of the plasma mass for the

Auroral Plasma Dynamics
Geophysical Monograph 80

magnetosphere [Chappell et al., 1987]. It is well known that energetic O^+ ions of energy below 12 keV are frequently the major ion (in number density) in the plasma sheet and ring current especially during periods of magnetic activity [Johnson et al., 1979]. Early measurements of the escape of ionospheric ions [Klumpar, 1979] showed that they received the necessary energy to do so initially in the direction transverse to the magnetic field at altitudes below 2000 km. The DE 2 satellite has measured transverse ion acceleration as low as 350 km altitude associated with strong sheared electron flow [Basu et al., 1988]. The accompanying turbulence in ambient ion density and electric fields studied by Basu et al. [1988] are now better understood in terms of the microphysics of the heating mechanism itself as revealed by sounding rocket data. Because this energization can take place in a collisional medium it is an important process in the coupling of auroral energy to the lower ionosphere.

Since the ISIS spacecraft, there has been little opportunity to make satellite measurements in or close to the actual ion acceleration region. Only recently has Whalen et al. [1991] reported direct observations of 90° TAI with instrumentation aboard the EXOS-D spacecraft and then the events occur in time spans small compared with the experiment resolution. Clearly sounding rockets moving at slower speeds and with much higher telemetry rates are more appropriate for studying TAI. The first observations of TAI at rocket altitudes were made by Canadians [Whalen et al., 1978; Yau et al., 1983] where it was demonstrated that the phenomenon occurs commonly above 500 km altitude in conjunction with nighttime aurora. These measurements revealed the extreme anistropy of the ions typically having a half-angle width of 4° about the peak at 90° pitch angle. Wave measurements made aboard these flights detected lower hybrid/H^+ Bernstein waves with wavelengths less than 20 m [Kintner et al., 1986] whose intensity displayed a rough linear relationship to the transverse ion flux.

During the past several years the UNH/Cornell sounding rocket program has concentrated on the nighttime auroral ionosphere where we have investigated auroral processes between 500 and 1100 km altitude. TOPAZ 2 was launched to an altitude of 926 km from Poker Flat, Alaska. Transversely accelerated ions were observed nearly continuously above 600 km altitude although the character of the phase space distributions depended on the intensity of the electron precipitation [Garbe et al., 1992]. In addition, the TAI appeared to exist in very narrow spatial regions sprinkled densely through the ionosphere. The payload typically traversed these narrow spatial regions in a time the order of the detector pitch and and energy sampling time (one second), making it difficult to resolve the bi-Maxwellian distributions. Nonetheless, partial distributions from which perpendicular characteristic energies could be calculated were obtained. Outside the regions of intense electron precipitation these distributions were characterized by a thermal core and a heated tail where the tail had a characteristic energy the order of a few tens of an eV while the core a characteristic energy of less than one eV. The tail heated ions generally co-existed with the lower hybrid/H^+ Bernstein waves. Within regions of intense auroral electron precipitation the character of the TAI measured aboard the TOPAZ 2 rocket changed. In these regions the core of the ion population was bulk heated to characteristic energies the order of several to 10 eV. If a tail distribution existed in this case, its characteristic energy was generally above 100 eV. With the

higher time resolution data we have received from the TOPAZ 3 flight, the microstructure of the low altitude transverse ion acceleration is becoming apparent.

The raw data from which the transversely accelerated ions were first identified aboard the TOPAZ 3 [Arnoldy et al., 1992] flight is given in the top panel of Figure 1. The top panel is the detector count rate summed over all pitch angles, while the bottom trace gives the energy sweep from 7 to 880 eV. Enhanced counts are measured between 576.7 and 577.7 seconds with two periods of large count rate when at the low energy end of its sweep at 577.1 and 577.6 seconds. The pitch angle image is given in the second panel where the three slanted light bars are dead zones in the detector due to structural

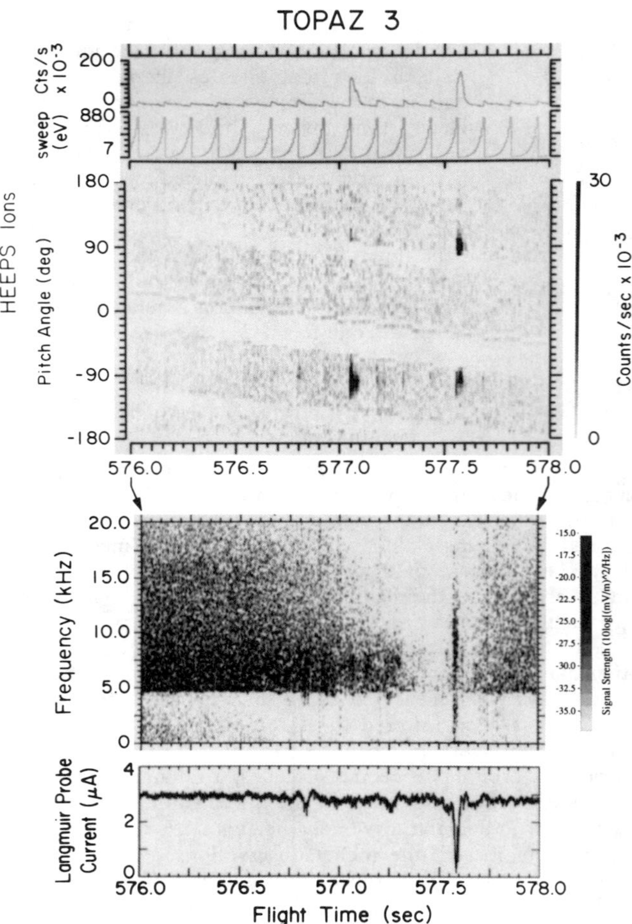

Fig. 1. This figure of TOPAZ 3 data is described thoroughly in the text. The top panel gives the raw count rate summed over all pitch angles and the energy sweep for the superthermal ion detector. The second panel from the top gives the pitch angle image of this detector. The light streaks are data gaps created by structural elements in the detector. The third panel from the top gives a frequency versus flight time spectrogram of the Cornell wave data. The bottom panel is the current measured by the Langmuir probe.

elements in its aperture. The enhanced ion fluxes are measured at or slightly greater than 90° pitch angle. Because of the energy sweep of this detector, it is not capable of resolving the temporal/spatial profile of the ion bursts. There were three other instruments aboard the flight which can be used to address this issue. The first is a fixed energy (13 eV) directional analyzer (always sensing 90° pitch angle ions) which measured the two bursts in Figure 1 as isolated events having durations between 50 and 100 ms. The second instrument is the Cornell electric field experiment from which the frequency-time spectrogram given in the third panel of Figure 1 is obtained. The wave power up to 20 kHz was measured by a one meter dipole antenna which was slowly spinning with respect to the magnetic field. At 577.5 s it was roughly aligned with the geomagnetic field which explains the minimum in received power at this time. The nearly continuous band of power with a low frequency cut-off at 4.5 kHz was auroral hiss which is typically produced by field-aligned electron beams with energies of 100 eV to a few keV [Maggs and Lotko, 1981]. The cut-off frequency of 4.5 kHz corresponds to the lower hybrid frequency. Coincident with the burst of low energy ions at 577.6 s there is an impulsive broadband increase in wave power. This appears to be a time when the payload was within the acceleration region as will be further discussed below. Kintner et al. [1992] discuss the clipped response of the electric field wave receiver to the coherent waves above 5 kHz contained within these bursts having amplitudes up to 300-400 mV/m which severely saturated the instrument. On the other hand, the large ion bursts at 577.1 s in Figure 4 has a very minimal coincident wave response. These ions, however, have pitch-angles centered around 100° rather than 90° as is the case for those ions at 577.6 s, therefore could be on the order of 100 kilometers along B removed from the region of intense waves and above their acceleration volume.

The third instrument that may tell us something about the acceleration volume is the Cornell Langmuir probe. A characteristic feature of the TAI events when simultaneously accompanied by wave impulse events is a decrease in the Langmuir probe current which is plotted in the bottom panel of Figure 1. Clearly the TAI at 577.1 seconds is devoid of such a decrease in peak current or density depletion, again supporting the model that these particular ions are above their acceleration volume as characterized by the wave impulses and density depletion. One might argue that the density depletions accompanying some TAI suggests that both are the result of some negative charging of the payload. However, the 577.1 event clearly dispels this interpretation since ions are coming from a region below the spacecraft where such a mechanism could not be operating.

It is inferred in the discussion above that the waves were the origin of the transverse ion heating rather than the TAI producing the waves. Over three hundred distinct TAI events similar to the two large events presented in Figure 1 were measured during the TOPAZ 3 flight when above 500 km altitude. Outside regions of auroral electron precipitation, where the VLF hiss was weaker, lower hybrid wave impulses with smaller amplitudes were frequently measured without TAI events; however, within regions of auroral precipitation above 500 km altitude, 85% of the TAI events were accompanied by large plasma wave impulses. If the TAI locally produced the wave events, one could not account for 15% of the events not being coincident with wave impulses.

Since the ions that are transversely accelerated (~ 10 eV) can spend several seconds to tens of seconds on a field line of a few hundred kilometers length, they are not a good indicator of duration of the acceleration mechanism; however, because of their attachment to field lines they can be used to obtain the scale size of the acceleration volume across B. Since about 85% of the few hundred ion bursts measured aboard TOPAZ 3, like those given in Figure 1, are associated with wave bursts, the acceleration must occur along an extended length of field line and last for several seconds or more. If this were not the case, given the relatively long duration of ions on a field line in the vicinity of the payload, it would not be possible to obtain the 85% correlation if the actual acceleration time was of very short duration. The measured short duration of ion and wave bursts is, therefore, a measure of the scale size of their acceleration volume across B. With a payload speed across B the order of 1 km/sec, the 100 ms duration corresponds to 100 m (a few hot O^+ gyroradii).

Two papers have recently been published presenting the TOPAZ 3 measurements of what appears to be the embryonic stage of transverse ion acceleration in the lower ionosphere [Kintner et al., 1992; Arnoldy et al., 1992] which will be summarized here. Even though the acceleration volume may be small, the frequency of occurrence of the wave/ion bursts (about once per second) provides for a sufficient average ion flux to reproduce the early ISIS satellite measurements [Klumpar, 1979] made just a few hundred kilometers above the rocket measurements. The Cornell measurements [Kintner et al., 1992; Vago et al., 1992] of the waves comprising the electric field bursts show that they are electrostatic, have a spectrum peaking just above the lower hybrid resonance, apparently propagate on the resonance cone, and from interferometric data have wavelengths between 2 and 20 m (lower hybrid solitary structures, LHSS). An example of a TOPAZ 3 measurement of the power spectrum within a lower hybrid solitary structure (LHSS) is given in Figure 2. A broad emission enhancement for the event of almost 15 dB is evident to the right of the background 5.8 kHz ambient lower hybrid frequency. The peak at 4.8 kHz corresponds to the lower hybrid frequency inside the medium (the LHSS) supporting the wave growth. The 1 kHz shift of the lower hybrid frequency between the event and the

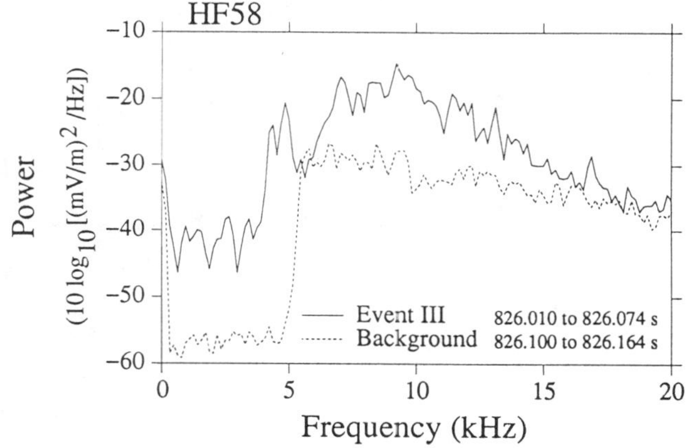

Fig. 2. Electric field spectrum before and during a LHSS event with accompanying transversely accelerated ions (TAI).

background data can be accounted for by a 20% decrease in density inside the LHSS from the background level as measured by the Langmuir probe.

A mass analysis instrument aboard the TOPAZ 3 flight separated protons from the heavier ions at rocket altitudes, presumably singly charged oxygen. In Figure 3 three phase space plots are given. The left and middle plots are data from the mass spectrometer having an energy threshold of 0.6 eV (dashed line) and about 20° pitch angle resolution, while the right panel gives data from an ion instrument which was not capable of mass discrimination but had 5° pitch angle resolution. The ion instrument data are consistent with the measurement of protons at this time. These three phase space plots clearly show that both the oxygen and proton tail distributions are simultaneously accelerated commencing at a transverse velocity of 18 km/sec comparable to the typical phase velocity of the intense lower hybrid waves in the LHSS. Remember both the wave number and the frequency of the waves

were measured so phase velocities are known. These data show that the resonant ion acceleration within the LHSS is clearly Landau (n=0), resulting in sixteen times more energy given to the oxygen ions than hydrogen. The hundreds of bursts of transversely accelerated ions measured aboard the TOPAZ 3 flight not only are consistent with conic fluxes measured by a low orbit satellite, like ISIS, but the associated density and electric field fluctuations of the LHSS also explains the turbulence in these parameters as recorded by the DE-2 spacecraft [Basu et al., 1988] during an ion conic.

The LHSS and the transversely accelerated ions are clearly auroral related, but the understanding of the physics is still something that future flights will be studying. The hypothesis is that the lower hybrid solitary structures condense, presumably via pondermotive force [Kintner et al., 1992], out of the whistler mode auroral hiss created by precipitating electrons. Current data shows that the structured aurora, rather than the steady isotropic component, is important, and, that the

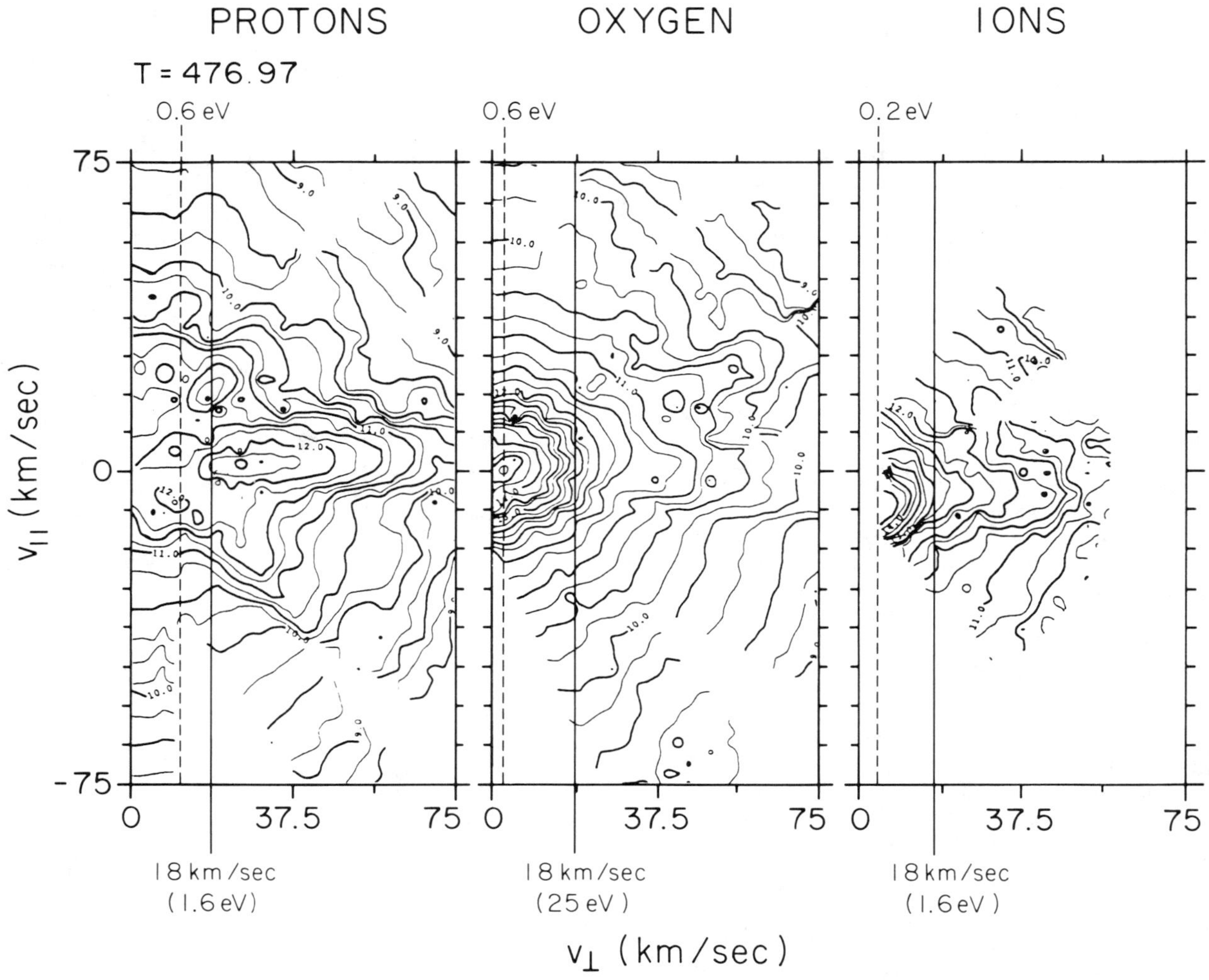

Fig. 3. Proton (left) space phase plot and oxygen (middle) phase phase plot made from data taken by a mass spectrometer during a TAI measured aboard the TOPAZ 3 flight. Each contour is a quarter of a decade change in the distribution function f. The right phase space plot is obtained by another ion instrument whose data at this time is consistent with the measurement of protons. Energy thresholds are indicated by dashed lines.

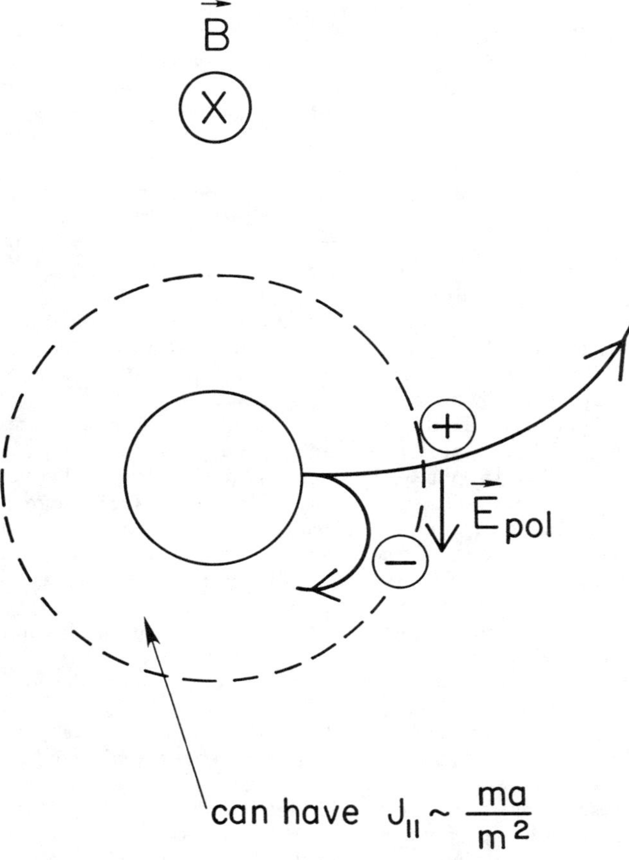

Fig. 4.   Cartoon showing the injection of argons ions transverse to the local magnetic field during the ARCS experiments.

beam precipitation which is plateaued is better correlated with periods of TAI than are the dispersed bursts.

ION BEAM ACTIVE EXPERIMENTS (ARCS ROCKET FLIGHTS)

In the ARCS 2, 3 and 4 flights, 100-200 eV argon ions (~80 ma) were injected both perpendicular and parallel to the local magnetic field. The argon beam and the injection payload were supposed to be neutralized by thermal/superthermal electrons pulled out of the argon plasma generator by space charge electric fields. For injection transverse to B, these electrons become magnetized and cannot follow the ions to neutralize the argon beam. As seen in the cartoon of Figure 4, the resulting polarization electric field could E x B drift the electrons away from the payload moving them with the injected ions. However, a significant result obtained from the Porcupine experiments, where xenon plasma was injected transverse to B at rocket altitudes, was that this polarization electric field was measured to be only 10% of the value needed to convect electrons along with the energetic beam ions [Hausler et al., 1986]. Under these circumstances, payload neutralization can then be achieved only if the superthermal gun electrons can leave the payload and its vicinity along the magnetic field, resulting in field-aligned currents on the order of milliamperes per meter squared. Likewise, ion beam neutralization must occur via ambient electron motion along the magnetic field. Such large field-aligned currents would be unstable to wave growth presumably producing the intense waves near the lower hybrid resonance which in turn provide energy to transversely accelerated ions as will be discussed next. For argon beam injections parallel to B (the injection cone full angle was 60°) the resulting field-aligned currents might be significantly reduced because now many electrons can move away from the injection payload with the ions. The waves and accelerated ions created by the ARCS argon beam injections will now be discussed.

Figure 5 gives the ARCS 3 measurement of ions on a payload removed from the beam injection payload. The ions

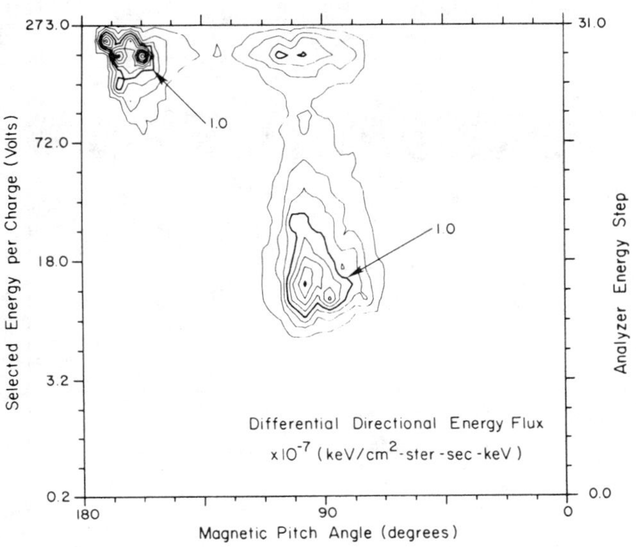

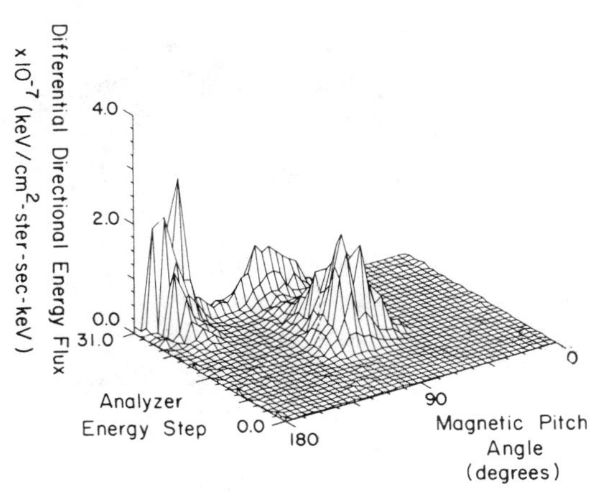

Fig. 5. Contour and three dimensional plots of the ARCS 3 ions measured during the second plasma injection.

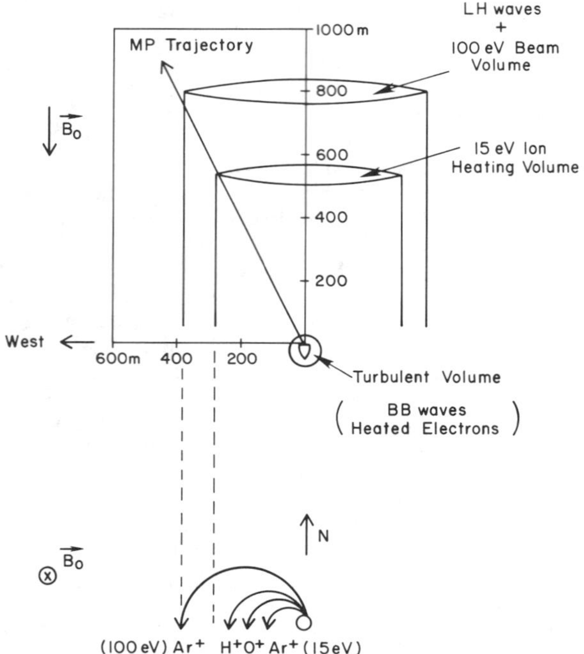

Fig. 6.   Cartoon depicting the three volumes of interest surrounding the ARCS beam injecting payload. Discussed in the text.

near 200 eV energy are beam ions from a plasma injection, while the ion peak between 10 and 30 eV is apparently ambient ions that have been transversely accelerated by the argon beam injection. These ~ 15 eV ions are not seen for all transverse beam injections, and for parallel injections are reduced in intensity by about a factor of a hundred. Figure 6 summarizes the ion and wave effects measured by the ARCS rockets as the diagnostic payload moved away from the argon beam injection payload along the line labeled MP trajectory in the top part of the figure. The injection payload is given by the circle in the middle of the figure. Three volumes of space measured with respect to the 100 eV beam injection payload can be identified [Arnoldy et al., 1990]. The first is called the "turbulent volume" having a radius on the order of several meters around the injection payload in which are found heated electrons with resulting light emission and intense broadband wave emissions. The intense field-aligned currents to neutralize the payload must pass through this volume and, in addition, there is apparently significant beam ion scattering in this volumes [Pollock et al., 1988]. The next volume is what is called the "heating volume" in Figure 6. As the diagnostic payload moved upward and across the field lines from the beam injection payload through this volume the ~ 15 eV heated ions discussed with regard to Figure 5 were measured Although the mass of these ions was not directly measured by instruments aboard the rocket payloads, they occupied a volume of space measured with respect to the beam injection field line consistent with being hydrogen ions as can be seen in the bottom of Figure 6 where the gyrodiameters are given for various 15 eV ions. In this volume, and in the third volume which has been identified ("beam volume"), namely that which contained the beam ions, intense waves near the lower hybrid frequency were measured.   Figure 7 gives wave spectrograms obtained from the ARCS 2 flight by dual probe and Δn/n wave instruments.   For the first perpendicular beam injection between 120 and 135 seconds flight time, the wave measurements were made within the "turbulent volume", hence the intense broadband emissions.   The electromagnetic hiss terminating at the lower hybrid frequency produced by the

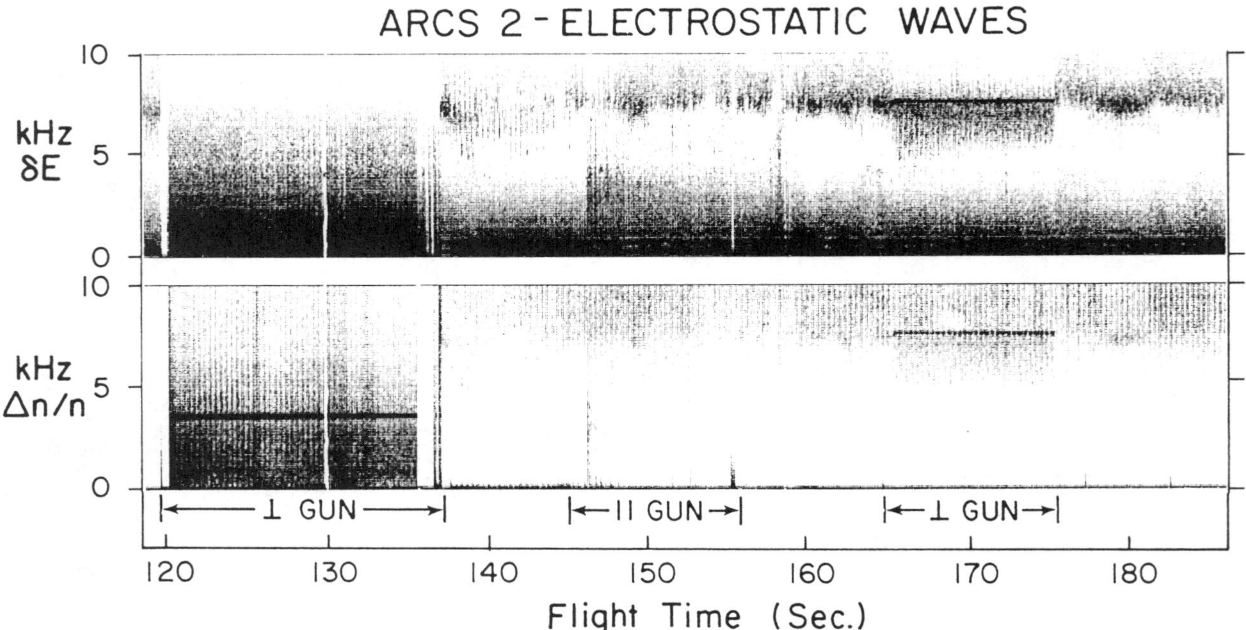

Fig. 7. Wave spectrograms obtained by dual probe and Δn/n instruments aboard the ARCS 2 flight. The type and duration of three beam injections are given near the bottom of the figure.

auroral precipitation during the flight is seen after the first perpendicular injection by the electric field antenna. During the second perpendicular beam injection between 165 and 175 seconds flight time when the payload was within the heating volume, intense narrow bandwidth waves near the lower hybrid frequency were seen by both wave instruments as well as transversely heated ions. The $\Delta n/n$ interferometer probe measured the wavelength of these electrostatic waves to be 6 meters and a phase velocity of 45 km/sec [Kintner et al., 1984]. Note the lack of significant wave emission created by the parallel beam injection between 145 and 155 seconds flight time. The plasma accelerator used aboard the ARCS rocket flight was first flown aboard the ECHO 1 flight. The creation of an intense band of waves near the lower hybrid resonance frequency was also observed [Cartwright and Kellogg, 1974].

The transversely accelerated, superthermal (~15 eV) ions were a feature of nearly every exercise of the beam injector aboard the ARCS 3 flight when the diagnostic payload was in the "heating volume" along with waves near the lower hybrid frequency just described above. Whether one causes the other is not clear at the present time, nor is the mass of the transversely accelerated ions. In a simulation of the ion beam-ionosphere interaction using an electrostatic particle code, Scales and Kintner [1990] find that for beam drift velocities much larger than the H^+ thermal velocity, the lower hybrid and ion-ion hybrid instabilities dominate with the lower hybrid instability having the largest growth rate. For this case, the perpendicular H^+ heating is more efficient than the perpendicular O^+ or parallel electron heating. Hydrogen ions, as discussed above, also give the correct volume of the heating region if the ions are accelerated near the field line containing the beam injection payload, presumably by the LH waves created by the large payload neutralization current on this field line.

ELECTRON BEAM ACTIVE EXPERIMENTS (ECHO ROCKETS)

In the Electron Echo Flights 5, 6 and 7, superthermal ions were measured which were apparently accelerated as a result of the injection of 80 ma of 10-40 keV electrons into the upper ionosphere [Arnoldy and Winckler, 1981; Arnoldy et al., 1984; and Arnoldy et al., 1985]. This report will just discuss the Echo 7 measurements since this was the first Echo flight which could determine the pitch angle distribution of the ions. A schematic of the orientation of the four rocket payloads of the Echo 7 flight is given in Figure 8. The electron beam was injected from the main payload while the ion measurements under discussion were made aboard the PDP payload. In this figure the PDP payload lies in the plane of the figure which contains the electron beam (spiral trace) and is the geomagnetic meridional plane. Over the duration of the flight the PDP moved about 100 meters magnetic south of the beam (north is to the right in the figure) and reached a separation of 600 m above the main (gun) payload. The flight reached an apogee of 290 km.

Figure 9 gives an example of the ions measured in the second trace from the bottom. Their energy between, 10 and 40 eV, is given by the bottom trace which is the energy sweep of the detector. These ions were measured during a 36 keV fixed energy electron injection at pitch angles around 90° (the pitch angle of the injected electrons is given in Figure 9 by the third trace up from the bottom). The auroral electron precipitation was measured during this flight by a directional electrostatic analyzer the data from which is presented in the top three traces

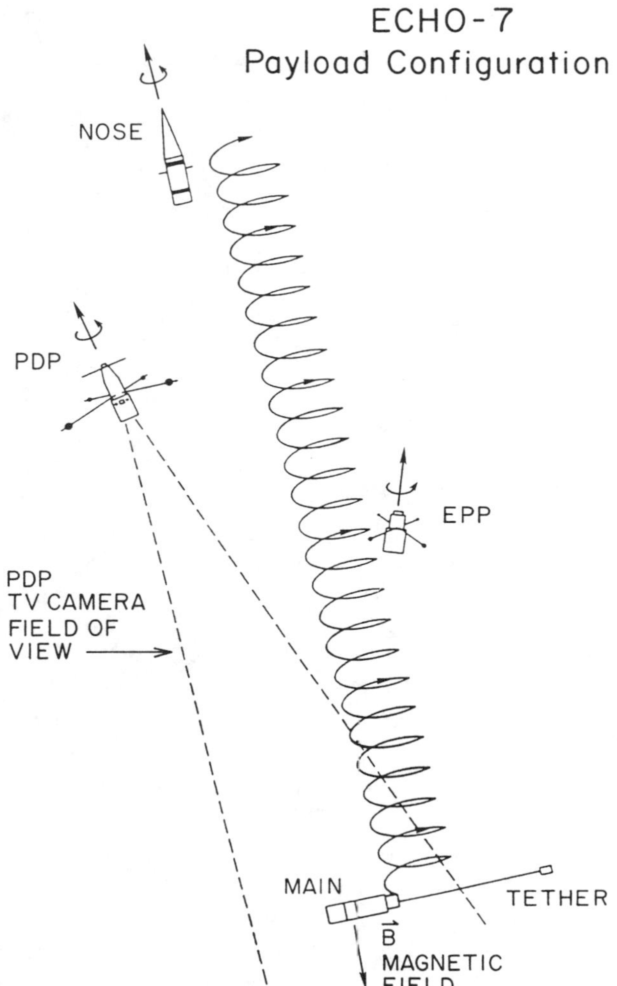

Fig. 8. Schematic of the ECHO 7 payload configuration.

of Figure 9. The top trace is the pitch angle of detected electrons, the second from the top is count rate, and the third from the top the energy sweep of the detector. During the time there was essentially no energetic electron precipitation, hence the measured ions were not auroral related. Figure 10 gives the pitch angle image of the ions with pitch angle as the ordinate and flight time as the abscissa. The bursts of ions between flight times 295.75 and 295.85 seconds are at pitch angles greater than 90° presumably having been transversely accelerated below the point of measurement. The magnetic azimuth (magnetic east is 90°) of the ions was 50°.

Over forty large bursts of ions, such as seen in Figure 9, were measured during the Echo 7 flight. There are several parameters that can be used to categorize the ion bursts. These are; pitch angle; azimuth angle, i.e. direction of arrival or location of guiding center with respect to the location of the electron beam moving past the PDP subpayload; energy; and, structure in the ion bursts. The structured bursts, such as seen in Figure 9, are the higher energy bursts (8-50 eV) and are modulated strongly at 50 Hz, which is apparently the O^+

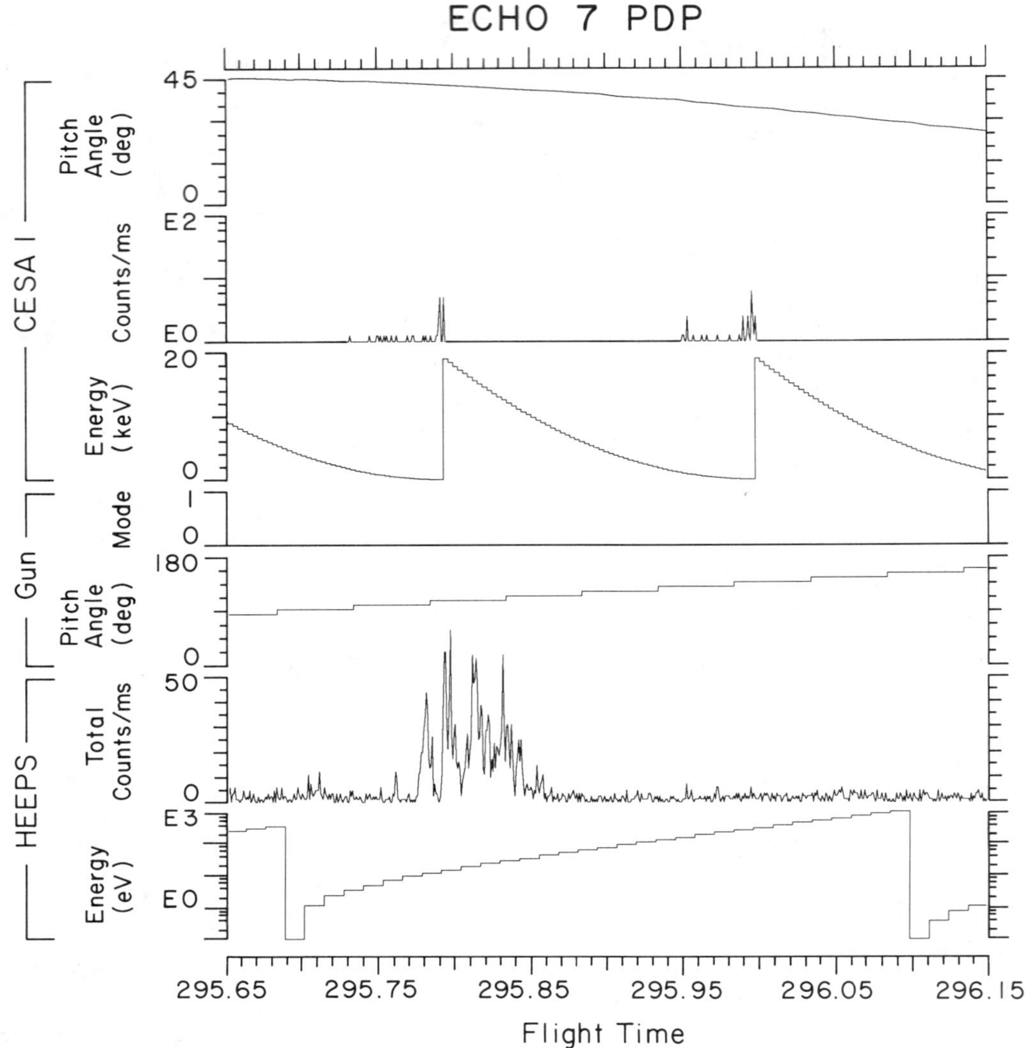

Fig. 9. Electron and ion data from the ECHO 7 flight. Described in the text. The panel labeled HEEPS gives data from the pitch-angle imaging, capped hemisphere electrostatic analyzer. HEEPS stands for hemispherical electrostatic particle spectrometer.

gyrofrequency at the rocket altitude. The meaning of this is not clear but could be related to the mechanism of acceleration, i.e. ion cyclotron wave resonance, or else Landau resonance in a very small volume of short duration which creates a phase-bunched beam. Figure 11 attempts to summarize all these parameters for the 44 ion bursts analyzed. The bottom panel plots the pitch angle versus the azimuth of the events. The events generally fall into two categories, detected in the ram or antiram direction. There is a peculiar separation of events also according to pitch angle, the ions are either downgoing or upcoming. Downgoing ion bursts means that the ions are not only accelerated in the direction transverse to B but also receive significant energy parallel to B. This is not understood at the present time.

The top panel of Figure 11 states that the ram bursts were low energy (between 2 and 6 eV) and unstructured, while the antiram bursts were structured and of higher energy (8-50 eV). The schematics in the top panel give a view looking down the magnetic field towards the Earth. The schematics are approximately to scale with the electron beam given by the open circle, the location of the PDP payload south of the electron beam by the open rectangular box covering the range of distances for all the events plotted in the bottom panel, and, the location of the guiding center for all the events within the shaded rectangular box. The large partial circles are representative gyro-trajectories for the average ion energy in each grouping of bursts. Clearly, the ram events do not originate in the vicinity of the electron beam, but appear to be just ions rammed into the detector by the payload motion, with possibly some energy increase due to a small negative payload potential. Contrary to this result, all the higher energy, structured ion bursts detected in the anitram direction have

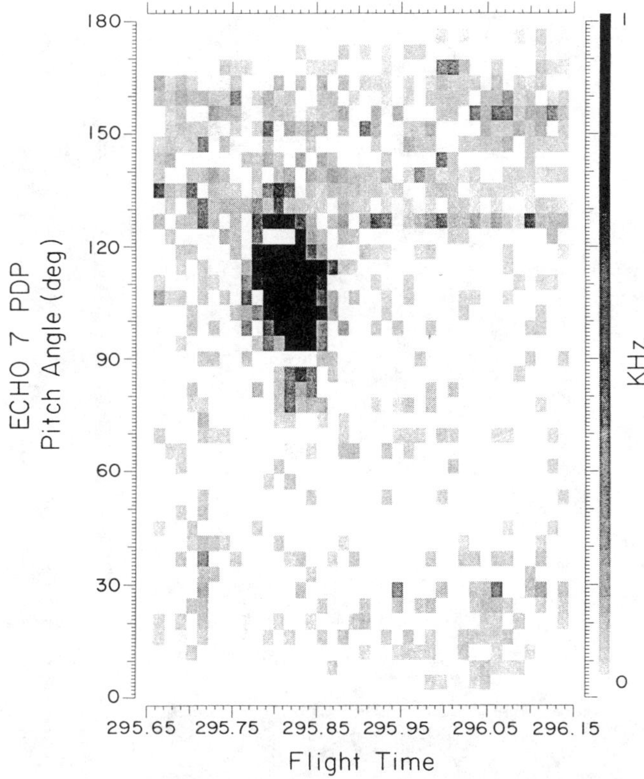

Fig. 10. Pitch angle spectrogram of the ions given in Figure 9.

trajectories which take them within 10 m of the electron beam, leading one to speculate that they are energized near the electron beam.

The injected electron beam does represent a column of negative charge. However, any electric field resulting from this space charge would have a duration some two to three orders of magnitude longer than the 0.02 second bursts comprising the 50 Hz modulated ion events. Presumably a wave-particle interaction is responsible for the ion energization. The injection of electron beams into the upper ionosphere by the ECHO rockets is known to create large amplitude plasma frequency waves, whistler mode waves down to receiver thresholds (50 kHz), and lesser amplitude electron cyclotron harmonic waves [Cartwright and Kellogg, 1974]. Indeed, one would expect the ECHO beams to replicate in some fashion the wave spectrum created by the natural aurora.

WAVE INJECTION EXPERIMENT (WISP ROCKET)

One of the major objectives of the WISP series of rockets is to answer the question, Does the injection of kilowatts of VLF electromagnetic wave power into the ionosphere result in the acceleration of ambient particles? In analyzing the results from the ARCS, ECHO and TOPAZ rocket flights we have speculated that waves created by the ARCS neutralization currents and waves created by the ECHO electron beam and the TOPAZ natural auroral field-aligned currents resulted in the transverse acceleration of ambient ions. The hope in the WISP program is that the wave frequency and power that will transversely accelerate ions can be identified as the injected wave frequency is varied from 30 Hz to 10 kHz and the antenna voltage is varied from 20 V to 10,000 V. The first WISP payload consisted of two parts; a main payload, containing a 2 KW VLF transmitter and a 60 m electric dipole antenna, and a diagnostic payload containing wave receivers, and electron and ion detectors.

The WISP 1 payload was launched in April, 1990 to an altitude of 586 km. All of the mechanical and electronic functions performed nominally except for the attitude control system which impacted the experiment in two ways. First, the subpayload was ejected at an angle of 35° from the geomagnetic field line instead of along it, and second, the transmitting antenna was severely warped. The transmitting antenna may have made frequent contact with the payload skin. The shape of the transmitting antenna was known to be changing since the main payload spin frequency increased and decreased by about 50% several times during the flight. Because both the ULF/VLF waves and accelerated particles should be constrained to a flux tube, we expected to see little if any evidence of the transmissions at the subpayload. However, both waves and the apparent transverse acceleration of ambient ions were seen for short durations during the flight.

Much of the WISP 1 data is still being analyzed, however, we can confidently state that ambient ions were accelerated transverse to B to energies of about 15 eV (surprisingly similar to the ARCS transversely accelerated ions). These ions were seen at just one period of time during the flight which was the first time at maximum radiated power when the antenna frequency was near the expected lower hybrid frequency resonance for the daytime F-region of the ionosphere. Two

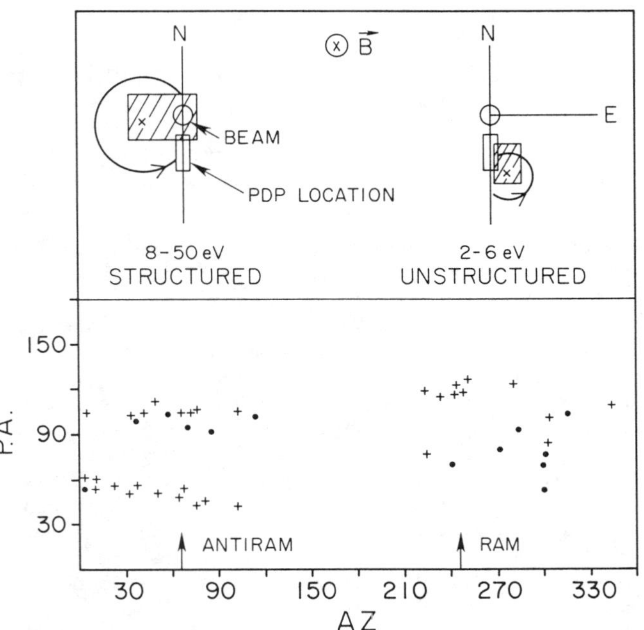

Fig. 11. Summary plot of all the large ion bursts measured aboard the ECHO 7 flight. Described in text. The crosses represent downleg data and the dots upleg data.

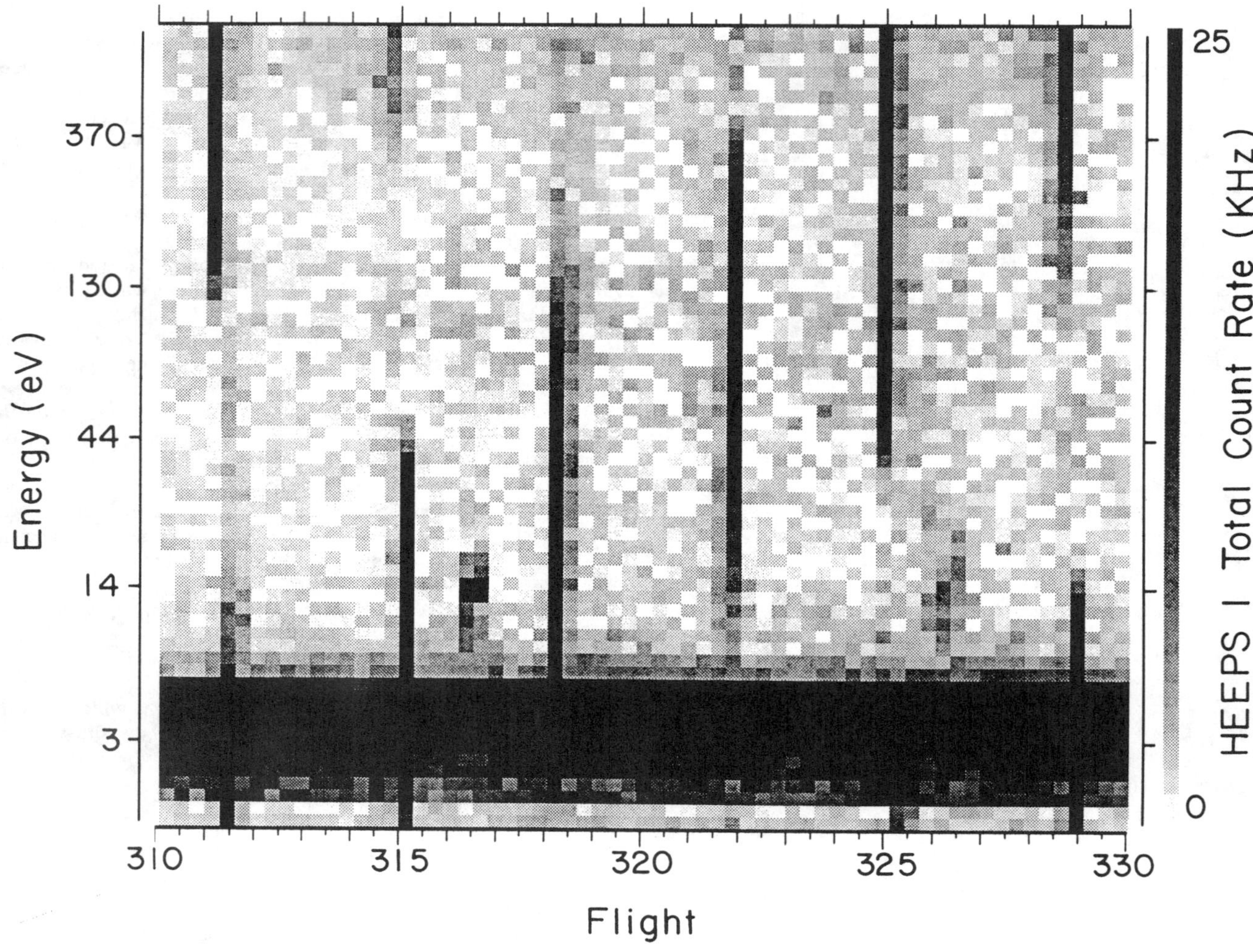

Fig. 12. Ion energy spectrogram obtained from the WISP 1 flight. Discussed in the text.

bursts of transversely accelerated ions are seen in the ion spectrogram given in Figure 12 between 316 and 317 seconds flight time and between 326 and 327 seconds flight time. The dark band centered on a few eV are the ambient thermal population. The dark streaks extending over a broad energy range and occurring once per spin of the payload are spurious counts due to sunlight leakage into the detector. The lack of proper payload orientation due to the ACS failure resulted in direct exposure of the instrument entrance aperture to the sun. The antenna frequency at 316 seconds flight time was 6750 Hz and at 326 seconds flight time 7000 Hz. The wave instrument on the diagnostic payload did not detect waves from the antenna for either of these events. The 15 eV ions moving transverse to B have a gyroradius of about 43 m, which is only about a seventh of the distance that the diagnostic payload was removed from the wave-injection payload field line at the time of their measurement. This seems to imply that the ions were accelerated by waves that have propagated off the wave-injection field line but not to the diagnostic payload since such

waves were not measured. If electrostatic lower hybrid waves are responsible for the ion acceleration then it should also be possible to detect the electron heating along the magnetic field that accompanies these waves. Such electrons were not measured during the WISP 1 flight, possibly because the electron detector was not on the wave-injection field line. The important result of this flight, that ions are accelerated apparently by waves of frequency near the lower hybrid frequency, have warranted a second attempt of the experiment with the WISP 2 mission scheduled for 1993

CONCLUSIONS

Ion acceleration has been observed at rocket altitudes between 500 and 1100 km due to: 1) the injection of 100-200 eV argon plasma; 2) the injection of 10-40 keV electron beams; 3) auroral electron precipitation; and, 4) the injection of electromagnetic waves. Recent high time resolution ion measurements show that the auroral TAI are accelerated in lower hybrid solitary wave structures of small dimension transverse to B (order of 100 m). The origin of these structures appears to be enhanced generation of whistler mode waves which on the resonance cone condense via the pondermotive force into the solitary structures. The ion heating is a damping mechanism for the collapse. Laboratory experiments have created such localized fields and density perturbations which result in ion heating due to self-focusing of nonlinear lower-hybrid waves [Gekelman and Stenzel, 1975]. The ion beams are known to be strong generators of electrostatic waves near the lower hybrid resonance apparently due to the intense field-aligned neutralization currents. The electron beam experiments also have large neutralization currents, and, moreover, inherently resemble the aurora presumably to the degree that there is an electron beam and that ion acceleration in short duration bursts take place. Since the scientific objectives of the ion and electron beam rocket flights did not include transverse ion acceleration, the complement of detectors, both wave and particle, and the temporal resolution of the particle detectors were inadequate to study the phenomena in the manner that the TOPAZ flights under natural auroral conditions did. One cannot, therefore, make more progress with these data sets other than that they all have a common similarity, i.e., electron beams, either directly injected or represented by neutralization current systems, that presumably result in intense wave generation near the lower hybrid resonance. The exciting active experiment is the WISP rocket program where electromagnetic waves are directly injected into space from a powerful (2 kW) rocket-borne transmitter. From the first flight, which was only partially successful due to the loss of attitude control of the payloads, injection of electromagnetic waves near the lower hybrid frequency did result in the transverse acceleration of ambient ions. Future work in this program with another launch scheduled for summer 1993 will attempt to determine if the acceleration mechanism is the lower hybrid solitary structure, under what conditions this structure forms out of the injected waves, and, whether the lower hybrid waves can heat electrons parallel to B.

Acknowledgements. The University of New Hampshire portion of these active experiments was supported by NASA under NASA grant NAG 6-12.

REFERENCES

Arnoldy, R.L., K. A. Lynch, P.M. Kintner, J. Vago, S. Chesney, T. E. Moore, and C. J. Pollock, Bursts of transverse ion acceleration at rocket altitudes, Geophys. Res. Lett., 19, 413, 1992.

Arnoldy, R.L., C.J. Pollock, L.J. Cahill, Jr., R.E. Erlandson, and P.M. Kintner, Observations of the plasma environment during an active ionospheric ion beam injection experiment, Adv. Space Res. 10, (7)1107, 1990.

Arnoldy, R.L., C.J. Pollock, and J.R. Winckler, The energization of electrons and ions by electron beams injected in the ionosphere, J. Geophys. Res., 90, 5197, 1985.

Arnoldy, R.L., C.J. Pollock, and J.R. Winckler, Suprathermal/energetic electron and ion populations resulting from the injection of 10-40 keV electron beams into the ionosphere, Physics of Space Plasmas, SPI Conference Proceedings and Reprint Series, Vol. 5, 107, 1984.

Arnoldy, R.L., and John R. Winckler, The hot plasma environment and floating potentials of an electron-beam-emitting rocket in the ionosphere, J. Geophys. Res., 86, 575, 1981.

Basu, S., S. Basu, E. MacKenzie, P.F. Fougere, W.R. Coley, N.C. Maynard, J.D. Winningham, M. Sugiura, W.B. Hanson, and W.R. Hoegy, Simultaneous density and electric field fluctuations spectra associated with velocity shears in the auroral oval, J. Geophys. Res., 93, 115, 1988.

Cartwright, D.G., and P.J. Kellogg, Observations of radiation from an electron beam artificially injected into the ionosphere, J. Geophys. Res., 79, 1439, 1974.

Chappell, C.R., T.E. Moore, and J.H. Waite, The ionosphere as a fully adequate source of plasma for the Earth's magnetosphere, J. Geophys. Res., 92, 5896, 1987.

Garbe, G.P., R.L. Arnoldy, T.E. Moore, P.M. Kintner, and J.L. Vago, Observations of transverse ion acceleration in the topside auroral ionosphere, J. Geophys. Res., 97, 1257, 1992.

Gekelman, W., and R.L. Stenzel, Localized fields and density perturbations due to self-focusing of nonlinear lower hybrid waves, Phys. Rev. Lett., 35(25), 1708, 1975.

Häusler, B., R.A. Treumann, O.H. Bauer, G. Haerendel, R. Bush, C.W. Carlson, B. Theile, M.C. Kelley, V.S. Dokukin, and Yu. Ya. Ruzhin, Observations of the artificially injected Porcupine xenon ion beam in the ionosphere, J. Geophys. Res., 91, 287, 1986.

Johnson, R.G., Energetic ion composition in the Earth's magnetosphere, Rev. Geophys. 17, 696, 1979.

Kintner, P.M., J. Vago, S. Chesney, R.L. Arnoldy, K.A. Lynch, C.J. Pollock, and T.E. Moore, Localized lower hybrid acceleration of ionospheric plasma, Phys. Rev. Lett., 68, 2448, 1992.

Kintner, P.M., J. LaBelle, W. Scales, A.W. Yau, and B.A. Whalen, Observations of plasma waves within regions of perpendicular ion acceleration, Geophys. Res. Lett., 13, 1113, 1986.

Kintner, P.M., J. LaBelle, M.C. Kelley, L.J. Cahill, Jr., T.E. Moore, and R.L. Arnoldy, Interferometer phase velocity measurements, Geophys. Res. Lett., 11, 19, 1984.

Klumpar, D.M., Transversely accelerated ions, An ionospheric source of hot magnetospheric ions, J. Geophys. Res., 84, 4229, 1979.

Maggs, J.E., and W. Lotko, Altitude dependent model of the auroral beam and beam-generated electrostatic noise, J. Geophys. Res., 86, 3439, 1981.

Musher, S.L. and B.I. Sturman, On the collapse of plasma waves near the lower-hybrid resonance, Sov. Phys. JETP Letts., Engl. Transl., 22, 265, 1975.

Pollock, C.J., R.L. Arnoldy, R.E. Erlandson, and L.J. Cahill, Jr., Observations of the plasma environment during an active ionospheric ion beam experiment, J. Geophys. Res., 93, 1988.

Retterer, J.M., T. Chang, and J.R. Jasperse, Ion acceleration by lowr hybrid waves in the suprauroral region, J. Geophys. Res., 91, 1609, 1986.

Retterer, J.M., T. Chang, and J.R. Jasperse, Lower hybrid collapse and charged particle acceleration, Research Trends in Nonlinear Space Plasma Physics, ed. R.Z. Sagdeev, 1992.

Scales, W.A., and P.M. Kintner, Artificial ion beam instabilities 2: Simulations, J. Geophys. Res., 95, 10,643, 1990.

Shapiro, V.D., V.I. Shevchenko, G.I. Solov'ev, V.P. Kalinin, R. Bingham, R.Z. Sagdeev, and M. Ashour-Abdalla, Wave collapse at the lower-hybrid resonance, in press, 1992.

Sotnikov, V.I., V.D. Shapiro, and V.I. Shevchenko, Macroscopic consequences of collapse at the lower hybrid resonance, Sov. J. Plasma Phys., 4, 252, 1978.

Vago, J.L., P.M. Kintner, S.W. Chesney, R.L. Arnoldy, K.A. Lynch, T.E. Moore, and C.J. Pollock, Transverse ion acceleration by localized lower hybrid waves in the topside auroral ionosphere, J. Geophys. Res., submitted 1992.

Whalen, B.A., S. Watanabe, and A.W. Yau, Thermal and suprathermal ion observations in the low altitude transverse ion energization region, preprint, 1991.

Whalen, B.A., W. Bernstein, and P.W. Daly, Low altitude acceleration of ionospheric ions, Geophys. Res. Lett., 5, 55, 1978.

Yau, A.W., B.A. Whalen, A.G. McNamara, P.J. Kellogg, and W. Bernstein, Particle and wave observation of low-altitude ionospheric ion acceleration events, J. Geophys. Res., 88, 341, 1983.

R.L. Arnoldy, Institute for the Study of Earth, Oceans and Space, University of New Hampshire, Space Science Center, Morse Hall, Durham, NH 03824.

Ion Heating by Low Frequency Waves

TOM CHANG AND MATS ANDRÉ[1]

Center for Space Research, Massachusetts Institute of Technology
Cambridge, Massachusetts

Recent theories of ion heating in the magnetosphere by low frequency turbulence in the range of the ion gyrofrequencies or below are briefly reviewed and assessed. It is concluded that the majority of observed ion heating events at altitudes above a few thousand kilometers in the auroral and polar cusp regions, that are locally heated or generated over extended regimes, may best be explained by the electromagnetic ion cyclotron resonance process.

INTRODUCTION

Rocket and polar orbiting satellites have detected transversely accelerated ions at altitudes ranging from a few hundred kilometers to several Earth radii at auroral and polar cusp altitudes. [See Lennartsson, 1983; Klumpar, 1986; Yau et al., 1987; Peterson et al., 1992; Arnoldy et al., 1992; and references contained therein.] The generally accepted scenario for transverse ion acceleration is some sort of plasma wave-particle interaction. Popular candidates include lower hybrid waves, ion-ion hybrid waves, electrostatic and electromagnetic waves in the ion cyclotron range of frequencies, as well as other low frequency waves. [See reviews by Lysak, 1986; Chang et al, 1988; and Ball and André, 1991a].

The observed wave amplitudes at frequencies near and below the local gyrofrequency are often much higher than those at other frequencies. [See, e.g., Gurnett et al., 1984, Chang et al., 1986, Temerin and Roth, 1986; Retterer et al., 1987; André et al., 1988; 1990; Hultqvist et al., 1988; Crew et al., 1990; Lundin et al., 1990; and Block and Fälthammar, 1990]. Consequently, a number of theories have been proposed to utilize parts or all of the electric field energy at these low frequencies to explain the phenomenon of transverse ion acceleration. In the following, we provide a brief review as well as an assessment of the essence of those theories which address ion conic events at altitudes above a few thousand kilometers, that are generated either locally or over extensive altitudinal regimes at auroral and polar cusp latitudes.

CYCLOTRON RESONANCE THEORY

Perhaps the simplest and most efficient of the proposed ion heating mechanisms involving low frequency waves is the electromagnetic ion cyclotron resonance theory [Chang et al., 1986]. Using reasonable assumptions, Retterer et al. [1987] applied this mechanism to observations in the central plasma sheet. In this study, a simultaneously observed wave spectrum (Fig. 1) together with a Monte Carlo simulation were used to produce an ion distribution quantitatively similar to that observed by Winningham and Burch [1984] at a geocentric height of 2 R_e, Fig. 2. Further tests of this mechanism, involving simultaneously observed wave and ion data in the auroral zone [Crew et al., 1990] and in the polar cusp [André et al., 1990] at high altitudes have also shown good agreement between theory and experimental observations.

The ion cyclotron resonance theory is based on a diffusion operator. This operator can be simplified by using the long wavelength approximation which should be valid, for example, for broad band Alfvén waves in most regions of the magnetosphere. The appropriate diffusion coefficient is then simply proportional to the electric field spectral density, S, at the local gyrofrequency, f_c, of the ion species of interest [Chang et al., 1986; Retterer et al., 1987]. The resulting average heating rate per ion, Q, can be shown to be also proportional to the spectral density (and thus proportional to the mean square of the electric field fluctuations) at the local ion gyrofrequency:

$$Q = (q^2/2m) \cdot S_L \qquad (1)$$

where q and m are the charge and mass of the ion, respectively. The spectral density, S_L, in (1) is the fraction of the spectral density due to left-hand polarized waves. As an illustration of

---

[1] Also at Swedish Institute of Space Physics, the University of Umeå, Umeå, Sweden.

Auroral Plasma Dynamics
Geophysical Monograph 80

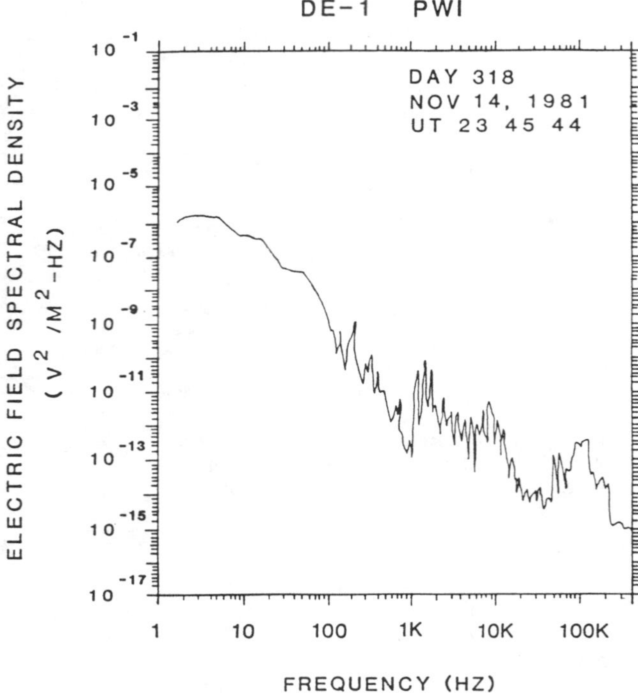

Fig. 1 Typical electric field spectral density in the central plasma sheet. Geocentric height = $2R_e$, invariant latitude = 60°. [Private communication, M. Mellott and D. Gurnett, 1986.]

the relative efficiency of this heating mechanism, we note that a spectral density of 10 $(mV/m)^2/Hz$ would yield an average heating rate of 30 eV/s for oxygen ions, [André et al., 1988]. Figure 3a depicts how an observed wave spectrum may be used to estimate the local ion heating rate. As the ions move up the field line, they interact with the spectral densities of different frequency bands as the local ion gyrofrequency changes due to the inhomogeneity of the geomagnetic field.

Cyclotron resonance heating represents the lowest order theory of ion heating for sufficiently low amplitude fluctuating electric fields using the low frequency portion of the wave spectrum. As it has been alluded earlier, there exist other theories of transverse ion acceleration using spectral densities at frequencies below the local gyrofrequency, f_c. It is tempting to use that portion of the spectra because the wave amplitudes are higher there. Below, we briefly describe a few of these theories and contrast them with that based on the ion cyclotron resonance mechanism.

Multiple Cyclotron Resonance

It has been suggested by Temerin and Roth [1986] that nonlinear resonance involving two modes of finite wavenumbers, with the sum of their frequencies equal to the fundamental cyclotron frequency, could also produce transverse ion heating. This process is often called "double cyclotron absorption (or resonance)." Assuming low amplitude electric field fluctuations, the average heating rate per ion for this process was shown to be proportional to the product of the local electric field spectral

densities at the two frequencies, $S_1 S_2$. Although this mechanism is less efficient than the ion cyclotron resonance theory (since it is of higher order in spectral densities), it can involve the entire frequency range (below the local gyrofrequency) of the spectrum and produce fair amount of transverse ion acceleration, Fig. 3b.

Ball [1989a; b], using a different approach, obtained similar results. It can be demonstrated that the maximum heating rate for such a process is obtained when the wavelength is roughly of the order of the ion gyroradius and that the heating rate vanishes for extremely long wavelength fluctuations (k = 0). A detailed comparison of double-cyclotron absorption with cyclotron resonance heating in the cusp/cleft region of the magnetosphere for typically observed spectra indicated that double-cyclotron absorption might give an appreciable contribution to O^+ energization, especially when the heating was accomplished locally (in altitude), [Ball and André, 1991a]. Nevertheless, the major conclusion of that study was that cyclotron resonance heating is still the more efficient mechanism for the observed oxygen ion conic events in the cusp/cleft region.

From a theoretical point of view, there exist obviously other higher order resonance processes involving spectral densities at more than two frequencies. However, the heating rates under these circumstances will be of even higher order in wave amplitudes (or spectral densities). For reasonably low wave amplitudes, these higher order effects will be much less efficient and can usually be neglected.

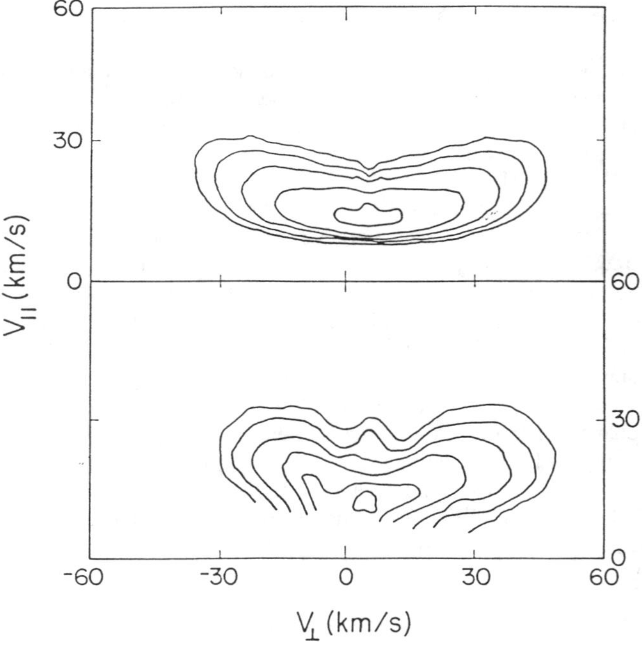

Fig. 2. Bottom panel is a contour diagram of the observed ion-conic distribution function, measured by HAPI instrument on DE-1 [Winningham and Burch, 1984]. Top panel is the calculated ion-velocity distribution [Retterer et al., 1987] using the electromagnetic ion cyclotron resonance theory [Chang et al., 1986], plotted in the same way as the observed distribution.

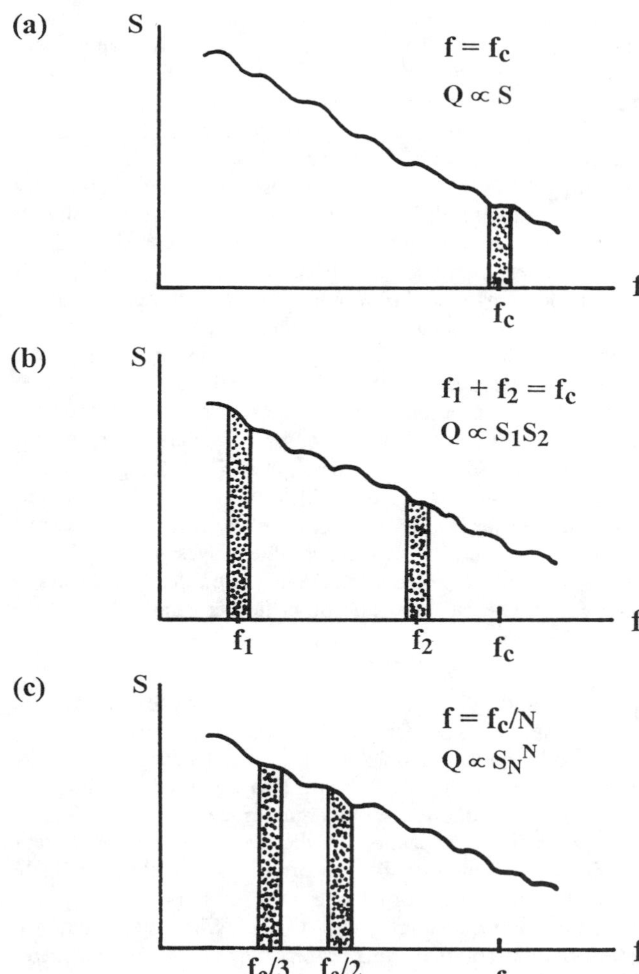

Fig. 3   Sketches depicting how an observed wave spectrum can be used to estimate local ion heating by the (a) cyclotron resonance, (b) double cyclotron resonance, and (c) subharmonic resonance mechanisms. Shaded areas depict bandwidths.

SUBHARMONIC HEATING

It follows from the discussion in the previous section that waves with frequencies f around the Nth order subharmonic ($f = f_c/N$, $N = 2,3,4,...$) can resonantly heat ions for finite wavelength electric field fluctuations, Fig. 3c [Roth and Temerin, 1986]. The average heating rate is then proportional to the Nth power of the electric field spectral density, S_N^N, at the Nth subharmonic. For sufficiently low amplitude fluctuating fields, this leads directly to the following conclusions: (1) Cyclotron resonance ($N = 1$) is generally more efficient than subharmonic resonances ($N > 1$) as has been demonstrated by Ball and André [1991a], (2) 2nd order subharmonic resonance heating (which is equivalent to the special case of double cyclotron resonance for $f_1 = f_2 = f_c/2$) is generally more efficient than those due to other higher order subharmonic resonances, and (3) For broad band

electric field spectra, it is more important to include all the double cyclotron resonance effects (in addition to the $N = 2$ subharmonic effect) than to include other higher order ($N > 2$) subharmonic effects.

We end this section by noting that recently, Terasawa and Nambu [1989] have considered subharmonic ion heating by magnetosonic waves. This process, however, requires magnetic fluctuations larger than those generally detected in the auroral and cusp regions.

NONRESONANT HEATING

It has also been suggested that nonresonant ion energization may be quite important. One possibility is acceleration of ions by an oblique double layer as suggested by Borovsky [1984]. Here the increase in perpendicular velocity is limited roughly by the $\mathbf{E} \times \mathbf{B}$ velocity corresponding to the perpendicular component of the electric field. This mechanism might give significant ion energization provided that strong enough localized electric fields exist. However, detailed comparisons between data and theory for this type of process are not possible because of the narrowness of the regions of intense electric fields and the lack of ion data with adequate temporal resolution and sufficient count rates.

Another possibility is nonresonant heating by waves with frequencies far below f_c. Since both the subharmonic heating and multiple cyclotron resonance heating will be of very high order in wave amplitudes there, they can generally be neglected. It has been suggested that such waves nevertheless might be significant for perpendicular ion energization in the magnetosphere [Lundin and Hultqvist, 1989; Lundin et al., 1990].

To test this idea, Ball and André [1991b] performed a numerical calculation based on a model of an observed broad band wave spectrum. The upper part of Fig. 4 depicts such a model spectrum, constructed by superimposing many sine waves. A spectrum similar to this was observed by the Viking satellite in conjunction with keV protons [Hultqvist et al., 1988]. The model waves are linearly polarized, and hence include a left-handed component. It was assumed in this calculation that the wavelengths of the fluctuations were very long, i.e., $k = 0$, so that multiple cyclotron resonance and subharmonic resonances are not present. To test the hypothesis of non-resonant heating, the model spectrum was subdivided into three parts: (1) one resonant part (RES) around the proton gyrofrequency at 30.0 Hz, (2) one low frequency (LF) nonresonant part, and (3) one high frequency (HF) nonresonant part. (Thus, the nonresonant part LF contains more than just the very low frequency part.) The effect of the different parts of the spectrum on ion energization was tested by simply solving the equation of motion for protons subject to the (known) time-varying electric field of the model in the presence of a static uniform magnetic field. The result obtained by averaging over many ions is displayed in the lower panel of Fig. 4, showing the mean proton energy vs. elapsed time. The difference between resonant heating by waves around the proton gyrofrequency and non-resonant energization by other parts of the spectrum is dramatically illustrated. It is seen that the low frequency part of the spectrum causes the perpendicular

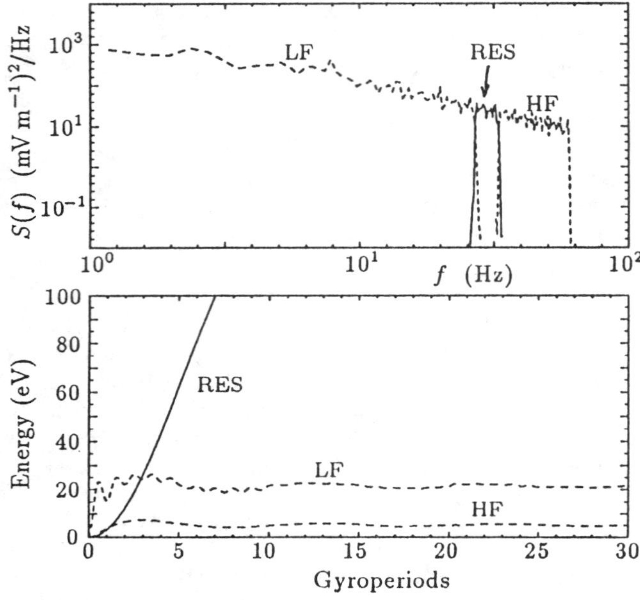

Fig. 4   Proton energization by cyclotron resonance and nonresonant effects due to electric field fluctuations consistent with a spectrum observed in conjunction with proton conics [Ball and André, 1991b]. The resonant part of the spectrum has a width of $f_c/5$ centered at f_c.

proton energy to increase rapidly to around 25 eV; after which, no further energy transfer occurs. The energization process due to the high frequency part of the spectrum is similar. The resulting energy is lower, however, since the spectral density is lower there. Estimates show that including wave power at even higher energies will not provide much additional ion energization.

On the other hand, after a few gyroperiods, the mean proton energy due to resonant heating by waves around f_c is higher than the total energy obtained from nonresonant energization processes. It is noted that no peak in the spectrum around the gyrofrequency is needed for efficient resonant heating, provided sufficient wave power is present near the gyrofrequency.

The numerically obtained resonant heating rate agrees well with the theoretically obtained value from Eq. (1) given by Chang et al. [1986], as discussed by Ball and André [1991b]. Such resonant heating can readily explain the observed keV protons reported by Hultqvist et al. [1988]. On the other hand, nonresonant energization processes can only account for a small fraction of the observed ion energy. We conclude that fundamental ion cyclotron resonance heating can readily explain the observed high energy proton conics (above 100 eV), which are hard to explain with any other mechanism.

For events involving heavier ions, the situation is less clear. As discussed above, detailed studies of O^+ heating events show that resonant heating by observed broadband waves can heat the ions to the observed energies. However, since the gyrofrequencies of heavy ions are rather low, it will be more difficult to differentiate the contributions of individual energization mechanisms. At least in some instances, double-

cyclotron absorption may significantly contribute to O^+ heating [Ball, 1989a,b]. Similarly, nonresonant interaction with waves below f_c may produce significant O^+ heating [Ball and André, 1991b]. Furthermore, such nonresonant energization may well give high enough energies to explain some ion outflow from the ionosphere [Lundin and Hultqvist, 1989]. Nevertheless, Ball and André [1991a,b] concluded that cyclotron resonance currently provides the best explanation for the majority of observed ion conic events (that are generated either locally or over an extended region in space) at altitudes above a few thousand kilometers in the auroral and cusp/cleft regions.

In passing, we point out that nonresonant heating due to stochastic acceleration could also play a role in ion heating if the wave field is coherent and sufficiently strong [Karney, 1978; 1979]. The observation of strong, narrow banded electrostatic ion cyclotron waves within the auroral acceleration region by Mozer et al. [1980] motivated Lysak et al. [1980], Papadopoulos et al. [1980], and Singh et al. [1981; 1983] to consider such type of ion heating processes. It is not clear, however, if there exist regions in the auroral zone where the waves are sufficiently coherent for such a process to operate. We refer the readers to the review article by Lysak [1986] for further details.

PONDEROMOTIVE EFFECTS

More recently, Li and Temerin [1992] suggested that the ponderomotive force derived from the very low frequency portion of the electric field wave spectra could provide field-aligned energization of ions. The idea utilizes the gradient of the $\mathbf{E} \times \mathbf{B}$ drift in the Earth's diverging magnetic field. (Similar concept of ion energization in the Earth's nonuniform static electric and magnetic fields was originally described by Cladis [1986].) This mechanism requires rather large low frequency fluctuating perpendicular electric fields. When this criterion is satisfied, this parallel heating process may need to be considered in conjunction with the perpendicular heating processes when considering ion evolution in the Earth's ionosphere and magnetosphere. Another mechanism that should be mentioned here is that advanced by Mauk (1986), who suggested that a sudden displacement of curved field lines resulting from a short-lived application of an intense electric field may violate the second adiabatic invariant of the lower energy ions, leading to significant parallel ion energization.

DISCUSSION

We have discussed a number of ion heating mechanisms applicable at high altitudes in the auroral and cusp regions. Amongst these, the electromagnetic ion cyclotron resonance theory seems to be the most efficient and relevant process for the formation of ion conics that are either energized locally in the high altitude region or over an extended regime.

The mechanisms discussed above provide bulk heating of the ion distributions. Consequently, the resulting conics due to the heating over extended regimes will generally exhibit a bowl-shaped geometry in velocity space. (See, e.g., Fig. 2). Other shapes of ion conics have also been detected in the suprauroral region, particularly below the auroral acceleration regime (See, e.g., Temerin et al., 1981). These ion distributions have the classical conic shape with their vertices located at the origin in velocity space, indicating that the source regions for heating are far below the points of observation. Suggested mechanisms for ion heating at low altitudes include resonant wave-particle interactions due to electrostatic ion cyclotron waves [Ashour Abdalla and Okuda, 1984; Lysak et al., 1980] and lower hybrid waves [Chang and Coppi, 1981]. It was suggested by Retterer et al. [1986] that nonlinear wave interactions near the lower hybrid frequency, such as the collapse of waves into cavitons might play an important role in the initial perpendicular energization of the ambient ions. Recent rocket observations of the simultaneous occurrence of lower hybrid spikelets and transverse ion acceleration events in the topside auroral ionosphere seem to confirm these predictions [LaBelle et al., 1986; Yau et al., 1986; Kintner et al., 1992; Vago et al., 1992; and Arnoldy et al., 1992]. Discussions describing the observations of this different class of conics can be found in a separate paper by Kintner et al. in this monograph.

We end this review by noting that ion energization on auroral field lines can also be the result of the ion two-stream instability. A review of this work is provided by Bergmann et al. (1993) in this volume.

Acknowledgments. The authors are indebted to R. Arnoldy, L. Ball, J. Burch, J. Cladis, G.B. Crew, D. Gurnett, J.R. Jasperse, P.M. Kintner, D. Klumpar, Xinlin Li, K.A. Lynch, R. Lundin, R. Lysak, W. Peterson, J.M. Retterer, I. Roth, S.W. Chesney, M. Temerin, J. Vago, D. Winningham, and A Yau for useful discussions. This research is partially supported by AFOSR, NASA, the Geophysics Directorate of the Phillips Laboratory, and the Swedish National Science Research Council.

References

André, M., H. Koskinen, L. Matson and R. Erlandson, Local transverse ion energization in and near the polar cusp, *Geophys. Res. Lett., 15*, 107, 1988.

André, M. G.B. Crew, W.K. Peterson, A.M. Persoon, C.J. Pollock and M.J. Engebretson, Ion heating by broadband low frequency waves in the cusp/cleft, *J. Geophys. Res., 95*, 20809, 1990.

Arnoldy, R.L., K.A. Lynch, P.M. Kintner, J.L. Vago, S.W. Chesney, T. Moore and C. Pollock, Bursts of Transverse Ion Acceleration at Rocket Altitudes, *Geophys. Res. Lett., 19*, 413, 1992.

Ashour-Abdalla, M. and H. Okuda, Heating of heavy ions on auroral field lines, *J. Geophys. Res., 89*, 2235, 1984.

Ball, L., Can ion acceleration by double-cyclotron absorption produce O^+ ion conics?, *J. Geophys. Res., 94*, 15257, 1989a.

Ball, L., Heavy ion acceleration by double-cyclotron absorption: some analytic approximations, *Aus. J. Phys., 42*, 493, 1989b.

Ball, L., and M. André, Heating of O^+ ions in the cusp/cleft: double-cyclotron absorption versus cyclotron resonance, *J. Geophys. Res., 96*, 1429, 1991a.

Ball, L. and M. André, What parts of broadband spectra are responsible for ion conic production?, *Geophys. Res. Lett., 18*, 1683, 1991b.

Bergmann, R., P.C. Gray, M.K. Hudson, and I. Roth, Interaction of ion beams in the auroral acceleration region, this volume.

Block, L.P. and C.-G. Fälthammar, The role of magnetic-field aligned electric fields in auroral acceleration, *J. Geophys. Res., 95*, 5877, 1990.

Borovsky, J.E., The production of ion conics by oblique double layers, *J. Geophys. Res., 89*, 2251, 1984.

Chang, Tom, G.B. Crew and J.M. Retterer, Electromagnetic tornadoes in space: ion conics along auroral field lines generated by lower hybrid waves and electromagnetic turbulence in the ion cyclotron range of frequencies, *Computer Phys. Comm., 49*, 61, 1988.

Chang, Tom, G.B. Crew, N. Hershkowitz, J.R. Jasperse, J.M. Retterer and J.D. Winningham, Transverse acceleration of oxygen ions by electromagnetic ion cyclotron resonance with broad band left-hand polarized waves, *Geophys. Res. Lett., 13*, 636, 1986.

Chang, Tom and B. Coppi, Lower hybrid acceleration and ion evolution in the supraauroral region, *Geophys. Res. Lett., 8*, 1253, 1981.

Cladis, J.B., Parallel acceleration and transport of ions from polar ionosphere to plasma sheet, *Geophys. Res. Lett., 13*, 893, 1986.

Crew, G.B., T. Chang, J.M. Retterer, W.K. Peterson, D.A. Gurnett and R.L. Huff, Ion cyclotron resonance heated conics: theory and observations, *J. Geophys. Res., 95*, 3959, 1990.

Gurnett, D.A., R.L. Huff, J.D. Menietti, J.L. Burch, J.D. Winningham and S.D. Shawhan, Correlated low-frequency electric and magnetic noise along the auroral field lines, *J. Geophys. Res., 89*, 8971, 1984.

Hultqvist, B., R. Lundin, K. Stasiewicz, L. Block, P.-A. Lindqvist, G. Gustafsson, H. Koskinen, A. Bahnsen, T.A. Potemra and L.J. Zanetti, Simultaneous observations of upward moving field-aligned energetic electrons and ions on auroral zone field lines, *J. Geophys. Res., 93*, 9765, 1988.

Karney, C.F.F., Stochastic ion heating by a lower hybrid wave, *Phys. Fluids, 21*, 1584, 1978.

Karney, C.F.F., Stochastic ion heating by a lower hybrid wave: II, Phys. Fluids, 22, 2188, 1979.

Kintner, P.M., J. Vago, S. Chesney, R.L. Arnoldy, K.A. Lynch, C.J. Pollock and T. Moore, Localized lower hybrid acceleration of ionospheric plasma, *Phys. Rev. Lett.*, 68, 2448, 1992.

Klumpar, D.M., A digest and comprehensive bibliography on transverse auroral ion acceleration, in *Ion Acceleration in the Magnetosphere and Ionosphere*, edited by Chang et al., AGU Monograph No. 38, p. 389 (American Geophysical Union, Washington, D.C., 1986).

LaBelle, J., P.M. Kintner, A.W. Yau and B.A. Whalen, Large amplitude wave packets observed in the ionosphere in association with transverse ion acceleration, *J. Geophys. Res., 91*, 7113, 1986.

Lennartsson, W., Ion acceleration mechanisms in the auroral regions: general principles, in *Energetic Ion Composition in the Earth's Magnetosphere*, edited by. R.G. Johnson, pp. 23-41, (Terra Publishing Company, Tokyo, 1983).

Li, Xinlin and M. Temerin, Ponderomotive effects on ion acceleration in the auroral zone, submitted to *Geophys. Res. Lett.*, 1992.

Lundin, R., G. Gustafsson, A.I. Eriksson and G. Marklund, On the importance of high-altitude low-frequency electric fluctuations for the escape of ionospheric ions, *J. Geophys. Res., 95*, 5905, 1990.

Lundin, R. and B. Hultqvist, Ionospheric plasma escape by high-altitude electric fields: magnetic moment "pumping", *J. Geophys. Res., 94*, 6665, 1989.

Lysak, R.L., Ion acceleration by wave-particle interaction, in *Ion Acceleration in the Magnetosphere and Ionosphere*, edited by Chang et al., AGU Monograph No. 38, p. 261 (American Geophysical Union, Washington, D.C., 1986).

Lysak, R.L., M.K. Hudson, and M. Temerin, Ion heating by strong electrostatic ion cyclotron wave turbulence, *J. Geophys. Res., 85*, 678, 1980.

Mauk, B.H., Quantitative modeling of the "convective surge" mechanism of ion acceleration, *J. Geophys. Res.*, 91, 13423, 1986.

Mozer, F.S., C.A. Cattell, M.K. Hudson, R.L. Lysak, M. Temerin and R.B. Torbert, Satellite measurements and theories of auroral acceleration mechanisms, *Space Sci. Rev., 27*, 155, 1980.

Papadopoulos, K., J.D. Gaffey and P.J. Palmadesso, Stochastic acceleration of large m/q ions by hydrogen cyclotron waves in the magnetosphere, *Geophys. Res. Lett., 7*, 1014, 1980.

Peterson, W.K., H.L. Collin, M.F. Doherty and C.M. Bjorklund, O^+ and He^+ restricted and extended (bi-modal) ion conic distributions, *Geophys. Res. Lett., 19*, 1439, 1992.

Retterer, J.M., Tom Chang, G.B. Crew, J.R. Jasperse and J.D. Winningham, Monte Carlo modeling of ionospheric oxygen acceleration by cyclotron resonance with broad-band electromagnetic turbulence, *Phys. Rev. Lett.,* *59,* 148, 1987.

Roth, I. and M. Temerin, Nonlinear ion heating in magnetized plasma by monochromatic low frequency waves, *IEEE Trans. on Plasma Science,* 616, PS14, 1986.

Singh, N., R.W. Schunk and J.J. Sojka, Energization of ionospheric ions by electrostatic hydrogen cyclotron waves, *Geophys. Res., Lett., 8,* 1249, 1981.

Singh, N., R.W. Schunk and J.J. Sojka, Preferential perpendicular acceleration of heavy ions by interactions with electrostatic hydrogen cyclotron waves, *J. Geophys. Res., 88,* 4055, 1983.

Temerin, M. and I. Roth, Ion heating by waves with frequencies below the ion gyrofrequency, *Geophys. Res. Lett., 13,* 1109, 1986.

Temerin, M., C. Cattell, R. Lysak, M. Hudson, R.B. Torbert, F.S. Mozer, R.D. Sharp and P.M. Kintner, The small scale structure of electrostatic shocks, *J. Geophys. Res., 86,* 11278, 1981.

Terasawa, T. and M. Nambu, Ion heating and acceleration by magnetosonic waves via cyclotron subharmonic resonance, *Geophys. Res. Lett., 16,* 357, 1989.

Vago, J.L., P.M. Kintner, S.W. Chesney, R.L. Arnoldy, K.A. Lynch, T.E. Moore and C. Pollock, Transverse ion acceleration by localized lower hybrid waves in the topside auroral ionosphere, *J. Geophys. Res., 97,* 16935, 1992.

Winningham, J.D. and J. Burch, Observation of large scale ion conic generation with DE-1, in *Physics of Space Plasmas (1982-4),* edited by J. Belcher, H. Bridge, Tom Chang, B. Coppi and J.R. Jasperse, SPI Conference Proceedings and Reprint Series, Vol. 5, p. 137 (Scientific Publishers, Inc., Cambridge, MA, 1984).

Yau, A, W., A. Whalen, F. Creutzberg and P.M. Kintner, Low altitude transverse ion acceleration: auroral morphology and in-situ plasma observations, in *Physics of Space Plasmas (1985-7),* edited by Chang et al., SPI Conference Proceedings and Reprint Series, Vol. 6, p. 77 (Scientific Publishers, Inc., Cambridge, MA 1987).

Tom Chang and Mats André, Center for Space Research. Massachusetts Institute of Technology, Cambridge, MA 02139, U.S.A.

Interaction of Ion Beams in the Auroral Acceleration Region

RACHELLE BERGMANN

Physics Department, Eastern Illinois University

PERRY C. GRAY

Bartol Research Institute, Sharpe Laboratory, U. of Delaware

MARY K. HUDSON

Dept. of Physics and Astronomy, Dartmouth College

ILAN ROTH

Space Sciences Laboratory, U. C. Berkeley

The interaction of relatively streaming ions accelerated within a quasistatic long spatial scale parallel electric field causes the ions to heat at a range of angles to the magnetic field. Linear stability analyses and plasma simulations indicate that the pitch angle distributions of the ion beams are related to the relative beam drifts and to the magnitude of the electric field. The larger the relative drift, the larger the angle of maximum wave growth; the larger the electric field, the smaller the time scale over which that wave mode will dominate the heating. These effects also constrain the type of analytic theories and plasma simulations that can reasonably model ion heating in the auroral acceleration region. Applicable theories and simulations will depend on the spatial scale of the electric field being modelled. For electrostatic potentials inferred from observations, the spatial scales are such that acceleration and heating should be treated simultaneously.

INTRODUCTION

The stability of relatively streaming ion beams has been studied by many investigators to explain the heating of upward accelerated ion distributions and the various wave spectra observed in the auroral acceleration region. Linear stability analyses [Bergmann and Lotko, 1986; Kaufmann et al., 1986; Dusenbery and Martin, 1988; Bergmann et al., 1988] and plasma simulations [Dusenbery et al., 1988; Roth et al., 1989; Winglee et al., 1989; Schriver and Ashour-Abdalla, 1990; Gray et al, 1990] have shown that this fluid instability has a large growth rate and is very effective in thermalizing the

Auroral Plasma Dynamics
Geophysical Monograph 80

distributions. Plasma simulations using various initial conditions have shown that the interaction can lead to substantial heating. This heating mechanism produces distributions that are in good qualitative agreement with observed ion beams. [Kaufmann et al., 1986; Roth et al., 1989; Winglee et al., 1989; Bergmann, 1990; Gray et al, 1990].

Of the various sites in the magnetosphere where the ion-ion two-stream interaction is important the acceleration region is made unique by the presence of a quasistatic parallel electric field. If this field is of long spatial extent, then the beams are accelerating and heating simultaneously. Since the instability depends on the relative drifts and temperatures of the beams, the heating process is influenced by the electric field.

The typical model of the acceleration region used to explore the ion-ion interaction includes a quasistatic accelerating electric field, $E_{\parallel}$, directed parallel to the magnetic field, **B**. Quasistatic refers to the time scale for ionospheric ions to be accelerated through the region. Observations of accelerated electrons and ions indicate that the field-aligned scale of the acceleration region may be several thousands of kilometers [Mozer et al., 1980; Reiff et al., 1988]. An O^+ ion, starting from rest, would take about one minute to travel 1000 km in a 0.1 mV/m field. If the acceleration region extends over several thousand kilometers, the region must be static on the order of minutes to apply this model to the whole of the field line.

The electric field is not included explicitly in linear stability calculations or in the simulations, with the exception of one simulation by Gray et al.[1990]. (The difficulty with including a static electric field in a linear stability analysis is that there is no equilibrium about which to perform a perturbation expansion.) On a local scale, the plasma is assumed to be homogeneous and uniformly magnetized. The parallel electric field is represented implicitly by the drift of the ion beams. The plasma is taken to be uniform perpendicular to **B**.

The assumption of uniformity in calculating linear growth rates requires that length scales be much larger than the wavelengths of the unstable modes. The transverse dimension of the density cavities associated with the acceleration region can be tens of hydrogen gyroradii. The perpendicular wavelengths of a two-stream instability are on the order of an H^+ gyroradius. Parallel to **B**, wavelengths of the unstable modes depend on the relative beam drifts. The limit of the spatial scale depends on the most unstable mode. This scale is much less than thousands of kilometers. The plasma is homogeneous when these conditions are used to define homogeneity.

This paper does not address the effect of weak double layers, ion holes, localized or solitary waves on ion distributions. Such potential structures are not monotonic, nor are they large scale. Discussions of these nonlinear forms can be found elsewhere in this volume. Further, the ion heating to be discussed is based only on excitation by the relative streaming of different mass ions. Microinstabilities stemming from other free energy sources, e.g. accelerated electrons, are not considered.

This paper starts with a review of important properties of the two-stream interaction in the context of the acceleration region based both on linear theory and simulations, and concludes that there should be a relationship between the pitch angle distribution of the beams and their relative drift. Further, recent simulations have shown that the assumption of spatial homogeneity is, under some circumstances, violated due to the properties of the instability. Though the spatial scales of the waves produced are much smaller than the plasma scale, the spatial scale over which the stability properties change is not. This limits the use of spatially homogeneous plasma simulations that do not implicitly include a static parallel electric field.

The next section briefly reviews the salient linear stability properties of the two-stream interaction in two dimensions.

Following that, these properties are related to ion heating as examined in the simulations of Gray et al. [1990]. In the fourth section these results are used to discuss the qualitative relationship between the observed pitch angle distribution of ions and the conditions in the auroral acceleration region. Spatial scales are derived in the section 5. Results are discussed and summarized in the conclusions.

LINEAR STABILITY

Models used to investigate the interaction between like and unlike ion beams have all contained at least stationary electrons and two ion beams. The focus here is on beams of different mass, in particular H^+ and O^+, with stationary electrons. As ionospheric ions are accelerated upward in a monotonically decreasing potential the ions develop different average drift speeds. The relative velocity between these beams, ΔV, is the free energy source for the two-stream instability.

In one dimension (along the magnetic field) the interaction has two characteristics important to the spatial development of the beam heating. First, the interaction is stable for relative drifts greater than

$$\Delta V_L \approx c_s \left(\frac{n_H}{n_e} \right)^{1/2} (1 + \delta^{-1})^{3/2} \qquad (1)$$

where $\delta^3 = n_H m_0 / n_0 m_H$ and $c_s = (T_e/m_H)^{1/2}$ is the acoustic speed. Second, there is a limit on `the electron-to-ion temperature ratio, T_e/T_i, below which the the plasma is stable to ion-ion interactions for all ΔV. (For H^+-H^+ and H^+-O^+ instability this limit is 3.5.) This is imposed by the real part of the dispersion relation [Koons et al, 1971].

When the analysis is extended to two dimensions, it is found that the beams are unstable for relative drifts above ΔV_L and that the growing waves are oblique to **B**. The angle of the fastest growing mode is proportional to ΔV. The value of this angle is determined by the electron temperature (which sets the acoustic speed) and the perpendicular temperature of the ions which, in a kinetic calculation, limits the maximum perpendicular wave vector to values on the order of the inverse of the lightest ion gyroradius. The trend is for the most unstable mode to move toward perpendicular with increased relative drift [Bergmann et al., 1988].

Putting this into the larger scale, ionospheric ions enter the lower altitude end of the acceleration region. As they move upward their relative drift increases until the kinetic damping due to the ion beams becomes small enough for the fluid instability to grow. This takes place a few to tens of kilometers into the acceleration region [Bergmann, 1990]. The beams are unstable to parallel wave modes. The parallel electric field, however, continues to increase ΔV. The unstable modes become oblique as ΔV_L is surpassed heating the ions at non-zero pitch angles. The angle of greatest wave growth (the angle where heating dominates) becomes progressively more oblique forming a broad pitch angle distribution as seen in the data. Figure 1 shows a schematic of this scenario.

The obvious caveat is that the heating of the beams has not been included. The heating changes the relative drift and T_e/T_i. The relative drift between the beams is decreased by the two-stream interaction. The velocity space valley between the beams fills. When the beams are sufficiently heated, they will kinetically damp the two-stream instability. Also, the increased T_i will decrease T_e/T_i possibly stabilizing the plasma.

There is a competition between the increase in the relative drift due to the accelerating potential and the decrease in drift and growth rate due to the heating. Complicating this is the shifting of the wave vector at maximum growth with the change in ΔV. These competing effects were examined in simulations.

SIMULATION WITH A PARALLEL ELECTRIC FIELD

Gray et al. [1990] performed a series of two-dimensional simulations. The first set used initial conditions similar to others simulations cited in the introduction, starting the H^+ and O^+ distributions at a particular relative drifts below and above ΔV_L. The last simulation started the ions at rest but included a spatially uniform $E_{||}$ of order 1 mV/m . The electrons were treated as a stationary Boltzmann distribution (ie. fluid electrons with finite inertia) with a thermal spread much larger than highest acquired H^+ beam drift. This mimics the conditions in the lower acceleration region and allowed the investigators to focus on the ion-ion two-stream interaction.

The first set of simulations showed that the ions initially heat along the angle of the most unstable mode. These results are consistent with other simulations. The simulations were not run until the heating process ceased, however, because the increase in the beam drift due to a parallel electric field was not included. (The simulation is not a valid representation of the model acceleration region after some time.) The form of the distributions were conic-like if $\Delta V > \Delta V_L$, Figures 2a-c, having peak fluxes at the angle of the most unstable mode.

With the parallel electric field included, the beams became unstable first to parallel propagating modes. The rate at which

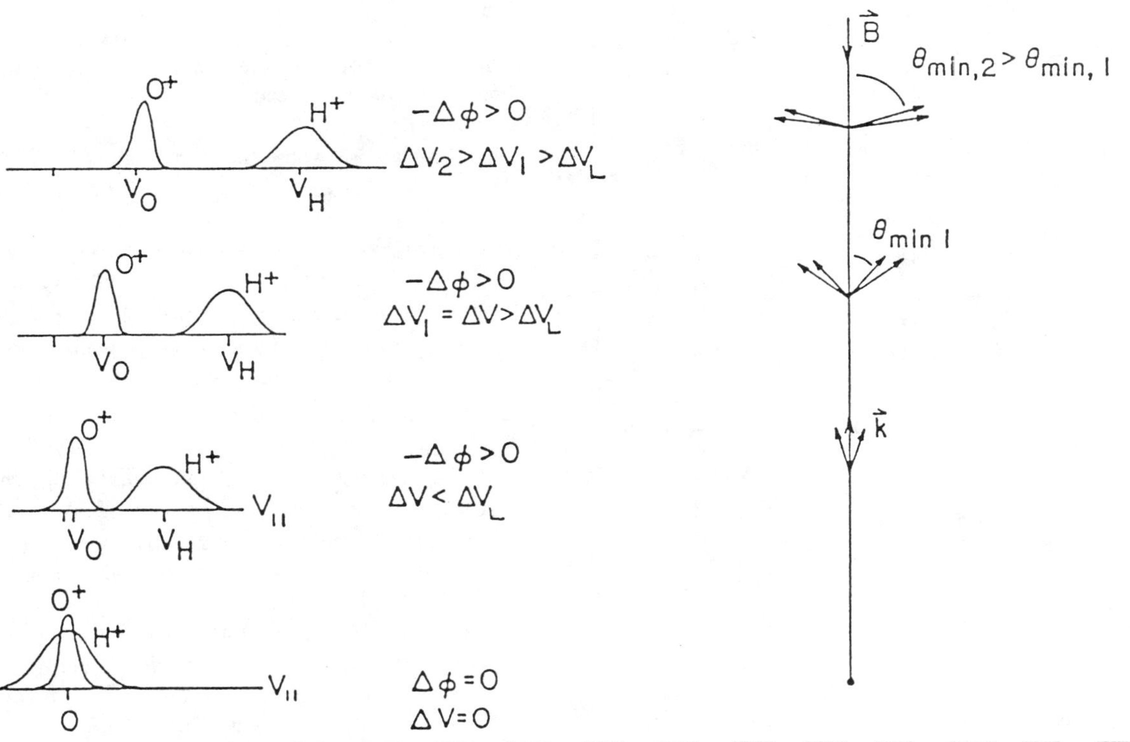

Fig. 1. Schematic of the acceleration region. The vertical axis represents altitude. On the left; reduced ion distributions. On the right; increase in the minimum unstable angle θ_{min} with altitude.

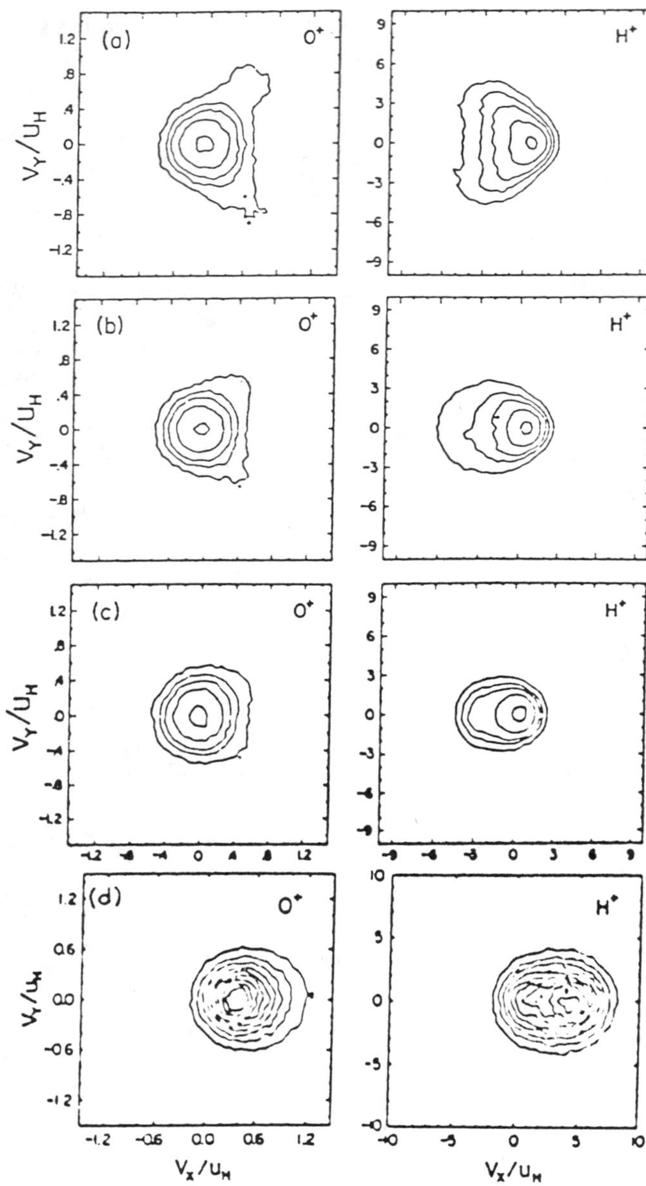

Fig. 2. H^+ and O^+ distributions t = 120 $\omega_H{}^{-1}$ with initial relative drifts (a) 1.6 ΔV_L, (b) 1.2 ΔV_L, and (c) 0.8 ΔV_L. (d) Distributions initialized at rest and $E_{\parallel} \approx 1$ mV/m. [Figures 2 and 4 from Gray et al., 1990].

the beam drift increased due to the parallel field was larger than the rate of decrease due to the heating, and wave modes oblique to **B** became unstable. The beams developed heating that broadened in pitch angle as the unstable mode swept away from **B**. However, there was no conic-like feature to the distributions. Rather, they appeared to be more or less bulk heated, with an elongation in the magnetic field direction. (Figure 2d). This is similar to observations, though it must be remembered that the simulations are run for a limited time; representing only 10-100 km. (It should be noted also that the ratio of the ion beam speeds is initially near $m_O/m_H = 16$ rather than the typically observed value of four. The simulation examined the temporal evolution of the beams following them up the field line, rather than the spatial evolution of beams.)

Based on the relationship between the angle of the linearly most unstable mode and the relative drift between the beams determined by the parallel electric field strength, Gray et al. found that the angle of this mode changed by at least 5° in one e-folding of the wave amplitude. Heating at any particular angle is limited in time. There is no conic-like feature in the distribution. The ions heat from parallel to increasingly oblique.

The form of the distribution functions and the time development of the wave energy in the case where beams are initialized with a relative drift and where an electric field accelerates them from rest are distinct. A finer understanding of this may lead to the identification of certain ion distributions and the potential structure along the magnetic field.

POTENTIAL STRUCTURE AND BEAM DISTRIBUTIONS

The results cited above and those from other simulations suggest that the ion distribution functions depend on the spatial structure of the potential not just the value of the potential difference between the local altitude and the bottom of the acceleration region $\Delta\phi(s)$.

Consider the case where the electric field is approximately constant. Let τ be the time scale over which a wave remains in the mode with the maximum growth rate. This can be approximated from the value of the electric field and the dependence of the most unstable wave vector on ΔV [Gray et al., 1990]. The quantity γ^{-1} is the e-folding time for this wave's amplitude. One can divide the heating process into three categories; (I) $\tau >> \gamma^{-1}$, (II) $\tau \approx \gamma^{-1}$, and (III) $\tau << \gamma^{-1}$. Each of these is related to the spatial extend of $\Delta\phi(s)$ as follows:

(I) $\Delta\phi(s)$ is extended over a large spatial scale. The electric field is relatively weak so ΔV changes relatively slowly as does the frequency and wave vector at maximum growth. The wave undergoes several e-foldings, heating the beams, before it becomes linearly stable due to the change in ΔV. The heating in such a potential might be predominantly parallel to the field, perhaps with the decrease in ΔV below that determined solely from acceleration. If the heating went to oblique angles the distributions may look much like those produced in (II) below, though with parallel to perpendicular temperature ratios larger than in (II).

(II) This intermediate range leads to the fanning out of the heating as shown in Figure 2d where the electric field was about 1 mV/m. The maximally growing mode undergoes about one e-folding before the wave vector with the largest γ shifts to a larger angle.

(III) The beams do not have time to heat via the two-stream interaction. The waves do not go through even one e-folding before the change in the relative drift causes them to stabilize. The other low frequency electrostatic instabilities, the electron-ion acoustic and electrostatic ion cyclotron, would not heat the beams; they have even longer e-folding times than the fluid two-stream.) Once the ions leave the potential structure heating they will have a relative drift and heating would proceed. The angle of dominant heating depends on the potential drop encountered. The resulting distributions would look like Figures 2a-c.

Simulations that start the beams with a relative drift are comparable to letting the beams fall through a potential satisfying condition (III) and then being released into a region of zero electric field where the two-stream interaction proceeds. The different relative drifts used would correspond to different potential drops. The extreme case of the ions falling through a field-aligned potential of several thousand volts before interacting was examined by Schriver et al. [1990]. This resulted in nearly perpendicular conic formation.

Spatial Scales and Homogeneous Plasma Models

What would be typical spatial scales for the potential structures in the three categories? To derive order of magnitude scale lengths consider an $E_{||}$ that is constant for a distance Δs. Then the time t spent within a structure could be related to Δs by $\Delta s = (eE_{||}t^2)/(2m)$. The value of the growth rate for a fluid instability is comparable to the cyclotron frequency $\gamma = \Omega_H = 2\pi(100\ Hz)$. Setting τ equal to t, one gets spatial scales for each of the interaction conditions above of

$$(I)\quad \Delta s \gg 350\ [e\Delta\phi]^{1/2}\ m, \qquad (2a)$$

$$(II)\quad \Delta s \approx 350\ [e\Delta\phi]^{1/2}\ m, \qquad (2b)$$

$$(III)\quad \Delta s \ll 350\ [e\Delta\phi]^{1/2}\ m. \qquad (2c)$$

where $e\Delta\phi$ is the potential energy difference in keV. Typical potential drops inferred in the acceleration regions are a few keV. The Debye length is on the order of 10 -100 m depending on the electron temperature assumed. The more important result is (III). In order for the beams to be in a potential structure and not heat by an ion-ion two-stream interaction (that is $\tau \ll \gamma^{-1}$), that structure must be on the order of Debye lengths; a potential structure similar to a strong double layer. Separating (I) and (II) requires that one carefully address the balance between the decrease in ΔV due to heating and increase due to the electric field.

Parallel scale lengths determine the applicability of linear theory based on a homogeneous plasma. The parallel wavelength of the most unstable mode is roughly

$$\lambda \approx \frac{\Delta V}{\omega_H} \qquad (3)$$

where ω_H is the hydrogen plasma frequency. The larger the drift the longer the spatial scale of the wave. Again assume a constant electric field, let $\Delta\phi$ and Δs be the total potential difference and length of the structure, and $\delta\phi$ be the potential difference at some distance $\delta s < \Delta s$ into the structure. At what $\delta\phi$ does $\lambda = \Delta s$? If $\delta\phi \ll \Delta\phi$, then the beams travel a short distance before λ becomes larger that Δs and uniformity is violated. For adiabatic motion $\Delta V = (\ e\delta\phi/\ m_H)^{1/2}$. This yields

$$\frac{\delta\phi}{\Delta\phi} \approx \frac{m_H\omega_H^2\,(\Delta s)^2}{e\Delta\phi} \rightarrow \begin{cases} \gg 1 & \text{for (I)} \\ \approx 1 & \text{for (II)} \\ \ll 1 & \text{for (III)} \end{cases} \qquad (4)$$

where (2a)-(2c) have been used for Δs, and $n_H = 1\ cm^{-3}$. The beams can be accelerated through the full potential before nonuniformity is violated with the exception of (III). This would be a shock structure where the gradients in the plasma are not small and homogeneous linear theory should not be applied to this situation. Nonlinear and spatial gradient effects within and at the boundaries of the shock structure may well be most important for determining the distribution functions.

The plasma may be treated as homogeneous for (I) and (II) in a local sense but not globally. Two spatial variations besides the potential structure affect the instability. First, the electron temperature is a function of altitude. There are several electron components in the acceleration region; ionospheric, backscattered, trapped and magnetospheric [Chiu and Schulz, 1978]. At lower altitudes the cooler ionospheric, backscattered and trapped electrons dominate the electron susceptibility, leading to lower effective electron temperatures at lower altitudes. With increasing altitude, the cooler electrons are reflected below the potential and the warmer magnetospheric and trapped electrons determine the electron temperature. The electron temperature affects two aspects of the linear stability of the beams. The electron temperature sets the acoustic speed and so ΔV_L. Also there is a possible lower limit on T_e/T_i where the plasma is two-stream stable. While the beams may become stable at lower altitudes due to this effect, they may destabilize at higher altitudes where the electron temperature has increased. The second nonuniformity is the diverging magnetic field. This tends to cause the distributions to become more field aligned as they accelerate into the magnetosphere. These preclude using homogeneous plasma simulations to model the whole of the acceleration region.

Conclusions

The ion-ion two-stream instability is qualitatively successful in explaining many features of H^+ and O^+ ion beam distributions in the acceleration region. The combination of linear theory and simulations have shown that there should be a relationship between the pitch angle spread of the distributions, the relative drifts, and the potential structure. Large spatial scale potential drops (small electric fields) allow the unstable modes to undergo several e-foldings before the increase in relative beam drift changes the most unstable wave vector. Since parallel propagating modes are unstable at the smallest potentials differences, the beam heating may be predominantly field aligned. For larger electric fields the unstable mode has limited time to grow before it stabilizes due to the change in ΔV. The angle of maximum growth becomes increasingly oblique and the beams are heated over a range of pitch angles, producing broad pitch angle distributions. For electric fields large enough that the two-stream unstable modes undergo less than one e-folding during the time the beams are within the potential structure, heating by the two-stream interaction will not occur until the beams enter a region of $E_{||} = 0$ (or at least small). Then the beams have a large relative drift and the angle of heating is determined by the magnitude of ΔV. However, this condition exists only if the potential drop is in the form of a strong double layer.

Average potential drops inferred from observations range up to a few keV. A study by Reiff et al. [1988] indicated that the relationship between the beam and electron distributions at high and low altitudes was not consistent with the acceleration of the species through a double layer with subsequent heating after the particles exited the layer. Rather the distributions were consistent with a potential distributed between the lower altitude DE-2 satellite and and DE-1 at higher altitude, several thousand kilometers. Scale lengths based on the work of Gray et al. [1990], and derived here, show that under such conditions a reasonable model of ion heating via an ion-ion two-stream interaction must simultaneously address the acceleration due to the field and heating due to the instability.
Simulations having zero $E_{||}$ and initialized with relatively drifting ions are applicable to the acceleration region only if the simulation is viewed as modelling the interaction that would take place after the ions exit the acceleration region with no heating occuring within the region.

For models with extended potential drops and modest electric field strengths, the heating associated with a continuous shift in the most unstable wave mode should be examined. Because one wave mode does not undergo several e-foldings before it stabilizes, the heating might best be modeled by quasilinear theory. Extended potentials also introduce spatial gradient considerations in the form of the electron temperature profile and magnetic field divergence. Because of the limitations on computer speed and memory it would be quite difficult to simulate several thousand kilometers of field line. The simulations are important for developing an appropriate local heating theory. Then global effects can be included by joining a series of local uniform plasma calculations.

REFERENCES

Bergmann, R. and W. Lotko, Transition to unstable ion flow in parallel electric fields, *J. Geophys. Res., 91,* 7033, 1986.

Bergmann, R., I. Roth, and M. K. Hudson, Linear stability of the H^+–O^+ two-stream interaction in a magnetized plasma, *J. Geophys., Res, 93,* 4005, 1988.

Bergmann, R., H^+–O^+ Two-Stream Interaction on auroral field lines, in *Physics of Space Plasmas (1989), SPI Conf. Proc. and Reprint Series, vol. 9,* edited by T. Chang, G. B. Crew and J. R. Jasperse, p. 229, Scientific Pub., Inc., Cambridge, MA, 1990.

Chiu, Y. T. and M. Schulz, Self-consistent particle and parallel electric field distributions in the magnetospheric-ionospheric auroral region, *J. Geophys. Res, 83,* 629, 1978.

Dusenbery, P. B. and R. F. Martin, Generation of Broadband turbulence in the accelerated auroral ions, 1. parallel propagation, *J. Geophys. Res., 92,* 3261, 1987.

Dusenbery, P. B., R. F. Martin, and R. M. Winglee, Ion-ion waves in the auroral acceleration region; wave excitation and ion heating, *J. Geophys. Res, 93,* 5655, 1988.

Gray, P. C., M. K. Hudson, R. Bergmann and I. Roth, Simulation study of ion two-stream instability in the auroral acceleration region, *Geophys. Res. Lett., 17,* 1609, 1990.

Kaufmann, R. L., G. R. Ludlow, H. L. Collin, W. K. Peterson and J. L. Burch, Interaction of upgoing H^+ and O^+ beams, *J. Geophys. Res, 91,* 10080, 1986.

Koons, H. C., D. A. McPherson, and M. Schulz, Ion-wave instabilities at VLF in the polar wind, *J. Geophys. Res.,76,* 6122, 1971.

Mozer, F. S., C. A. Cattell, M. K. Hudson, R. L. Lysak, M. Temerin, and R. B. Torbert, Satellite, measurements and theories of low altitude particle acceleration, *Space Sci. Rev.,27,* 155, 1980.

Reiff, P. H., H. L. Collin, E. G. J. D. Craven, J. L. Burch, J. D. Winningham, E. G. Shelley, L. A. Frank and M. A. Friedman, Determination of auroral electrostatic potential using high and low altitude particle distributions, *J. Geophys. Res., 93,* 7441, 1988.

Roth, I., M. K. Hudson, R. Bergmann, Effects of ion two-stream instability on auroral ion heating, *J. Geophys. Res, 94,* 348, 1989.

Schriver, David, M. Ashour-Abdalla, H. Collin, and N. Lallande, Ion beam heating in the auroral acceleration region, *J. Geophys. Res, 95,* 1015, 1990.

Winglee, R. M., P. B. Dusenbery, H. L. Collin, C. S. Lin and A. M. Persoon, Simulation and observations of heating of auroral ion beams, *J. Geophys. Res., 94,* 8943, 1989.

Rachelle Bergmann, Physics Department, Eastern Illinois University, Charleston, IL 61920.

Perry C. Gray, Bartol Research Institute, Sharpe Laboratory, U. of Delaware, Newark DE, 19716.

Mary K. Hudson, Department of Physics and Astronomy, Dartmouth College, Hanover, NH, 03755.

Ilan Roth, Space Sciences Laboratory, U. C. Berkeley, Berkeley, CA, 94720.

Effects of Solar Cycle on Auroral Particle Acceleration

C. A. Cattell, T. Nguyen[1], and M. Temerin

Space Sciences Laboratory, University of California, Berkeley

W. Lennartsson and W. Peterson

Lockheed Palo Alto Research Laboratory, Palo Alto, California

We present the results of two studies designed to assess the effect of solar cycle on auroral particle acceleration processes. The first study is of data obtained from the S3-3 satellite at geocentric distances from 1.3 to 2.2 R_e during solar minimum and the rising phase of the solar cycle. The S3-3 study included electrostatic shocks, ion beams and conics, and electron beams, but did not include ion composition. The second study is of data obtained by the ISEE-1 satellite when the ion mass spectrometer was operating in a special mode designed to study auroral ion distributions. This study covers geocentric distances from ~2.5 to 7 R_e during solar maximum and includes O^+ and H^+ beams and conics, and electrostatic shocks and low frequency turbulence. The S3-3 statistics for shocks with ion beams imply that the auroral parallel potential drop usually occurs above ~2.2 R_e at solar maximum, while the ISEE-1 observations suggest the potential drop, at solar maximum, may be located at ~2.5 to 4 R_e. The ISEE-1 beam observations (usually at energies less than or equal to 1 keV) imply that the potential drop is less, on average, for solar maximum than for solar minimum. Both data sets provide evidence for increased perpendicular acceleration of ions during solar maximum, associated with electrostatic shocks and enhanced low frequency turbulence. The ISEE-1 data show that the flux of upflowing O^+ continues to increase as $F_{10.7}$ increases from "low" to "high" solar maximum. These data also provide evidence that the ion two-stream instability is important in modifying the ion distributions. Most of these observations may be understood as being due to increased heating of the atmosphere during solar maximum which results in an increase in the H^+ - O^+ charge exchange altitude and an increase in the ionospheric density and temperature. Some effects, including the $F_{10.7}$ dependence of the local time distribution of electrostatic shocks and the apparent decrease in the magnitude of the auroral potential drop at solar maximum, are still not understood.

1. Introduction

The effect of solar cycle on ion composition in the magnetosphere was first recognized by Young et al. [1982], who showed that the percentage of oxygen observed near geosynchronous orbit was higher during solar maximum than during solar minimum. Similar increases were observed in auroral particles by the S3-3 and DE spacecraft [Yau et al., 1985; Collin et al., 1987] and in the plasmasheet by ISEE-1 [Lennartsson, 1989]. Possible effects of solar cycle on auroral particle acceleration processes have also been examined. Ghielmetti and Johnson [1983] first presented evidence that the occurrence frequency of ion beams decreased during the rising phase of the solar cycle. Yau et al. [1985], in a DE-1 study of upflowing H^+ and O^+ ions, showed that at solar maximum: (1) the flux of O^+ was higher than H^+; (2) the occurrence probability of conics increased; and (3) the occurrence probability of O^+ at angles of 160° - 180° (i.e., field aligned) increased at high altitudes. Utilizing ISEE-1 electric field and ion composition data on auroral field lines, Nguyen et al. [1986] showed that ion beams were less common and generally had lower energies during periods of high $F_{10.7}$, and that O^+ ions (both beams and conics) were more common than H^+ ions. Collin et al. [1987, 1988] utilizing S3-3 and DE-1 data, provided evidence that: (1) the average energy of O^+ beams was greater than that of H^+ beams at solar minimum, and comparable during solar maximum; and (2) the flux of O^+ beams was often greater than that of H^+ beams. Both they and Reiff et al. [1988] suggested that the observations were consistent with acceleration of the ions in parallel electric fields followed by modification of the ion distributions via the ion two-stream instability.

In this paper, we review and extend the results of two different studies. The first was a study of electrostatic shocks, ion beams and conics, and electron beams using data from the S3-3 spacecraft

---

[1] Also at Center for X-ray Optics, Lawrence Berkeley Laboratory, Berkeley, California

Auroral Plasma Dynamics
Geophysical Monograph 80

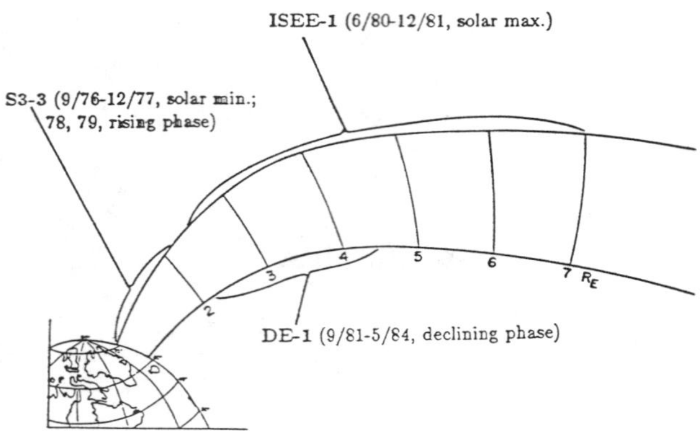

Fig. 1. The radial ranges (geocentric) and time periods of satellite coverage discussed herein. This study uses data from S3-3 and ISEE-1.

[Cattell et al., 1991b]. Composition information was not obtained by the ion detectors utilized for this study. The second study was a study of electrostatic shocks, low frequency turbulence, and O^+ and H^+ beams and conics using data from the University of California at Berkeley electric field detector and the Lockheed Palo Alto Research Laboratory ion composition instrument on ISEE-1 [Nguyen et al., 1986]. The radial ranges of the various data sets utilized herein and in studies referenced above are shown in Figure 1. S3-3 obtained data from 1.3 to 2.2 R_e during solar minimum and the rising phase of the solar cycle. DE-1 covered geocentric distances from ~2.3 to 4.5 R_e. ISEE-1 obtained data on auroral field lines at altitudes from 2.5 to 7 R_e. The coverage in monthly $F_{10.7}$ for different studies is indicated in Figure 2. $F_{10.7}$ was used as an indicator of the solar EUV flux which is correlated with solar cycle. Although daily $F_{10.7}$ was used herein, some studies have suggested that long term averages might correlate better with atmospheric parameters [Hedin, 1984; Gorney, 1990].

The questions to be addressed by the S3-3 study are the following: (1) Is there a difference in the occurrence probability for

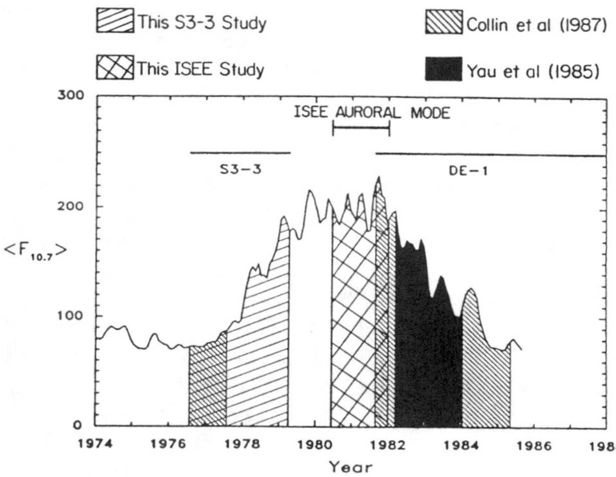

Fig. 2. The coverage in monthly $F_{10.7}$ of various referenced studies.

shocks, shocks with beams, and shocks with conics between periods of high $F_{10.7}$ and periods of low $F_{10.7}$?; (2) Is there evidence for a change in the altitude and/or magnitude of the parallel potential drop which produces auroral ion and electron beams and electrons?; and (3) Is there increased perpendicular acceleration of ions for high $F_{10.7}$? The ISEE-1 study examines the following three questions, viewing the ISEE-1 data set as a solar maximum data set: (1) How do characteristics, i.e., energy, composition and type of upflowing ions at ISEE-1 altitudes compare to previous studies?; (2) How do shocks, beams and conics compare to S3-3 observations at solar minimum and the rising phase of the solar cycle?; and (3) Is there evidence for parallel and/or perpendicular acceleration of ions at these altitudes? In addition, the ISEE-1 data set was divided according to $F_{10.7}$ to determine whether trends in the occurrence of shocks and ions continue to the highest values of $F_{10.7}$ and whether there is evidence for the O^+-H^+ two-stream interaction.

The methodology for the S3-3 study is presented in Section 2.1 and the S3-3 statistical results are shown in 2.2. The ISEE-1 study methodology is described in Section 3.1, including an example of the type of event studied. The statistical results for the ISEE-1 dataset, considered as an example of solar maximum data, are shown and compared to prior studies in Section 3.2. The comparison of high and low $F_{10.7}$ events in the ISEE-1 data is described in Section 3.3. In Section 4, the results of the two studies are combined and conclusions are presented.

2. S3-3 STUDY OF ELECTROSTATIC SHOCKS, ION BEAMS AND CONICS, AND ELECTRON BEAMS

Methodology

The S3-3 study is based on a previous statistical database which was compiled by Redsun et al. [1985] and Bennett et al. [1983]. The database contains 2196 crossings of the auroral zone. Electric fields were obtained by the University of California at Berkeley double probe; electrostatic shocks were defined as the occurrence of a field > 90 mV/m. Ions (0.09 to 3.9 keV) and electrons (0.17 to 9 keV) were measured by the Aerospace Corporation detectors [Mizera and Fennell, 1977]. There was no composition information available from this detector. Daily $F_{10.7}$ was utilized as an indicator of solar activity and solar EUV. Details can be found in the above referenced papers, as well as in Cattell et al. [1991b]. Most of the data was obtained during solar minimum. Good coverage in altitude and magnetic local time was obtained except near noon as can be seen in Figure 3a in which each auroral zone crossing is plotted versus altitude and magnetic local time. The points correspond to a model auroral zone latitude [see Bennett et al., 1983, p. 7108 for details]. Points with $F_{10.7} < 80$ are shown as "x's" and those with $F_{10.7} > 100$ are open squares. Another possible source of bias in the data set is magnetic activity. However, as shown in Figure 3b, both high and low $F_{10.7}$ cases cover almost the complete range of Kp values.

Statistical Results

The occurrence probability of an electrostatic shock for low and high $F_{10.7}$ is shown versus altitude (summed over local time) in

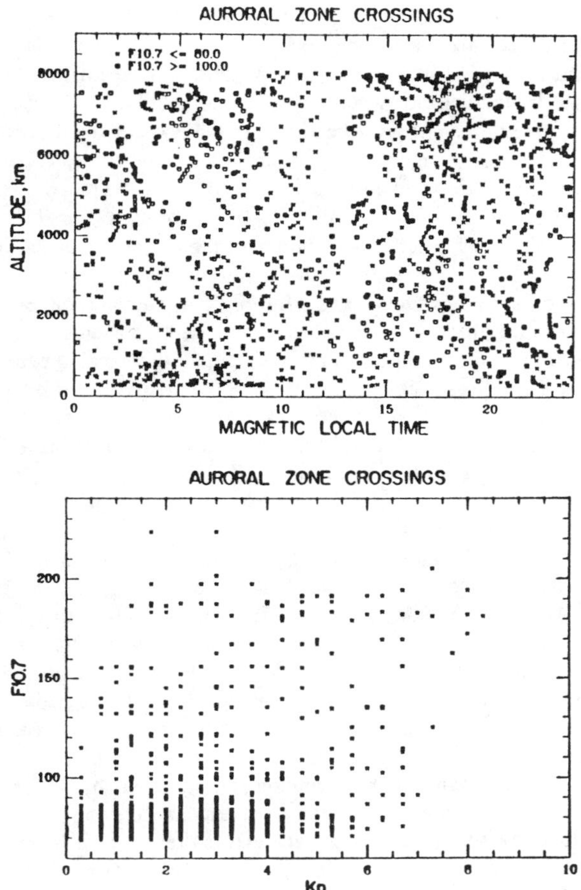

Fig. 3. (a) Scatter plot of S3-3 auroral zone crossings versus altitude and magnetic local time with high (low) $F_{10.7}$ plotted as squares (x's); (b) scatter plot of S3-3 crossings versus $F_{10.7}$ and Kp.

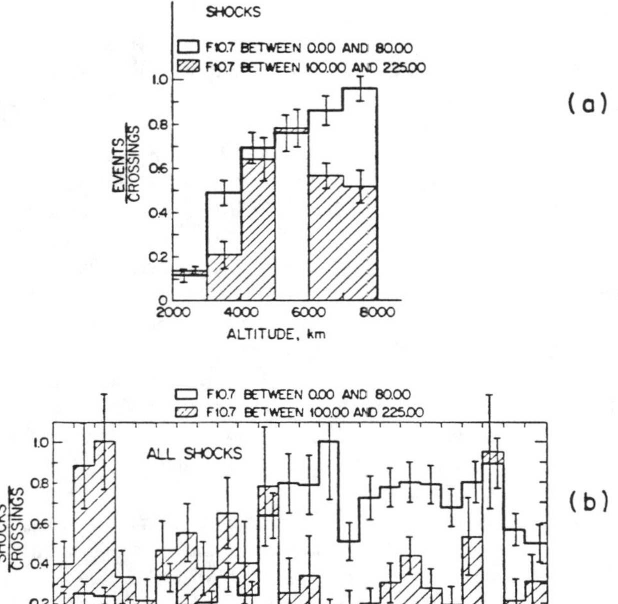

Fig. 4. The occurrence probability of S3-3 electrostatic shocks plotted versus (a) altitude and (b) magnetic local time for high and low $F_{10.7}$.

common in the evening, ~1600–2200 MLT, during low $F_{10.7}$ (Figure 5b). The occurrence frequency of ion beams was much higher during low $F_{10.7}$ except during the hours of 0000–0400 MLT. Since electrostatic shocks with ion beams were less common during peri-

Figure 4a and versus magnetic local time (summed over altitude) in Figure 4b. In each bin, the total number of shocks observed has been normalized by the number of auroral zone crossings in that bin. Multiple shocks could occur in a given auroral zone crossing [see Bennett et al., 1983]. During periods of low $F_{10.7}$, there was a clear increase with altitude of the probability of observing an electrostatic shock (see Figure 4a). This increase was also observed during periods of high $F_{10.7}$, but there was a leveling off at higher altitudes. The magnetic local time dependence was more complicated. The occurrence frequency of electrostatic shocks was ~2.5–3 times higher for low $F_{10.7}$ than for high $F_{10.7}$ in the postnoon sector. During the prenoon sector, however, shocks had a higher occurrence frequency during high $F_{10.7}$. On average, shocks were more commonly observed during low $F_{10.7}$ than during high $F_{10.7}$.

The distribution of shocks with ion beams is shown in Figure 5. (A shock was defined as having a beam if a beam was observed within one spin period of the shock.) At all altitudes, electrostatic shocks with ion beams are at least twice as likely to occur during low $F_{10.7}$ than during high $F_{10.7}$. Shocks with ion beams were most

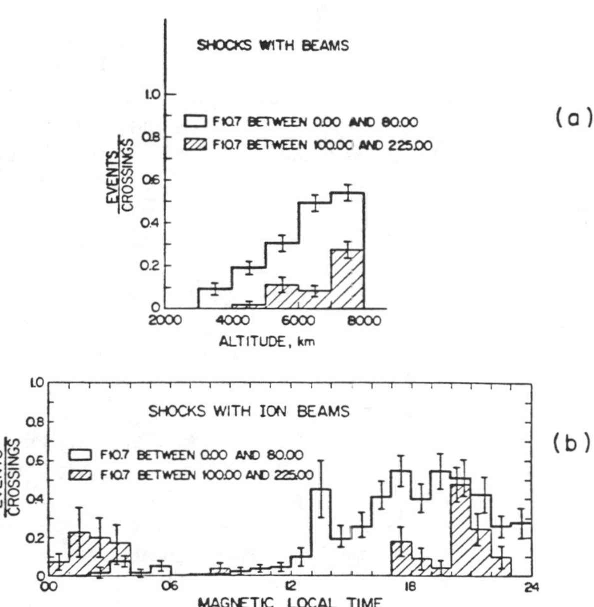

Fig. 5. Same as Figure 4 for electrostatic shocks with ion beams.

ods of high $F_{10.7}$ than during periods of low $F_{10.7}$, it is important to determine whether the remaining shocks were associated with ion conics or with ion distributions which are not indicators of either perpendicular or parallel ion energization. Approximately 18% of the high $F_{10.7}$ shocks were associated with ion beams and ~64% with ion conics. For the periods of low $F_{10.7}$, in contrast, 43% of shocks were observed with ion beams and 43% with conics, while ~14% of the shocks were associated with neither type of distribution.

To examine the question of whether there is a statistically significant difference between the altitude dependence of the magnitude of shocks for solar maximum and solar minimum, we plotted the average magnitude of the shocks in each altitude bin, scaled to the average magnitude at 8000 km. No clear difference was seen. Similarly, the average of the peak energy of the ion beams observed in a given altitude bin increased with altitude for both high and low $F_{10.7}$, but the difference for the two $F_{10.7}$ ranges was not statistically significant.

The main conclusions of the above study are the following: (1) Shocks, in general, are more common during solar minimum than during solar maximum (except from 0 - 5 MLT); (2) Shocks with beams are also more common during solar minimum; and (3) A larger percentage of shocks are associated with conics during solar maximum. It can be inferred that, on average, the auroral parallel potential drop occurs at higher altitudes during solar maximum compared to solar minimum. In addition, there was more perpendicular ion acceleration at altitudes up to 8000 km during solar maximum.

3. ISEE-1 STUDY OF ELECTRIC FIELDS AND IONS

Methodology

The data for this study were obtained from instruments on board the ISEE-1 satellite during 1980 and 1981 as the satellite encountered auroral field lines at distances of 2.5 to 7 R_e. During this period, the Lockheed Ion Mass Spectrometer [Shelley et al., 1978] was operated in a special Auroral Scan Mode [Sharp et al., 1983] which looked at a restricted set of energies and masses so that complete pitch angle spectra could be obtained in approximately 1 minute. In this mode, which was designed for study of auroral phenomena at high altitudes, a complete pitch angle scan for O^+ and H^+ was obtained in 8 energy channels with center energies ranging from 0.21 to 17.4 keV in 52 s (high bit rate) or 64 s (low bit rate). For a sample to be included in this study, the following criteria had to be met: (1) The center angle of the pitch angle bin closest to the magnetic field had to be within 20° of the field direction (~90% of the samples were within 10°, and no qualitative differences in the results were observed between a 10° cutoff and a 20° cutoff); and (2) There was a minimum of 9 counts per sample (corresponding to a flux of 2 to 6×10^4 /cm²-sec-ster-keV for H^+ and 4 to 7×10^4 for O^+ depending only weakly on energy due to pre-acceleration). A beam was defined as the existence of a maximum in the flux along the magnetic field in at least 1 energy channel. A conic was defined as the existence of a maximum in the flux in at least 1 energy channel at an angle between 0° and 90° (or 90° and 180° for the northern hemisphere). Note that a 90° conic at a radial distance

of 2.5 R_e would be observed as a beam within the instrument resolution at a radial distance of ~5-6 R_e, assuming no further energization. It is also possible that some very narrow ion beams would be missed by the detector when the magnetic field is out of the spin plane. Due to the energy dependence of the vertical angular width of the detectors, this would be more likely to occur for energies above ~2 keV. It would also be more likely in low bit rate orbits.

One minute samples (actually 52 s [64 s] for high [low] bit rate orbits) of the raw electric field data from the University of California at Berkeley double probe [Mozer et al., 1978] were utilized. Electric field data were available for ~85% of the ion samples. An electrostatic shock was defined to occur when the average magnitude of the largest 10 points, mapped to 2.0 R_e assuming equipotential field lines, was greater than 100 mV/m. Enhanced low frequency turbulence was defined as the occurrence of one-minute averaged power in 2 out of 3 broadband filters (centered at 4, 32, and 256 Hz) greater than a set value (18, 2, 0.5 $\mu V/mHz^{1/2}$ respectively).

Both ion and electric field data were sorted into 648 three-dimensional bins with nine 0.5-R_e bins by twelve 2-hour bins by six 2-degree bins. The invariant latitude ranged from 64° to 76°. Data were also sorted by $F_{10.7}$ and AE. There were more than 7000 individual samples of ~1 minute each obtained during the ~1 1/2 years of auroral mode operation. Error bars shown in the plots discussed below are statistical and do not reflect systematic errors due to orbital constraints. In particular, there was not much coverage obtained from ~6 to 12 LT for low AE or from 4 to 14 LT for high AE. Since ion beams are not common at those local times [Gorney et al., 1981], this does not strongly influence our conclusions. The distribution of samples in $F_{10.7}$ and AE is presented in Figure 6. For AE $\lesssim 800$, the coverage in $F_{10.7}$ is relatively uniform. The highest values of AE occurred only for the lower values of $F_{10.7}$. Note that all the data were obtained at solar maximum so that the range of $F_{10.7}$ was 140 to 310 in comparison to the S3-3 data which had a range of 60 to 220. For this reason, the ISEE-1 statistical study will be analyzed in two ways: (1) with the data viewed as a

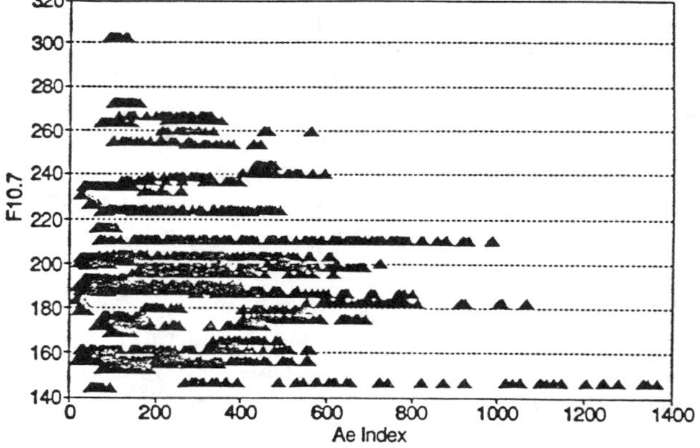

Fig. 6. Scatter plot of ISEE-1 samples, with simultaneous auroral scan mode ion data and electric field data, versus AE and $F_{10.7}$.

whole to study solar maximum effects and (2) with the data divided by $F_{10.7}$ value into two groups (called low maximum and high maximum) to determine whether trends previously observed to occur as $F_{10.7}$ increased continue to the highest values of $F_{10.7}$.

An example of the type of event utilized in this study is shown in Figure 7 in which 70 minutes of data obtained as ISEE-1 moved up along auroral field lines from a radial distance of ~2.5 R_e to ~4.2 R_e near midnight local time. In panel a, the spin-averaged duskward electric field is plotted. Large spiky electric fields, which were identified as electrostatic shocks, occurred around 2120 and 2145 UT. Note that the raw electric field (the component along the spinning boom; not shown) had magnitudes up to ~200 mV/m at those times. The electric field in three filters is shown in panel b. Regions of enhanced power can be seen at ~2110, ~2112-2125, ~2130, ~2141-2155 and ~2200. The negative of the spacecraft potential, panel c, indicates that there was a density depletion in the region between the electrostatic shocks. Selected pitch angle distributions of O^+ and H^+ show that outside the shocks and density depletion region, there were conical ions, whereas inside there were upflowing ion beams which became broader at higher altitudes. It can be seen that the isotropic H^+ background was usually higher than the O^+ background. This may have affected selection of events. These data are consistent with observations made on auroral field lines at lower altitudes [see, for example, Mozer et al., 1980]. Another example of the events used in this study can be seen in Figure 2 of Cattell et al. [1991a].

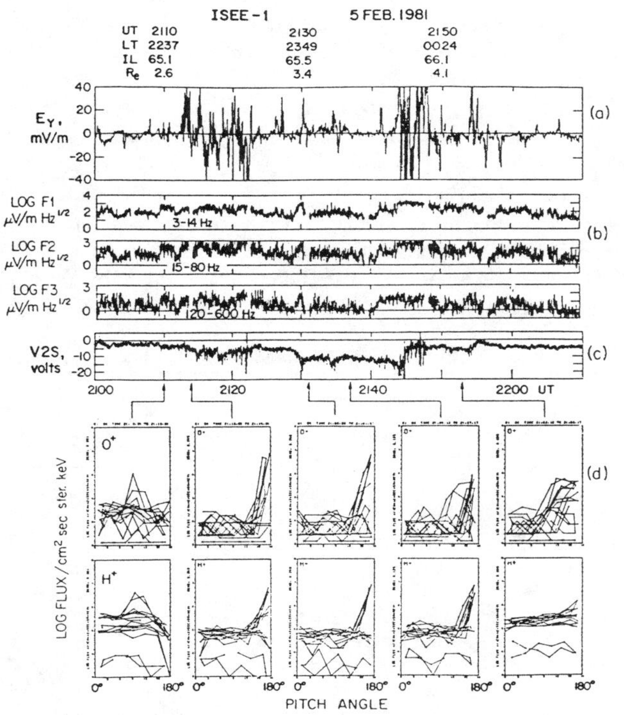

Fig. 7. Electric field and ion data obtained as ISEE-1 moved up along auroral field lines from ~2.5-4.5 R_e. (a) spin-averaged duskward electric field, E_y; (b) electric field in 3 filters; (c) negative of the spacecraft potential, V2S; and (c) selected O^+ and H^+ pitch angle plots.

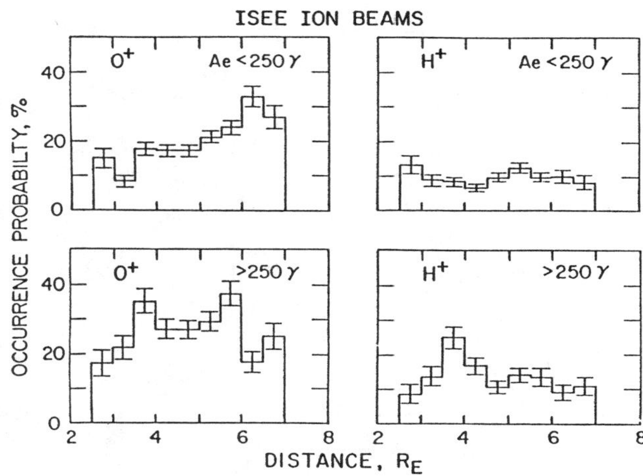

Fig. 8. Occurrence probability of ISEE-1 ion beams for O^+ and H^+ versus altitude for low and high AE.

Statistical Results for the Study of ISEE-1 Data as a Solar Maximum Data Set

In Figure 8, the occurrence probability of an ion beam at a given radial distance is plotted for both O^+ and H^+ and for periods of high and low magnetic activity. At these radial distances, both H^+ and O^+ beams are more common during periods of high magnetic activity than low magnetic activity. The occurrence probability of O^+ beams increases with altitude for AE < 250; the probability for H^+ does not show a similar increase. This change can be interpreted in the following way: the O^+ conics observed at ~3 R_e (see Figure 9) fold up into beams by ~5-6 R_e accounting for the increase in beam occurrence. This would require local acceleration of conics at the higher altitudes to maintain the conic occurrence probability. Very few H^+ conics were observed, consistent with the constant occurrence of H^+ beams with altitude. During periods of high AE, the probability of observing a beam increases with altitudes up to a radial distance of ~4 R_e. In this case, in order to

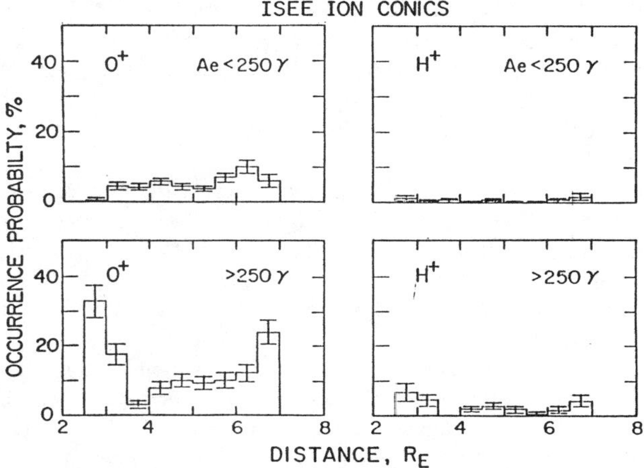

Fig. 9. Same as Figure 8 for ion conics.

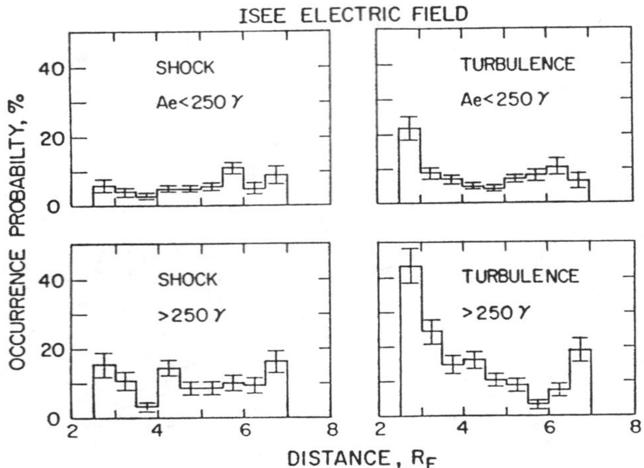

Fig. 10. Same as Figure 8 for electrostatic shocks and samples with enhanced low-frequency wave power.

explain the increased probability of beams at ~3.5 - 4 R_e from the folding up of the conics at ~2.5 - 3 R_e, a parallel potential drop would be required. This suggests that a potential drop exists at these altitudes (~3 - 4 R_e) during active times at solar maximum. The data are consistent with an approximately constant occurrence for O^+ beams between ~4 R_e and 6 R_e. The decrease observed above 6 R_e may be due to: (1) the fact that the LT time coverage above 6 R_e is less uniform; and/or (2) the detector missing very narrow beams, as discussed above. For all AE, O^+ beams are approximately twice as common as H^+ beams. The results for ion conics are shown in Figure 9. O^+ conics are much more common than H^+ conics for all levels of magnetic activity. There is no altitude dependence for quiet times (AE < 250), whereas there is a strong altitude dependence for active times (AE > 250). Note that for high AE the conic probability has an altitude minimum at the same radial distance where the beam probability peaks, consistent with the explanation that the beam peak is due to higher altitude parallel acceleration of the conics. The statistics for electrostatic shocks and enhanced low frequency turbulence are presented in Figure 10. The occurrence probability of shocks is almost constant with altitude for high and low AE. Intense waves are more likely to occur at low altitudes. Both intense waves and electrostatic shocks are approximately twice as likely to occur during high AE than low AE.

The conclusions which can be inferred from the above figures are the following:
(1) For all AE, O^+ dominates upflowing ions (UFI) by a factor of ~3 for conics and ~2 for beams. The distribution in altitude and local time (not shown) is similar for O^+ and H^+. Approximately 85% of the ion beams and 90% of the conics are at energies ≤ 1 keV. These results confirm the suggestions of Ghielmetti and Johnson [1983] and Yau et al. [1985] about UFI at solar maximum.
(2) A significant number of conics is accelerated at all radial distances. There is evidence for a source of O^+ conics at 2.5 - 3.5 R_e for active times. Conics are strongly correlated with shocks and low frequency turbulence which suggests that the conics are produced by these electric fields.

(3) The radial distribution of beams suggests the possibility of parallel electric field acceleration at radial distances of 2.5 - 4.0 R_e during active times. Comparison with the electrostatic shock data implies that these parallel electric fields are associated with electrostatic shocks, as was found at S3-3 altitudes.
(4) Just as the S3-3 studies have shown, most electrostatic shocks are associated with either ion conics or ion beams. For the rising phase of the solar cycle ($F_{10.7}$ >100), the S3-3 data showed that ~20% (~65%) of the shocks were observed with ion beams (conics). The ISEE-1 data at solar maximum had ~40% (~30%) associated with beams (conics). Note that a percentage of the ion distributions which are identified as beams in upper range of altitudes in the ISEE-1 data set may actually be conics whose opening angle is less than the resolution of the measurement. As discussed above, a 90° conic at 2.5 R_e would fold up and be identified as a beam at ~5 R_e; smaller angle conics would fold up earlier.

$F_{10.7}$ Effects in the ISEE-1 Data Set

Because the range of $F_{10.7}$ values observed in the ISEE-1 auroral scan mode data set was as large as that observed in the S3-3 data set (even though the lowest $F_{10.7}$ was 140 compared to 60 for S3-3), we decided to examine the question of whether previously observed trends, such as the increase in the flux of O^+ and in the occurrence probability of conics and the decrease in the occurrence of beams, continued to the highest values of $F_{10.7}$. In addition, we wished to examine the evidence for the operation of the O^+-H^+ streaming instability at high altitudes. The local time sampling in the ISEE-1 study is not completely uniform; the low $F_{10.7}$ data set has less data at dusk and the high $F_{10.7}$ set has less data postmidnight. Table 1 summarizes the ISEE-1 results for low and high values of $F_{10.7}$. The probability of observing an ion beam decreased with $F_{10.7}$ for both O^+ and H^+. The occurrence frequency of O^+ conics increased with $F_{10.7}$, whereas that of H^+ remains approximately the same. There was no altitude dependence for the O^+ conic events. The ratio of O^+/H^+ conics also increased with $F_{10.7}$, although the ratio for beams remained approximately constant.

Because the methodology used herein and that used by Yau et al. [1985] differed, we have averaged the probability for the 2.2 - 3.2

TABLE 1. ISEE Auroral Scan Mode Data

	(less data at dusk) $F_{10.7} < 180$		(less data post-midnight) $F_{10.7} > 180$	
	$R < 4.5\,R_E$	$R > 4.5\,R_E$	$R < 4.5\,R_E$	$R > 4.5\,R_E$
		Ion Data		
H^+ Beams	0.17	0.11	0.1	0.11
O^+ Beams	0.27	0.27	0.17	0.24
O^+/H^+ Beams	1.6	2.5	1.7	2.2
H^+ Conics	0.03	0.01	0.02	0.01
O^+ Conics	0.06	0.06	0.09	0.09
O^+/H^+ Conics	2	6	4.5	9
		Electric Field Data Restricted to 20-04MLT		
Shocks	0.12	0.02	0.2	0.14
Turbulence	0.11	0.02	0.3	0.16

TABLE 2. Yau et al. (1985)

	$100 < F_{10.7} < 170$	$F_{10.7} > 170$
H^+ Beams	.15	.14
O^+ Beams	.14	.17
H^+ Conics	.06	.11
O^+ Conics	.06	.13

R_e bin and the 4.5 - 4.6 R_e bin from their tables to produce the Yau et al. entry in Table 2. Nevertheless, the numbers in the two tables can not be directly compared because of differences in the $F_{10.7}$ and altitude ranges and instrumentation; instead, trends observed in the two studies can be compared. It can be seen that the number of H^+ beams decreased slightly with $F_{10.7}$ in the Yau et al. study in agreement with the ISEE study. The number of O^+ beams, however, increased slightly. This may actually reflect an increase in small angle conics since the Yau et al. study did not distinguish between conics at angles of 160° - 180° and beams. The "conics" entry is for angles from 100° - 160°. It indicates an increase in the occurrence of both H^+ and O^+ conics.

Studies of DE-1 and S3-3 data by Reiff et al. [1988] and Collin et al. [1987, 1988] have provided evidence for the occurrence of the H^+-O^+ streaming instability which mediates exchange of energy between O^+ and H^+ beams. By comparing the energy of simultaneously observed O^+ and H^+ beams, they showed that the instability was suppressed when the ratio of O^+/H^+ flux increased just as had been theoretically predicted [Bergmann et al., 1988]. In Figure 11, both the ratio of the O^+/H^+ average energy and the ratio of the O^+/H^+ flux for the ion beam events are plotted versus altitude for the two ranges of $F_{10.7}$ utilized in this section. The ratio of the flux of O^+ to the flux of H^+ is larger for the higher $F_{10.7}$ range, whereas the ratio of the average energy of O^+ to the average energy of H^+ is smaller. The ratio of the average energy for DE-1 data at solar minimum (maximum) was ~1.4 (1.1). These numbers are in good agreement with the ISEE-1 results (averaged over the altitude range below DE-1 apogee) which were ~1.3 (1.15) for low (high) maximum.

In summary, the trends observed in the occurrence probability of ions from solar minimum to solar maximum (i.e., decrease in beams, increase in conics and increase in O^+) continue from low to high solar maximum. In addition, ratio of the flux of O^+ to that of H^+ was larger for high maximum, whereas the ratio of the average energy of O^+ was smaller. This is consistent with the operation of the ion two-stream instability and the results of Collin et al. [1987, 1988].

4. DISCUSSION AND CONCLUSIONS

The results of the S3-3 study can be combined with those of the ISEE-1 auroral scan mode study to infer some effects of solar cycle on auroral acceleration processes. The S3-3 observations of electrostatic shocks with ion beams suggest that, at solar maxi-

mum, most of the auroral parallel potential drop occurs above ~2.2 R_e (S3-3 apogee). The ISEE-1 observations of the occurrence probability of ion beams suggest that the parallel potential occurs between ~2.5 and ~4 R_e during solar maximum. In addition, the ISEE-1 study suggests that the magnitude of the parallel potential may be smaller at solar maximum since beams were rarely observed with energies above ~1 keV. The S3-3 data indicate that there is an increase in perpendicular acceleration relative to parallel acceleration from altitudes of ~1.3 to 2.2 R_e from solar maximum to solar minimum. The ISEE-1 data suggest that this trend extends at least to 3.5 or 4 R_e. Both data sets provide evidence that the perpendicular acceleration process is associated with electrostatic shocks and enhanced low-frequency turbulence. The ISEE-1 data indicate that trends (such as the decrease in beams, increase in conics and increase in O^+, and evidence for the effects of the ion two-stream instability) observed when solar minimum data were compared to solar maximum data are also seen from low solar maximum (as indicated by $F_{10.7}$) to high solar maximum.

Several of these varied effects of solar cycle on auroral acceleration processes may be explained by the effects of solar EUV on the atmosphere and ionosphere. Many researchers have postulated that the heating of the atmosphere due to the increased EUV flux at solar maximum raises the O^+-H^+ charge exchange altitude and the O^+ scale height so that more oxygen can escape [Klumpar, 1979; Young et al., 1982; Moore, 1980; Yau et al., 1985]. Ionospheric temperature and density also increase during solar maximum [Kelley, 1989]. Since many mechanisms which have been proposed to heat ions and to provide the parallel potential drop are current driven, the altitude at which they would become unstable would increase (for a given current density) when the ionospheric temperature and density increased. The increased oxygen density also has an effect on some instabilities including acoustic double layers [Gray et al., 1992], ion cyclotron waves [Cattell et al., 1991a], and ion two-stream waves [Bergmann et al., 1988]. If electrostatic shocks are the nonlinear evolution of a current driven instability [Witt and Lotko, 1983], then the decrease in shock occurrence probability at low altitudes may be understood. If the bulk of the auroral parallel potential drop occurs in shocks, then the $F_{10.7}$ dependence of beams observed by DE-1 and ISEE-1 and

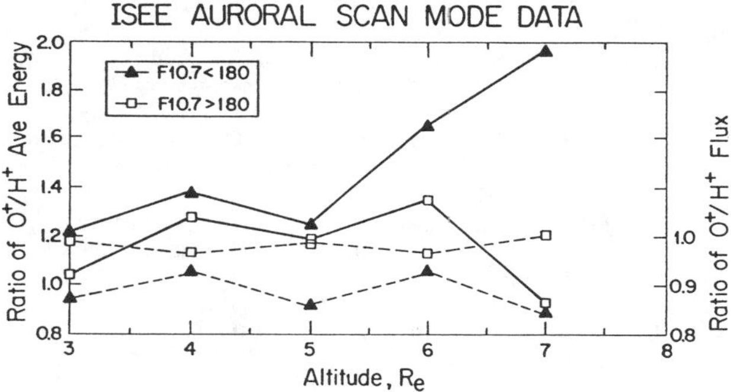

Fig. 11. The ratio of the average energy of O^+ to that of H^+ (left-hand scale, solid lines) and the ratio of the average flux of O^+ to that of H^+ (right-hand scale, dotted line) versus altitude for low $F_{10.7}$ (triangles) and high $F_{10.7}$ (open boxes).

of shocks with beams observed at S3-3 altitudes may also be explained.

There are several remaining puzzles in the data sets. It is not yet understood why electrostatic shocks are more common post-midnight for high $F_{10.7}$ than for low $F_{10.7}$. It may be due to the local time dependence of both the heating of the atmosphere by solar EUV and the ionization by solar EUV and precipitating particles [Cattell et al., 1991b]. The dependence of conic acceleration on altitude and $F_{10.7}$ is also not fully understood. This study suggests that the acceleration is due to both shocks and waves. If these mechanisms are current driven, then we would not expect enhanced perpendicular acceleration at low altitudes during solar maximum. Some of the enhanced perpendicular acceleration of oxygen at solar maximum may be due to the fact that, when the O^+/H^+ density ratio increases, oxygen cyclotron waves are more linearly unstable than hydrogen cyclotron waves [Kindel and Kennel, 1971; Cattell et al., 1991a]. Finally, the fact that most of the observed ion beams during solar maximum had energies of less than ~1 keV implies a reduction in the total auroral potential drop during solar maximum. Although this conclusion would seem to contradict observations of more intense auroral displays at solar maximum, the fact that the auroral luminosity depends on other factors as well may mitigate this problem.

Acknowledgments: This work was supported by NASA grants NAG5-375 and NAG5-1098 and NSF grant ATM-8918774 at UCB, and NAS5-33047 at Lockheed Palo Alto Research Laboratory. We thank J. Bonnell for assistance in programming and production of figures, D. Gorney and A. Ghielmetti for useful discussions, and the referees for helpful comments.

REFERENCES

Bennet, E. L., M. Temerin, and F. S. Mozer, The distribution of auroral electrostatic shocks below 8000-km altitude, *J. Geophys. Res.*, 88, 7107, 1983.

Bergmann, R., I. Roth and M. K. Hudson, Linear stability of the H^+-O^+ two stream interaction in a magnetized plasma, *J. Geophys. Res.*, 93, 4005, 1988.

Cattell, C., F. S. Mozer, I. Roth, R. R. Anderson, R. C. Elphic, W. Lennartsson and E. Ungstrup, ISEE-1 Observations of EIC waves in association with ion beams on auroral field lines from 2.5 - 4.5 R_e, *J. Geophys. Res*, 96, 11421, 1991a.

Cattell, C., S. Chari, and M. Temerin, An S3-3 Satellite Study of the Effects of Solar Cycle on the Auroral Acceleration Process, *J. Geophys. Res.*, 96, 17903, 1991b.

Collin, H. L., W. K. Peterson and E. G. Shelley, Solar cycle variation of some mass dependent characteristic of upflowing beams of terrestrial ions, *J. Geophys. Res.*, 92, 4757, 1987.

Collin, H., W. Peterson, and E. Shelley, The helium components of Energetic Terrestrial ion upflows: Their occurrence, morphology and intensity, *J. Geophys. Res.*, 93, 7558, 1988.

Ghielmetti, A. G., and R. G. Johnson, Variations in the occurrence frequency of UFI events from July 1976 to April 1979 (abstract), *EOS Trans AGU*, 64, 807, 1983.

Gorney, D. J, A. Clarke, D. Croley, J. Fennell, J. Luhmann, and P. Mizera, The Distribution of Ion Beams and Conics Below 8000 km, *J. Geophys. Res.*, 86, 83, 1981.

Gorney, D. J., Solar cycle effects of the near-earth space environment, *Rev. Geophys.*, 28, 315, 1990.

Gray, P., M. Hudson, R. Bergmann, and I. Roth, Simulation study of the ion two-stream instability in the auroral acceleration region, *Geophys. Res. Lett.*, 17, 1609, 1990.

Gray, P., M. K. Hudson, and W. Lotko, Acoustic double layers in multi-species plasma, *IEEE Trans. Plasma Sci.*, in press, 1992.

Hedin, A. E., Correlations between thermospheric density and temperature, solar EUV flux and $F_{10.7}$ cm flux variations, *J. Geophys. Res.*, 89, 9828, 1984.

Kelley, M., in *The Earth's Ionosphere—Plasma Physics and Electrodynamics*, Academic Press, Inc., San Diego, 1989.

Kindel, J. M., and C. F. Kennel, Topside current instabilities, *J. Geophys. Res.*, 76, 3055, 1971.

Klumpar, D., Transversely accelerated ions: An ionospheric source of hot magnetospheric ions, *J. Geophys. Res.*, 84, 4229, 1979.

Lennartsson, W., Energetic (0.1- to 16 keV/E) Magnetospheric ion composition at different levels of solar $F_{10.7}$, *J. Geophys. Res.*, 94, 3600, 1989.

Lysak, R. L. and M. K. Hudson, Coherent anomalous resistivity in the region of electrostatic shocks, *Geophys. Res. Lett.*, 6, 661, 1979.

Mizera, P. F., and J. F. Fennell, Signatures of electric fields from high and low altitude particle distributions, *Geophys. Res. Lett.*, 4, 311, 1977.

Moore, T. E., Modulations of terrestrial escape flux composition, *J. Geophys. Res.*, 85, 2011, 1980.

Mozer, F. S., R. Torbert, U. V. Fahleson, C.-G. Falthammar, A. Gonfalone, and A. Pedersen, Measurements of quasistatic and low-frequency electric fields with spherical double probes on ISEE-1 spacecraft, *IEEE Trans. Geosci. Electron.*, GE-16, 258, 1978.

Mozer, F. S., C. A. Cattell, R. L. Lysak, M. K. Hudson, M. Temerin, and R. B. Torbert, *Spac. Sci. Rev.*, 27, 155, 1980.

Nguyen, T., C. A. Cattell, F. S. Mozer, E. Ungstrup, and W. Lennartsson, A statistical study of upflowing ions and electric fields on auroral field lines from ~3 to 7 R_e, *EOS Trans AGU*, 67, 1159, 1986.

Redsun, M. S., M. Temerin, and F. S. Mozer, Classification of auroral electrostatic shocks by their ion and electron associations, *J. Geophys. Res.*, 90, 9615, 1985.

Reiff, P. H., H. L. Collin, J. D. Craven, J. L. Burch, J. D. Winningham, E. G. Shelley, L. A. Frank, and M. A. Friedman, Determination of auroral electrostatic potentials using high and low altitude particle distributions, *J. Geophys. Res.*, 93, 7441, 1988.

Sharp, R., W. Lennartsson, W. Peterson and E. Ungstrup, The Mass Dependence of Wave Particle Interactions as Observed with the ISEE-1 Energetic Ion Mass Spectrometer, *Geophys. Res. Lett.*, 10, 651, 1983.

Shelley, E. G., R. D. Sharp, R. G. Johnson, J. Geiss, P. Eberhardt, H. Balsiger, G. Haerendel and H. Rosenbauer, Plasma Composition Experiment on ISEE-A, *IEEE Geosci. Electron.*, GE-16, 266, 1978.

Witt, E. and W. Lotko, Ion-acoustic solitary waves in a magnetized plasma with arbitrary electron equation of state, *Phys. Fluids*, 26, 2176, 1983.

Yau, A.W., P. H. Beckwith, W. K. Peterson, and E. G. Shelley, Long term (solar cycle) and seasonal variations of upflowing ionospheric ion events at DE-1 altitudes, *J. Geophys. Res.*, 90, 6395, 1985.

Young, D. T., H. Balsinger, and J. Geiss, Correlations of magnetospheric ion composition with geomagnetic and solar activity, *J. Geophys. Res.*, 87, 9077, 1982.

C. Cattell, T. Nguyen, and M. Temerin, Space Sciences Laboratory, University of California, Berkeley, California 94720

W. Lennartsson and W. Peterson, Lockheed Palo Alto Research Laboratory, Palo Alto, California

Sounding Rocket Observations of Ion Injections in the Morning Convection Reversal Region

J. H. CLEMMONS[1] AND C. W. CARLSON

Space Sciences Laboratory, University of California, Berkeley, California

Energetic ions which were impulsively injected onto high-altitude magnetic field lines have been measured by instruments borne by an ionospheric sounding rocket launched into the morning auroral region from Sondre Stromfjord, Greenland at 0926 UT on January 23, 1985. The measured ion fluxes exhibit an energy-time dispersion due to a time-of-flight effect in which more energetic particles arrive from a distant source before those of lesser energies. Analysis of these signatures shows the source to be ~20-50 R_E distant, consistent with a source located in the flank region. Measurements of energetic electrons associated with the injections show the presence of a mixture of two electron populations, one having a temperature of ~100 eV and a density of ~3-8 cm^{-3} and the other having a temperature of ~1 keV and a density of 10^{-2}-10^{-1} cm^{-3}. These values are consistent with the mixture of magnetosheath and plasma sheet electron populations commonly found in the boundary layer. Electric field measurements indicate that the ion injections were observed in a region of sunward ionospheric convection and a pass of the HILAT satellite shows that the convection reversal boundary was near the rocket trajectory. A magnetic field model used to trace field lines which intersected the rocket reveals that the field lines mapped tailward into the morning flank region. These results are consistent with a source in the flank boundary layer, contrasting with reports of similar measurements from earlier sounding rockets and recent results from the VIKING satellite, both of which inferred sources near the dayside magnetopause.

INTRODUCTION

The regions in and near the geomagnetic cusp have been studied to develop a deeper understanding of the interaction between the solar wind with the magnetic field of the earth. Due to its magnetic geometry, the cusp provides relatively little hindrance to the entry of magnetosheath plasma into the magnetosphere, permitting low-altitude observations of phenomena related to that interaction. Observations of charged particles precipitating in or near the geomagnetic cusp have been reported by many researchers. Cusp ion signatures have been studied in detail, and the observations may be grouped into two categories with respect to the inferred temporal structure. Quasi-steady fluxes were first observed by Heikkila and Winningham [1971] and by Frank [1971]. Shelley et al. [1976] and Reiff et al. [1977] explained the observed dispersion in ion velocity with latitude in terms of an $E \times B$ velocity filter model. A steady source was assumed in their model. On the other hand, observations have been made in which the ion signatures were interpreted as being due to an inherently time-dependent source. These include observations by Carlson and Torbert [1980], Lundin et al. [1991], Mukai et al. [1991], Woch and Lundin [1991], and Woch and Lundin [1992]. The results of a sounding rocket experiment which was similar to that of Carlson and Torbert [1980] are presented here. These are further measurements of the transient phenomena.

MEASUREMENTS

Rocket Flight

A Black Brant X sounding rocket was launched in an eastward direction into the morning auroral zone from Sondre Stromfjord, Greenland at 0926 UT on January 23, 1985. The payload consisted of instruments designed to measure DC and wave electric and magnetic fields and fluxes of energetic particles. (Further descriptions of the instruments may be found in Boehm [1987] and in Clemmons [1992] and the references therein.) Included were pitch-angle imaging electrostatic analyzers to measure fluxes of ions and electrons. Apogee was reached 520 s after launch at an altitude of 770 km. Figure 1 indicates that the rocket traveled along a path which covered a range in MLT of 0730-1200. The range from 8 MLT to 10 MLT, corresponding to flight times of 200-600 s, was covered at a nearly constant latitude before the payload turned southward. The flight occurred during a disturbed time, as indicated by the reported K_P value of 5 [Coffey, 1985].

---

[1]Now at Max-Planck-Institut für extraterrestrische Physik, Garching, Germany.

Auroral Plasma Dynamics
Geophysical Monograph 80

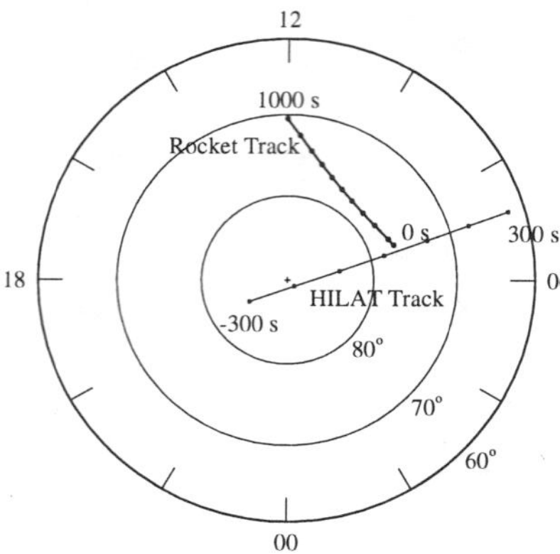

Fig. 1. Sounding rocket flight path. Also shown is a nearby pass of the HILAT satellite. Tick marks indicate intervals of 100 s, and times are relative to the time of the rocket launch at 0926 UT on January 23, 1985.

Measurements from Other Experiments

Figure 2 shows measurements of the magnetosheath conditions returned by the IMP-8 spacecraft for a four-hour period including the rocket flight. Its position in GSE coordinates at the time of the rocket launch was (-10.18, -24.66, -13.01) R_E. The magnetic field was strongly southward prior to the northward turning of B_z at 0735 UT. It then remained rather steady for the next three hours, making only small excursions to the south such as the one near 0925. At 0805 the plasma density began an increase from 15 cm^{-3} to 60 cm^{-3} ten minutes later. The plasma flow speed increased from 450 km/s to 600 km/s in the same period. The combined effect of these changes was to increase the plasma dynamic pressure from 6 nPa to 36 nPa, a six-fold increase. By the time the rocket was launched, the pressure had tailed off to

16 nPa due to the density falling to 25 cm^{-3}. Considering both the proximity in time and the magnitude of change, the changes in dynamic pressure seem to be more important than the changes in B_z in terms of effects in the magnetosphere during the rocket flight.

The trajectory of the HILAT satellite is shown in Figure 1 for a pass which was nearly over the rocket launch site at the time of launch. HILAT provided information about the primarily latitude-dependent features of the auroral region during the rocket flight, complementing the rocket measurements. The data from the cross-track drift meter in the middle panel of Figure 3 and the measurements made by the electric field instruments on the rocket (not shown here) show that the rocket was launched into a region of sunward (eastward) convection. The HILAT data also show that there was a mixed flow pattern very near the rocket trajectory, indicating that the payload was near the convection-reversal boundary. The cross-track component of the HILAT magnetic field deviation shown in the top panel of Figure 3 indicates the presence of field-aligned currents in this region. (The approximately sinusoidal signal with a period of about 40 s appears superposed on the signal as an artifact of an imperfect spacecraft attitude solution.)

Rocket Ion Measurements

A stackplot of the rocket measurements of downgoing ion fluxes is shown in Figure 4. Particle flux has been plotted against velocity (assuming the proton mass and charge) for 16-s averages. Each subsequent trace has been offset by a factor of 10^{14} for clarity. The small vertical bars represent uncertainties in the data due to counting statistics. The traces exhibit several well-defined peaks in the spectra, and many of the peaks were present for several traces, progressing to lower speeds with increasing time. This is similar to the behavior reported by Carlson and Torbert [1980], and their method of analysis was adapted to characterize the movement of the peaks. A computer fitting procedure was

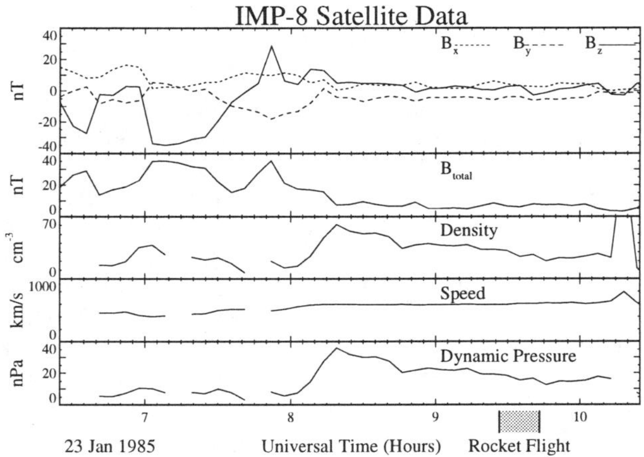

Fig. 2. Measurements of magnetosheath conditions returned by the IMP-8 spacecraft. Shown are the magnetic field, plasma density, flow speed, and dynamic pressure. The interval during which the rocket flew is indicated near the bottom of the plot.

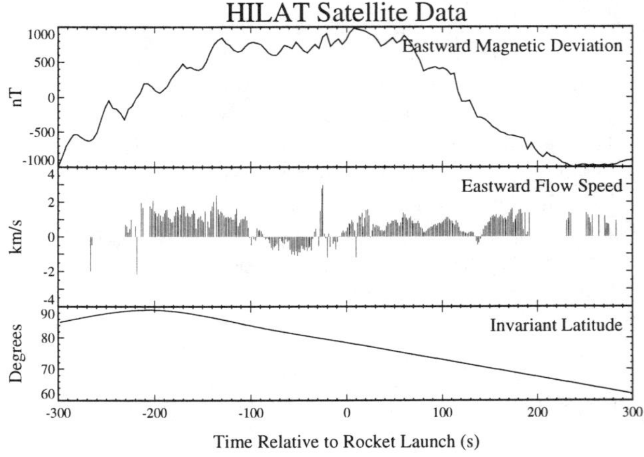

Fig. 3. Data returned by the magnetometer and drift meter aboard the HILAT spacecraft. Field-aligned currents appear as a deviation of the magnetic field from a model field. Flow reversal is apparent near -100 s and 0 s. Also shown is the invariant latitude of the spacecraft.

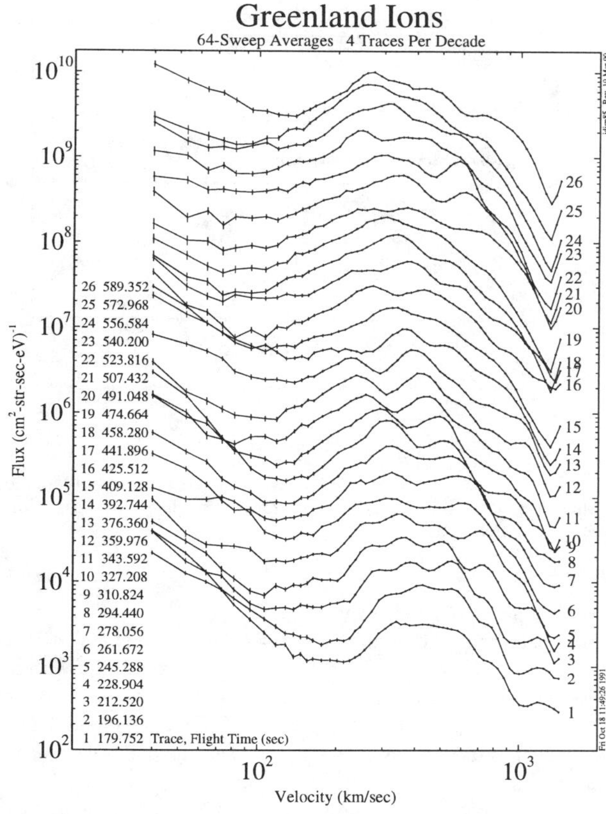

Fig. 4. Stackplot of precipitating ion fluxes measured by the rocket. Each trace represents an average over 16 s for the flight times listed, with earlier times nearer the bottom of the plot.

developed to identify peaks by fitting individual 2.048-s spectra to a function consisting of a sum of Gaussians:

$$F(v) = \sum_{i=1}^{N} A_i \exp\frac{(v-V_i)^2}{W_i^2} \qquad (1)$$

Here $F(v)$ is the differential ion flux at the speed v and N the number of Gaussians included in the fit, while A_i, V_i, and W_i are respectively the amplitudes, average speeds and widths of the component functions. The Gaussian function was used because of its computational simplicity and its flexibility, not because it follows from the assumption of a particular physical model which predicts the functional form for the spectral peaks. Figure 5 depicts an example of the results of the fitting procedure for one of the spectra. The spectra were fitted for the interval from 180 s to 650 s flight time, after which the intensity of the ion fluxes diminished. Figure 6 shows how the reciprocals of the center velocities of the peaks changed as a function of flight time. If the peaks were due to a distant, transient source which "turned on" for a brief period, the peaks would be expected to be observed at lower speeds at later times as the slower source particles required longer travel times than the faster ones. This "time-of-flight" dispersion may be represented by:

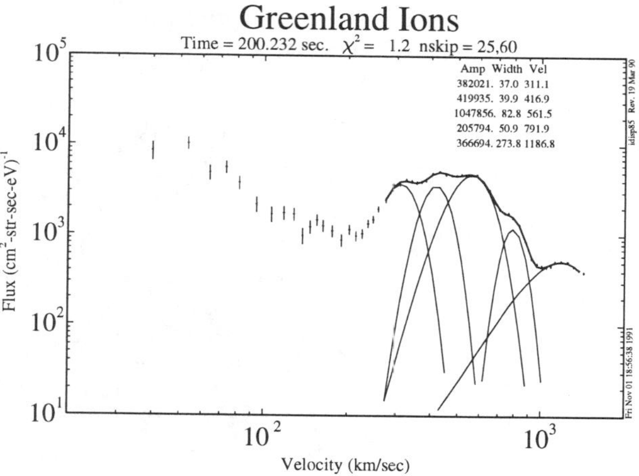

Fig. 5. Results of the peak-identification procedure for one of the ion spectra averaged over 2.048 s. The heavy curve indicates the fitted function, while the light curves represent the individual component Gaussians.

$$\frac{1}{V} = \frac{t - \tau}{L} \qquad (2)$$

Here L is the path length from the source to the detector and τ is the time of the source injection. V represents the component of the velocity parallel to the magnetic field. Since there was little dependence of the downgoing fluxes on pitch angle, it was assumed that the particles were highly field-aligned for most of their flights, and V was taken to be the total velocity of the particles. Equation (2) indicates that the reciprocal of the slope of a straight line on Figure 6 represents the source distance, while the horizontal intercept represents the time at which the source injection occurred. In light of this idea, points from Figure 6 were placed into groups and straight lines were fitted to them, with Figure 7 showing the results. The fitted parameters have been

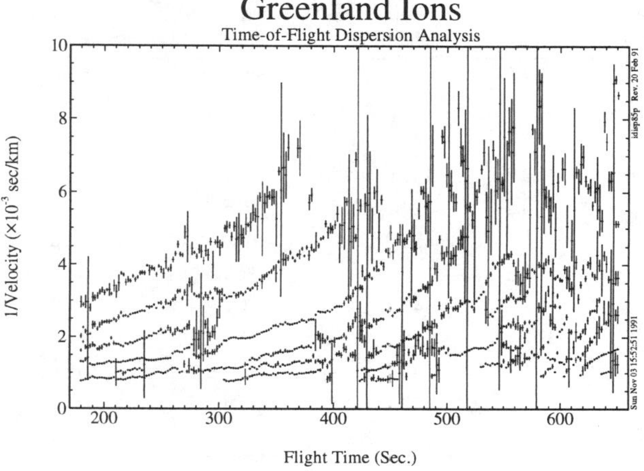

Fig. 6. Scatterplot of parameters from the fits of individual ion spectra. The center velocities of the individual peaks have been plotted versus flight time as $1/V_i$. The vertical bars represent uncertainties.

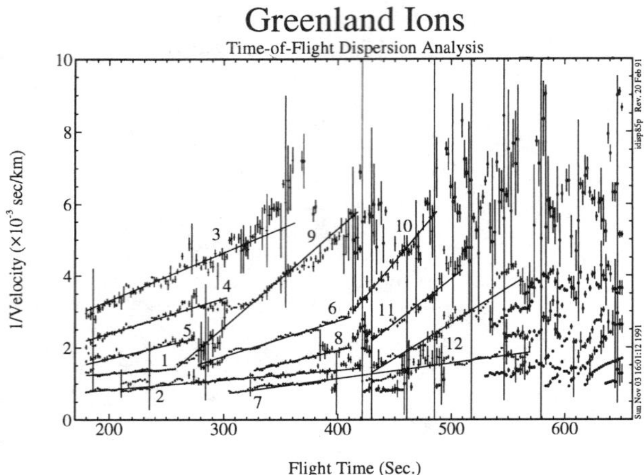

Fig. 7. Same as Figure 6 with dispersion lines described by Equation (2). Fitted dispersion parameters may be found in Table 1.

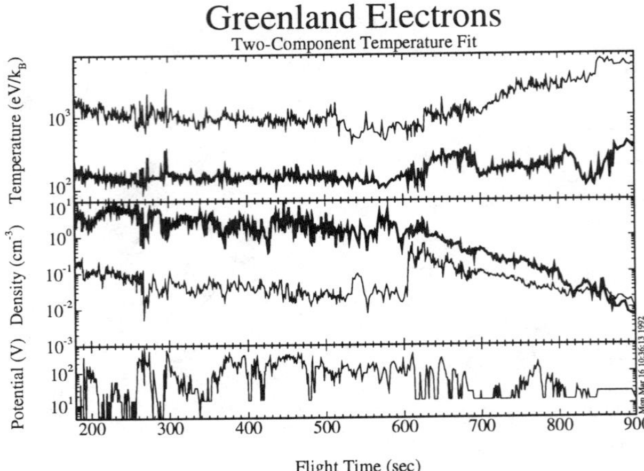

Fig. 8. Fitted electron temperatures, densities and estimated field-aligned potential drop from the rocket electron measurements. The heavy lines represent the cool component.

tabulated in Table 1. The fitted source distances have a wide range of values, from 4 to 70 R_E. This range is much wider than the range of 7-19 R_E reported by Carlson and Torbert [1980] and it is difficult to attribute such disparate distances to a single phenomenon. The applicability of the linear dispersion of (2) is further called into question by examining the loci of points in Figure 6. They seem to be described much better by a few curves than by many short line segments. The linear dispersion treatment does provide an estimate of the distances involved, however. The longer distances (>20 R_E) are deemed to be relatively accurate because they were observed at higher energies for which field-aligned potentials have little effect. However, due to the inadequacy of the linear dispersion model, other avenues were explored to determine the origin of the observed ion fluxes.

Rocket Electron Measurements

The spectra returned by the rocket electron electrostatic analyzer were fit to two thermal components throughout the flight, with the results shown in Figure 8. The temperatures and densities of the two components and the estimated field-aligned potential drop have been plotted versus flight time. During the time interval before 600 s the cool component exhibited temperatures of about 100 eV and densities of 3-8 cm⁻³, while the warm component had temperatures of about 1 keV and densities of 10^{-2}-10^{-1} cm⁻³. The values for the cool component are consistent with those found in the magnetosheath and/or low-latitude boundary layer, while the values for the warm component are consistent with a source in the plasma sheet. The mixture of these components suggests a source in or near the magnetopause boundary layer. The sharp drop in the density of the magnetosheath and/or LLBL component after

TABLE 1. Parameters for the Fitted Lines of Figure 7.

Line No.	Observation Time (s)	Injection Time (s)	Apparent Source Distance (R_E)	Correlation Coefficient
1	77.8	-368.0 ± 19.2	69.8 ± 2.2	0.90
2	239.6	-112.5 ± 2.7	58.8 ± 0.4	0.95
3	182.3	-46.8 ± 4.5	11.7 ± 0.2	0.91
4	122.9	-45.0 ± 5.0	16.1 ± 0.3	0.96
5	94.2	-24.3 ± 6.5	20.8 ± 0.5	0.90
6	131.1	121.2 ± 1.6	16.0 ± 0.1	0.98
7	264.2	130.0 ± 0.8	37.1 ± 0.1	0.99
8	84.0	142.3 ± 4.4	21.0 ± 0.4	0.98
9	157.7	206.2 ± 0.3	5.7 ± 0.1	0.99
10	73.7	333.8 ± 2.9	4.2 ± 0.1	0.95
11	79.9	340.5 ± 2.6	6.4 ± 0.1	0.94
12	133.1	362.8 ± 0.9	8.0 ± 0.1	0.98

Times are relative to the time of the rocket launch. The correlation coefficient represents the degree of correlation between the data and the fitted lines.

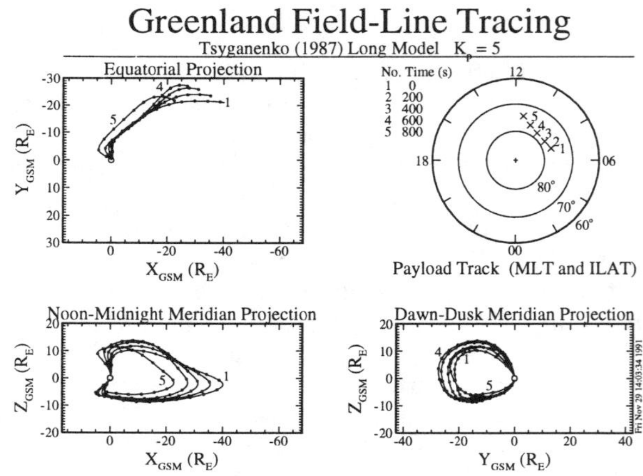

Fig. 9. Magnetic field line tracings using the model of Tsyganenko [1987]. The traces begin at the indicated points on the flight path. Tick marks have been placed at 5 R_E intervals along the traces.

600 s, combined with the density increase observed in the plasma sheet component, is coincident with the southward turn of the payload and with the decrease in the ion fluxes, indicating entry into a different plasma domain.

Tracing of Magnetic Field Lines

Another method used to determine the region from which the ions were injected was the use of a magnetic field model to trace the field lines upon which the ions were observed. Figure 9 shows three projections of field lines traced using the Tsyganenko [1987] model. Field lines were traced beginning at five points along the flight path as indicated. It is apparent that the field lines map to the nightside flank region, but it should also be pointed out that model limitations at these distances do not allow a more precise interpretation of the mapping. Although the magnetic field model does not contain a magnetopause, the mapping results are consistent with the results inferred from the electron data, i.e. that the source was in or near the magnetopause boundary layer.

CONCLUSIONS AND FURTHER WORK

The main conclusion of this work is that the observed ion signatures are consistent with a source located in or near the nightside magnetopause boundary layer about 20-50 R_E from the rocket. The idea is supported by the electron measurements and the use of the model magnetic field. This conclusion is in contrast to the previous work of Carlson and Torbert [1980] who found a source in the subsolar region, and the work of Lundin et al. [1991] who inferred sources in or near the dayside magnetopause boundary layer. A possible explanation for the difference is that the level of disturbance was much higher during the experiment described here, as indicated by a larger value of K_p. This implies a much more distended magnetosphere and a more tailward mapping of field lines in this region. The IMP-8 measurements indicate that the change in solar wind dynamic pressure may be implicated in triggering the ion injections. Work remains to be done to narrow the possible scope of the source region. More work will also be done to identify the mechanism by which the injections were triggered.

Acknowledgments. The satellite data have been presented here through the courtesy of J. Vickrey (HILAT) and P. Song (IMP-8). The authors gratefully acknowledge these courtesies. This work was supported by NASA through grants NAG6-10 and NGT-50313.

REFERENCES

Boehm, M. H., Waves and static electric fields in the auroral acceleration region, Ph.D. dissertation, 266 pp., University of California, Berkeley, April, 1987.

Carlson, C. W. and R. B. Torbert, Solar wind ion injections in the morning auroral oval, *J. Geophys. Res., 85,* 2903-2908, 1980.

Clemmons, J. H., Sounding rocket observations of precipitating ions in the morning auroral region, Ph.D. dissertation, 135 pp., University of California, Berkeley, April, 1992.

Coffey, H. E. (ed.), Geomagnetic and solar data, *J. Geophys. Res., 90,* 5365-5368, 1985.

Frank, L. A., Plasma in the earth's polar magnetosphere, *J. Geophys. Res., 76,* 5202-5219, 1971.

Heikkila, W. J. and J. D. Winningham, Penetration of magnetosheath plasma to low altitudes through the dayside magnetospheric cusps, *J. Geophys. Res., 76,* 883-891, 1971.

Lundin, R., I. Sandahl, J. Woch and R. Elphinstone, The contribution of the boundary layer EMF to magnetospheric substorms, in *Magnetospheric Substorms,* ed. by J. R. Kan, T. A. Potemra, S. Kokubun, and T. Iijima, 355-373, American Geophysical Union, Washington, DC, 1991.

Mukai, T., A. Matsuoka, H. Hayakawa, S. Machida, K. Tsuruda, A. Nishida and N. Kaya, Signatures of solar wind injection and transport in the dayside cusp: EXOS-D observations, *Geophys. Res. Lett., 18,* 333-336, 1991.

Reiff, P. H., T. W. Hill and J. L. Burch, Solar wind plasma injection at the dayside magnetospheric cusp, *J. Geophys. Res., 82,* 479-491, 1977.

Shelley, E. G., R. D. Sharp, and R. G. Johnson, He^{++} and H$^+$ flux measurements in the day side cusp: Estimates of convection electric field, *J. Geophys. Res., 81,* 2363-2370, 1976.

Tsyganenko, N. A., Global quantitative models of the geomagnetic field in the cislunar magnetosphere for different disturbance levels, *Planet. Space Sci., 35,* 1347-1358, 1987.

Woch, J. and R. Lundin, Temporal magnetosheath plasma injection observed with Viking: A case study, *Ann. Geophys., 9,* 133-142, 1991.

Woch, J. and R. Lundin, Signatures of transient boundary layer processes observed with Viking, *J. Geophys. Res., 97,* 1431-1447, 1992.

C. W. Carlson and J. H. Clemmons, Space Sciences Laboratory, University of California, Berkeley, CA 94720.

Centrifugal Flow Reversal in the Equatorial Magnetosphere

D.C. DELCOURT

CRPE/CNET-CNRS, Saint-Maur des Fossés, France

J.A. SAUVAUD

CESR/CNRS, Toulouse, France

T.E. MOORE

NASA/Marshall Space Flight Center, Huntsville, Alabama

A change in the magnetic field orientation yields a centrifugal acceleration in the course of a particle motion, which in return contributes a drift to its guiding center. Particles traveling in the equatorial magnetosphere experience the well-known curvature drift due to the local stretching of the magnetic field lines. A less explored curvature drift is that obtained when particles are rapidly transported across the field lines. Such a drift occurs when the particles are acted upon by fast ExB drifts, like during the expansion phase of substorms or in the steady state magnetotail. The role of this latter drift is examined by means of single-particle codes. It is shown that it yields the confinement of populations with relatively low parallel speeds near the tail midplane, until these are convected into the inner magnetosphere. This effect is here referred to as "centrifugal trapping" as, in contrast to magnetic trapping, it does not follow from an intensification of the field magnitude but from its rotation in the course of the ExB transport. This flow reversal mechanism is independent of nonadiabatic interaction with the neutral sheet. It is effective at all pitch angles and solely depends upon parallel speed. It prevents less energetic particles of the distant plasma sheet from precipitating.

INTRODUCTION

Due to the local stretching of the magnetic field lines, the dynamics of charged particles in the equatorial magnetosphere forms a key aspect of the large scale circulation of magnetospheric plasma. Provided that the field line curvature radius largely exceeds the particle Larmor radius, the adiabatic (guiding center) theory can be applied. In this case, the particles are known to experience curvature drifts which drive them across the equipotentials of the dawn-dusk convection electric field and thus lead to significant energy changes. This behavior likely characterizes particle motions in inner magnetospheric regions, but it does not hold further out into the geotail owing to the locally abrupt field reversal at the equator. Here, the breaking of adiabaticity at neutral sheet crossing has been increasingly investigated since the early work of Speiser [1965]. Different classes of particle orbits

have been uncovered, reflecting distinct changes of magnetic moment [e.g., Chen and Palmadesso, 1986; Büchner and Zelenyi, 1989]. Among these, quasi-trapped ("cucumber-type") particles display back and forth motions and experience repeated interactions with the neutral sheet, while transient ("Speiser-type") particles interact only once and exit toward the ionosphere (see Chen [1992] for a review).

In the present report, we investigate a trajectory feature which, though it occurs in taillike field geometries, does not result from nonadiabatic motion but from a specific parallel behavior of particles with relatively low parallel speeds. It consists in a reversal of the flow if these particles attempt to leave the low-latitude region while simultaneously acted upon by fast ExB drifts. This effect (termed "centrifugal trapping" hereinafter) was already noted in the time-dependent trajectory study of Delcourt et al. [1990], which was intended to explore the transport of near-Earth plasma sheet particles during storm-time dipolarization of the magnetospheric field lines. Here, we examine it in steady state conditions. Section 2 describes the underlying reason for such a behavior, while results of numerical simulations are presented in Section 3. For comparison, these results are shown both during dipolarization

Auroral Plasma Dynamics
Geophysical Monograph 80

events and in the steady state magnetotail. Since the particle magnetic moment is at stake in both cases, the trajectory calculations were carried out using the full equation of motion. The "centrifugal trapping" itself however can be approached either via guiding center or full particle treatment since, as stated above, it relates to the very parallel motion. This will be made more apparent in the following section.

CENTRIFUGAL TRAPPING

When developing the equation of motion of the guiding center, Northrop [1963] obtained a term which describes the acceleration resulting from the change in the magnetic field orientation. Taking the $\mathbf{E} \times \mathbf{B}$ drift velocity (noted $\mathbf{U_E}$ hereinafter) as the zero order velocity component perpendicular to $\mathbf{B}$, this term is:

$$\ddot{\mathbf{R}} = V_{/\!/} d\mathbf{b}/dt = V_{/\!/} \partial\mathbf{b}/\partial t + V_{/\!/}^2 \partial\mathbf{b}/\partial s + V_{/\!/}(\mathbf{U_E}.\nabla)\mathbf{b} \quad (1)$$

where $V_{/\!/}$ is the parallel velocity component, $\mathbf{b}$, a unit vector along $\mathbf{B}$, and s, the curvilinear coordinate along the field line. Accordingly, in the general expression of the drift velocity (equation (1.17) of Northrop [1963]), this yields a component of the form:

$$\dot{\mathbf{R}}_\perp = (m/qB)\, \mathbf{b} \times [V_{/\!/} \partial\mathbf{b}/\partial t + V_{/\!/}^2 \partial\mathbf{b}/\partial s + V_{/\!/}(\mathbf{U_E}.\nabla)\mathbf{b}] \quad (2)$$

(m being the particle mass and q, its charge). An additional second order drift of the form $\mathbf{b} \times d\mathbf{U_E}/dt$ is also apparent in equation (1.17) of Northrop [1963], which is not relevant to the present analysis as can be seen from examination of the guiding center parallel equation of motion. In the absence of parallel electric fields and neglecting the gravitational acceleration, this latter equation is (equation (1.20) of Northrop [1963]):

$$\dot{V}_{/\!/} = -(\mu/m)\partial B/\partial s + \mathbf{U_E} . [\partial\mathbf{b}/\partial t + V_{/\!/}\partial\mathbf{b}/\partial s + (\mathbf{U_E}.\nabla)\mathbf{b}] \quad (3)$$

(noting μ the particle magnetic moment). Note that, with the exception of the first term on the right side of (3), a similar equation can be obtained for the full parallel motion as shown in Delcourt et al. [1990]. While an explicit time dependence of the field may at times (such as during substorms) substantially affect the particle motion, the $\dot{\mathbf{R}}.\nabla$ part of the total time derivative of $\mathbf{b}$ is the component of interest here. It yields the last two terms in (2) and (3), the first one of these accounting for the particle centrifugal acceleration in the course of its longitudinal motion. It yields the well-known drift due to field line curvature, of particular importance in the equatorial vicinity [e.g., Cladis, 1986]. The last term in (2) and (3) has comparatively received lesser attention. It results from the direction-changing $\mathbf{B}$ as the particle travels across the magnetic field lines, i.e., in first approximation, as it drifts in $\mathbf{E} \times \mathbf{B}$. When this transverse motion is slow as compared to the parallel one (for instance, in the innermost ground state magnetosphere or for relatively fast particles) or in regions where little field structure exists (e.g., in the magnetospheric lobes), the corresponding change in the field orientation is small and this additional centrifugal term needs not be retained. However, in some instances, rapid $\mathbf{E} \times \mathbf{B}$ drifts can affect the

particles and drastically alter their propagation path and energization. Such drifts are known to develop, for example, during storm-time dipolarization events as a result of large (a few millivolts per meter) though short-lived induced electric fields. In the steady state magnetotail, particles also experience fast $\mathbf{E} \times \mathbf{B}$ drifts, in this case due to the weak $\mathbf{B}$ magnitude rather than to intense transient electric fields.

The respective influence of the last two terms in (2) and (3) can be better visualized in Figure 1 which presents a schematic illustration of the change in the $\mathbf{B}$ orientation for distinct particle motions near the equator. In this figure, it can be seen that, for particles approaching the field reversal at the tail midplane, the $\Delta\mathbf{B}$ implied by the longitudinal motion (case a-to-b) and that due to the $\mathbf{E} \times \mathbf{B}$ drift across the field lines (case a-to-c) occur in similar directions. The former $\mathbf{B}$ change varies with field line curvature, while the latter one depends upon curvature of the normal (as defined by the $\mathbf{E} \times \mathbf{B}$ drift orientation) to the magnetic field. These combined motions result into an enhanced outward oriented centrifugal impulse; hence, a curvature drift which points toward dusk (in the midnight meridian plane) and gives rise to systematic acceleration (via crossing of the electric field equipotentials). In contrast, for particles leaving the equator, the longitudinal motion (case d-to-e in Figure 1) and the $\mathbf{E} \times \mathbf{B}$ drift (case d-to-f) yield $\Delta\mathbf{B}$s in opposite directions and thus a less trivial energization. Particularly, in this latter case, slow particles subjected to a fast $\mathbf{E} \times \mathbf{B}$ transport will experience a centrifugal impulse which points inward (case d-to-f) or, equivalently, an eastward oriented curvature drift (in the midnight meridian plane). Consequently, as will be seen in the following section, rapid decelerations will affect these particles until they mirror and return to the equator. This centrifugal trapping mechanism at low latitudes differs from magnetic trapping as it does not follow from an intensification of the $\mathbf{B}$ magnitude, but from the field rotation in the course of the $\mathbf{E} \times \mathbf{B}$ transport.

TRAJECTORY RESULTS

Flow reversal during storm times

During the expansion of substorms, as the magnetospheric field lines rapidly evolve from a taillike geometry to a more dipolar one, large electric fields develop [e.g., Aggson et al., 1983], which trigger abrupt plasma injections into low L-shells. Recent studies [e.g., Mauk, 1986; Delcourt et al., 1990; Lewis et al., 1990] have demonstrated that, as a result of the

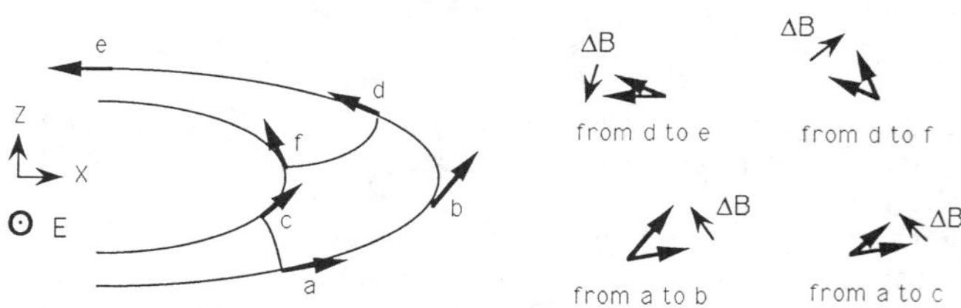

Fig. 1. Schematic illustration of the field orientation change in the course of a particle motion in the equatorial region. Heavy arrows indicate the $\mathbf{B}$ vectors at various positions (labeled from a to f) along the field lines. The $\Delta\mathbf{B}$ obtained for distinct motions (primarily parallel or primarily perpendicular) toward and away from the equator are summarized at right.

local field line curvature, these short-lived electric fields result into impulsive parallel energization of the particles at equatorial crossing. Furthermore, it was shown in Delcourt et al. [1990] that, during such events, conflicting centrifugal effects as described in Section 2 can strongly sever the ion flow. This is illustrated in Figure 2 which presents the storm-time trajectories of protons in the near-Earth plasma sheet. These trajectories were obtained by means of full motion calculations, considering a two-minute relaxation of the field from disturbed to quiet Mead and Fairfield [1975] configurations (the reader can consult Delcourt et al. [1990] for further details on the time-dependent model). At substorm onset, the particles were assumed to flow along the magnetic field ($\alpha_0 = 0°$), initialized in the midnight meridian plane with 10 eV energy (solid and dashed lines) and at symmetrical positions (5° and -5° latitude) about the midplane. The left panel in Figure 2 shows the trajectory projection in the X-Z plane (with X axis oriented toward the tail, and the Z axis from south to north), while the two panels at right display the respective kinetic energy and pitch angle variations as a function of time.

It is apparent from this figure that, while the H$^+$ initiated below the center plane (dashed line) experiences a large acceleration at equatorial crossing (from 10 eV up to 600 eV) and, not surprisingly, proceeds toward the northern ionosphere, the proton started above the equator (solid line) is rapidly reflected toward the opposite hemisphere. This H$^+$ subsequently travels in a direction anti-parallel to **B** ($\alpha_f = 180°$), exhibiting a lesser net energy gain (viz., from 10 eV up to 120 eV). This behavior contrasts with that of a proton started at a similar position but with an energy of 1 keV (dotted line). This latter ion does travel toward the northern auroral zone and remains essentially unaffected by the dipolarization process. The flow reversal portrayed here results from the conflicting contributions of the last two terms in (2) and (3), as discussed in the previous section. For the particle leaving the equator with "insufficient" parallel speed (solid line), the

($\mathbf{U_E}.\nabla$)**b** component prevails, resulting into centrifugal deceleration and ultimately a new mirror point at high altitude.

Flow reversal in the steady tail

In the magnetotail, particles are subjected to large **ExB** drifts owing the weak magnitude of the magnetic field. At relatively low latitudes, one expects a rapid rotation of the particle gyration plane in the course of the **ExB** transport since, in the quasi-neutral sheet geometry with a small B_z component at Z = 0, the **ExB** drift velocity points toward the center plane somewhat above the equator and earthward at the equator. Centrifugal effects as described in Section 2 are thus expected to act upon the particles, prior to or after nonadiabatic interaction with the neutral sheet depending upon the flow direction. Figure 3 shows H$^+$ trajectories representative of these effects. These trajectories were calculated using the low Kp version of the Tsyganenko [1987] model which features a current sheet thickness scale of the order of 5 R$_E$ and a somewhat uniform Z component of the tail field of the order of 1-2 nT. As for the large scale convection electric field, a Volland [1978] potential distribution was adopted in the ionosphere, with a potential drop of 25 kV across the polar cap (or, equivalently, a dawn-dusk convection electric field of the order of 0.15 mV/m at 60 R$_E$ geodistance). In Figure 3, test protons were launched from an arbitrary radial distance of 25 R$_E$. These H$^+$ were prescribed to initially flow along the magnetic field (zero magnetic moment), initiated with 10 eV energy (solid and dashed lines) at symmetrical positions about the center plane (5° and -5° latitude in the midnight meridian plane). In a format similar to that of Figure 2, Figure 3 presents the trajectory projections in the X-Z plane (left panel), together with the respective kinetic energy and pitch angle variations versus time (right panels).

It is apparent from the left panel of Figure 3 that the path of the H$^+$ initiated above the equator (solid line) is severely altered by the action of the fast **ExB** drift, namely: instead of traveling toward the ionosphere (initially positive parallel speed), this

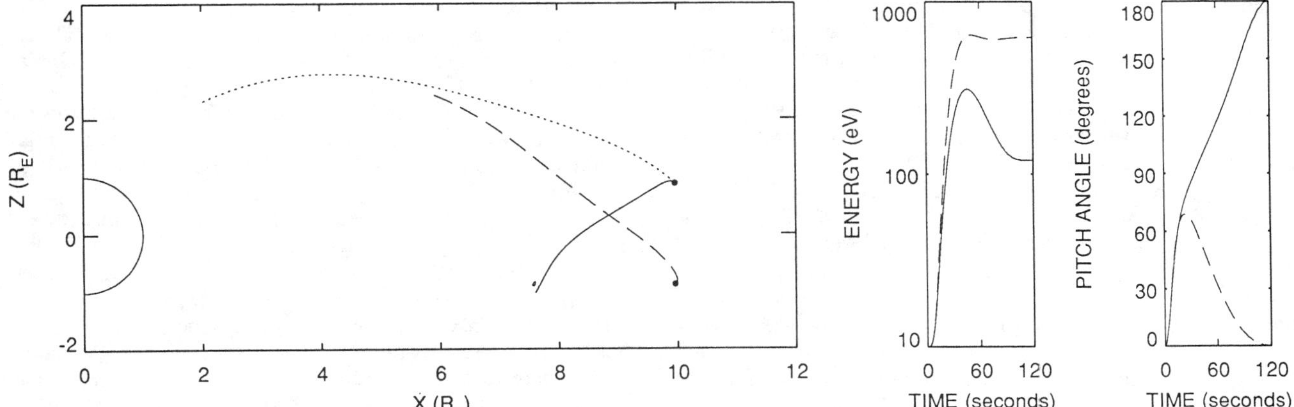

Fig. 2. Proton trajectories during storm-time dipolarization of the magnetic field lines: trajectory projection in the midnight meridian plane (left panel), kinetic energy versus time (middle panel) and pitch angle versus time (right panel). These simulations relate to a two-minute transition within the Mead and Fairfield [1975] model. At substorm onset, the test H$^+$ are located at symmetrical positions about the equator (10 R$_E$ geocentric distance, 5° and -5° magnetic latitude), initiated with 10 eV energy along the magnetic field ($\alpha_0 = 0°$) (solid and dashed lines). The dotted line in the left panel relates to a proton started above the center plane with 1 keV energy.

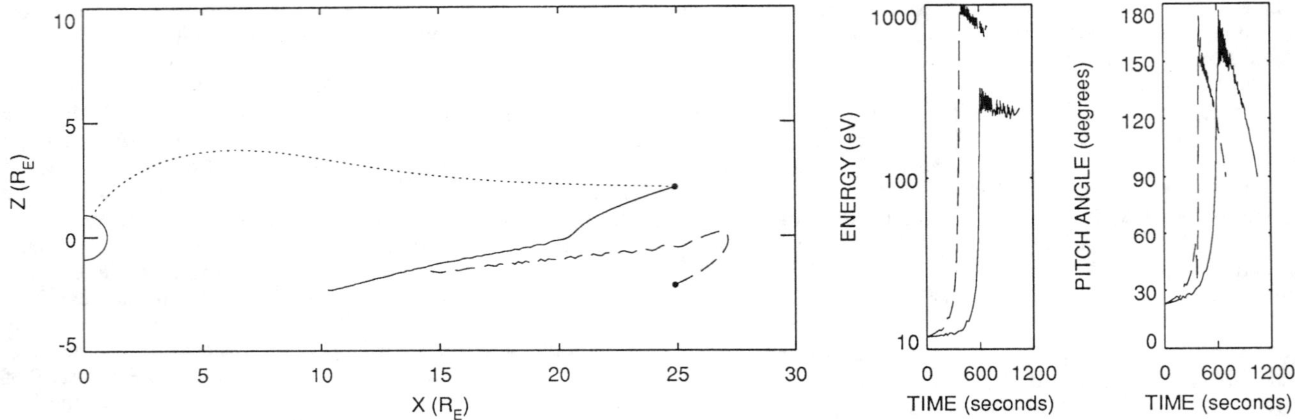

Fig. 3. Segments of H$^+$ trajectories in the low Kp model of Tsyganenko [1987]: trajectory projection in the midnight meridian plane (left panel), kinetic energy versus time (middle panel) and pitch angle versus time (right panel). The cross-tail potentail drop is set to 25 kV. The ions are injected at symmetrical positions about the center plane (25R$_E$ geocentric distance, 5° and -5° magnetic latitude), traveling from south to north with 10 eV energy along the magnetic field (solid and dashed lines). The dotted line in the left panel relates to a proton started above the tail midplane with 1 keV energy.

ion is rapidly reflected toward the equator. After nonadiabatic traversal of the neutral sheet [e.g., Speiser, 1965; Chen and Palmadesso, 1986; Büchner and Zelenyi, 1989], it leaves the equatorial region with a substantial magnetic moment and subsequently experiences magnetic mirroring at a geocentric distance of the order of 10 R$_E$ (at which point the trajectory tracing was interrupted). This behavior contrasts with that of a proton injected at a similar position but with an energy of 1 keV (dotted line). This latter particle does proceed toward the ionosphere and, as expected, ultimately precipitates over the auroral zone. The path of the H$^+$ started below the center plane (dashed line) is also of interest. Indeed, as schematically shown in Figure 1, the fast **ExB** drift and the longitudinal transport yield concurrent centrifugal accelerations during the equatorward sequence of the motion. However, after nonadiabatic crossing of the tail midplane, this ion exits with "insufficient" parallel energy (viz., of the order of 150 eV at t = 370 seconds). Accordingly, centrifugal trapping rapidly interrupts its outbound motion, and the particles goes through another traversal of the neutral sheet. As a result of its magnetic moment breakdown, it subsequently experiences magnetic mirroring at an altitude of 15 R$_E$.

A more general view of this parallel transport feature can be obtained from Figure 4 which presents the results of systematic trajectory calculations. Here, test protons were started with distinct parallel speeds (varied from 50 to 250 km/s by steps of 50 km/s) and at distinct positions (separated by equidistant steps of 0.2 R$_E$) along a given field line (that passing through Z = 0 at X = 25 R$_E$). These ions were initiated above the equator (positive s values) and prescribed to flow toward the ionosphere (positive parallel speed) aligned with the magnetic field (zero magnetic moment). Figure 4 presents their computed trajectories in the one-dimensional phase space (s,V$_\parallel$), until they reach the equator or exit toward the ionosphere. Not surprisingly, it can be seen on the right side of this figure that particles injected with large parallel speeds propagate earthward (increasing s values). Note their further energization

due field line curvature (the V$_\parallel \partial$b/∂s term in equations (2) and (3)) if injection occurs close enough to the tail midplane. On the other hand, a centrifugal reversal of the flow at low energies is readily noticeable at left. Indeed, it can be seen that, regardless of their initial distance, particles with velocities of the order of 50 km/s (and below) travel toward lower s values and ultimately display negative parallel speeds. Between these two outer regimes, a more erratic behavior is apparent for initial speeds of the order of 100-150 km/s, which results from the conflicting contributions of the last two terms in (2) and (3). Clearly, this latter behavior delineates the boundary between escape and centrifugal trapping modes. Note particularly the apparent discontinuity near s = 0.3 R$_E$ for an initial speed of 150 km/s since, as mentioned above, the H$^+$ originating closer to the equator experiences a substantial energization due to field line curvature.

It should be stressed here that, at exit from the neutral sheet, only particles displaying small pitch angles (i.e., "Speiser-type" orbits as opposed to "cucumber-type" ones) are susceptible of precipitating over the auroral zone. However, what Figure 4 demonstrates is that, even though highly collimated (the test H$^+$ are taken to be field-aligned in Figure 4), these particles will be subjected to centrifugal mirorring and return to the equator if their parallel energy is "insufficient". This in fact corresponds to the behavior seen near X = 26 R$_E$ in Figure 3 for the particle started below the equator (dashed line). In other terms, ejection onto untrapped ("Speiser-type") orbits after interaction with the neutral sheet appears as a necessary but not sufficient condition to precipitate over the auroral zone. There exists in addition a "critical" parallel speed below which particles remain trapped at low latitudes. As this critical speed is a straightforward consequence of the last term in (2) and (3), it depends upon the magnitude of the **ExB** drift as well as the curvature of the normals to the magnetic field. It thus significantly varies with depth into the magnetotail. As a matter of fact, as examined in Delcourt et al. [1993], similar H$^+$ trajectory calculations at L = 50 reveal substantially higher

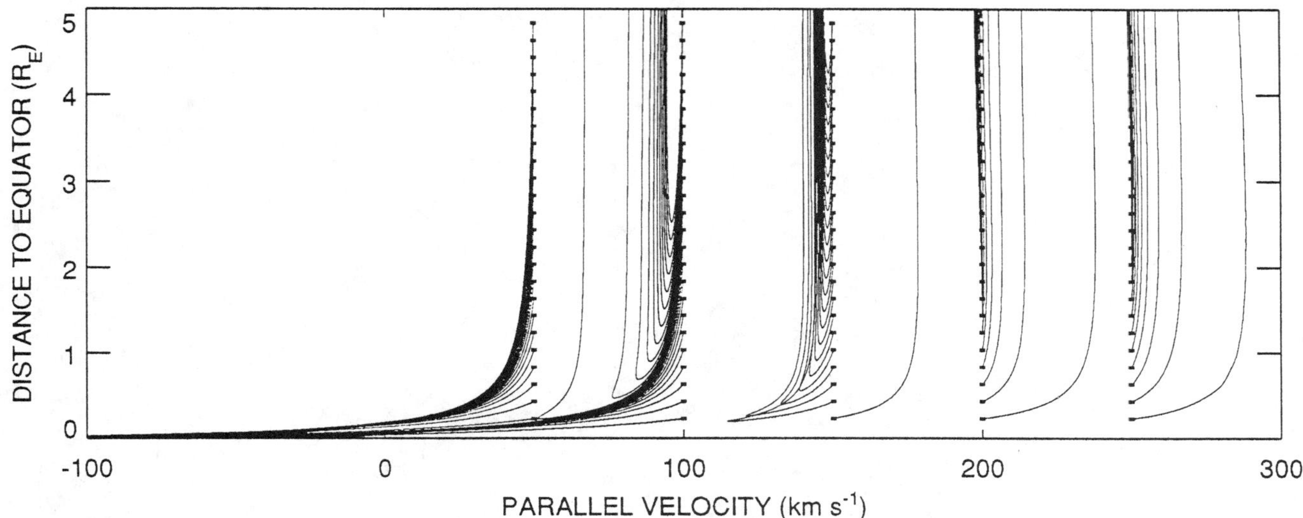

Fig. 4. Proton trajectories in the one-dimensional phase space $(s, V_\parallel)$. The test particles are initiated at equidistant steps of 0.2 R_E along the $L = 25$ field line and with distinct parallel speeds (from 50 to 250 km/s by steps of 50 km/s). Heavy dots depict the injection positions.

critical speeds, viz., of the order of 100-150 km/s and 200-250 km/s during quiet and disturbed conditions respectively (adopting, in the latter case, a cross-tail potential drop of 80 kV). On the other hand, by imposing a confinement of less energetic populations to low latitudes, centrifugal trapping will result into local density enhancements. This confinement vanishes when the particles gain access to inner magnetospheric regions, as a result of the weakening **ExB** and the more dipolar field configuration (see for instance Figure 1 of Delcourt et al. [1993]). Note at last that this parallel transport feature may be in relation with the low-speed cutoffs observed near the lobeward edge of the plasma sheet. A thorough examination of this latter assertion will require a more comprehensive trajectory analysis.

SUMMARY

Under the effect of fast **ExB** drifts such as during the expansion phase of substorms (large transient electric fields) or in the geomagnetic tail (weak field magnitude), particles traveling at low latitudes are subjected to rapid rotations of the magnetic field. For inbound (with regard to the tail midplane) particles, this re-orientation of the field results into a centrifugal acceleration which adds to that due to the strictly longitudinal motion. In contrast, longitudinal and **ExB** motions give rise to conflicting centrifugal accelerations for outbound particles. For these latter particles, a reversal of the flow is obtained when the transverse motion prevails over the longitudinal one (i.e., for relatively low parallel speeds as compared to the **ExB** drift velocity). While fast particles of the distant plasma sheet can gain access to high latitudes and possibly the auroral zone, this "centrifugal trapping" mechanism yields the confinement of less energetic populations to the equatorial vicinity until these are convected into inner magnetospheric regions.

REFERENCES

Aggson, T.L., J.P. Heppner, and N.C. Maynard, Observations of large magnetospheric electric fields during the onset phase of a substorm, *J. Geophys. Res.*, 88, 3981, 1983.

Büchner, J., and L.M. Zelenyi, Regular and chaotic charged particle motion in magnetotaillike field reversals: 1. Basic theory of trapped motion, *J. Geophys. Res.*, 94, 11821, 1989.

Chen, J., Nonlinear dynamics of charged particles in the magnetotail, *J. Geophys. Res.*, 97, 15011, 1992.

Chen, J., and P.J. Palmadesso, Chaos and nonlinear dynamics of single-particle orbits in magnetotaillike magnetic field, *J. Geophys. Res.*, 91, 1499, 1986.

Cladis, J.B., Parallel acceleration and transport of ions from polar ionosphere to plasma sheet, *Geophys. Res. Lett.*, 13, 893, 1986.

Delcourt, D.C., J.A. Sauvaud, and A. Pedersen, Dynamics of single-particle orbits during substorm expansion phase, *J. Geophys. Res.*, 95, 20853, 1990.

Delcourt, D.C., J.A. Sauvaud, and T.E. Moore, Polar wind ion dynamics in the magnetotail, *J. Geophys. Res.*, in press, 1993.

Lewis, Z.V., S.W.H. Cowley, and D.J. Southwood, Impulsive energization of ions in the near-Earth magnetotail during substorms, *Planet. Space Sci.*, 38, 491, 1990.

Mauk, B.H., Quantitative modeling of the "convection surge" mechanism of ion acceleration, *J. Geophys. Res.*, 91, 13423, 1986.

Mead, G.D., and D.H. Fairfield, A quantitative magnetospheric model derived from spacecraft magnetometer data, *J. Geophys. Res.*, 80, 523, 1975.

Northrop, T.G., The Adiabatic Motion of Charged Particles, *Wiley Interscience*, New York, 1963.

Speiser, T.W., Particle trajectories in model current sheets: 1. Analytical solutions, *J. Geophys. Res.*, 70, 4219, 1965.

Tsyganenko, N., Global quantitative models of the geomagnetic field in the cislunar magnetosphere for different disturbance levels, *Planet. Space Sci.*, 35, 1347, 1987.

Volland, H., A model of the magnetospheric electric convection field, *J. Geophys. Res.*, 83, 2695, 1978.

D.C. Delcourt, CRPE/CNET-CNRS, Saint-Maur des Fossés, France

J.A. Sauvaud, CESR/CNRS, Toulouse, France

T.E. Moore, NASA/Marshall Space Flight Center, Huntsville, Alabama

DE 1 Particle and Wave Observations in an AKR Source Region

J. D. MENIETTI AND J. L. BURCH

Department of Space Sciences, Southwest Research Institute, San Antonio, Texas

Near the AKR source region wave-particle interactions appear to have modified the observed electron distributions. We compare the observations to those predicted by recently published numerical simulations. Observations of electron distributions indicate a region of perpendicular heating ($T_\perp/T_\parallel > 10$) adjacent to and within the source region. Loss cones, trapped particles, beams, and electron conical distributions are also observed near and within the source region, which extends perpendicular to the magnetic field line for at least 20 km in a density cavity. The high altitude plasma instrument on board the DE 1 satellite was operating during a near crossing of the AKR source in the nightside auroral region.

INTRODUCTION

The theory of the cyclotron maser instability has had much success in explaining observations of auroral kilometric radation (AKR). Wu and Lee [1979] have demonstrated that a loss cone distribution of electrons within a region of depleted plasma density (with ratio of plasma frequency to gyrofrequency $f_p/f_g < 0.3$) can be an effective free-energy source for the generation of auroral kilometric radiation. This distribution provides the requirement of the cyclotron maser instability (CMI) that $\partial f/\partial v_\perp > 0$. As pointed out by Omidi and Gurnett [1982] and Omidi et al. [1984], in addition to the loss cone, the "hole" distribution and the "bump" or trapped particle distribution can also act as free-energy sources of AKR. Dusenbury and Lyons [1982], LeQueau et al. [1984], and Lin et al. [1986] have also shown that the hole distribution can be significant under certain conditions in the generation of AKR. Louarn et al. [1990] have shown that the trapped electron population may also be an effective free-energy source.

The AKR source region is associated with inverted V events [Gurnett, 1974; Green et al., 1979; Benson and Calvert, 1979]. Sharp et al. [1979] have shown that energetic upward flowing ion beams observed in association with inverted V events may be due to parallel electric fields beneath the satellite. Such electric fields are hypothesized as the source of the inverted V events and are believed to play an important role in the generation of AKR [Wu et al., 1982; Wu, 1985]. Benson and Calvert [1979] have shown that AKR has a low frequency cutoff at the R-X cutoff, which for a low density plasma is very near the electron gyrofrequency.

To date, no satellite has been capable of obtaining a complete distribution in phase space on a time scale adequate to resolve the question of the most effective free energy source of AKR.

Typically, the best satellites require several seconds to obtain a complete distribution in phase space. During this time, within an AKR source region, the waves would have grown and already altered the existing plasma distribution.

Observations of distributions in the AKR source region have recently been reported by investigators of the Viking satellite data. Ungstrup et al. [1990] have reported wide loss cones adjacent to the AKR source region, but field-aligned potentials and nearly filled loss cones appear within the source region. Ungstrup et al. [1990] attribute the filling of the loss cone to the AKR instability and fast diffusion of electrons as predicted by the cyclotron maser theory [Wu et al., 1982]. Louarn et al. [1990] have reported simultaneous measurements of EM fields and particle distributions measured by the Viking satellite during a crossing of the AKR source region. They have determined that trapped electrons may play an important role in the generation of AKR. Using high resolution measurements, Hilgers [1992] has subsequently shown that density cavities of dimensions < 20 km are associated with AKR sources, and that AKR seems to exist in regions where the ratio of energetic to thermal electron number densities is greater than 0.3 and $f_p/f_g < 0.15$. These results are consistent with earlier findings of Benson and Calvert [1979] and Benson [1985] that $f_p/f_g < 0.2$ in X mode AKR source regions. During 1981, DE 1 flew through the nightside auroral region at altitudes > 8000 km, a region that is believed to be at the upper extent of the source region of AKR. We here report DE 1 particle and wave data for a pass of the nightside auroral region near the AKR source center. We will compare these results to those reported by Viking and also to recent numerical simulation studies of Winglee and Pritchett [1986] which show a great similarity to the observations.

OBSERVATIONS

The high altitude plasma instrument (HAPI) and the plasma wave instrument (PWI) have been described by many authors and

Auroral Plasma Dynamics
Geophysical Monograph 80

will not be described here [cf. Burch et al., 1981; Shawhan et al., 1981]. In Plate 1 we show an energy-vs-time spectrogram with energy flux gray-coded according to the gray-bar at the right. The pass is of the nightside auroral region on October 12, 1981 (day 285). Precipitating electrons are shown in the top panel, and upward-ions are in the second panel. In the bottom panel we display the PWI AC electric field data, with intense AKR extending down below 100 kHz to frequencies near the local gyrofrequency f_g (indicated by the black dotted line), in the time interval $0028 < t < 0030$. In fact, the AKR extends to frequencies $< f_g$ which is to be expected when relativistic electrons are the dominant component. This follows directly from the relativistic gyroresonance condition for propagation at wave normal angles near 90° [cf. Wu, 1985]. The boundary plasma sheet (BPS) electrons are most energetic near 0028 UT, and the electron average energy decreases after 0029 UT. The upper energy extent of the upward flowing ions observed at the

same time reach a maximum in the interval $0029 < t < 0030$ UT. Thus, the potential below the spacecraft peaks at this time.

We have replotted the electron and PWI data at higher resolution in Plate 2. In this plot the electron data are shown at all pitch angles in the top panel, and the AC magnetometer data from PWI are plotted in the bottom panel on a linear scale over a reduced frequency range. On this plot the proximity of the satellite to the AKR source region is more clearly noted. The usual criterion for determining if the satellite is near the AKR source center is that intense AKR is observed near the local gyrofrequency. We also indicate on this plot a region of temperature anisotropy ($T_\perp/T_\parallel > 1$) which corresponds to times between about 0027:30 and 0029 as identified from the electron distribution discussed later. This region is adjacent to and extends into the AKR source region.

In Figure 1 we have plotted some of the calculated plasma parameters for an interval of this pass including the times near the

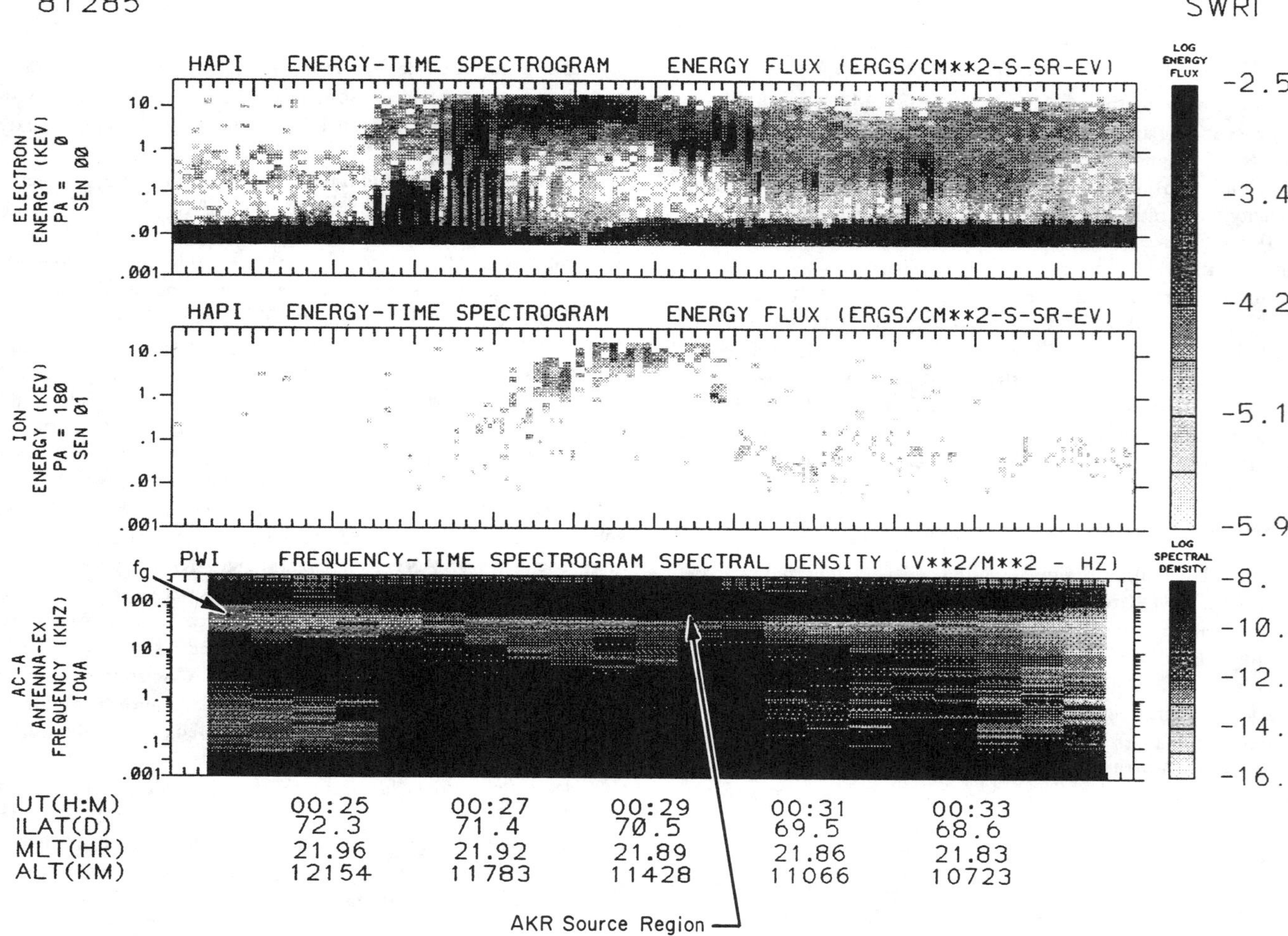

Plate 1. Energy-versus-time spectrogram with energy flux gray-coded for precipitating electrons (top panel) and upgoing ions (second panel). The bottom panel is a frequency-versus-time spectrogram with the electric field intensity gray-coded. The black dots indicate the local gyrofrequency for electrons (highest frequency) and ions.

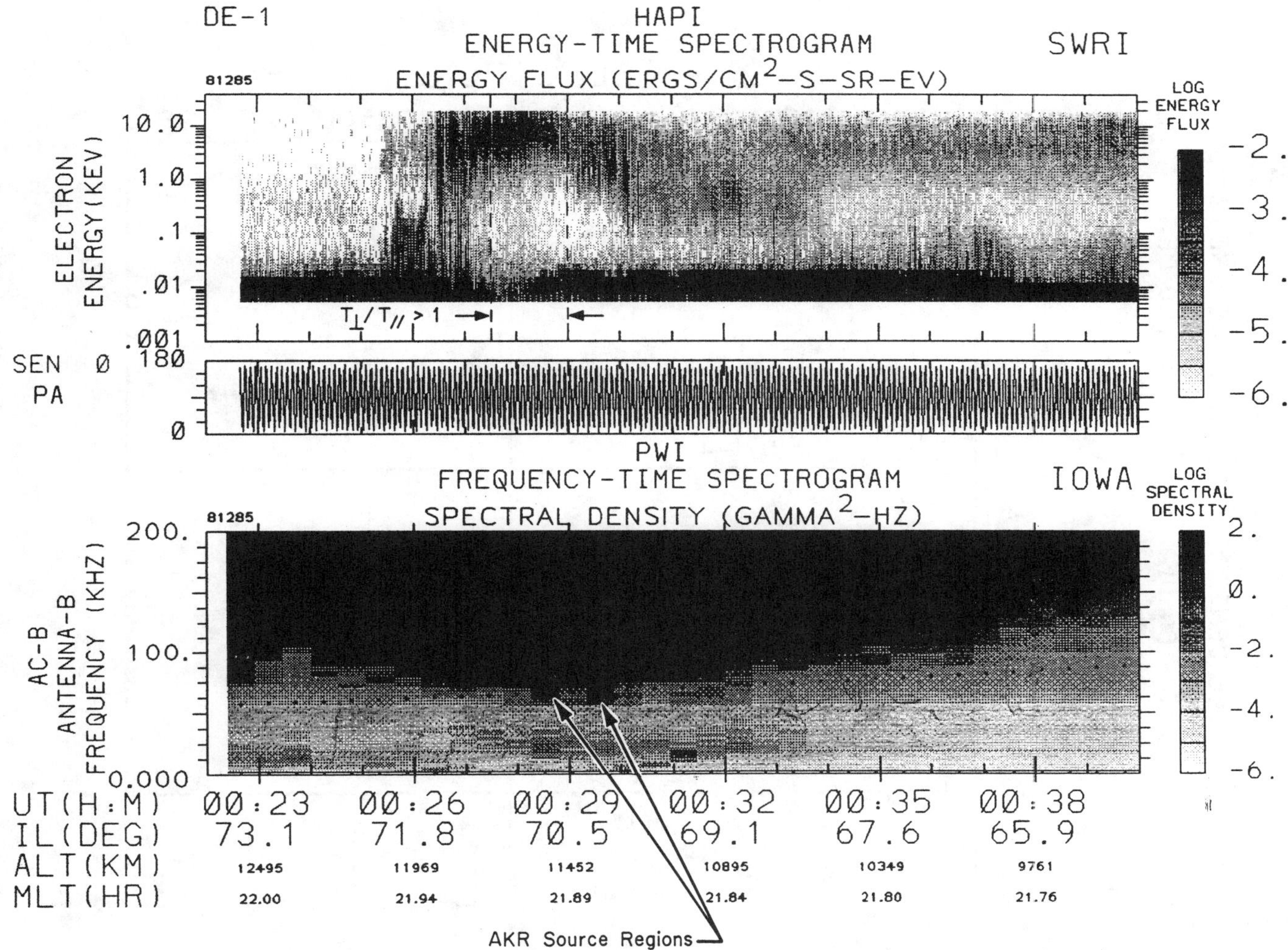

Plate 2. A higher resolution spectrogram of the same pass as Plate 1. The electrons are displayed at all pitch angles in the top panel, and the magnetic field intensity is gray-coded in the bottom panel with the frequency plotted on a linear scale.

AKR source region. We see in Figure 1 that when the satellite is closest to the AKR source region, 0028 < t < 0030, a density cavity was encountered and the average electron energy was decreasing. The plasma density was determined by integration of the electron data for E > 20 eV (to avoid contamination due to spacecraft photo electrons), so it only indicates a lower limit to the actual density. The plasma frequency estimated near the density cavity from the plasma wave instrument indicates an actual density of less than 2 cm^{-3}. In Figure 2 we plot the temperature of the electrons versus invariant latitude. The temperature was calculated assuming a linear fit to a Maxwellian, and shows a distinct peak just poleward of the AKR source center, in the region where $T_\perp/T_\parallel > 1$.

In Plate 3 we show the electron and ion data in the region nearest the AKR source center in higher resolution. Wide loss cones (that become semifilled for t > 0028:30) are observed throughout this time period, and the electric potential beneath the

satellite reaches its peak in the range 0028:20 < t < 0029 as indicated by the upward ion beams. These beams have maximum energies equal to ~ 13 keV, which was the programmed high energy limit of the instrument at the time of the observations.

We now examine these data with the aid of phase space contour plots of the distribution function. As noted in Plate 2, a region adjacent to and overlapping the AKR source region contained distributions with $T_\perp/T_\parallel > 1$. We show two examples of such distributions in Figure 3. In both of these figures the electron distributions show enhancements in the perpendicular direction continuously from the lowest energies to E > 2 keV. It is possible that such distributions are the result of heating by wave-particle interactions. In this case the waves could be the AKR emissions. We have estimated that for Figure 3b, $T_\perp/T_\parallel$ ~ 10. Also seen in both distributions of Figure 3, are field-aligned beams. This is an example of a beam with temperature anisotropy of the kind Newman et al. [1988], Wong

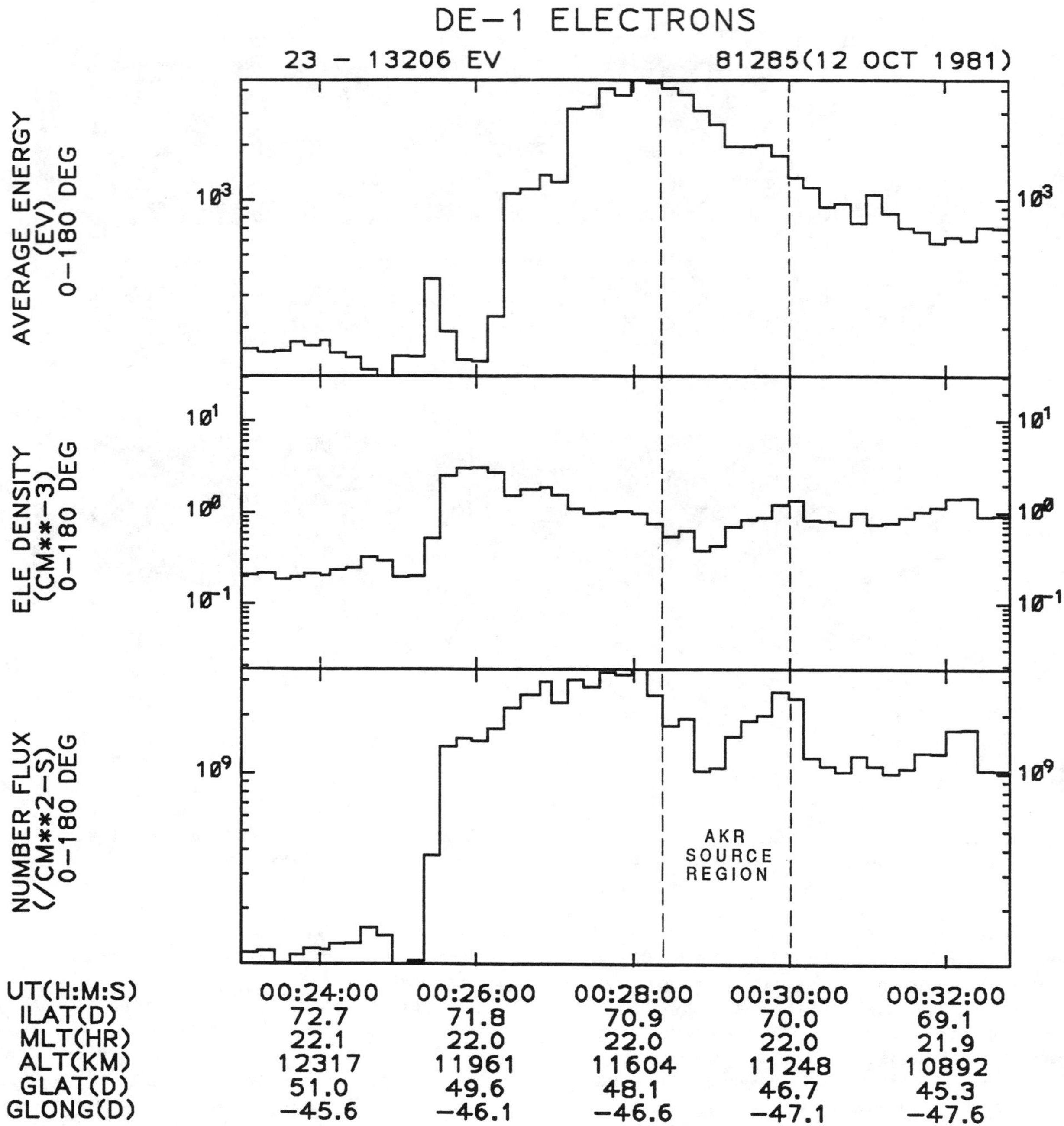

UT(H:M:S)	00:24:00	00:26:00	00:28:00	00:30:00	00:32:00
ILAT(D)	72.7	71.8	70.9	70.0	69.1
MLT(HR)	22.1	22.0	22.0	22.0	21.9
ALT(KM)	12317	11961	11604	11248	10892
GLAT(D)	51.0	49.6	48.1	46.7	45.3
GLONG(D)	−45.6	−46.1	−46.6	−47.1	−47.6

Fig. 1. Average energy, electron density, and number flux versus time for the pass of October 12, 1981. The times when the satellite was closest to the AKR source center are indicated by dotted lines.

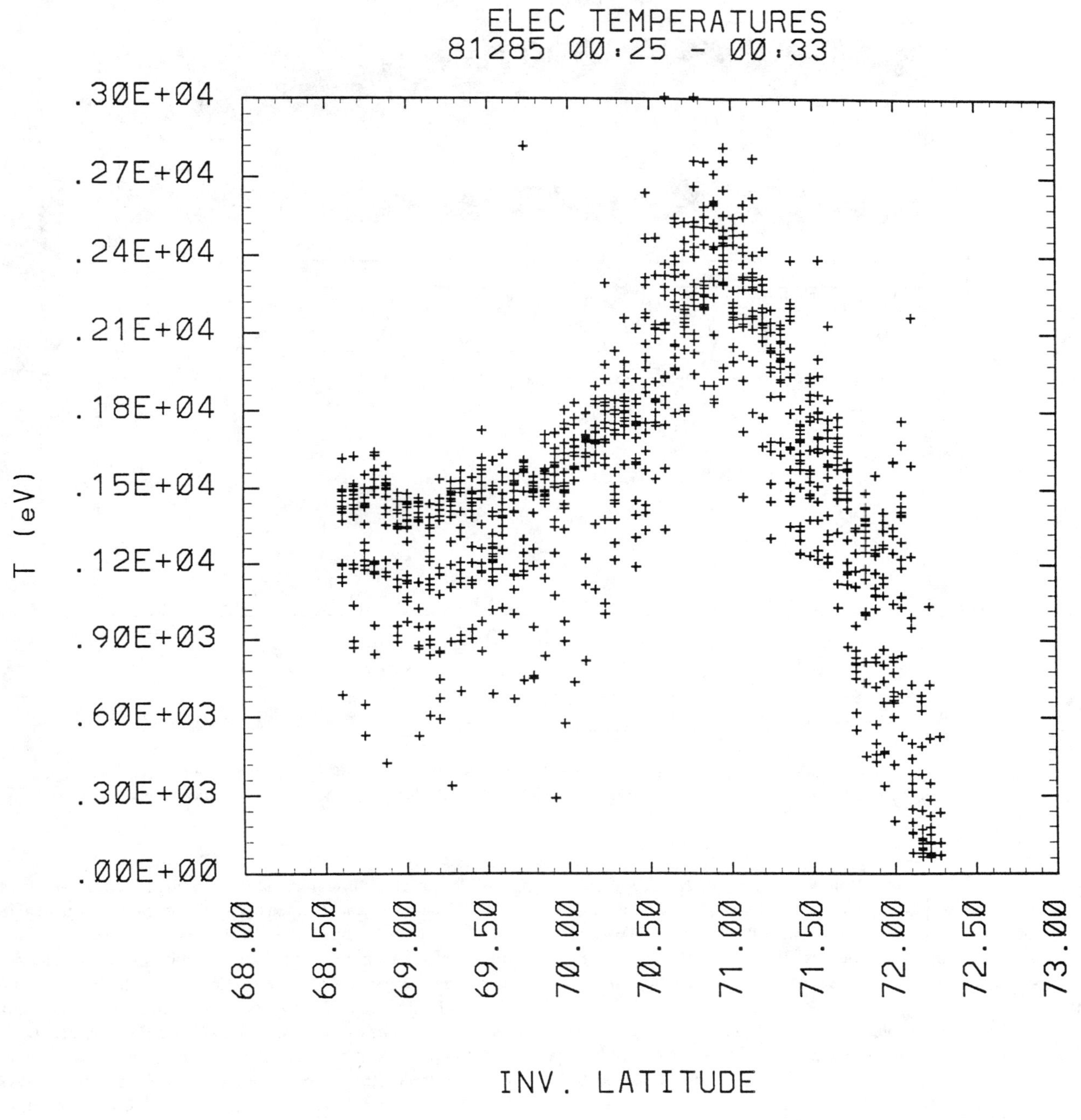

Fig. 2. Temperature versus invariant latitude for the pass of October 12, 1981.

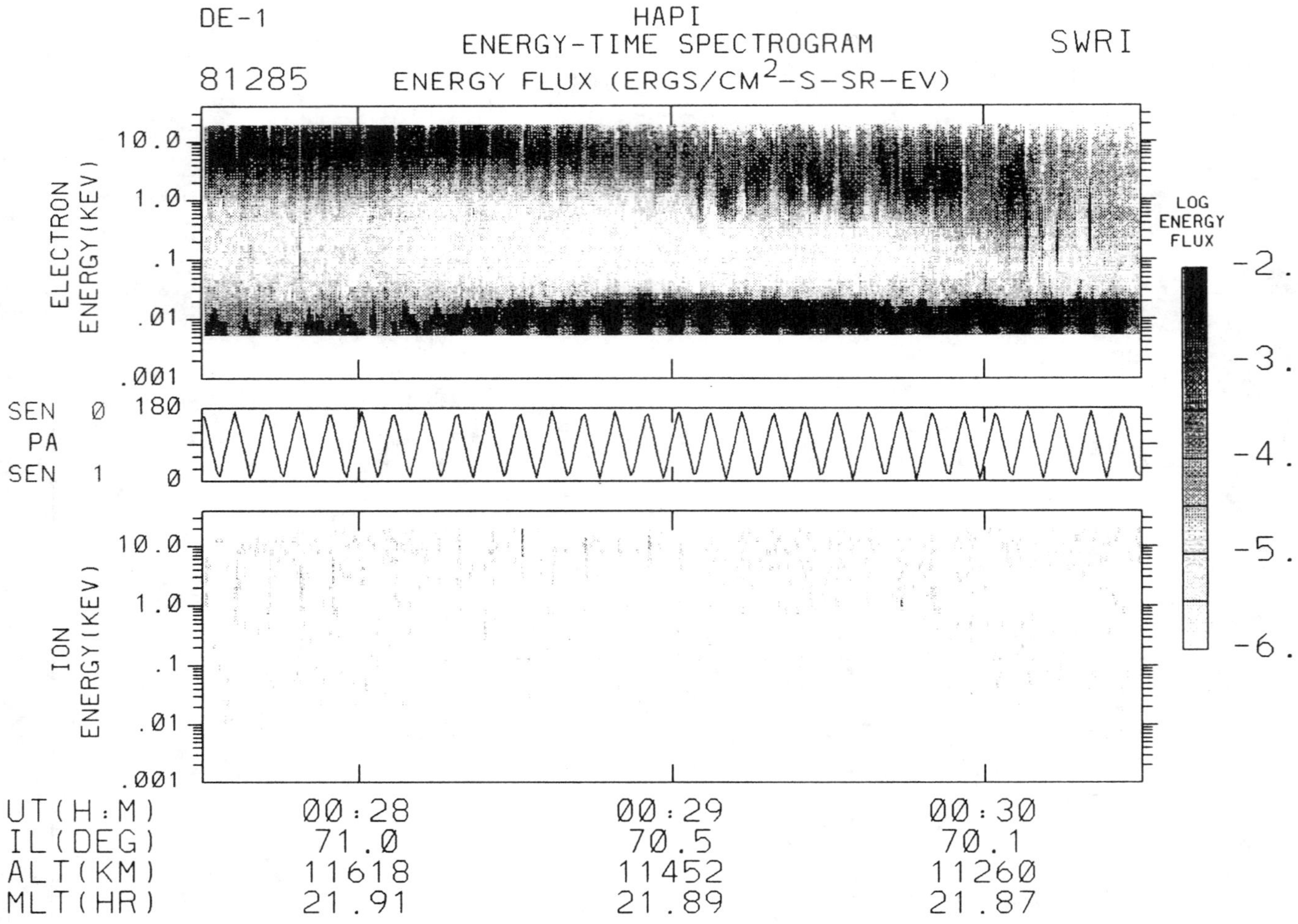

Plate 3. A high resolution spectrogram of electron (top panel) and ion (bottom) energy flux. The upward ion beams indicate a field-aligned potential beneath the satellite. For times greater than about 0029 UT the energy flux and average energy of the electrons decreased.

and Goldstein [1990] and Winglee et al. [1992] have shown is unstable to bursty radio emission via a modified kinetic Weibel or temperature anistropic beam instability (TABI). We suggest this distribution may also be responsible for the broadbanded electromagnetic bursts recently reported by Sonwalkar et al. [1990].

The AKR source center in Plate 2 is seen to consist of two separated regions of intense emission. Region I and III are separated by region II, from about 0028:50 to about 0029:20 where the AKR intensity slightly decreases. We do not know if this decrease is a temporal or spatial effect. Within Region I the distribution of electrons includes semifilled loss cones and trapped particles, and temperature anisotropies as shown in Figures 4a-4d. In region II the distributions include semifilled loss cones and an electron conic (5c) as shown in Figures 5a-5d. We hypothesize the electron conic may be the result of wave-particle heating resulting from the AKR emission. Wave particle interactions

resulting in electron conics have been suggested by Wong et al. [1988], Temerin and Cravens [1990] and Menietti et al. [1992] among others. Within region III the loss cone features are not as prevalent, but more beam-like distributions are observed as displayed in Figures 6a-6d. Also observed in both regions II and III are "banana" distributions (so named because of the shape of the beam) as shown in Figure 5b, 5d, 6c, and Figure 7. Such distributions are the result of beams that have been spread in pitch angle due to the conservation of the 1st adiabatic invariant [cf. Winglee and Pritchett, 1986]. In Figure 7 we show in greater resolution a distribution that appears to be a combination of a banana distribution and a "trapped" population. It takes 6 seconds to sample the data required for the contour plot. The data sampling began at the data point indicated and proceeded in the counterclockwise direction. Thus, the "banana" was sampled in the early part of the satellite spin and the "trapped" distribution was sampled near the end of the spin. It is possible that the

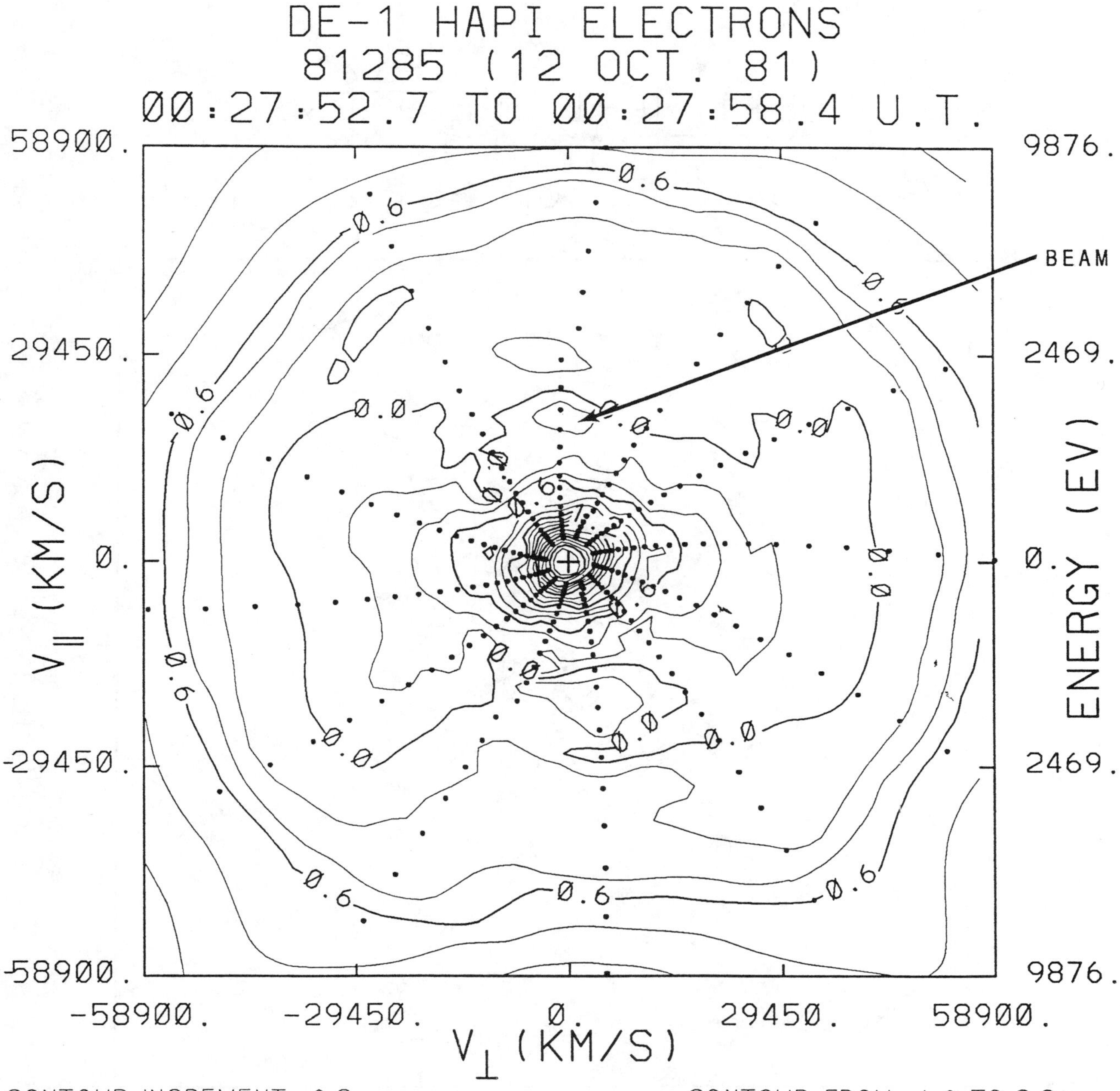

Fig. 3. Contours of the $\log_{10}$ of the distribution function in phase space obtained adjacent to and poleward of the AKR source region. The units are $\sec^3/km^6$. Both of these examples show strong anisotropies with $T_\perp > T_\parallel$.

banana distribution may have been unstable to cyclotron maser emission which heated the electron population, causing diffusion that produced the "trapped" distribution observed later in the same satellite spin.

Winglee and Pritchett [1986] have examined with the aid of numerical simulations how such distributions lead to the generation of electrostatic as well as AKR emissions. In their Figure 1 they show schematically how a field-aligned beam of electrons, as produced in an acceleration region, is spread in pitch angle due to the conservation of the first adiabatic invariant. In their Figure 7, Winglee and Pritchett schematically show how such banana distributions contain both $\partial f/\partial v_\parallel > 0$ and $\partial f/\partial v_\perp$

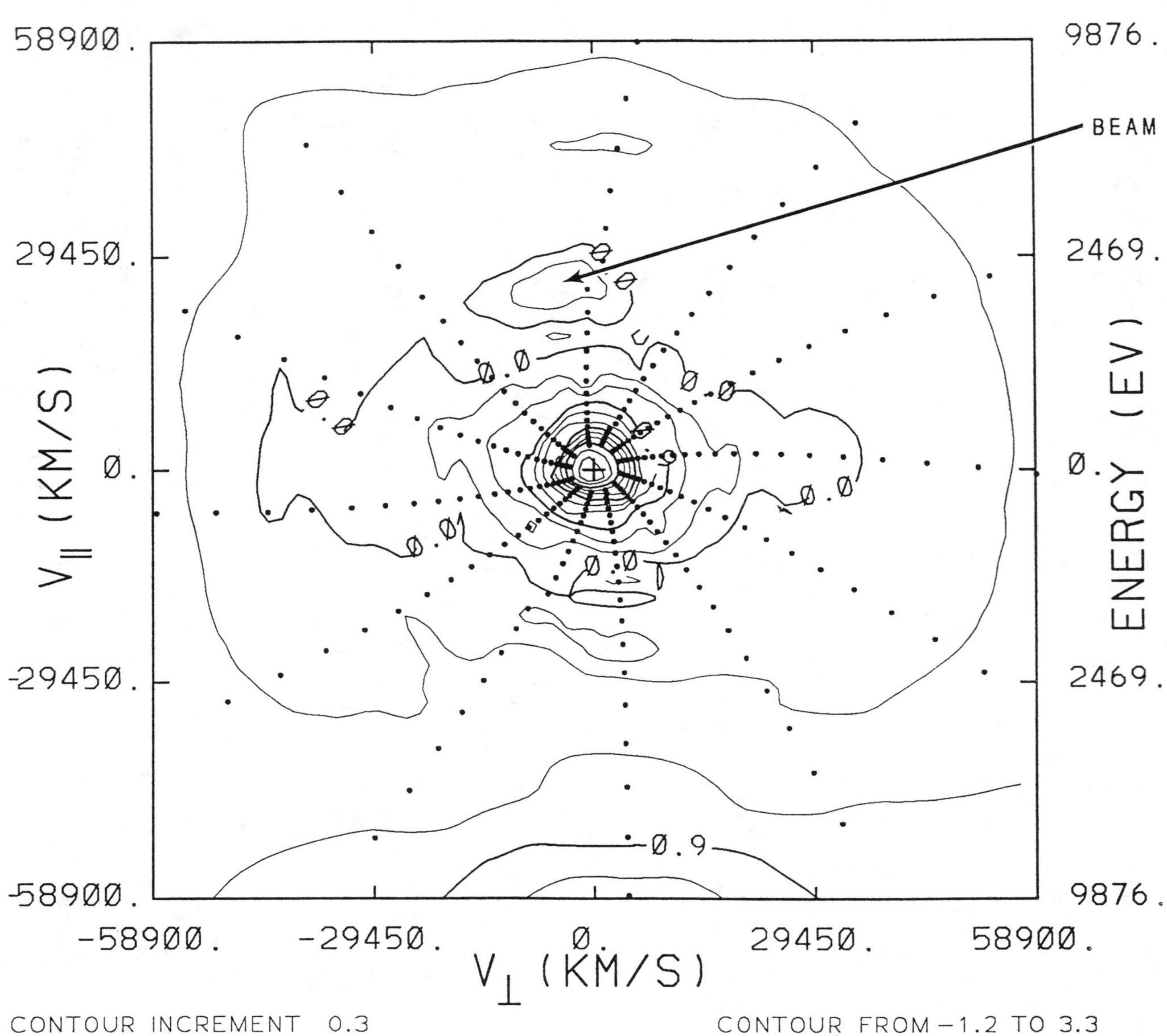

Fig. 3. (continued)

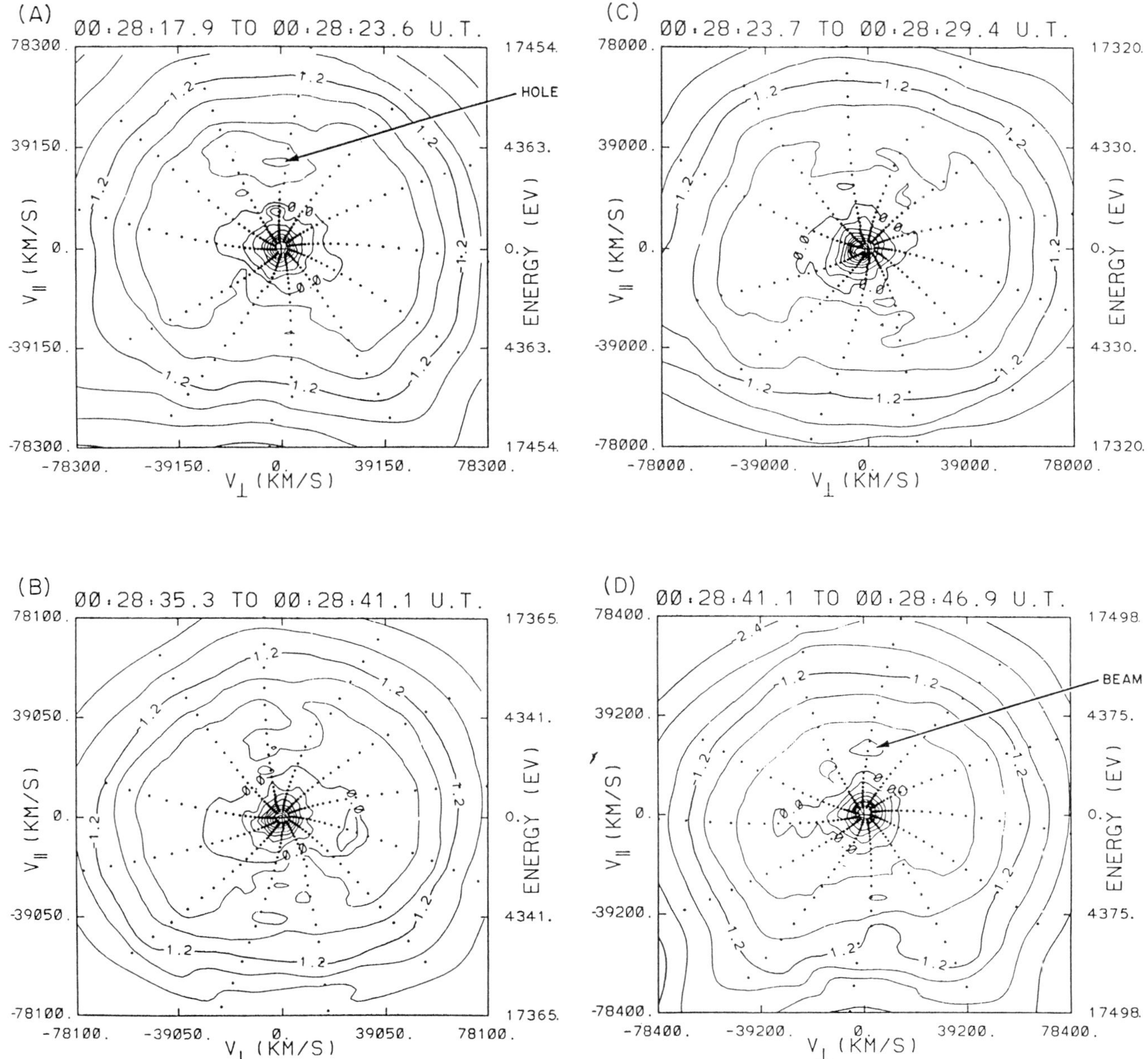

Fig. 4. A series of contours of the distribution function obtained when the satellite was in region I (0028:20 < t < 0028:50). Note the presence of semifilled loss cones (4a-4d), temperature anisotropies with $T_\perp > T_\parallel$ (4a-4d), and an electron conic (4d).

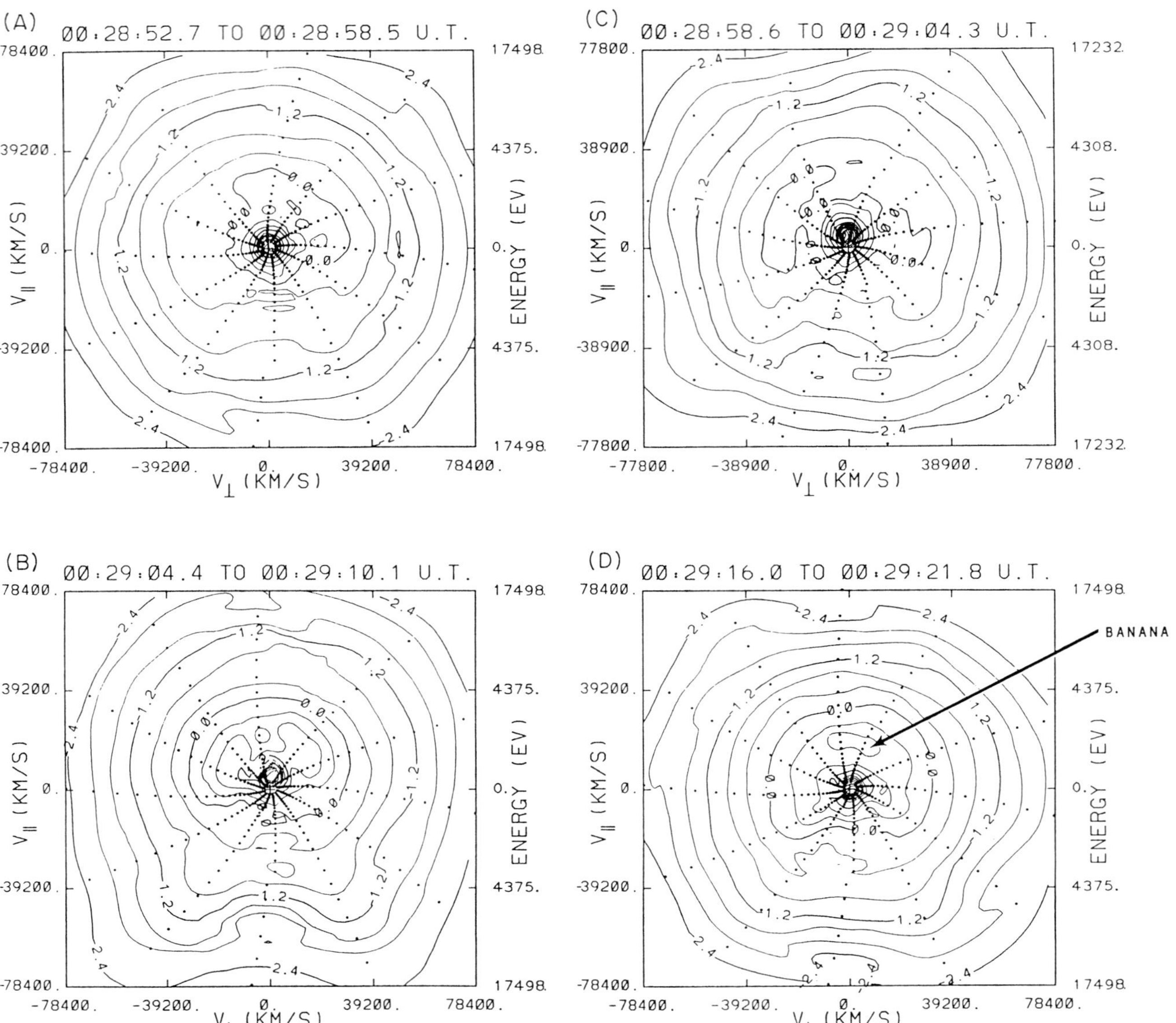

Fig. 5. A series of contours of the distribution function obtained when the satellite was in region II (0028:50 < t < 0029:20). The salient features are semifilled loss cones (5a–5d) and a banana distribution (5d), but the temperature anisotropies are not as large as in region I.

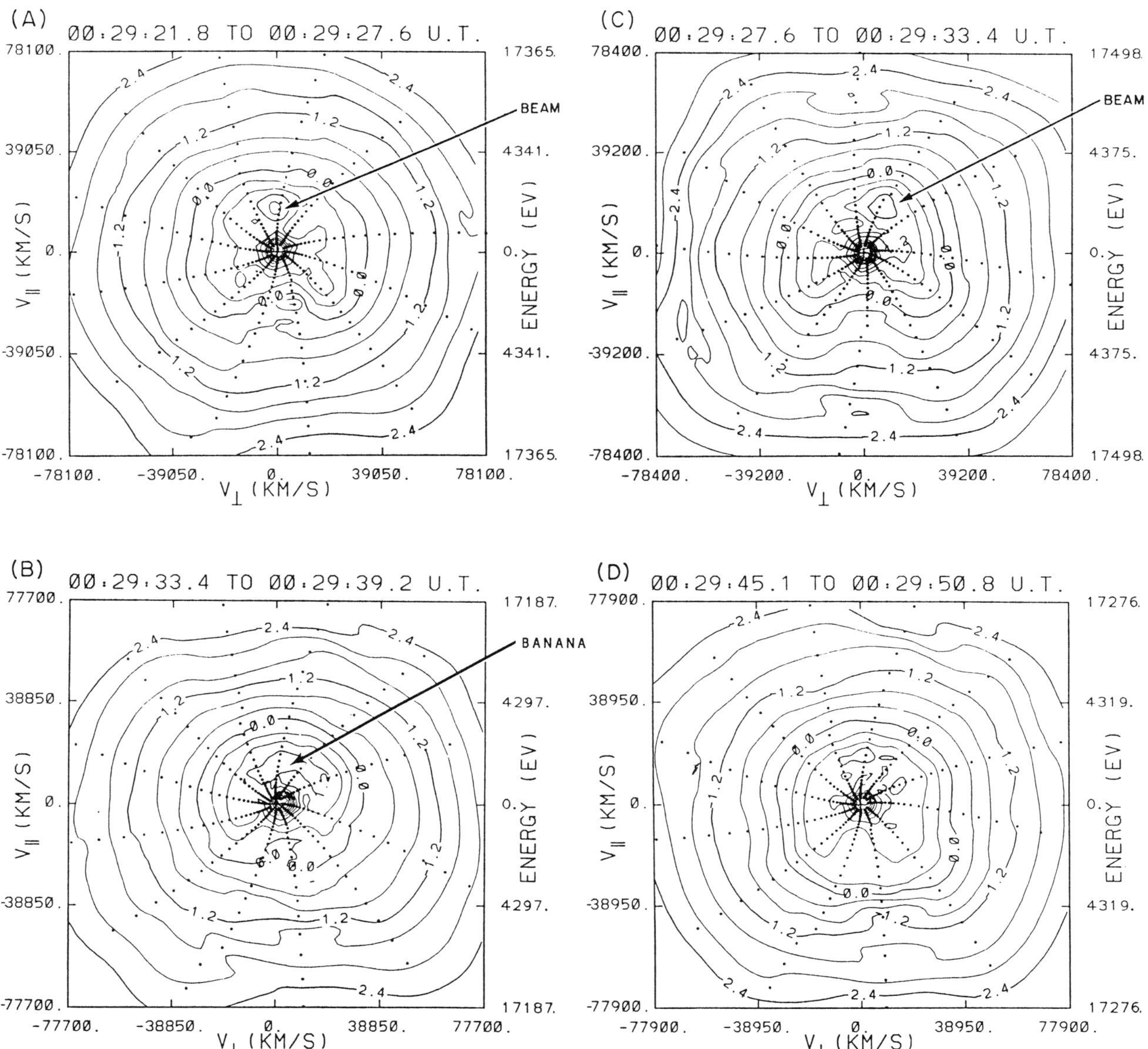

Fig. 6. Same as Figures 4 and 5, but now for region III (0029:20 < t < 0030). A banana distribution appears in 6c, and an electron conic is observed in 6d.

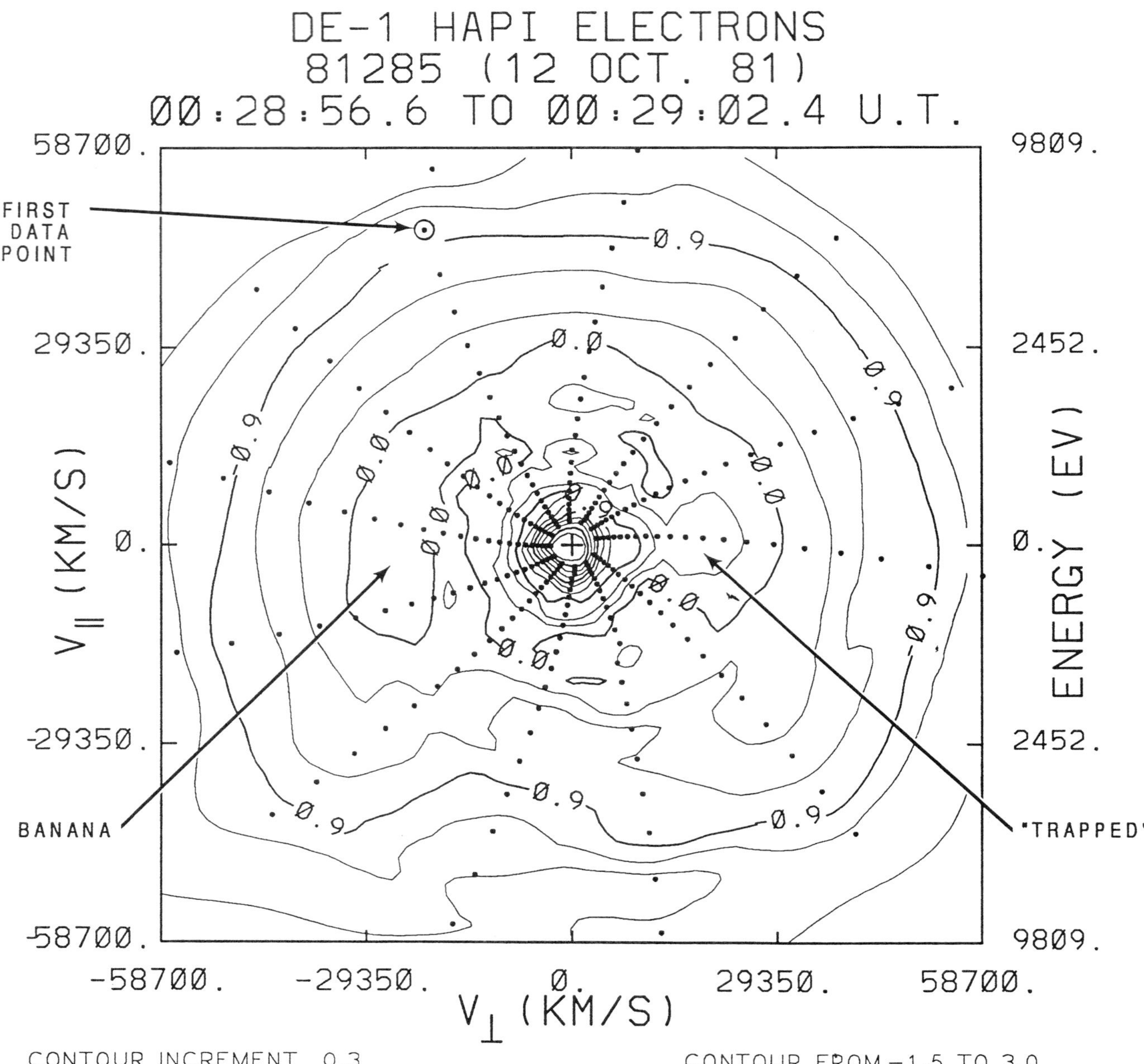

Fig. 7. A high-resolution plot of a time almost overlapping that of Figure 5b. The banana distribution has apparently been extended in pitch angle due to the conservation of the first adiabatic invariant. Perhaps local heating due to wave-particle interactions has produced the "trapped" distribution seen (see text).

> 0 and are thus unstable to both the bump-in-tail and the cyclotron maser instability. In Figure 8 we reproduce contours of the distribution function as calculated from 2-D simulations [cf. Winglee and Pritchett, 1986, Figure 14]. Note that diffusion of the distribution function resulting from the bump-in-tail instability as well as the cyclotron maser instability drive the distribution to a more isotropic state. The similarity of some of the distributions shown in Figure 8 to those of Figures 5 and 6 is apparent. We must conclude from this comparison that it is probable that the observed distributions are the result of wave-particle interactions.

Summary and Conclusions

We have presented examples of electron distributions as observed by the high altitude plasma instrument as the DE 1

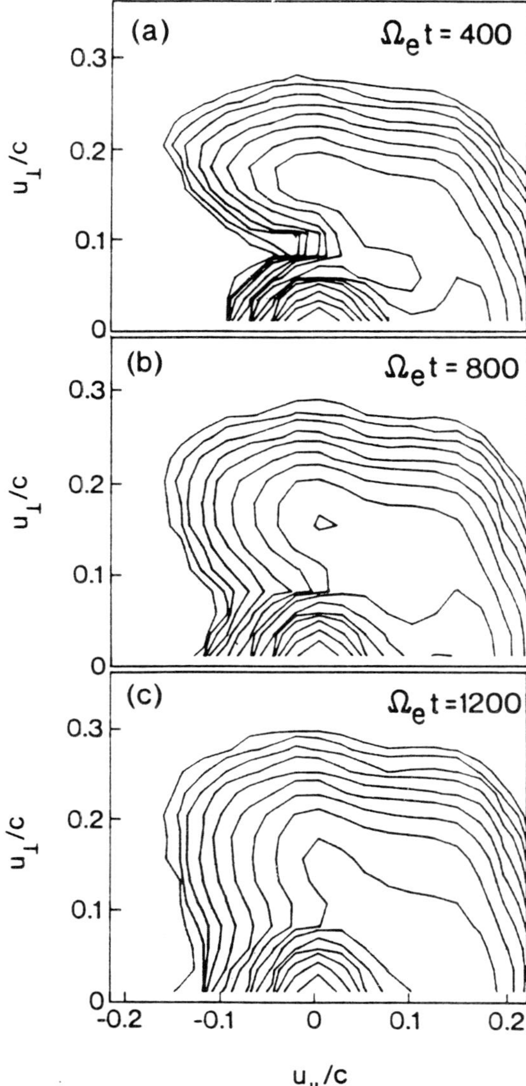

Fig. 8. A reproduction of Figure 14 of Winglee and Pritchett [1986]. This series of contours of the distribution function are the results of numerical simulation of an initial banana distribution. The isotropization is due to wave-particle interactions and diffusion of the electrons that have been heated both perpendicular and parallel to the magnetic field direction. These figures resemble many of the contours of Figures 4, 5, and 6.

satellite passed near an AKR source region. For this example, loss cones (some semifilled) were observed when the satellite was closest to the AKR source region, and temperature anisotropies in the distribution and electron conics were observed sometimes within and sometimes adjacent to the source region. These data are consistent with the hypothesis that perpendicular heating due to wave-particle interactions occurs near the AKR source region, thus modifying the "source" distribution.

The observations indicate that it is difficult to determine whether trapped particles are the source of AKR emission as has been suggested by Louarn et al. [1990] or the result of wave/particle interactions. Our results are in agreement with Ungstrup et al. [1990], who find the electron distributions are considerably modified by the waves near the AKR source region. The results of Winglee and Pritchett [1986] indicate that "banana" distributions commonly observed in acceleration regions of the nightside auroral region are unstable to AKR emission, which in turn heats the electron distribution and results in an isotropization of the electron population, which could lead to distributions that resemble trapped distributions (cf. Figure 7). To resolve the outstanding problems, it will be necessary to obtain multiple and simultaneous measurements both outside and inside the AKR source region by both particle and wave instruments. More detailed modelling including particle-in-cell simulations which incorporate the structure of the source region as described above is also needed to help identify sources and sinks for the observed waves.

Acknowledgements. We gratefully acknowledge many useful discussions of these observations with Kit Wong. We wish to thank Tricia Becker for clerical assistance and Tony Sawka for drafting some of the figures. This research was supported at SwRI by NASA grants NAS5-28711 (JDM and JLB) and NAG5-1552, NAG5-2102, and NAGW-3143 (JDM).

REFERENCES

Benson, R. F., Auroral kilometric radiation: Wave modes, harmonics, and source region electron density structures, *J. Geophys. Res.*, *90*, 2753, 1985.

Benson, R. F., and W. Calvert, ISIS I observations at the source of auroral kilometric radiation, *Geophys. Res. Lett.*, *6*, 479, 1979.

Burch, J. L., J. D. Winningham, V. A. Blevins, N. Eaker, W. C. Gibson, and R. A. Hoffman, High-altitude plasma instrument for Dynamics Explorer-A, *Space Sci. Instrum.*, *5*, 455, 1981.

Dusenbury, P. B., and L. R. Lyons, General concepts on the generation of auroral kilometric radiation, *J. Geophys. Res.*, *87*, 7467, 1982.

Green, J. L., D. A. Gurnett, and R. A. Hoffman, A correlation between auroral kilometric radiation and inverted-V electron precipitation, *J. Geophys. Res.*, *84*, 5216, 1979.

Gurnett, D. A., The Earth as a radio source: Terrestrial kilometric radiation, *J. Geophys. Res.*, *79*, 4227, 1974.

Hilgers, A., The auroral radiating plasma cavities, *Geophys. Res. Lett.*, *19*, 237, 1992.

LeQueau, D., R. Pellat, and A. Roux, *Phys. Fluids*, *27*, 247, 1984.

Lin, C. S., J. L. Burch, C. Gurgiolo, and C. S. Wu, DE-1 observations of hole electron distribution functions and the cyclotron maser resonance, *Annales Geophys.*, *4*, 33, 1986.

Louarn, P., A. Roux, H. de Feraudy, D. Le Queau, M. Andre, and L. Matson, Trapped electrons as a free energy source for the auroral kilometric radiation, *J. Geophys. Res.*, *95*, 5983, 1990.

Menietti, J. D., C. S. Lin, H. K. Wong, A. Bahnsen, and D. A. Gurnett, Association of electron conical distributions with upper hybrid waves, *J. Geophys. Res.*, *97*, 1353, 1992.

Newman, D., R. M. Winglee, and M. V. Goldman, Theory and simulation of electromagnetic beam modes and whistlers, *Phys. Fluids*, *31*, 1515, 1988.

Omidi, N., and D. A. Gurnett, Growth rate calculations of auroral kilometric radiation using the relativistic resonance condition, *J. Geophys. Res.*, *87*, 2377, 1982.

Omidi, N. A., C. S. Wu, and D. A. Gurnett, Generation of auroral kilometric and Z-mode radiation by the cyclotron maser mechanism, *J. Geophys. Res.*, *89*, 833, 1984.

Sharp, R. D., R. G. Johnson, and E. G. Shelley, Energetic particle measurements from within ionospheric structures responsible for auroral acceleration processes, *J. Geophys. Res.*, *84*, 480, 1979.

Shawhan, S. D., D. A. Gurnett, D. L. Odem, R. A. Hoffman, and C. G. Park, The plasma wave and quasi-static electric electric field experiment (PWI) for Dynamics Explorer-A, *Space Sci. Instrum.*, *5*, 535, 1981.

Sonwalkar, V. S., R. A. Helliwell, and U. S. Inan, Wideband VLF electromagnetic bursts observed on the DE 1 satellite, *Geophys. Res. Lett.*, *17*, 1861, 1990.

Temerin, M. A., and D. Cravens, Production of electron conics by stochastic acceleration parallel to the magnetic field, *J. Geophys. Res.*, *95*, 4285, 1990.

Ungstrup, E., A. Bahnsen, H. K. Wong, M. Andre, and L. Matson, Energy source and generation mechanism for auroral kilometric radiation, *J. Geophys. Res.*, *95*, 5973, 1990.

Winglee, R. M., and P. L. Pritchett, The generation of low-frequency electrostatic waves in association with auroral kilometric radiation, *J. Geophys. Res.*, *91*, 13,531, 1986.

Winglee, R. M., J. D. Menietti, and H. K. Wong, Numerical simulations of bursty radio emissions from planetary magnetospheres, *J. Geophys. Res.*, *97*, 17,131, 1992.

Wong, H. K., and M. L. Goldstein, A mechanism for bursty radio emission in planetary magnetospheres, *Geophys. Res. Lett.*, *17*, 2229, 1990.

Wong, H. K., J. D. Menietti, C. S. Lin, and J. L. Burch, Generation of electron conical distributions by upper hybrid waves in the Earth's polar region, *J. Geophys. Res.*, *93*, 10,025, 1988.

Wu, C. S., Kinetic cyclotron and synchrotron maser instabilities: Radio emission processes by direct amplification of radiation, *Space Sci. Rev.*, *41*, 215, 1985.

Wu, C. S., and L. C. Lee, A theory of the terrestrial kilometric radiation, *Astrophys. J.*, *230*, 621, 1979.

Wu, C. S., H. K. Wong, D. J. Gorney, and L.C. Lee, Generation of the auroral kilometric radiation, *J. Geophys. Res.*, *87*, 4476, 1982.

J.D. Menietti and J.L. Burch, Department of Space Sciences, 6220 Culebra Rd., POD 28510, Southwest Research Institute, San Antonio, TX 78228-0510.

Acceleration and Radiation From Auroral Cavitons

R. POTTELETTE

Centre de Recherche en Physique de l'Environnement, Saint-Maur-des-Fossés, France.

R.A. TREUMANN

Max-Planck-Institut für Extraterrestrische Physik, Garching bei München, Germany

G. HOLMGREN

Swedish Institute of Space Physics, Uppsala, Sweden

N. DUBOULOZ AND M. MALINGRE

Centre de Recherche en Physique de l'Environnement, Saint-Maur-des-Fossés, France

Emphasis is put on the possible role played by lower hybrid cavitons on the structuring of the auroral plasma on a fine scale of the order of one kilometer and below. Intense lower hybrid waves, evolving into cavitons, have been observed on the Viking satellite in close correlation with steep thermal density gradients and with strong Auroral Kilometric Radiation bursts. The cavitons are shown to accelerate background electrons producing an energetic tail on the distribution function. At the same time, the cavitons act as sources for the kilometer wave radiation when interacting with electron beam excited upper hybrid waves.

INTRODUCTION

One of the main challenges for the auroral space plasma community consists in the requirement to understand interacting global and microscopic nonlinear processes on a hierarchy of scales. Due to the high sampling and telemetry rates implemented on the Swedish Viking satellite, it has become possible to determine the role of transient nonlinear structures - possibly associated with large amplitude waves - in the basic microscopic processes taking place in the magnetospheric auroral regions. Recently, great progress has been achieved in probing the nonlinear turbulent plasma processes which accelerate energetic particles and generate the Auroral Kilometric Radiation (AKR).

It is now widely acknowledged that the generation of AKR is directly connected with the acceleration processes of charged particles along the auroral magnetic field lines. Consequently, an unified theory applying to these closely related physical processes should reflect the basic properties of both the observed radiation and acceleration mechanisms. In this paper, we are going to put some emphasis on the possible role played by nonlinear coherent plasma structures with regards to these processes. Strong localized high frequency electrostatic fields - associated with theses nonlinear structures - may act as powerful particles accelerators that generate suprathermal tails. In addition, they produce a modification of the density profile via their associated ponderomotive forces. The scenario leading to the formation and to the related effects of the nonlinear structures is the following:

When the electrostatic waves involved have large growth rates, they can readily enter into the nonlinear regime, and under certain circumstances can evolve into solitons or cavitons.

The electrostatic waves become localized and the interaction is restricted to small volumes in space. Under these circumstances, the wave energy of a large amplitude wave is confined to the small volume of the caviton yielding intense interaction.

The soliton or envelope caviton may collapse thereby enhancing this effect due to increase of the energy density. In restricting the acceleration to the caviton volume, the caviton starts acting as an antenna.

Electromagnetic emission from caviton should yield narrow band radiation because of its finite spatial extension along magnetic field lines. Drifting tones of the radiation are caused by the upward and downward motion of the caviton along field

Auroral Plasma Dynamics
Geophysical Monograph 80

lines, while discrete narrow-band short bursts reflect the various temporal stages of evolution of the caviton: formation, saturation, and collapse.

Let us apply this scenario to the experimental results obtained in probing the auroral region. One fundamental property of AKR is that it does not appear as a continuous broadband emission. The radiation consists of many discrete narrowband emissions with typical bandwidth of ~1 kHz; most often the center frequency of these discrete emissions sweeps either upward or downward across the spectrum [Gurnett and Anderson, 1981]. Moreover, the drifting features are not continuous but consist of short bursts occuring on ~1 s time scale. These observations suggest the following properties for the acceleration mechanisms:

Since the 1 kHz bandwidth corresponds to a vertical source thickness of about 20 km in the AKR source region [Pottelette et al., 1992], significant acceleration of charged particles should occur on this spatial scale.

The variability of the radiation (on a 1 s time scale) should reflect the equivalent variability of the acceleration mechanisms. In this respect, it should be mentioned that 1 s is the typical time scale associated with the energization of both ions and electrons in the auroral magnetosphere [Hultqvist et al., 1988].

In this paper we are going to present high time resolution data acquired by the Viking satellite which enligthen the fact that AKR bursts can be produced in the presence of nonlinear lower hybrid waves which may lead to the formation of cavitons and subsequent collapse. In the following, emphasis will be put on the filamentation of the auroral plasma by lower hybrid waves, and on the sensitive role played by lower hybrid cavitons in the acceleration and radiation mechanisms.

OBSERVATIONS

The Figure 1 shows a typical spectrogram measured in the AKR source region by the step frequency analyser on Viking.

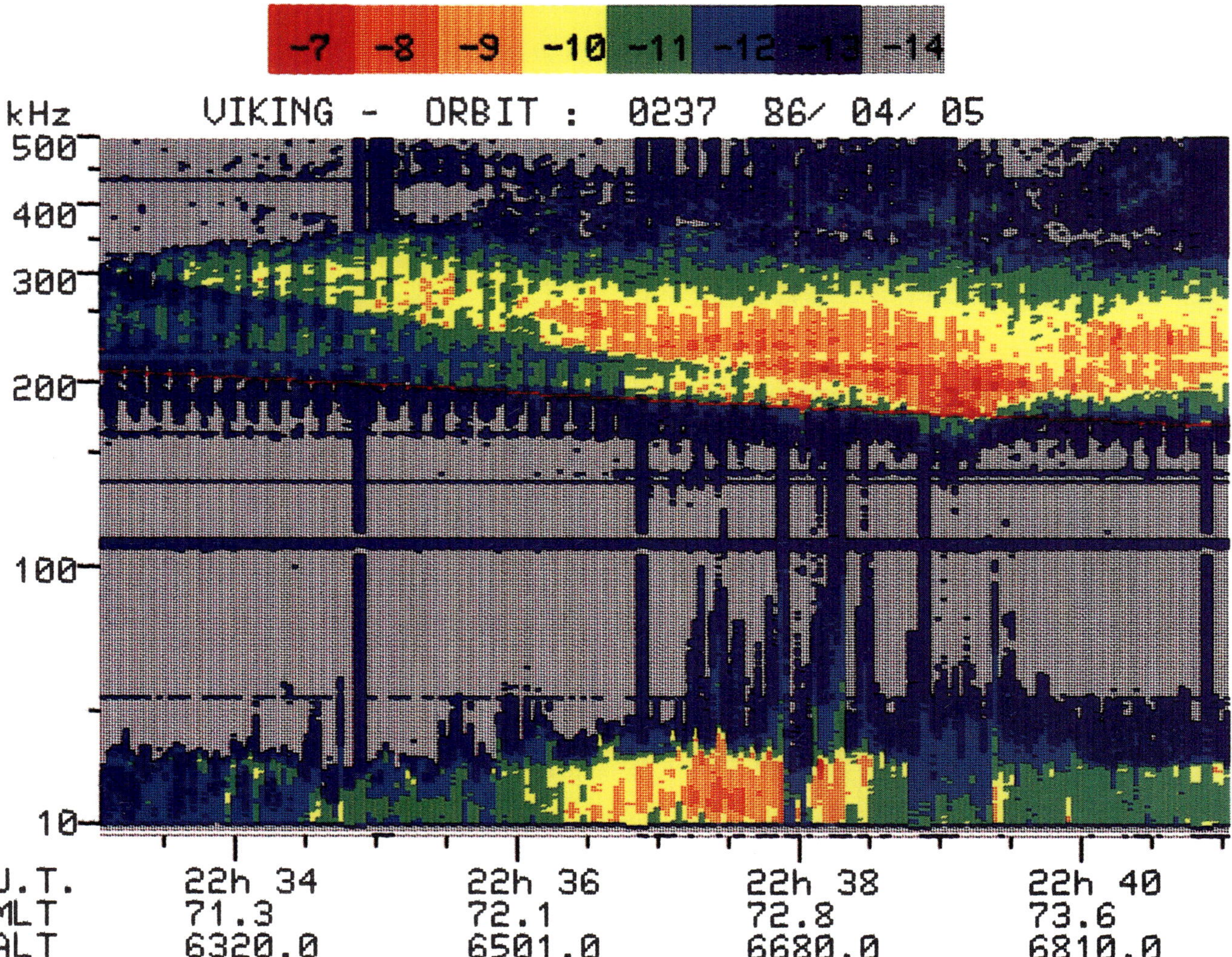

Fig. 1. Frequency-time spectrogram of the high-frequency electric field data during AKR source crossings. The electron gyrofrequency f_{ce} is plotted as the solid red line. The frequency scale inside each frequency interval 10-128, 128-256, 256-512 kHz is linear. Horizontal lines are interferences due to the on-board converter. The vertical lines appearing every 2 min are caused by an active wave experiment.

Only the electric component is displayed and the full frequency sweep between 9 and 500 kHz is performed within 2.4 s. The red line is the local electron gyrofrequency f_{ce} and traces above f_{ce} belong to electromagnetic emissions of AKR. The crossing of the AKR generation region is identified from the spectrogram as the position along the spacecraft trajectory where the strongest AKR signals are generated just above f_{ce}. In Figure 1 this occurs two times: a narrow localized source takes place around 22:38 UT, while a much more extended source is crossed around 22:39 UT. At this time the Viking satellite was near 2010 MLT at an altitude of 6650 km , and at an invariant latitude of 72.8°. The local electron gyrofrequency was $f_{ce} \sim 170$ kHz. In this paper we characteristize wave acceleration of particles along auroral field lines by studying the localized source of electromagnetic radiation. In order to identify the basic properties of this acceleration, we concentrate on high time resolution measurements.

The coulor spectrogram displayed in Figure 2 illustrates the different wave modes recorded during the first source crossing at about 22:38 UT. The three panels of this spectrogram show the variation of the high-frequency fluctuations as measured by the eight separate electric filters of the high frequency V4-H experiment with center frequencies spaced one octave from 4 kHz up to 512 kHz (top panel, frequency axis in log scale), the wave spectrum in the 5 to 428-Hz (middle panel), and the density variation as deduced from the Langmuir probe experiment (bottom panel), respectively. The time resolution was 37.5 ms. The most spectacular event takes place between 2237:50 and 2237:55 UT when very intense and sporadic electrostatic hiss signals are detected in the 3-10 kHz range in close connection with sporadic AKR emissions in the 128-kHz and 256-kHz channels, and with very steep density gradients (lower panel). The solid dark line in the top panel of Figure 2 marks the variation of the electron plasma frequency f_{pe}, corresponding to the variations in electron density, and indicates that the strongest hiss emissions (with amplitudes reaching 100 mV/m) are recorded for waves propagating at large angles with respect to the magnetic field. These waves are fairly broadband. Resonant lower hybrid waves are restricted to the lower part of the spectrum while the higher frequency part overlaps with the whistler mode hiss band. It is well known that these modes are closely related because lower hybrid waves for oblique propagation smoothly connect to the whistler mode thereby attaining an electromagnetic component.

The 256-kHz channel (top panel of Figure 2) detected mainly radiation propagating perpendicular to the magnetic field from the second AKR source located on the field lines crossed by Viking around 22:39 UT (compare Figure 1). The perpendicular polarisation of the radiation is obvious from the spin modulation of the signal yielding the intensity maximum at 90 degrees of the angle between the y-boom and the magnetic field. Consequently, the main signature of the local AKR source appears in the 128-kHz channel, which is close to f_{ce}.

The event we are interested in takes place at 22:37:50 UT. It consists itself of two sub-events at 2237:50 - 37:53.5 UT, and at 2237:54 - 37:55.5 UT. The AKR source region is crossed during the second sub-event when strong electromagnetic bursts are detected at 128 kHz. At this time, the density drops to its lowest value of about 2 cm^{-3}. Correspondingly the lower hybrid frequency falls below the range of this figure. Intense

lower hybrid electrostatic noise (reaching amplitudes of 100 mV/m) has been detected during this case in the filter banks of the low frequency V4-L experiment and has been found similar to the data given in Pottelette et al. (1992). Due to the low density the wave to thermal energy ratio maximizes at this time so that one can expect nonlinear processes to take place here which in itself is probably the reason for the appearance of the steep drop in densitiy. Since such steep density drops are known to be related to electrostatic shocks, it is concluded that in the source region the lower hybrid waves and AKR appear right in the electrostatic shock transition. One should also note the ~ 1 s time modulation of the AKR bursts around 2237:55 UT which indicate that the source itself is not stationary. On the other hand the first sub-event is less violent in the radiation. Weak sporadic intensification of AKR is detected during this event in the 128 kHz-channel but the main contribution is in the 256 kHz-channel and therefore comes from the second remote source. At the same time the density decrease is less violent while the low frequency noise is broader band and less concentrated at the lower hybrid frequency. It nevertheless includes the lower hybrid and hiss bands, maximizing at the former frequency, but the wave to thermal energy ratio is less than in the second sub-event which prevents strongly nonlinear processes and may be responsible for the absence of strong locally excited radiation in this sub-event. Presence of hiss during this time is confirmed by the detection of a magnetic component in the magnetic filter bank data (not shown). As already pointed out by Mozer et al. (1980), a distinction between lower hybrid waves and whistler mode waves (hiss) must be made in the auroral region. Indeed waves at the lower hybrid resonance frequently appear together with hiss emission, however the former waves constitute distinct emission with little accompanying higher frequency hiss [see bottom portion of Figure 8 of Mozer et al. (1980) where S3-3 data, similar to those presented here, are displayed].

The central panel of Figure 2 exhibits waveform data below 400 Hz. The black solid line indicates the local singly charged hydrogen cyclotron frequency f_{cp}. Intense ion cyclotron waves are recorded after the radiation event. Before this time, very low wave activity is observed below the lower hybrid frequency.

Particle data for the same time period obtained by the electron (ESP1, ESP5) and ion (PISP) spectrometers are shown in Figure 3. ESP1 and ESP5 look always into opposite directions. ESP5 is a high time resolution magnetic detector for electrons with wide energy bandwidth $\Delta E/E \sim 70\%$. It shows the presence of downgoing 5-8 keV electron beams at 2237:53 UT (top panel). Two bursts separated by ~ 1s occur. Note that the second event is detected $\sim 20°$ off the magnetic field direction. ESP1 having high energy ($\Delta E/E \sim 5\%$) but low time resolution observed an upgoing electron component in the 8.17 keV channel (central panel). The presence of these counterstreaming energetic electron conics may be taken as evidence for stochastic electron acceleration by lower hybrid waves. Electron conics are also known to interact with upper hybrid waves [Wong et al., 1988; Menietti et al., 1992]. Around 2237:55 UT PISP shows the presence of ion conics with energy extending up to ~ 1 keV; the cone angle is about 55° (lower panel). These particles observations are similar to those previously discussed by Pottelette et al. (1988) for the same type of events.

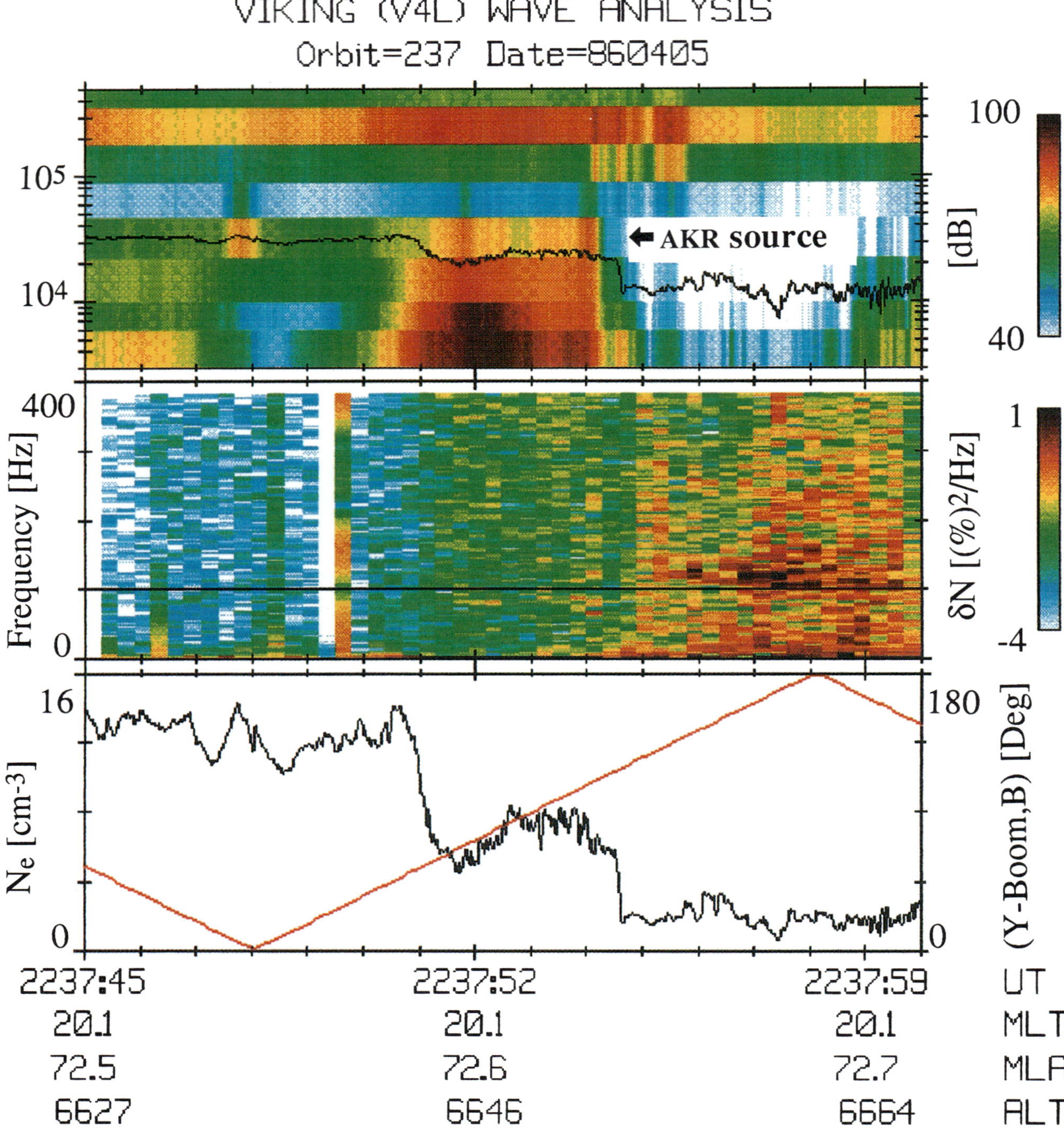

Fig. 2. Coulor spectrogram of the high and low frequency waves measured during the localized AKR source crossing. *Upper panel*: Electric filterbank data of the high frequency fluctuations. The variation of the electron plasma frequency f_{pe} is plotted as the solid dark line. The color coding is such that 100 dB corresponds to 100 mV/m *Central panel*: Low frequency spectrogram performed in the $\delta n/n$ mode (5-428 Hz). The dark lines exibits the value of the hydrogen gyrofrequency f_{cp}. *Bottom panel*: Density variation measures by the Langmuir probe. The orientation of the electric antenna measuring the waves is indicated by the solid red line.

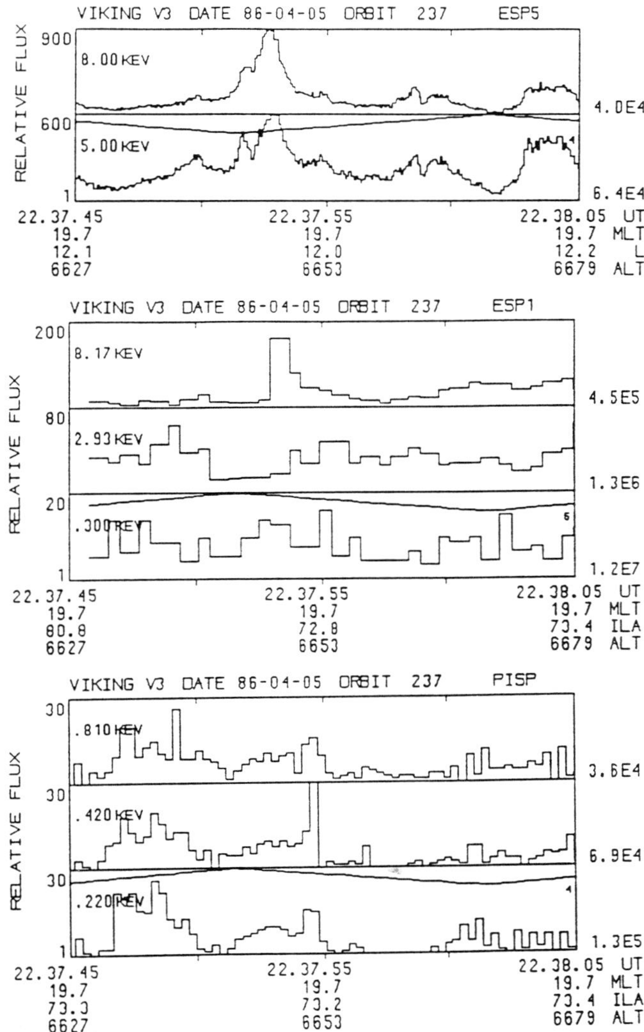

Fig. 3. Particle data obtained by the electron (ESP5,ESP1) and ion (PISP) spectrometers. For each spectrometer the pitch angle is shown in graphical form at the bottom of the panels. For each panel, counts accumulated at different energy levels are displayed versus time.

Unfortunately the time to construct a pitch angle distribution function is limited to the spin rate and requires about 10 s of observational time which is long in comparison with the duration of the wave events. Usually electrostatic ion cyclotron waves are found in association with upwardly accelerated ion beams, which could not be detected in the present case because the ion spectrometer was not looking in the right direction. Two electron energy distributions for different pitch angles (Figure 4) and 0.6 s time resolution taken at different times during the event (2237:51-2238:01 UT) show the flux increases in the downgoing beam case at 7° pitch angle at 2238:01 UT and 10 s earlier in the conic case at 156° pitch angle. In the former case the beam energy is lower and the flux enhancement is restricted to a narrow energy range.

The DC electric field measurements between 2237:50 and 22:38 UT are plotted in Figure 5. A well defined and highly structured electrostatic shock of the so-called S type [Mozer et al., 1980], with amplitude of 400 mV/m is measured between 2237:54 and 2237:56 UT, in the transition region between lower hybrid and ion cyclotron waves. Note however that the latter waves are detected somewhat later than the shock and are not correlated with AKR bursts. Consequently, lower hybrid waves seem more efficient for structuring the plasma.

From these experimental data it results that the type of waves that is measured by Viking depends on whether the measurement is made above or below the acceleration region. Above the region of parallel electric field the dominant mode appears to be the electrostatic hydrogen cyclotron wave (plus some incoherent noise below f_{cp}). Below the parallel electric field region the dominant waves are in the auroral hiss mode and occur in association with anisotropic energetic electron beams. Within the parallel field region where the electrons are accelerated, AKR bursts and intense electrostatic lower hybrid waves are observed in close association with electrostatic shocks. The relation between electrostatic shocks and lower hybrid waves has first been mentioned from S3-3 data by Mozer et al. (1980) and Temerin (1981). The new results obtained by Viking deal with the additional presence of AKR bursts and small scale density depressions. This is schematically illustrated in Figure 6.

The correlation between AKR bursts and shock structure implies that most of the particle acceleration takes place inside or close to this structure. The strong localization of the structure argues for its spatial character. As a matter of fact, taking into account the spacecraft velocity, the ~1 s duration of the shock corresponds to a transverse spatial extension of the order of ~5/6 km. It cannot be excluded that the entire growth structure is related to a kinetic Alfvén-wave which propagates in from the outer magnetosphere (for instance from the plasma sheet boundary layer) parallel to the magnetic field and has a slow perpendicular component of its velocity. For a wave frequency of ~1 Hz, and the local Alfvén velocity (which is approximately c/3) the local parallel wavelength of the Alfvén wave amounts to roughly 60,000 km, putting the outer magnetospheric source region at an approximate one parallel wavelength distance of 9-10 R_E. The perpendicular wavelength is c/f_{pe} ~30 km, corresponding to a transverse phase velocity of 30 km/s. In the spacecraft frame this corresponds to transition time of roughly 1 s for one wavelength. This requires very narrow transverse structuring of the Alfvén mode and contradicts the observed even much narrower stratifications (see Figure 5) that can be produced by the action of lower hybrid waves. The simultaneous presence of intense lower hybrid waves reaching their nonlinear level may therefore also play an important role in the generation of the shock. The shock could result from space charge fields resulting from the acceleration mechanisms involved with these waves as observed in laboratory experiments [Gekelman and Stenzel, 1975]. One can argue that kinetic Alfvén waves provide the coupling between the plasma sheet boundary layer and the auroral magnetosphere feeding energy into it on a large transverse scale corresponding to the entire source region of the auroral processes (particle acceleration and AKR), while the fine structure and the local processes are dominated by lower hybrid waves interaction. In the following sections we

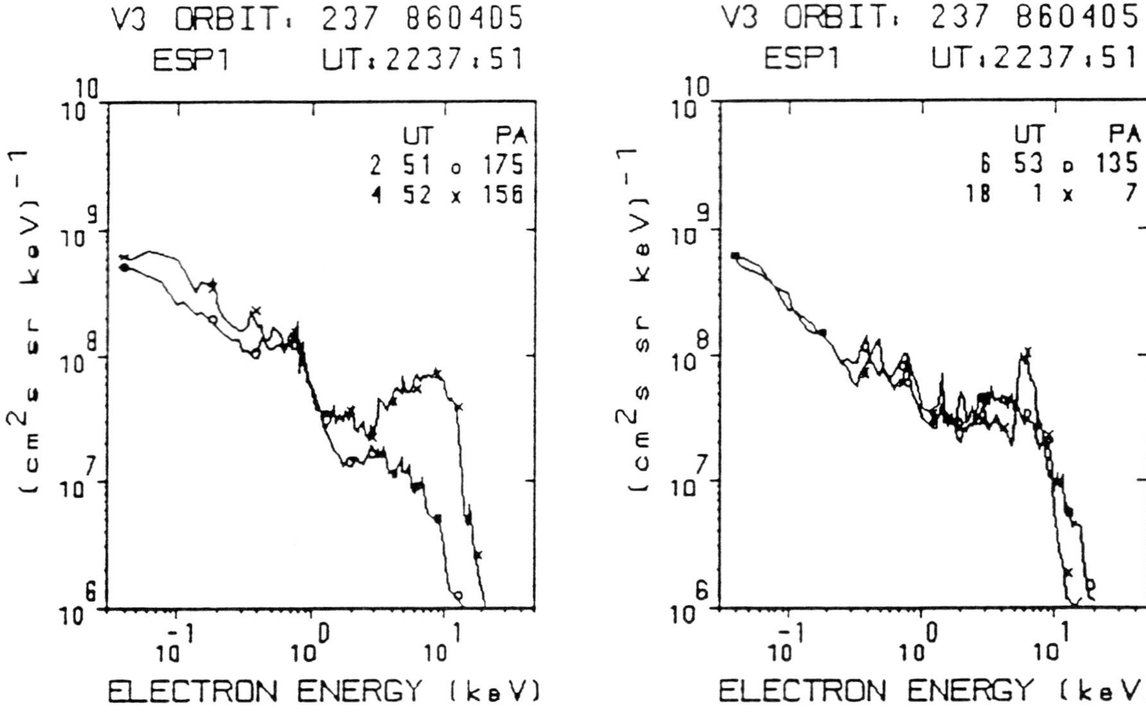

Fig. 4. Electron spectra showing that the peak for electron conics (pitch angle = 156°, taken at 2237:53 UT) is broader and more energetic than peak for downgoing electrons (pitch angle = 7°, taken at 2238:01 UT).

study the role of lower hybrid cavitons in the elaboration of basic physical processes taking place in the auroral magnetosphere: Acceleration and Radiation.

The ion cyclotron waves observed are found outside the region of steep density gradients and therefore appear less efficient for structuring the plasma than the lower hybrid turbulence. From a theoretical point of view this is due to the fact that their dispersion depends on the plasma density only to second order.

FILAMENTATION of the PLASMA by LOWER HYBRID WAVES

Since electrostatic emissions near the lower hybrid frequency are undoubtedly very large (≥ 100 mVm^{-1}), they may

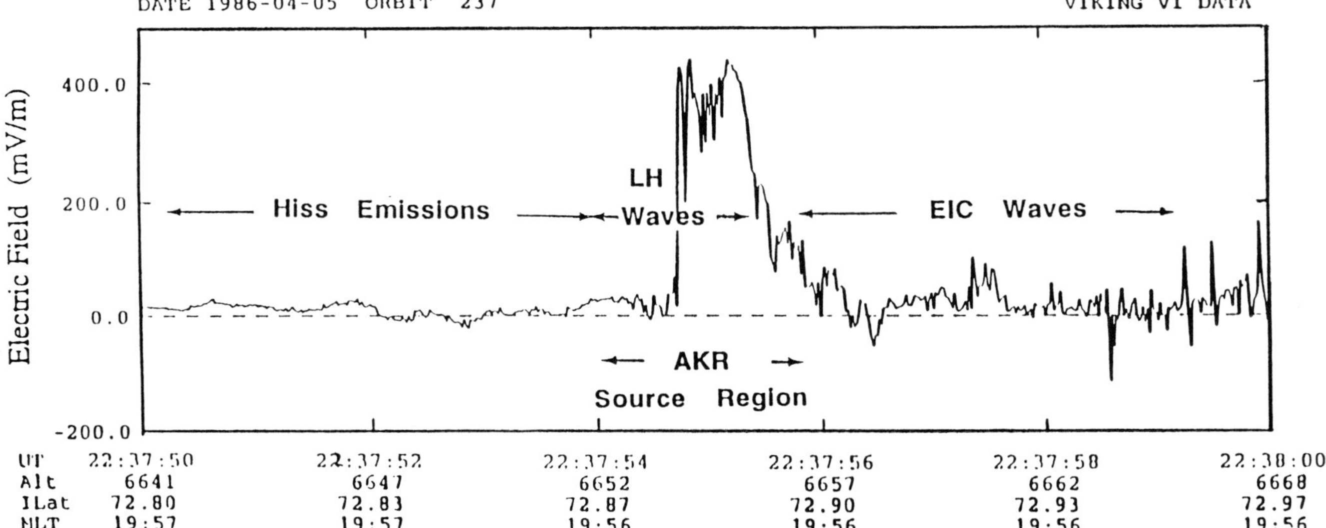

Fig. 5. DC electric field measurements performed during the localized AKR source crossing.

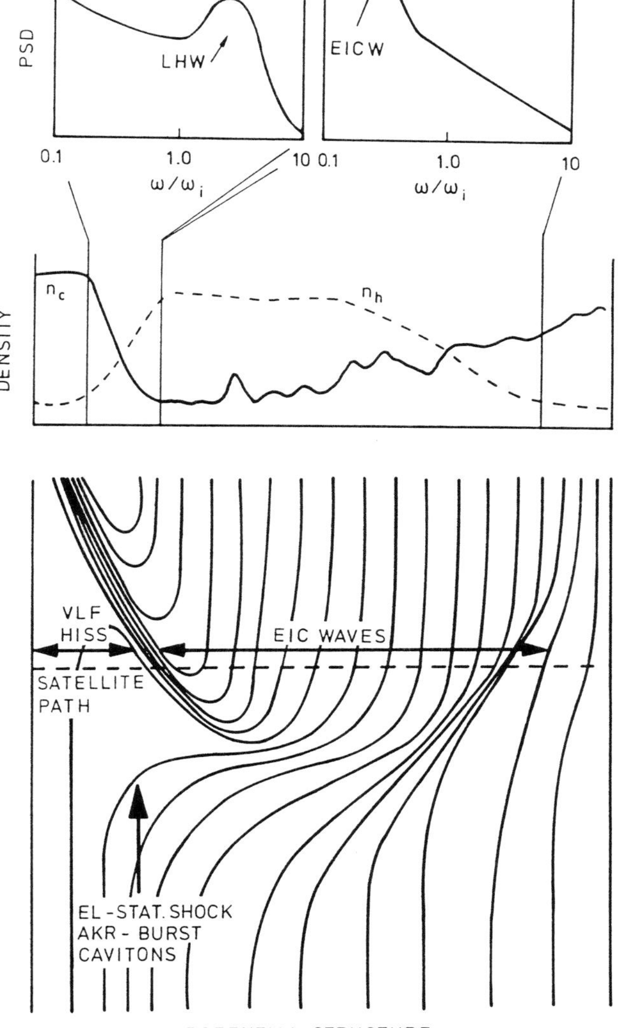

Fig. 6. Schematic of the characteristic waves emissions recorded by a satellite during crossing of an elementary cell of acceleration along the auroral field lines. Lower hybrid waves (LHW) are detected in association with steep cold plasma density (n_c) gradients, while lectrostatic ion cyclotron waves (EICW) occur in the hot plasma density (n_h) region. AKR bursts and shocks are recorded in the transition region between hot and cold plasmas.

significantly modify the local plasma density. Indeed, an electric field strength of 100 mV/m corresponds to an energy density of $4.5 \ 10^{-14}$ J m^{-3}. In an auroral plasma with an electron density $n = 2 \ 10^6$ m^{-3}, and an electron temperature $T_e = 8$ eV, this corresponds to an energy density ratio $W/nT_e = 3 \ 10^{-2}$ which conventionally is considered as large and in the non-linear regime. Under the conditions of high altitude auroral plasma, where the electron cyclotron frequency $f_{ce} >$

f_{pe} exceeds the plasma frequency, the dispersion relation of lower hybrid waves becomes

$$f_k^2 = f_{lh}^2\left(1 + \frac{m_i}{m_e}\frac{k_{//}^2}{k^2} + 3k^2\lambda_i^2\right) \qquad (1)$$

Here m_e and m_i are the respective electron and ion masses, and $f_{lh} = \left[f_{cp}^2 + f_i^2/\left(1 + f_{pe}^2/f_{ce}^2\right)\right]^{1/2}$ is the lower hybrid resonance frequency which can be expressed in terms of the ion plasma frequency f_i and of the ion gyrofrequency f_{cp}. The component of the wavenumber k in the magnetic field direction is $k_{//}$, while λ_i is the ion Debye length. When in addition $f_i > f_{cp}$, one obtains $f_{lh} \sim f_i$. The lower hybrid frequency is thus very sensitive to density variations, and the dispersive properties of the waves can be altered by a nonlinear response which causes a local perturbation in the plasma frequency. Let $\delta n \ll n$ be a small initial variation of the background plasma density, causing a variation $(\delta n/2n) f_{lh}$ in the lower hybrid frequency. The linear dispersion relation is replaced by the nonlinear dispersion relation

$$f_k^2 = f_{lh}^2(1 + \delta n/n)\left(1 + \frac{m_i}{m_e}\frac{k_{//}^2}{k^2}\right) + \frac{3k^2v_i^2}{4\pi^2} \qquad (2)$$

with v_i the ion thermal velocity. It is the result of the modulation of the density due to nonresonant interaction of the lower hybrid wave with the background plasma. Since the frequency of the lower hybrid plasmon is conserved during slow modulation, the decrease in plasma density ($\delta n < 0$) caused by the lower hybrid wave pressure must be compensated by an increase of k [Pottelette et al., 1992]. During this process the wavelength of the waves trapped in the density dips decreases again. The caviton starts deepening (amplitude of δn increasing) and shrinking in size, i.e., it collapses, and at the same time transfers the waves back into the short-wavelength domain where they can become damped at the particles. It is clear that for large density depletions the above dispersion relation is not valid anymore. It is basically the final stage of this process which is observed as large density depletion in the measurements. The final stage of the collapse process can be treated only by numerical simulation. It is analogous to self-focussing in nonlinear optics [Melrose, 1986]. Lower hybrid waves refract into the region of increasing refraction indices (i.e., the region of decreasing plasma density), enhancing the electric energy there and providing the feedback which drives the instability. During caviton formation the third term in the above dispersion relation dominates, so that the lower hybrid waves inside the nonlinear coherent structure move with velocity of the order of the ion thermal velocity v_i in the auroral plasma. As the caviton is formed due to the interaction of the trapped waves with ion waves which are mainly excited in the parallel direction, it results that the caviton essentially travels up and down along the auroral magnetic field lines.

Breakdown of the dispersion during collapse transports the wave energy back into the high phase velocity range where resonant collisionless damping yields dissipation of wave energy, limits the further collapse and accelerates particles. It has previously been demonstrated [Pottelette et al., 1992], that the cavitons produced are elongated along the magnetic field forming cigar-shaped cavitons of dimensions $L_{//}/L_\perp \sim (m_i/m_e)^{1/2}$, where $L_{//}$ is the parallel extension of the caviton. An estimate for the transverse scale $L_\perp$ is

$$L_\perp \approx 12 T_e / \left(e|E_{lh}| \right) \qquad (3)$$

where T_e is the electron thermal temperature and e the elementary charge, while $|E_{lh}|$ is the amplitude of the lower hybrid waves. For typical values of the auroral plasma, $|E_{lh}| \approx$ 100 mV/m (parallel wave field ~ 2.5 mV/m) and $T_e \approx 8$ eV, one obtains $L_\perp \sim 1$ km for the transverse size of the caviton.

One caviton is obvious from the data during the crossing of the first elementary AKR event around 22:38 UT. High time resolution density data measured by the two 80 m separated Langmuir probes are plotted in Figure 7. The upper panel shows the density dropping from 6 to 2 cm^{-3}, at 2237: 54.600 UT. At this time a density variation of 25% appears lasting for about 80 ms on both probes, as illustrated in the second panel. Taking into account the transverse satellite velocity of 6 km/s and assuming that lower hybrid cavitons propagate parallel to the magnetic field, the transverse scale of the caviton is 0.5 km. Correspondingly its longitudinal extension $L_{//}$ becomes about 20 km.

Several density modulations of similar amplitudes and durations are observed during the AKR event around 2239:22 UT (Figure 8). The cavitons appear on the low density side of the density gradients where the wave to plasma energy density ratio maximizes. The increase of energetic electron fluxes indicates a causal relationship to the formation of the cavitons.

PARTICLE ACCELERATION

Acceleration is frequently attributed to double layers, electrostatic shocks, or simply to wave fields [Wu et al. 1981; Bryant et al., 1992]. Since double layers are laminar structures not containing large fluctuation fields, the acceleration by double layers is mainly by the excess potential drop. This drop is in most cases small, hence a large number of double layers is required to yield particle energies the order of the observed few keV in the electron component. Electrostatic shocks [Mozer et al., 1980] are believed to contain larger (several 100 mV/m) electric fields and therefore larger potential drops which may yield effective acceleration. At the same time they contain different modes of plasma turbulence : hiss near the lower hybrid frequency is of particular strength, some very low frequency modes appear as broadband waves, and ion cyclotron waves of moderate to large amplitudes are also measured in the neighbourhood of the shock. Particles may be accelerated in those fields by different mechanisms which are either stochastic or due to intrinsic stochasticity implying that the space orbit of the particles becomes distorted when the wave amplitude becomes large. Stochastic acceleration applies to the electron component and acts along the magnetic field. Intrinsic acceleration in these low frequency fields is important for the ions [Karney, 1978; Lysak, 1986]. One therefore distinguishes between electron and ion acceleration.

Electron acceleration

Heating by ponderomotive potentials

A particularly interesting case is the acceleration in envelope solitons which arise due to the action of the

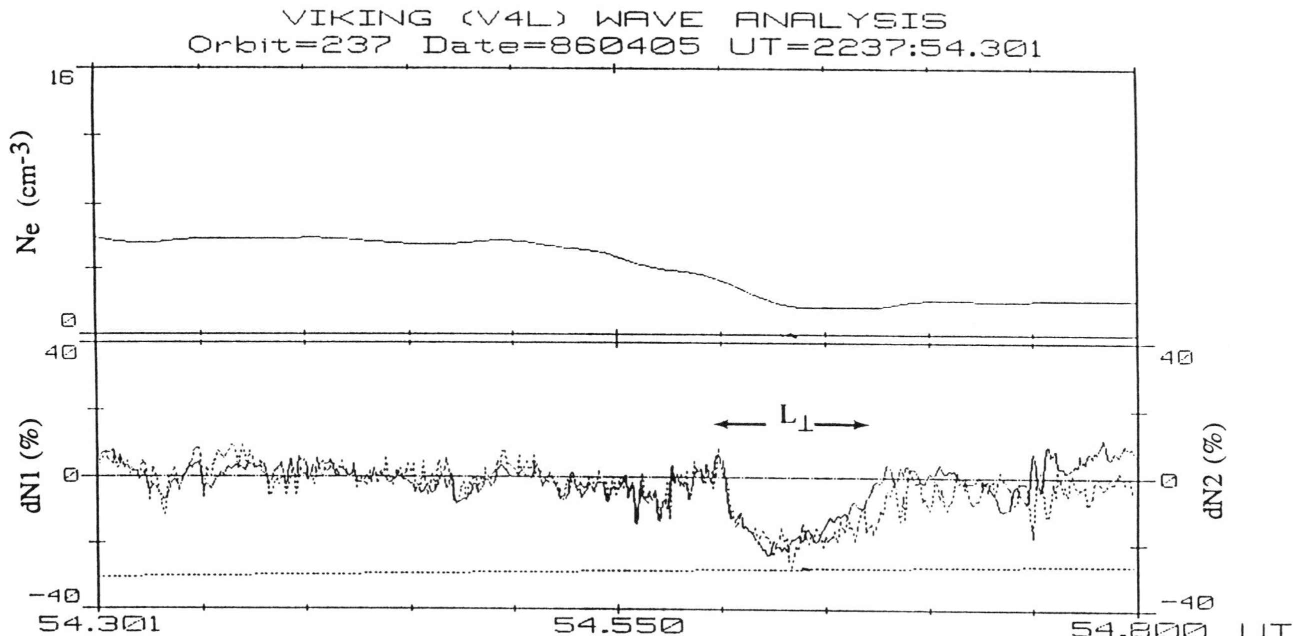

Fig. 7. High time resolution density data measured by the two 80 m separated Langmuir probes during the crossing of the localized AKR source.

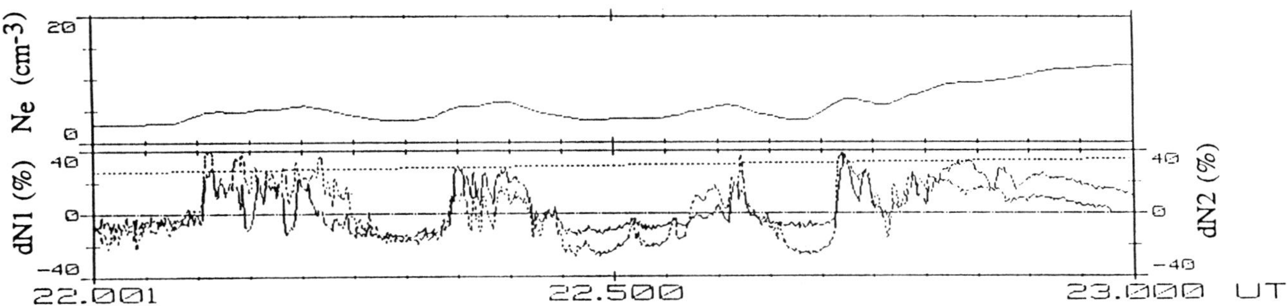

Fig. 8. Same as in Fig. 7. for the extended AKR source crossing.

ponderomotive force on the particles. The ponderomotive force potential can be easily calculated from the nonlinear coupling term in the particle (fluid) equation of motion $f_{pm}= -m\langle v_h \cdot \nabla v_h \rangle = -\nabla \phi_{pm}$ assuming that the high frequency wave field (index h) modulates the plasma on a slower time scale [Morales and Lee, 1975]. When the high-frequency field is identified with the lower-hybrid wave field, the potential becomes for electrons and ions, respectively

$$\phi_\parallel^e \equiv \frac{e|E_\parallel|^2}{4m_e \omega_{LH}^2} \qquad \phi_\perp^i \equiv \frac{e|E_\perp|^2}{4m_i \omega_{LH}^2} \qquad (4)$$

where ω_{LH} is the angular lower hybrid frequency. Electrons are accelerated parallel to the magnetic field B, while ions are affected only perpendicular to B. For lower hybrid waves the ratio of the perpendicular to the parallel electric wave field is equal to the root of the mass ratio. Therefore, even though the parallel electric field is less than the perpendicular by the root of the mass ratio, the above perpendicular and parallel potentials are equal in magnitude, and the ponderomotive electron and ion heatings are comparable. However, the parallel electron force pushes the electrons out of the lower hybrid field maximum region. Since this process produces charge separation, one must introduce an ambipolar electric field $E_{a\parallel} = -\nabla_\parallel \phi_a$ which accelerates the ions parallel to B. In the stationary state, however, this ambipolar potential is given by

$$e\phi_a = -\phi_\parallel^e / \left(1 + T_e / T_i\right) \qquad (5)$$

which for $T_e \gg T_i$ is negligible. In this case the ions stay at rest parallel to B, while for $T_e \sim T_i$ the potential is simply $\phi_a = -\phi_\parallel^e / 2e$.

The acceleration of the electrons can be up to the potential $\phi_\parallel^e$ and corresponds to a broadening of the distribution or heating. Taking $E_\parallel \sim 1mV/m$ the heating amounts to $\Delta T_e \sim 0.6$ eV. One hence requires unreasonably large parallel fields for ponderomotive force heating. In the late stage of lower hybrid collapse when the trapped wave fields intensify and isotropize, the heating by ponderomotive force may be more effective.

Quasi-Stochastic Acceleration

Stochastic electron heating in high-frequency wave fields is usually treated on the basis of quasilinear acceleration theory for the particles in a given wave field. Quasilinear theory does not apply to the evolution of the waves during modulation and collapse, however. The trapping of the wave field during this process favours energy transfer from waves to particles. This has been realized in treatments of pure transit time damping [Robinson, 1989]. The reason is that in a finite size wave packet the accelerated electron (damping the wave) spends shorter time in the wave field than the retarded electron (amplifying the wave) and therefore experiences less phase mixing. Phase mixing diminishes the efficiency of acceleration. Consequently, space limited wave fields (cavitons) should be much more efficient than extend wave fields in accelerating particles. This effect has been demonstrated using the Born approximation for transit time damping [Robinson, 1989]. For simplicity we choose a different, somewhat more relaxed treatment in which "stochastic" interaction is limited to the time the electron spends in the caviton. We use the test particle approximation where the wave field is assumed given and the particles experience quasi-stochastic acceleration when exposed to it. In a preliminary treatment we use a quasilinear (Fokker-Planck) equation for the evolution of the electron distribution function [Davidson , 1972] assuming that the wave spectrum is given:

$$\partial_t f_e + v_\parallel \nabla_\parallel f_e = \partial_{v_\parallel} D_\parallel (v_\parallel) \partial_{v_\parallel} f - f/\tau_l \qquad (6)$$

Since electron dynamics is restricted to parallel motion, only the parallel spatial dimension has to be considered. The last term on the right accounts for electron losses from the finite spatial acceleration region with τ_l the characteristic loss time.

The form of the diffusion coefficient arises from the quasilinear velocity space diffusion tensor $\mathbf{D} = \left(D_\parallel, D_{\parallel\perp}, D_{\perp\perp}\right)$ where for the electrons only the parallel component survives. Wu et al. (1981) considered solutions of this equation for a power law lower hybrid spectrum and concluded that stochastic acceleration generates a long suprathermal tail on the distribution function. This, however, assumes that the particles stay indefinitely long in the wave field. In the case of lower-hybrid cavitons the form of the diffusion coefficient changes according to the finite size of the cavitons and the finite interaction time between the electrons and the waves [Papadopoulos, 1992]. Defining $\lambda_\parallel$ as the parallel wavelength of waves resonant with electrons trapped in the caviton, one obtains:

$$D\left(v_\parallel\right)=\begin{cases} \dfrac{2e^2\left|E_\parallel\right|^2}{m^2 v_\parallel^2}\dfrac{\lambda_\parallel^2}{\tau_s} & v_\parallel \geq \lambda_\parallel\, f_k \\[2ex] 0 & v_\parallel < \lambda_\parallel\, f_k \end{cases} \qquad (7)$$

Slow electrons, slower than the characteristic parallel resonant velocity, simply oscillate in the wave field inside the caviton and are not affected, while the energetic electrons observe a stationary wave field during their transition and either gain or loose energy. It is clear that lowest energies are effected by the shortest trapped waves. These have perpendicular wavelengths $\lambda_\perp$ of a few Debye lengths ($k_\perp \lambda D \leq$ 1) or $\lambda_\perp \sim$ a few 50 m in a hydrogen dominated plasma of 8 eV temperature with a plasma frequency of about 20 kHz, corresponding to shortest parallel wavelengths of $\lambda_\parallel \sim$ 2 km. Since $f_{lh} \sim$ 500 Hz this implies that $v_\parallel >$ 1000 km/s. Hence particles of more than about 8 eV resonate and may become accelerated.

The "collision time" with the caviton can be approximated by $\tau_s = 1\,/N_s v_\parallel$ where N_s is the number of solitons per unit length along the particle path (in our case the magnetic field line). This may roughly be given by the linear initial parallel wave number of the lower-hybrid waves $N_s \sim k_{\parallel 0} = 2\pi/L_\parallel$. With these assumptions one finds that the diffusion equation coincides with one of the equations which have been treated by Wu et al. (1981). Its long time solution (if electron losses are neglected) is

$$F_e\left(u,t\right)=\frac{n_0\, 3^{1/3}}{\Gamma\!\left(\dfrac{1}{3}\right)\tau^{1/3}}\exp\!\left(-\frac{u^3}{9\tau}\right) \qquad (8)$$

Here $u = v_\parallel/v_e$ is the ratio of parallel to thermal electron velocities, n_0 the initial plasma density, and

$$\tau \equiv t\left(\frac{v_e e^2\left|E_\parallel\right|^2}{2T_e^2}\lambda_\parallel^2 k_{\parallel 0}\right) \qquad (9)$$

is the normalized time. The parallel wavelength $\lambda_\parallel$ of the waves trapped in the caviton is a function of time and changes during modulation. But assuming that this is a slow process, we take it as a constant ranging between 2-20 km, according to the conditions of the caviton and the stage of the collapse of the caviton.

The above distribution function can be used to find the increase in the energy K_{eT} of the accelerated particles in the tail of the distribution :

$$K_{eT} = 2T_e \int_{u_{min}}^{u_{max}} F(u)u^2\,du \qquad (10)$$

The boundaries of the integral are determined by the conditions that the particle velocity must be large enough for the particle to resonate with the wave and that the caviton has finite size. One then finds that

$$K_{eT} = \frac{2T_e}{\Gamma\!\left(\dfrac{1}{3}\right)}3^{4/3}\,\tau^{2/3}\left(\exp\!\left[-x_{min}\right]-\exp\!\left[-x_{max}\right]\right) \qquad (11)$$

Here $x_{min} = (\omega/k_{max}v_e)^3/9\tau$ and $x_{max} = (\omega L_\parallel/\pi v_e)^3/9\tau$. Since $x_{max} \gg x_{min}$ the second term can be neglected. The change in energy is found to be given by

$$\Delta K_{eT} = K_{eT}\left(\tau_2\right)-K_{eT}\left(\tau_1\right)$$

$$\approx 2T_e\,\frac{3^{4/3}}{\Gamma\!\left(\dfrac{1}{3}\right)}\tau_2^{2/3}\exp\!\left\{-\frac{1}{9\tau_2}\left(\frac{\omega_{LH}}{k_{max}v_e}\right)^3\right\} \qquad (12)$$

The above discussion suggest for an estimate we take $f_{LH} = 500$ s^{-1}, $L_\parallel = 20$ km, $\lambda_\parallel = 4$ km, $k_{max} \sim (2\pi/4 \times 10^3)$m^{-1}. Then for a parallel electric field of $E_\parallel \sim 2$ mV/m (corresponding to $E_\perp \sim$ 80 mV/m) numerical calculations show that sligthly suprathermal electrons can be accelerated to energies of the order of a few hundreds of eV in a time $t_2 \sim 1$ s. During this time the electrons travel a few thousand kilometers and traverse a large number of cavitons ($\sim$ 100) in which they are stochastically accelerated. These quoted numbers must however be looked after as very rough estimates because the theory does not account for the evolution of the waves during modulation and collapse. An understanding of electron acceleration in the late stages of collapse can only be accomplished with a three dimensional kinetic simulation, which, to this date has not be made.

One can make the following considerations. In a weak parallel electric wave field of 2 mV/m a resonant electron will be accelerated along the field lines up to about 40 eV in a caviton measuring 20 km, which is insufficient for auroral electrons. One therefore requires that the electron stays in resonance with many cavitons. However, an electron is unlikely to interact resonantly with each caviton. An optimistic estimate would be that the average electron interacts with 10% of the $\sim$ 100 cavitons. The electron is just as likely to be accelerated as decelerated. Then, the average heating would be the square root of the number of cavitons times the energy change. This amounts to $\sim$ 100 eV. Since the caviton wave fields are probably not coherent in different cavitons, this number might still be an overestimate. This fact leads for the conclusion that for the weak parallel wave fields considered here the electron acceleration may be at the lower end of the auroral electrons. However, lower hybrid wave fields up to 400 mV/m have been measured in the auroral acceleration regions [Mozer et al, 1980; Vago et al., 1992], leading to parallel wave fields $E_\parallel \sim 10$ mV/m and thus to a much larger electron acceleration. These measurements are confirmed by the fact that, quite frequently, in these regions the electric filters of the Viking wave experiment are oversaturated at 100 mV/m in the lower hybrid frequency range [Pottelette et al., 1988]. Taking into account the effects of collapse will also make the acceleration more efficient. Thus, it is expected that stochastic acceleration is an important source of hot parallel electrons escaping from the lower hybrid cavitons. A definite conclusion would require development of a more elaborate theory and/or numerical simulations.

Ion Acceleration

As electrons, ions can also be heated in two ways: by ponderomotive forces and stochastically. First, they react to the ponderomotive heating of the electrons parallel to the field via the ambipolar potential, but this effect is negligible for low ion temperature. Second, ponderomotive forces heat ions also in the perpendicular direction.

It has been shown first by Karney (1978) that ions encountering a lower hybrid wave may become accelerated in the direction perpendicular to the magnetic field. This acceleration is basically due to intrinsic stochasticity when the particle is kicked out of its gyro-orbit by the presence of the wave field. Lysak (1986) has shown that a similar mechanism acts in the presence of ion-cyclotron waves. More simple approaches have been presented by Chang and Coppi (1981) who assume resonant interaction between lower hybrid waves and ions. The stochastic heating is relevant above a certain threshold for the lower hybrid wave amplitude [Karney, 1978] which for the high measured wave amplitudes is satisfied. However, the interaction may be changed in the presence of lower hybrid cavitons similar to the case of electron acceleration discussed above. The concentration of the wave field to the region of the caviton limits the interaction to the transition or diffusion time of the ion across the soliton which will restrict the available acceleration. On the other hand, the high wave intensities lead to stronger acceleration than in the case of a distributed wave spectrum. Another point relates to the spatial distribution of the cavitons. One generally assumes that they are located in chains along the magnetic field. Accelerating an ion perpendicular to the field therefore will in general bring the ion out of the interaction region and will further limit the available heating if the ion is not moving a long distance along the magnetic field and encounters many cavitons. This was a natural assumption in the case of the electrons which may propagate only in the parallel direction, but for ions this assumption is relaxed. One therefore does not expect very strong stochastic ion heating in the presence of lower hybrid cavitons. However, this qualitative discussion has to be supplied in future by numerical calculations and simulations of the ion heating in spatially limited caviton wave fields before a definite conclusion can be drawn about the efficiency of transverse stochastic ion heating in lower hybrid cavitons. It is worth to mention that intense lower hybrid waves (100-400 mV/m) have recently been found responsible for transversely accelerating ions up to 6 eV in the lower auroral ionosphere [Vago et al., 1992]. It was possible to show that the wave-particle interaction takes place in thin filamentary cavities oriented along the magnetic field lines.

RADIATION

A large number of theories have been proposed in the past to explain the non thermal radiation from the Earth. A review of various radio emission mechanism can be found in Grabbe (1981) and in Goldstein and Goertz (1983).

Here we propose that lower hybrid cavitons in combination with upper hybrid waves and fast electrons can serve as emitters of high frequency radio radiation at the fundamental and harmonic upper hybrid frequencies. The mechanism proposed [Pottelette et al., 1992] is based on the nonlinear evolution of lower hybrid waves in the dilute plasma of the auroral magnetosphere. It results from this theory that the broad AKR radio emission must be caused by a large number of elementary cavitons moving on the background of a distribution of upper hybrid waves as illustrated in Figure 9. Assuming a caviton volume of the order of some ten km^3, an upper hybrid wave field amplitude of approximately 3 mV/m and a lower hybrid amplitude $|E_{lh}| \sim 100$ mV/m, we find a radiated power $P \approx 13$ W per caviton. Because of the uncertainties involved this power may range between 1-100 W, corresponding to energy fluxes of $10^{-11} - 10^{-9}$ $Wm^{-2} Hz^{-1}$ in agreement with Viking observations. To account for the total observed AKR power, about 10^6 cavitons are required to exist filling 10% of the auroral acceleration volume. It is worth to point out that the caviton mechanism allows for participation of a broad spectrum of waves in the radiation process.

The radiated energy is taken from the high frequency electrostatic upper hybrid waves. Hence the mechanism heavily depends on the simultaneous presence of upper hybrid and lower hybrid waves in the source volume. Theory [Wong et al., 1988; Maggs and Lotko, 1981; McFadden et al., 1986] suggests that both types of waves are excited by anisotropic energetic auroral electrons as have been observed on Viking.

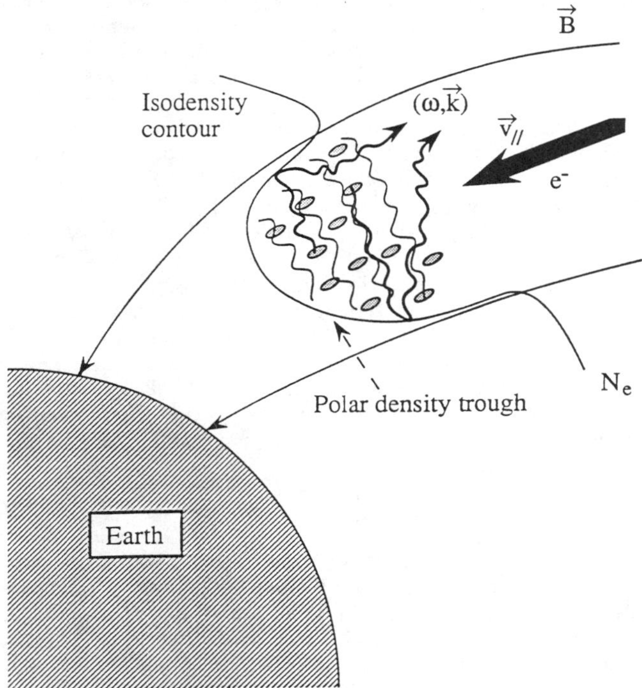

Fig. 9. Sketch of the AKR source region for caviton antenna radiation. The isodensity contour indicates presence of the auroral cavity filled with upper hybrid waves and lower hybrid cavitons. AKR emission from the caviton is shown to take place. It may become reflected at the through boundary before leaving the source region. The heavy arrow designates fast electrons flowing down the field lines.

The anisotropic energetic electron fluxes measured during the intense lower hybrid events are shown in Figure 3. Unfortunately, the Viking wave measurements do not permit to distinguish between the upper hybrid and electromagnetic contributions because of the too large bandwidth of the SFA receiver (2 kHz). Moreover, only one electric and one magnetic wave field component have been measured. In the source region the refraction index based on these measurements turns out to be sometimes larger than one. However, since the radiation is generated near the resonance, this fact cannot be used to determine the contribution of the electrostatic waves to the electric field amplitude. Morioka et al. (1990) as well as Menietti et al. (1992) have provided experimental evidence for the presence of upper hybrid noise during AKR emission events.

At the first glance there seems to be a problem with the high intensity emitted by cavitons when interacting with the weak upper hybrid wave spectrum. This problem is resolved by observing that the upper hybrid waves propagate at high velocity of the order of the hot electron thermal velocity with respect to the caviton [Pottelette et al., 1992]. Thereby they provide an intense high frequency energy flux per unit of time into the caviton which is transformed into radiation.The relatively high power a single caviton emits is hence due to its motion across this background of upper hybrid waves. Since interaction takes place only inside the caviton, the moving caviton works like a vacuum cleaner eating up the wave energy in the upper hybrid waves crossing the caviton at high speed (Figure 9). The mechanism has been explicated in Pottelette et al. (1992) and Treumann et al. (1992). Most of the radiation is in the X-mode. However, there are also contributions to O-mode emissions. O-mode emission is, however, a higher order process in which the parallel wave fields are involved which are smaller in intensity by the root of the electron to ion mass ratio. The intensity of the O-mode is therefore much weaker than that of the X-mode. It should be noted, however, that weak O-mode emissions have occasionally been reported in the literature [e.g. Shawhan and Gurnett, 1982].

The theory shows that the emission of the radio radiation actually consists of several different radiation bands which are separated by small contributions to the emitted frequency provided by the different frequency and wave number shifts involved in the caviton formation. The radiation therefore has a distinct fine structure which sometimes has observationally been resolved [Gurnett and Anderson, 1981]. The lifetime of ~1s of a caviton reflects itself in the characteristic modulations of the observed AKR bursts which vary on the same time scale.

CONCLUSIONS

In the present paper we have rewied the effects of caviton formation in the lower hybrid mode on the acceleration of electrons and ions as well as on the generation of electromagnetic radiation. Intense lower hybrid waves are found to be correlated with steep density gradients and high potential drops in the low density regions of the thermal electron component in the auroral magnetosphere. We have mainly focussed on acceleration of electrons. Acceleration of ions is found to take place mainly in the perpendicular direction by the localized ponderomotive potential inside the caviton and by the effect of intrinsic stochasticity in the lower

hybrid wave fields trapped in the caviton. For very strong wave fields there is also a parallel acceleration of ions due to space charge effects, which when combined with perpendicular heating produce ion conics. Heating by intrinsic stochasticity has so far only been considered for spatially extended lower hybrid waves [Karney, 1978]. The localisation of lower hybrid waves and cavitons should limit this kind of heating due to transition time effects.

Acceleration of electrons takes place only parallel to the magnetic field. The nonlinear ponderomotive potentials may accelerate the electrons, but acceleration is in general weak. In addition, interaction with the lower hybrid waves trapped in the caviton accelerates the electrons quasi-stochastically during transit time producing an energetic tail on the distribution function. This process has been shown to be efficient. These findings may be important in the auroral acceleration mechanisms for redistributing the bulk energy fed into the plasma by outer magnetospheric sources as for instance kinetic Alfvén waves, warm electrons fluxes or parallel electric currents. Large chains of lower hybrid cavitons generating narrow energetic electron beams of the dimension of the order of the transverse width of the cavitons - which is of the order of 1 km in the acceleration regions - might contribute to the very narrow structures observed in the aurora.

Cavitons also serve as sources of Auroral Kilometric Radiation. The process of radio wave emission has been found to be similar to the radiation from an antenna of kilometric length. The role of the antenna is played by the caviton which triggers interaction between auroral electrons, trapped lower hybrid waves, and upper hybrid waves which are excited in the same region. Excitation of lower and upper hybrid waves is known to be due to warm anisotropic auroral electron fluxes. The energy of the radiation is of course taken from the high frequency upper hybrid waves. Because of the low plasma density, their frequency is closed to the electron cyclotron frequency, and the emission is in the fundamental of the latter and propagates in the X-mode. The interesting point of this process is that the large difference in the propagation velocities of upper hybrid waves and cavitons has the effect of a vacuum cleaner, in which the caviton eats up the energy of all upper hybrid waves crossed by it. This produces the high radiation efficiency. Several other effects as narrow bandedness, time variability, and faint O-mode components can be attributed to this mechanism as well.

Acknowledgments. This work has been supported by the French-German foundation PROCOPE for scientific cooperation. The Centre National d'Etudes Spatiales sponsored this research and helped us with the data analysis. The Viking project is managed by the Swedish Space Corporation under contract from the Swedish Board for Space Activities. Thanks are due to P.A. Lindqvist (Royal Institute of Technology, Stockholm) and K. Stasiewicz (SISP, Uppsala) for providing wave and particle data. The effort of J. Paris, who developed the V4-H data processing software is thankfully acknowledged.

REFERENCES

Bryant, D. A., D. S. Hall, and R. Bingham, Auroral electron acceleration: a case for the stochastic alternative, in *Auroral Physics*, edited by C-I Meng, M. J. Rycroft and L. A. Frank, pp. 119-128, Cambridge University Press , 1991.

Chang, T., and B. Coppi, Lower hybrid acceleration and ion evolution in the suprauroral region, *Geophys. Res. Lett.*, 8, 1253-1256, 1981.

Davidson, R. C., *Methods in Nonlinear Plasma Theory,* pp. 166, Academic, San Diego, Calif., 1972.

Garbe, G. L., Auroral kilometric radiation: A theoretical review, *Rev. Geophys. Space Phys.*, 19, p 627-637, 1981.

Gekelman, R. L., and R. L. Stenzel, Localized fields and density perturbations due to nonlinear lower hybrid waves, *Phys. Rev. Lett.*, 35, 1708-1711, 1975.

Goldstein, M. L. and C. K. Goertz, Theories of radio emissions and plasma waves, in *Physics of the Jovian Magnetosphere*, edited by A. J. Dessler, Cambridge University Press, pp.198, Cambridge, 1983.

Gurnett, D. A., and R. R. Anderson, The kilometric radio emission spectrum: Relationship to auroral acceleration processes, in *Physics of Auroral Arc Formation*, Geophysical Monograph 25, edited by S.-I. Akasofu and J. R. Kan, pp. 341-350, AGU, Washington, D. C., 1981.

Hultqvist, B., R. Lundin, K. Stasiewicz, L. Block, P. A. Lindqvist, G. Gustafsson, H. Koskinen, A. Bahnsen, T. A. Potemra, and L. J. Zanetti, Simultaneous observation of upward moving field aligned energetic electrons and ions on auroral zone field lines, J. *Geophys. Res.*, 93, 9765-9774, 1988.

Karney, C.F.F., Stochastic ion heating by a lower hybrid wave, *Phys. Fluids*, 21, 1584-1602, 1978.

Lysak, R. L., Ion acceleration by wave-particle interaction, in *Ion Acceleration in the Magnetosphereand Ionosphere*, edited by T. Chang, Geophysical Monograph 38, p. 261-270, AGU, Washington, D.C., 1986.

McFadden, J. P., M. H. Boehm, and C. W. Carlson, High-frequency waves generated by auroral electrons, J. *Geophys. Res.*, 91, 12078-12085, 1986.

Maggs, J. E., and W. Lotko, Altitude dependent model of the auroral beam and beam generated electrostatic noise, J. *Geophys. Res.*, 86, 3439-3452, 1981.

Melrose, D. B., *Instabilities in space and laboratory plasmas*, pp. 112, Cambridge University Press, 1986.

Menietti, J. D., C. S. Lin, H. K. Wong, A. Bahnsen, and D. A. Gurnett, Association of electron conical distribution with upper hybrid waves, J. *Geophys. Res*, 91, 1353-1361, 1992.

Morales, G. J., and Y. C. Lee, Nonlinear filamentation of lower hybrid cones, *Phys. Rev. Lett.*, 35, 930-933, 1975.

Morioka, A., H. Oya, and K. Kobayashi, Polarization and mode identification of auroral kilometric radiation by PWS system onboard the Akebono (EXOS-D) satellite, *J. Geomagn. Geoelectr.*, 42, 443, 1990.

Mozer, F. S., C. A. Cattell, M. K. Hudson, R. L. Lysak, M. Temerin, and R. B. Torbert, Satellite measurements and theories of low altitude auroral particle acceleration, *Space Sci. Rev.*, 27, 156-213, 1980.

Papadopoulos, K., The CIV process in the CRIT experiments, *Geophys. Res. Lett.*, 19, 605-609, 1992.

Pottelette, R., M. Malingre, A. Bahnsen, L. Eliasson, K. Stasiewicz, R. E. Erlandson, and G. Marklund, Viking observations of bursts of intense broadband noise in the source regions of auroral kilometric radiation, *Ann. Geophys.*, 6, 573-586, 1988.

Pottelette, R., R. A. Treumann, and N. Dubouloz, Generation of auroral kilometric radiation in upper-hybrid wave-lower hybrid soliton interaction, *J. Geophys. Res.*, 97, 12029, 1992.

Robinson, P. A., New contributions to transit-time damping in multidimensional systems, Phys. Fluids B1, 490-498, 1989.

Temerin, M. A., Plasma waves on auroral field lines, in *Physics of Auroral Arc Formation*, Geophysical Monograph 25, edited by S.-I. Akasofu and J. R. Kan, pp. 351-358, AGU, Washington, D. C., 1981.

Treumann, R. A., R. Pottelette, N. Dubouloz, and J. Labelle, Lower hybrid soliton mediated radio emission, in *Proceedings of the 3rd International Workshop of Planetary Radio Emissions*, edited by H. O. Rucker, S. J. Bauer and M. L. Kaiser, pp. 391-403, Graz, 1991.

Vago, J. L., P. M. Kintner, S. W. Chesney, R. L. Arnoldy, K. A. Lynch, T. E. Moore, and C. J. Pollock, Transverse ion acceleration by localized lower hybrid waves in the topside auroral ionosphere, *J. Geophys. Res.*, 97, 16935-16957, 1992.

Wong, H. K., J. D. Menietti, C. S. Lin, and J. L. Burch, Generation of electron conical distribution by upper hybrid waves in the earth's polar region, *J. Geophys. Res.*, 93, 10025-10033,-1988.

Wu, C. S., J. D. Gaffey, Jr., and B. Liberman, Statistical acceleration of electrons by lower-hybrid turbulence, *J. Plasma Physics*, 25, 391-401 1981.

R. Potellette, Centre en Physique de l'Environnement, Saint-Maur-des-Fossé, France.

R.A. Treumann, Max-Planck-Institut für Extraterrestrische Physik, Garching bei München, Germany.

N. Dubouloz and M. Malingre, Center de Recherche en Physique de l'Environnement, Saint-Maur-des-Fossé, France.

Elusive Upper Hybrid Waves in the Auroral Topside Ionosphere

Robert F. Benson

Interplanetary Physics Branch, Laboratory for Extraterrestrial Physics
NASA/Goddard Space Flight Center, Greenbelt, Maryland

Swift [*J. Geophys. Res.*, *93*, 9815, 1988] suggested that upper hybrid waves may, under certain conditions, be a significant source of heating of auroral electrons in the direction perpendicular to the earth's magnetic field in the topside ionosphere. An investigation of emissions observed by the ISIS 2 topside sounder receiver indicates that upper hybrid emissions are only observed on about 3% of the passes through the auroral region, and that these emissions are often very weak. The strongest of these emissions, however, are observed under plasma conditions similar to those required by Swift. The observations are from a study involving ISIS 2 satellite passes that crossed 70° invariant latitude when the sounder was in the on/off mode of operation. In this mode, the sounder operates normally for 2 ionograms and then produces 2 ionograms with the transmitter off and the swept frequency receiver on. The data set was further restricted to include only passes where soft particle spectrometer data were also available; 550 ISIS 2 passes satisfying these conditions were investigated for upper hybrid emissions.

Introduction

Upper hybrid waves are waves associated with the upper hybrid frequency $f_T = (f_N^2 + f_H^2)^{1/2}$ where f_N and f_H are the electron plasma and gyrofrequencies, respectively. The ionospheric notation f_T is used to emphasize that the wave vector $\mathbf{k}$ is predominately perpendicular to the earth's magnetic field vector $\mathbf{B}$ for these waves. They have been the subject of considerable interest because of their role in electrostatic-to-electromagnetic wave-mode coupling, wave/particle interactions and planetary magnetospheric emissions. Their wave mode coupling role was vividly illustrated by Oya [1971] and applied to wave generation mechanisms for Jovian decametric radiation [Oya, 1974], terrestrial auroral kilometric radiation [Benson, 1975] and terrestrial continuum radiation [Jones, 1976]. Their importance in wave/particle interactions has been suggested by the observed periodicity of precipitating auroral electrons [Evans, 1967; Perkins, 1968] with a favored amplification when $f_N > f_H$ [Maggs, 1978] and a possible significance in the perpendicular heating of auroral electrons [Swift, 1988]. Their occurrence in nature has been demonstrated in the earth's magnetosphere by a number of spacecraft, e.g., near the plasmapause by IMP 6 and Hawkeye [Kurth et al., 1979] and ISEE 1 [Kurth, 1982] with a strong equatorial enhancement observed by EXOS D [Oya et al., 1990] and in Jupiter's Io plasma torus by Voyager 1 and 2 [Warwick et al., 1979a, 1979b].

Often, emissions in the entire frequency range from f_T down to the greater of f_N, f_H are described as upper hybrid waves. In this frequency range, the refractive index of the Z mode (low frequency slow branch of the extraordinary mode) becomes very large corresponding to short wavelength slow waves that can enter into strong wave/particle interactions [see, e.g., Oya, 1971]. Z mode emissions have been observed in this frequency range by a large number of spacecraft over a wide variety of plasma conditions, e.g., rocket flights from Wallops at night to 1,691 km [Walsh et al., 1964], daytime flights from Cape Kennedy (to 11,100 km) [Huguenin et al., 1964] and Wallops (to 1,068 km) [Stone et al., 1966] and an evening flight from Fort Churchill (to 328 km) [Kellogg et al., 1978]; satellite observations with UK-2 [Harvey, 1965], ATS 2 (to 8,000 km, 30° inclination) [Bauer and Stone, 1968], Alouette 1 and 2 (500 - 3,000 km, polar) [Hartz, 1969], Ariel 3 (500-600 km, 80° inclination) [Gregory, 1971] and Aureol/Arcad 3 (400 - 2,000 km, polar) [Beghin et al., 1989].

Waves in this frequency range have often been observed in space with a peak amplitude at or just above f_N, e.g., during rocket flights from Fort Churchill (into an intense auroral arc) [Kellogg et al., 1978], Cape Parry (to 640 km through a midday auroral arc [McFadden et al., 1986] and Poker Flat (to 1107 km through evening auroral arcs) [Ergun et al., 1991]; and at higher altitudes from satellites, e.g., from ISIS 1 (near 2,700 km and 55° invariant latitude Λ) [Benson, 1985; Figure 12] and ISEE 3 (observations of a type III solar radio burst from within the solar wind, 253 R_e upstream) [Lin et al., 1981].

Auroral Plasma Dynamics
Geophysical Monograph 80

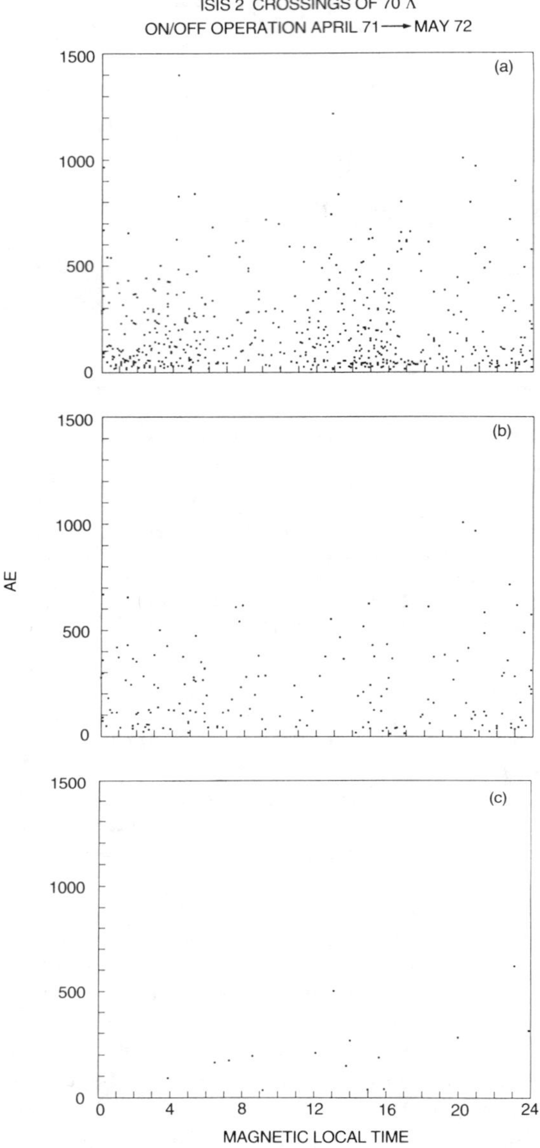

Fig. 1. (a) Magnetic activity and local time corresponding to 550 crossings of 70 invariant latitude by the ISIS 2 satellite during sounder on/off operation from 21 April 1971 to 16 May 1972. (b) Subset of the above passes that revealed wave emissions at f_N (171 passes), (c) subset of (a) that revealed wave emissions at f_T (15 passes).

The motivation for this paper was to search for waves in this frequency range with a peak amplitude at or just below f_T using ISIS 2 topside sounder data where the condition $f_N > f_H$ is often encountered in the auroral ionosphere since Swift [1988] predicted that f_T waves could then be a significant source of electron heating perpendicular to **B,** and evidence suggested to be in support of such heating was presented by Swift and Gorney [1989].

OBSERVATIONS

The ISIS topside sounders were designed to obtain the vertical electron density profile between the altitude of the satellite and the altitude of the F region peak density [Franklin and Maclean, 1969]. These sounders transmit short duration R. F. pulses (typically 0.1 ms) and record the reflected signal from the denser ionospheric region below the satellite as the frequency

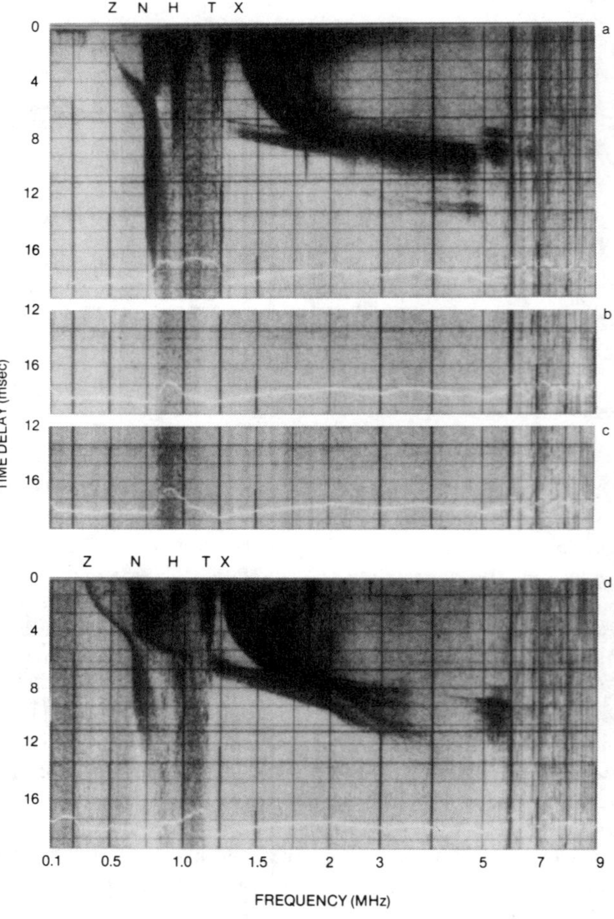

Fig. 2. ISIS 2 ionograms collected during the on/off mode. The off ionograms (b, c, e, and h) correspond to sounder receiver only operation and are conspicuous by their lack of resonances and wave cutoffs which appear as dark traces on the normal ionograms (a, d, f, and g). These resonances and cutoffs are identified by Z, N, H, T, and X for f_z, f_N, f_H, f_T, and f_x where f_z, and f_x are the Z and X wave cutoffs. The natural wave emissions also appear as dark signals (with darkness proportional to signal intensity) on both the on and off ionograms as does the interference from ground-based transmitters which is present when $f > 6$ MHz. Signal intensity is also indicated by the automatic gain control (AGC) level (the white trace at the bottom of each ionogram which saturates just below the 16 ns time delay level as seen during the whistler mode emissions from about 0.1 to 0.8 MHz in f, g, and h).

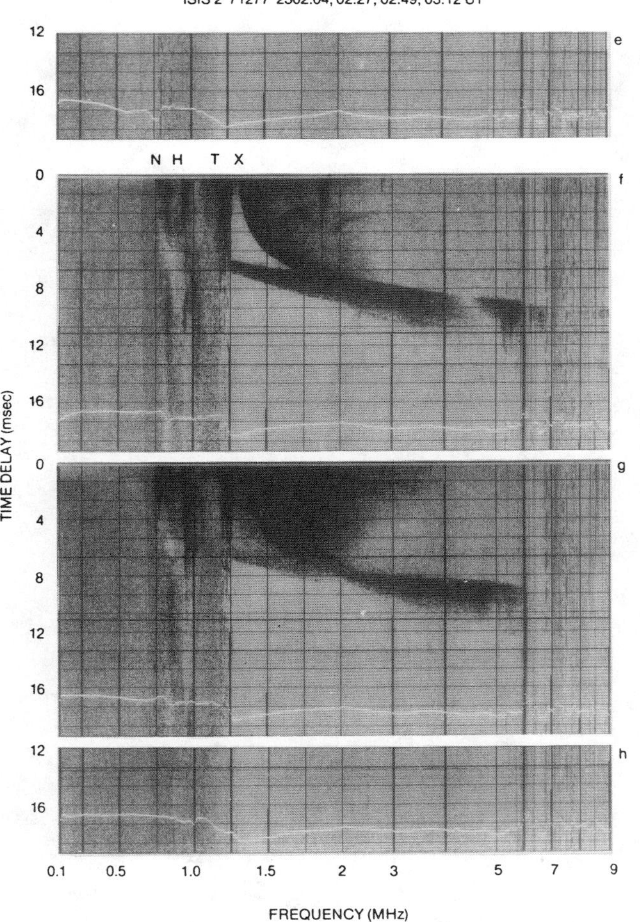

Fig. 2. (continued)

of the transmitter and receiver are continuously increased in a swept-frequency mode of operation. The data record, called an ionogram, displays the time delay of the reflected signal as a function of the transmitted frequency. Inversion techniques are used to convert the recorded frequency-time delay curves into electron density- true height profiles [Jackson, 1969].

In addition to the electromagnetic ionospheric reflection traces mentioned above, the ionograms reveal intense signals which appear as vertical traces resembling stalacites. They have been described as "resonances" because they appear as long duration signals (many milliseconds) which are received immediately after the transmission of the RF sounder pulse. They are attributed to sounder-stimulated electrostatic waves propagating in the near vicinity of the satellite (hundreds of meters to kilometers), and can be used to determine ambient plasma parameters (see review by Benson [1977])

In the present investigation, topside sounder data from ISIS 2 were used (rather than from ISIS 1) because there is a large ISIS 2 data base corresponding to the on/off mode of operation, i.e., normal sounder operation for 2 ionograms followed by 2 ionograms with only the swept-frequency receiver (no transmitter). The active ionograms were used to determine ambient values for f_N, f_H and f_T; the adjacent passive ionograms were used to identify wave emission of natural origin. Data from the first year of ISIS 2 operation were used so as to also have access to the soft particle spectrometer data; 550 satellite passes over the Ottawa telemetry station that crossed 70° Λ, under conditions that included a variety of magnetic activity and local times (see Figure 1a) were investigated. A pass was clasified as one detecting natural signals if any of the (typically) 20 to 30 ionograms on one of these satellite passes contained video noise signals above the background cosmic noise level. All of these passses contained ionograms that revealed either whistler (W) or Z mode emissions. It was sometimes difficult to distinguish between them in the frequency region below the minimum of f_N, f_H [Benson, 1991]. The Z mode emissions in the frequency range from f_N to f_T often showed enhancements at either f_N or f_T, the former being more common than the latter as illustrated by comparing Figures 1b and 1c. When $f_N \ll f_H$, any enhancements of the Z mode emission (between f_H and f_T) near f_T were of much smaller amplitude than similar enhancements when f_N and f_H were comparable. Of the 15 f_T enhancements represented by Figure 1c, 8 corresponded to the conditions $f_N < f_H$, 6 to $f_N > f_H$ and one to $f_N \cong f_H$.

Figure 2 illustrates the common occurrence of Z mode radiation in the frequency range from f_N (or f_H) to f_T with a large decrease in signal intensity near f_T. Here the natural emissions on the passive ionograms b and c are to be compared with the adjacent active ionograms a and d (that determine the ambient plasma conditions). Similarly, the emissions on e and h are to be compared with the plasma parameters determined from the intervening active ionograms f and g. In all cases, $f_N < f_H$. Relative signal intensity is indicated by the darkness of the video signal and by the level of the AGC trace which covers about 80 dB of dynamic range, on a log scale, to a saturation level of -30 dBm (relative to 1 mW CW input power) corresponding to an electric field intensity of approximately 7 mV/m integrated over the 50 kHz ISIS 2 receiver bandwidth (see caption). Note the build-up of W mode emission with $f < f_N$ in the latter 4 ionograms. The corresponding soft particle spectrometer (SPS) electron data, presented in Figure 3, indicates that the most intense electron precipitation corresponded to the ionogram h with the most intense natural W and Z mode emissions. On this ionogram, there is also a hint of a signal enhancement at f_T.

Figure 4 presents a case where $f_N > f_H$ and an enhancement of the Z mode signals near f_T are observed in c; the intensity level here is still much weaker than the Z mode emission near f_N. The corresponding SPS record (Figure 5) indicates the presence of energetic electrons with pitch angles near 90°, as would be expected from the mechanism proposed by Swift [1988], in the time interval between ionograms b and c.

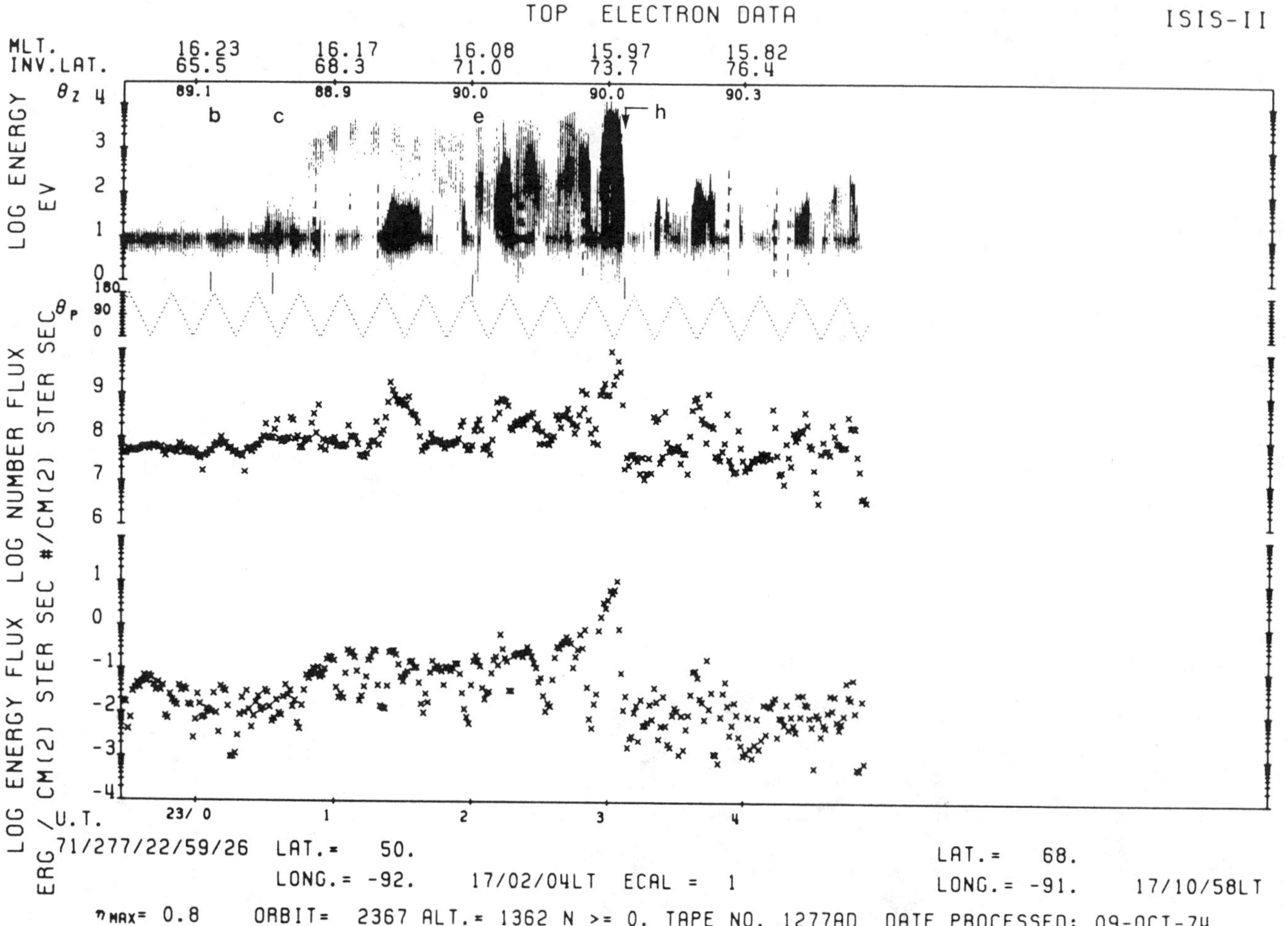

Figure 3. SPS spectrogram showing the electron data corresponding to the ionogram data of Figure 1 with the passive ionogram times indicated by the appropriate letters above the electron energy channel (top) and by vertical bars above the pitch angle channel (2nd from top).

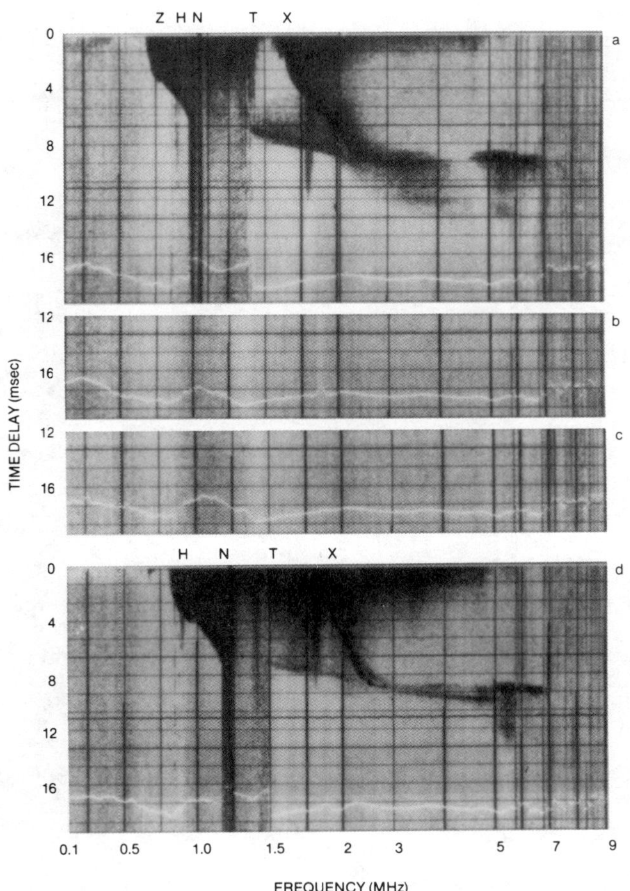

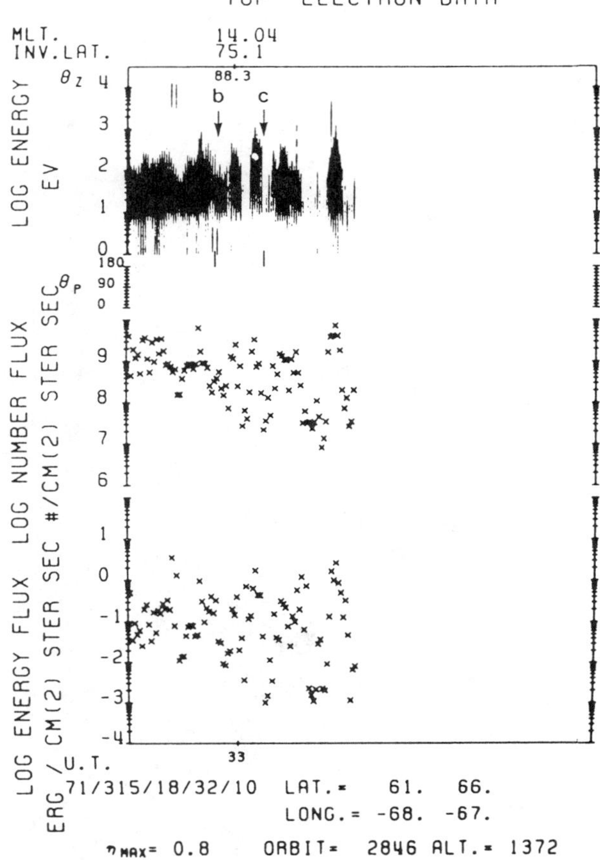

Figure 5. SPS spectrogram corresponding to the ionogram data of Figure 4. See Figure 3 caption.

Figure 4. Similar to Figure 2 except that the off ionograms correspond to b and c and the normal ionograms correspond to a and d.

Figure 6 presents another case with $f_N > f_H$ and emission at f_T. In this case the emission at f_T is greater than the emission at f_N. The corresponding SPS record (Figure 7) reveals electrons at all pitch angles but the peak electron number flux appears at 90° near the time of recording ionogram a.

Figure 8 presents a 4 ionogram consecutive sequence where the plasma conditions change from $f_N < f_H$ (in a) to $f_N > f_H$ (in d) with the onset of f_T emissions in (c) that are much more intense than the Z mode emissions near f_N. The corresponding SPS record (Figure 9) indicates a 90° peak in the number flux of energetic electrons near the time when ionogram c was recorded.

DISCUSSION

Even though upper hybrid emissions were only observed on about 3% of the satellite passes investigated, their appearance with greatest intensity when $f_N > f_H$ and, in particular, in the

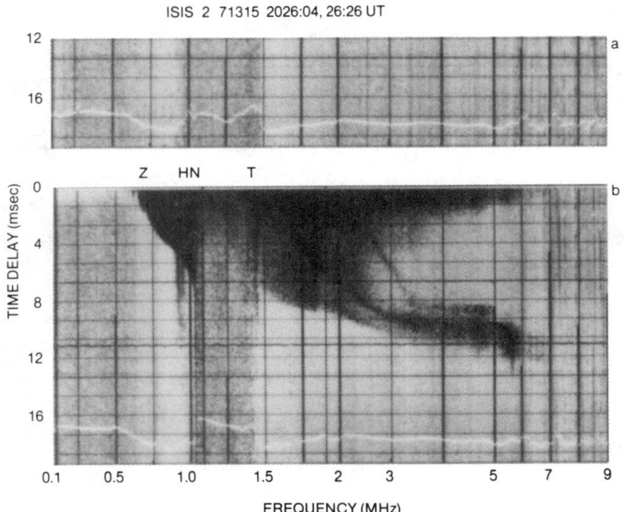

Figure 6. Similar to Figure 2 except that a and b correspond to the off and normal ionograms, respectively.

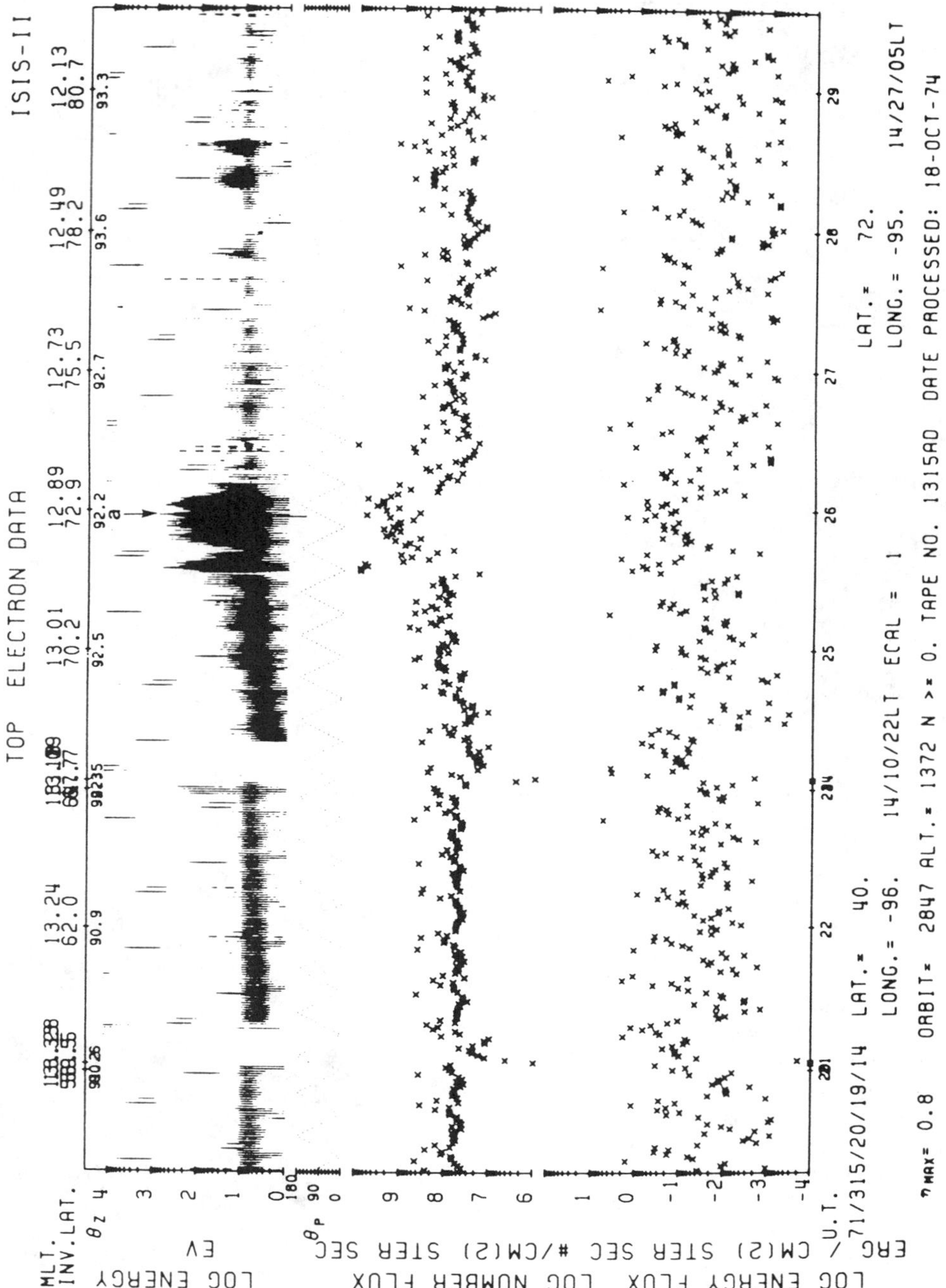

Figure 7. SPS spectrogram corresponding to the ionogram data of Figure 6.   See Figure 3 caption.

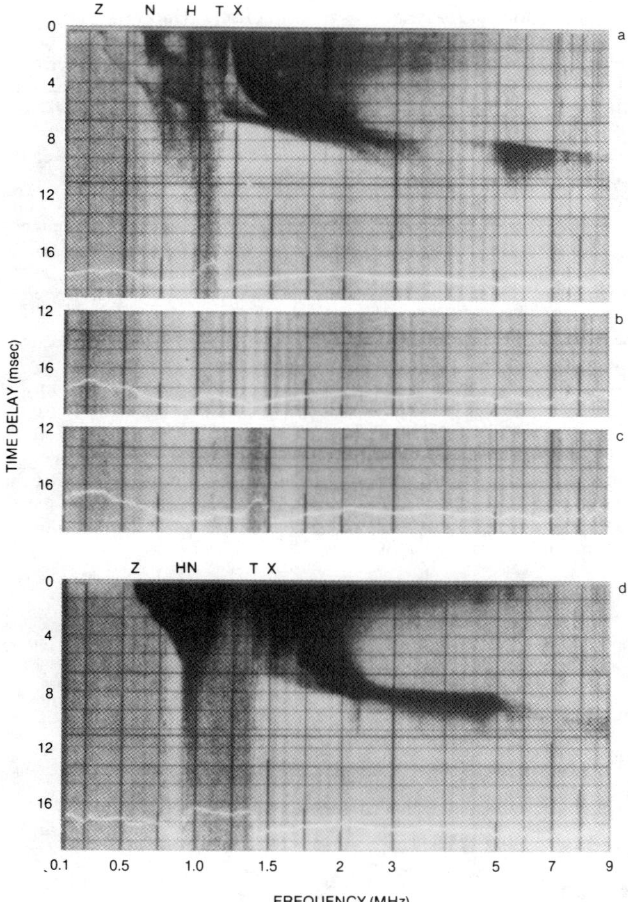

Figure 8. Similar to Figure 2 except that the off ionograms correspond to b and c and the normal iongrams correspond to a and d.

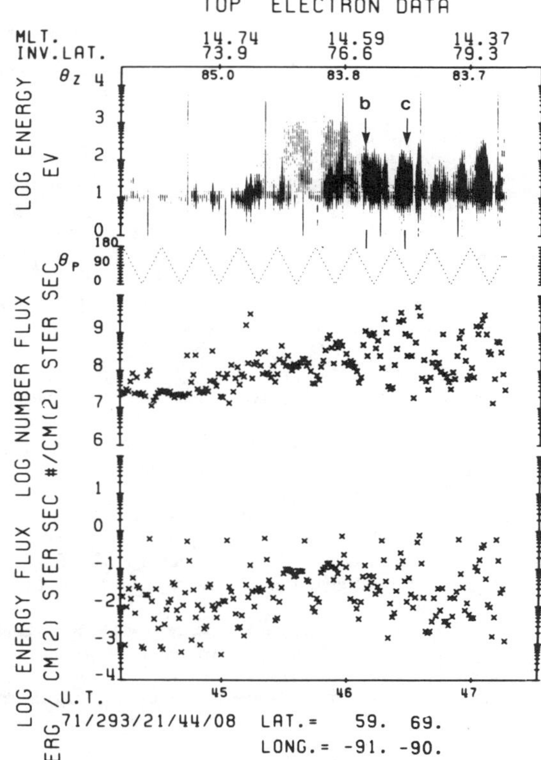

Figure 9. SPS spectrogram corresponding to the ionogram data of Figure 8. See Figure 3 caption.

presence of steep gradients in f_N (as in Figure 8) seem very consistent with the predictions of Swift [1988]. Even though it is not possible to separate between pitch angle, temporal and spatial variations on a spinning spacecraft such as ISIS 2, the presence of energetic electrons with pitch angles near 90° is consistent with the electron heating model proposed by Swift [1988]. These results may warrant a more thorough investigation that would determine the range of f_N (relative to f_H) on all of the passes rather than just on those that revealed f_T emissions (as was done here). Indeed, an investigation restricted to satellite passes where the condition $f_N > f_H$ is encountered near 70° Λ may yield a considerably higher percentage occurrence for f_T emissions.

Acknowledgments. I am grateful to W. Schar for assistance with the data analysis and to the National Space Science data Center at the Goddard Space Flight Center for the photographic reproductions of the ISIS 2 data.

REFERENCES

Bauer, S. J., and R. G. Stone, Satellite observations of radio noise in the magnetosphere, *Nature, 128,* 1145-1147, 1968.

Beghin, C., J. L. Rauch, and J. M. Bosqued, Electrostatic plasma waves and HF auroral hiss generated at low altitude, *J. Geophys. Res., 94,* 1359-1379, 1989.

Benson, R. F., Source mechanism for terrestrial kilometric radiation, *Geophys. Res. Lett., 2,* 52-55, 1975.

Benson, R. F., Stimulated plasma waves in the ionosphere, *Radio Sci., 12,* 861-878, 1977.

Benson, R. F., Auroral kilometric radiation: Wave modes, harmonics and source region electron density structures, *J. Geophys. Res., 90,* 2753-2784, 1985.

Benson, R. F., Z and whistler mode look-alikes in space plasma emissions (abstract), *Eos Trans. AGU, 72 (17),* Spring Meeting Suppl., 258, 1991.

Ergun, R. E., C. W. Carlson, J. P. McFadden, J. H. Clemmons, and M. H. Boehm, Langmuir wave growth and electron bunching: Results from a wave-particle correlator, *J. Geophys. Res., 96,* 225-238, 1991.

Evans, D. S., A 10-cps periodicity in the precipitation of auroral-zone electrons, *J. Geophys. Res., 72,* 4281-4291, 1967.

Franklin, C. A., and M. A. Maclean, The design of swept-frequency topside sounders, *Proc. IEEE, 57,* 897-929, 1969.

Gregory, P. C., Satellite observations of magnetospheric radio noise - I, *Planet. Space Sci., 19,* 13-820, 1971.

Hartz, T. R., Radio noise levels within and above the ionosphere, *Proc. IEEE, 57,* 1042-1050, 1969.

Harvey, C. C., Results from the UK-2 satellite (VII-5), *Ann.d' Astrophys., 28*, 248-254, 1965.

Huguenin, G. R., A. E. Lilley, W. H. McDonough, and M. D. Papagiannis, Measurements of radio noise at 0.700 Mc and 2.200 Mc from a high-altitude rocket probe, *Planet. Space Sci., 12*, 1157-1167, 1964.

Jackson, J. E., The reduction of topside ionograms to electron-density profiles, *Proc. IEEE, 57*, 960-976, 1969.

Jones, D., Source of terrestrial non-thermal radiation, *Nature, 260*, 686-689, 1976.

Kellogg, P. J., S. J. Monson, and B. A. Whalen, Rocket observation of high frequency waves over a strong aurora, *Geophys. Res. Lett., 5*, 47-50, 1978.

Kurth, W. S., Detailed observations of the source of terrestrial narrowband electromagnetic radiation, *Geophys. Res. Lett., 9*, 1341-1344, 1982.

Kurth, W. S., J. D. Craven, L. A. Frank, and D. A. Gurnett, Intense electrostatic waves near the upper hybrid resonance frequency, *J. Geophys. Res., 84*, 4145-4164, 1979.

Lin, R. P., D. W. Potter, D. A. Gurnett, and F. L. Scarf, Energetic electrons and plasma waves associated with a solar type III radio burst, *Astrophys. J., 251*, 364-373, 1981.

Maggs, J. E., Electrostatic noise generated by the auroral electron beam, *J. Geophys. Res., 83*, 3173-3188, 1978.

McFadden, J. P., C. W. Carlson, and M. H. Boehm, High-frequency waves generated by auroral electrons, *J. Geophys. Res., 91*, 12,079-12,088, 1986.

Oya, H., Conversion of electrostatic plasma waves into electromagnetic waves: numerical calculation of the dispersion relation for all wavelengths, *Radio Sci., 6*, 1131-1141, 1971.

Oya, H., Origin of Jovian decameter wave emissions--conversion from the electron cyclotron plasma wave to the ordinary mode electromagnetic wave, *Planet. Space Sci., 22*, 687-708, 1974.

Oya, H., A. Morioka, K. Kobayashi, M. Iizima, T. Ono, H. Miyaoka, T. Okada, and T. Obara, Plasma wave observation and sounder experiments (PWS) using the Akebono (EXOS-D) satellite - Instrumentation and initial results including discovery of the high altitude equatorial plasma turbulence, *J. Geomagn. Geoelectr., 42*, 411-442, 1990.

Perkins, F. W., Plasma-wave instabilities in the ionosphere over the aurora, *J. Geophys. Res., 73*, 6631-6648, 1968.

Stone, R. G., J. K. Alexander, and R. R. Weber, Measurements of antenna impedance in the ionosphere - II. Observing frequency greater than the electron gyro frequency, *Planet. Space Sci., 14*, 1007-1016, 1966.

Swift, D. W., A numerical model for auroral precipitation, *J. Geophys. Res., 93*, 9815-9830, 1988.

Swift, D. W., and D. J. Gorney, Production of very energetic electrons in discrete aurora, *J. Geophys. Res., 94*, 2696-2702, 1989.

Walsh, D., F. T. Haddock, and H. F. Schulte, Cosmic radio intensities at 1.225 and 2.0 Mc measured up to an altitude of 1700 km, in *Space Research, 4*, edited by P. Muller, pp. 935-959, North Holland Publishing Company, Amsterdam, 1964.

Warwick, J. W., et al., Planetary radio astronomy observations from Voyager 2 near Jupiter, *Science, 206*, 991, 1979.

Warwick, J. W., et al., Voyager 1 planetary radio astronomy observations near Jupiter, *Science, 204*, 955, 1979.

Robert F. Benson, Mail Code 692, NASA/Goddard Space Flight Center, Greenbelt, MD 20771

SCEX 3 Observations of HF Z-mode Emissions
From the Aurora

R. T. GOERKE,[1] P. J. KELLOGG, S. D. BALE AND S. J. MONSON

School of Physics and Astronomy, University of Minnesota, Minneapolis

H. R. ANDERSON AND D.W. POTTER

Science Applications International Corporation, Bellevue, WA

E. P. SZUSZCZEWICZ AND G. D. EARLE

Science Applications International Corporation, McLean, VA

We have observed moderately strong HF (high frequency ~ 4 MHz) plasma waves with amplitudes in excess of .1 mV/m(Hz)$^{1/2}$ at frequencies slightly less than the ambient plasma frequency and about 2 to 3 times the electron cyclotron frequency. These observations coincide with *in situ* measurements of broad band low energy (10 eV to 600 eV) auroral electrons and ground-based observations of auroral luminosity. These results were obtained from the rocket-borne experiment SCEX 3 (Several Compatible Experiments Using a Rocket-Borne Accelerator), NASA Flight 39.002 UE, launched February 1, 1990 from Poker Flat Research Range.

The frequency and intensity of the HF emissions are modulated at the spin period of the spacecraft in such a way that they appear like saucers on frequency-time spectrogram. This indicates that the source of these observations is a directional wave. If these waves are transverse (electromagnetic), then the k vector points in the north/south direction, which is consistent with an auroral source. Due to the lack of an observable interference pattern, we assume the waves have a wavelength that is much larger than the antenna length (2 m). These waves are seen consistently above the Z cutoff and persist over a region of ~100 km which is suggestive of transverse electromagnetic waves. Other features of the aurora which produced these emissions will be presented and discussed.

1. INTRODUCTION

The SCEX 3 experiment (Several Compatible EXperiments using a rocket-borne accelerator) was carried to ionospheric altitudes (375 km) by a Black Brant 11 rocket on February 1, 1990. The experiment was launched from Poker Flat Research Range (65.1°N, 147.5°W) at 1207 UT.

Preceding the launch of SCEX 3 an auroral arc along the southern boundary of the diffuse aurora suddenly brightened, split into two separate arcs and moved to a position north of the rocket's trajectory. SCEX 3 was launched into an active breakup aurora consisting of tall rays and diffuse patches.

---

[1]Now at International School of Minnesota, Eden Prairie.

Auroral Plasma Dynamics
Geophysical Monograph 80
Copyright 1993 by the American Geophysical Union.

During ascent the rocket trajectory passed very near a region of diffuse auroral luminosity. The closest approach to this region occurred at a flight time of ~130 s.

The emissions discussed here were observed prior to and between injections of the "artificial" electron beam and are therefore associated with natural processes. Because these observations were made during a time when auroral electron precipitation was observed both optically and by rocket-borne particle detectors, it is assumed that this radiation is produced by the auroral electrons. Following the observation of the energetic electrons (up to 600 eV—the limit of the instrument), the emissions fade gradually and eventually are eclipsed by emissions from the artificial electron beam. This observation supports the hypothesis that the source of these emissions is the aurora.

A wide variety of auroral emissions is commonly observed [Shawhan, 1979, Gurnett et al., 1983]. Auroral Kilometric

Radiation (AKR) is the most intense. It is commonly observed at altitudes greater than 2400 km [Benson, 1985]. The frequency of the most intense AKR is on the order of a few hundred kilohertz. This paper will focus on Z-mode emissions. The Z-mode emission frequency limits are always the Z-mode cutoff (f_Z) and the upper hybrid frequency (f_{UHR}). These frequencies are given by the following:

$$f_Z = 1/2[(f_{ce}^2 + 4f_{pe}^2)^{1/2} - f_{ce}] \tag{1}$$

and

$$f_{UHR} = (f_{pe}^2 + f_{ce}^2)^{1/2} \tag{2}$$

where f_{ce} and f_{pe} are the electron cyclotron frequency and plasma frequency respectively which are given by

$$f_{pe} = 9 \, (kHz) \, n_e^{1/2} \tag{3}$$

and

$$f_{ce} = 2.8 \, (MHz) \, |B| \tag{4}$$

The units of electron density (n_e) are cm^{-3} and the units of the magnetic induction $|B|$ are Gauss. The plasma frequency during the period of auroral observations varied from 6.4 MHz to 3.4 MHz while the cyclotron frequency was relatively constant at 1.4 MHz.

The Z-mode emissions in the auroral zone can be roughly grouped in two distinct categories. The high frequency emissions are observed at frequencies above ~1.7 MHz. It has been suggested that these emissions are produced at the second harmonic of f_{ce} by a cyclotron maser instability [Winglee, 1985; Benson and Wong, 1987]. The low frequency Z-mode emissions (<1.6 MHz) can easily be mistaken for Whistler mode because they are near or below the local f_{ce} in the topside auroral zone [Gurnett et al., 1983]. Ray-tracing techniques have been used extensively to estimate the lower boundary of the source region for the topside observations of Z-mode emissions at ~0.7-0.9 R_e. The lower frequency emissions have also been suggested to be produced at f_{ce} through the cyclotron maser instability [Melrose et al., 1984; Benson and Wong, 1987].

2. EXPERIMENT DESCRIPTION

The SCEX 3 payload was carried to a maximum altitude of 375 km by a Black Brant 11 rocket. In order to meet the scientific objectives the payload was planned to separate into six separate parts. Four Throw Away Detectors (TADs) were to be ejected normal to ambient magnetic field in four geographic directions from the aft payload (East, West, North and South respectively). The forward payload, which carried several diagnostic instruments to study the beam plasma interaction, was to be ejected from the aft payload in a direction parallel to ambient magnetic field so that it could pass through the beam region. The aft payload carried the electron accelerators and several diagnostic instruments.

The actual experiment configuration is shown in Figure 1. Due to a shorted relay in the pyrotechnic system of the aft payload, the pyrotechnic battery failed 68 seconds after launch causing the failure of several planned deployments. The most notable instruments that failed to deploy were: TAD 3, TAD 4, the tether and the aft payload Langmuir probe booms. TAD 4 blocked the view of the aft payload TV camera that was intended to view the electron beam injection directly. The forward payload, which carried several diagnostics instruments and the primary wave receiver package, was ejected late due to the battery failure. Hence it was ejected at the wrong azimuth but with nearly the correct elevation angle. It was ejected with a velocity of 2.7 m/s at 116 s, in a westerly direction (6° from the geomagnetic field). TAD 1 carried a second wave receiver package. It was ejected with a velocity of 0.97 m/s at 113 s, normal to the geomagnetic field in an easterly direction. TAD 2 carried electron and ion particle detectors for observing the beam electrons reflected from the auroral acceleration region and ions transversely accelerated by payload charging effects. It was ejected with a velocity of 0.96 m/s at 113 s in a westerly direction.

a. Plasma Wave Receivers

The plasma wave receivers were located on the forward payload and TAD 1, and were designed to detect electrostatic and electromagnetic waves associated with the electron beam injections as well as natural auroral emissions. The swept frequency receiver system is similar to those flown on previous sounding rocket experiments by the University of Minnesota [Cartwright and Kellogg, 1974; Kellogg and Monson, 1981; Kellogg et al., 1986; and Goerke et al., 1990]. The forward payload sweeping receiver had a bandwidth of 30 kHz and covered a range from 100 kHz to 12 MHz every 25.6 ms. The preamps were gain switched every two seconds with the low gain being attenuated by ~20 dB to check for spurious harmonics caused by possible saturation of the receiver. All the spectrogram plots presented in the results section have been normalized to the high gain response of the receiver. The gain switches are still apparent from the change in the noise level at the low frequency end of the sweeps. The low gain times show an increased noise level.

The forward payload also carried a wide band receiver which sampled the *in situ* electric field at a frequency of 60 kHz. The range of the instrument was maximized by telemetering the log of the observed signal. Both receivers were connected to the upper pair of antenna booms. The booms consisted of 10 cm diameter aluminum spheres extended on 81.3 cm fiberglass rods. The dipole length of the antenna was 1.93 m. The effective length of the dipole in the RF range was ~1 m due to the input impedance of the pre-amps. During the period of auroral emission observations the forward payload moved from ~10 to 400 m from the aft payload and ~0.6 m to 24 m from the field line containing the aft payload.

The first TAD also carried a swept-frequency receiver connected to a flexible stainless steel wire dipole antenna with a diameter of 1.4 mm. The dipole length of the antenna was 1.84 m and the effective length in the RF range was ~1 m also. This receiver also had a bandwidth of 30 kHz and covered the range from 100 kHz to 15 MHz. The TAD sweeping receiver

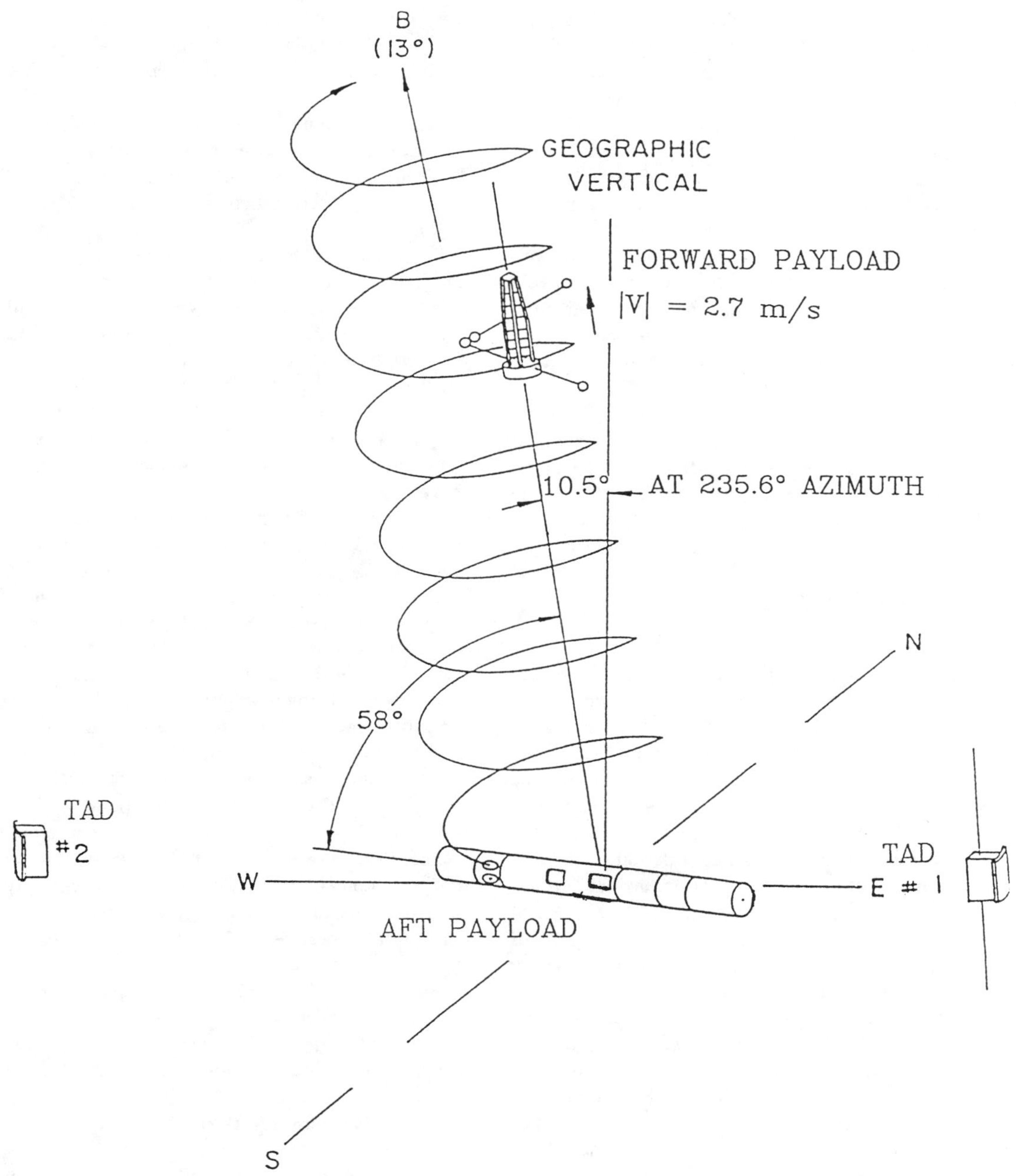

Fig. 1. The actual configuration for the SCEX 3 experiment. The forward payload is outside the beam spiral. The four booms on the forward payload are the antennas for the wave receiver package on board. The top pair was used by the swept-frequency and wide band receiver. The two wires on TAD 1 are the antennas for the TAD swept-frequency receiver.

was not gainswitched. The auroral emissions were only slightly above the threshold of this receiver and could not produce spurious harmonics. During the period of auroral emission observations the TAD moved from 7 to 80 m from the aft payload in a direction perpendicular to the magnetic field.

b. Particle Detectors

The forward payload retarding potential analyzer (RPA) measures the integral spectrum of electron flux from 600 eV to nearly 0 eV. The instrument sweeps the entire energy range with a 167 ms period and measures currents in the range of 10^{11}

A to 10^{-5} A. Current is collected through a 1 cm radius aperture perpendicular to the payload spin axis. Capacitive coupling between the sweeping voltage grid and the current collection plate requires the subtraction of a background signal. This is obtained from a quiet time and reveals itself as a series of bands on the high energy end of the plot.

The plasma density measurements were used to determine the plasma frequency during periods when there were no identifiable plasma frequency emissions. The density was scaled from the base current of the forward payload Langmuir probe (a.k.a. Pulsed Plasma Probe, P^3) when the probe was held at a potential of +5 V which is ~50 times the expected thermal energy of the ambient electrons. To eliminate the apparent roll modulation in the Langmuir probe data the maximum for each spin period was taken to be the value at the center of each spin period and the intermediate values are determined by a linear interpolation between the two nearest maxima. The current of the probe is proportional to the electron density if the probe dimensions are much less than the gyro-radius of thermal electrons (a few cm). This is satisfied by the cylindrical P^3 whose diameter is 1.4 mm. The error in the absolute density measurements is estimated to be ±50%, while the differential error is presumably much less (±5%).

3. AURORAL EMISSION OBSERVATIONS

During ascent the rocket trajectory passed very near a region of diffuse auroral luminosity. The closest approach to this region occurred at a flight time of ~130 s. Saucer shaped emissions were observed by the forward payload mid frequency sweeping receiver at ~5 MHz from the start of data taking (120 sec) to ~ 250 s. Examples of these emissions are shown in Figures 2, 3 and 4. The period of repetition for the saucers is ~1 s, which is half the forward payload spin period. There is no apparent fingerprint pattern in the spectrogram data caused by interference nulls (when nl = antenna length ∥ to k). All of these observations lead one to the conclusion that the waves observed are directional and have a wavelength which is larger than the antenna length. The forward payload spectrogram (Figure 2, middle panel) show Z-mode emissions which apparently extend slightly below the local Z-mode cutoff. This apparent incongruity may be the result of overestimating the local plasma frequency. The frequency of these emissions follows the changing Z-mode cutoff frequency fairly well. The plasma frequency (and all other characteristic frequencies) are determined from the forward payload Langmuir probe data from 125 s to 144 s. The spin modulation of the Z-mode waves is consistent with a directional transverse (EM) wave whose source is either north or south of the forward payload.

From 144 to 156 s the plasma frequency is scaled from the TAD sweeping receiver data (refer to Figure 2, top panel), which show a relatively intense emission extending from the plasma frequency up to the upper hybrid frequency during that period. There is also a hint of Z-mode emission observed by this receiver but it is seen at a somewhat lower frequency and much lower amplitude than the forward payload. The TAD antennas were wire dipoles while the forward payload antennas were spherical probes. The TAD antennas were probably not straight. Prior to launch the antennas were wrapped around the TAD inside the aft payload for approximately three weeks, quite possibly causing them to become permanently bent. This bending may diminish the effective length of the antenna and limit it's ability to observe directional radiation. Further, the TAD antennas were nearly parallel to B, while the forward antennas were perpendicular.

The forward payload receiver did not observe the plasma frequency-upper hybrid frequency emissions. This fact is another peculiarity of these observations. The severe outgassing of the forward payload may have limited it's ability to observe the plasma frequency emissions. The ionization of the outgassed neutrals by auroral electrons may have set up small scale irregularities in the vicinity of the forward payload which could have shielded the forward payload from plasma frequency emissions. The Z-mode emissions would not be strongly effected by these irregularities because of their large wavelength.

Using the calculated characteristic frequencies as a guideline, one can determine that the low frequency cutoff of these emissions is consistently near the calculated Z-mode cutoff, and the high frequency cutoff is either the plasma frequency or a cyclotron harmonic. This leads one to identify the waves as Z-mode (transverse electromagnetic) waves. The wavelength of such waves in the observed frequency range is several hundred meters which is much larger than the antenna length (~2m). So the saucer shape is consistent with the Z-mode identification.

The bottom panel of Figure 2 is a sonogram-style plot of electron energy as a function of flight time with electron flux as intensity. The 2 s payload spin period evidences itself as a modulated low energy electron flux. This is thought to be due to the ram/wake effect of the payload's motion through the ambient ionosphere. The payload's velocity through the ionosphere is of the same order as the ion thermal velocity. Ions are thus preferentially excluded from the wake side of the payload, and, since the Debye length is small, electrons are excluded also. So, the low energy spin-modulated peaks in the electron flux are indicative of the ram/wake and are maximum in the direction of the payload motion. The low energy flux seems to reach a maximum at about 138 s and decays, only to increase again near 150 s. This seems to imply that the payload is passing through regions of local density enhancement.

Higher energy electrons (10-600 eV) are visible, first in the ram direction and then in all directions at times which coincide with Z-mode emission characteristic of the aurora. The all-sky TV, at this time, shows the rocket moving northward into a region of diffuse auroral patches. The broad energy spectrum of electrons are most likely backscattered electrons from the nearby aurora.

One peculiarity in these emissions is the vertical stripes which are observed in Figure 3 at 131.6-133 s, and perhaps also at 130 s. The striped pattern was caused by a change in the local plasma density having a periodicity of about 12 Hz. This density oscillation is apparent in both the Langmuir probe data

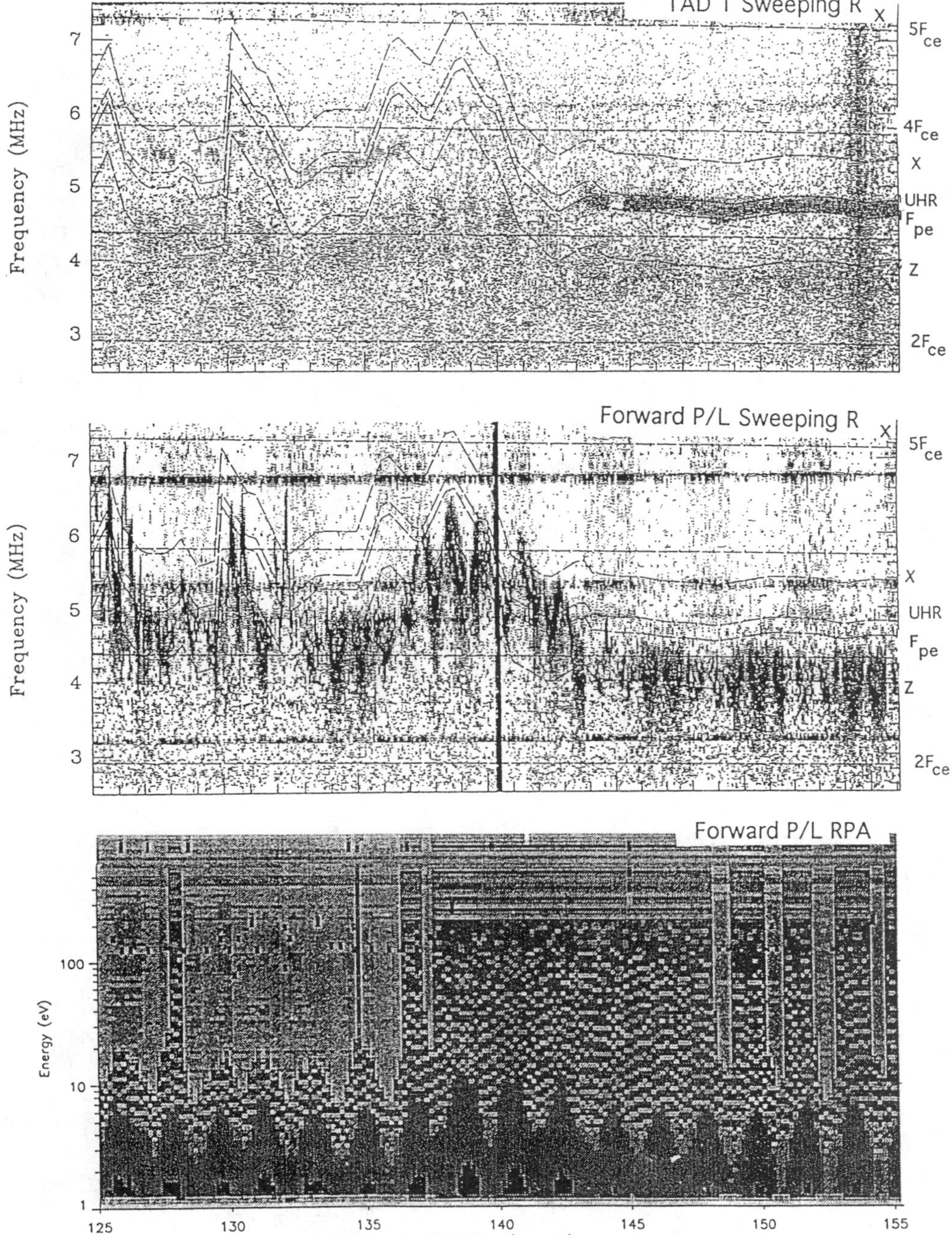

Fig. 2. The SCEX 3 auroral data summary plot is shown here. The top panel is a spectragram of the TAD 1 sweeping receiver data. The limits of the grey scale are 10^{-4} to 10^{0} mV/m(Hz)$^{1/2}$. The strongest auroral emissions are observed between the plasma frequency and the upper hybrid frequency from ~142 s to 155 s.

The middle panel is a spectrogram of the forward payload data covering the same period with the same grey scale limits. The vertical line at 140.4 s is a calibration signal. The Z-mode emissions follow the measured plasma density profile fairly well. Because of the slow sampling rate of the density (~2 Hz) rapid density changes which occur at ~126 and ~132 s are not reflected in the calculated characteristic frequencies.

The bottom panel is a sonogram-style plot of electron energy as a function of flight time with the log of electron flux as intensity. The two second payload spin period evidences itself as a modulated low energy electron flux. presumably caused by wake and ram effects.

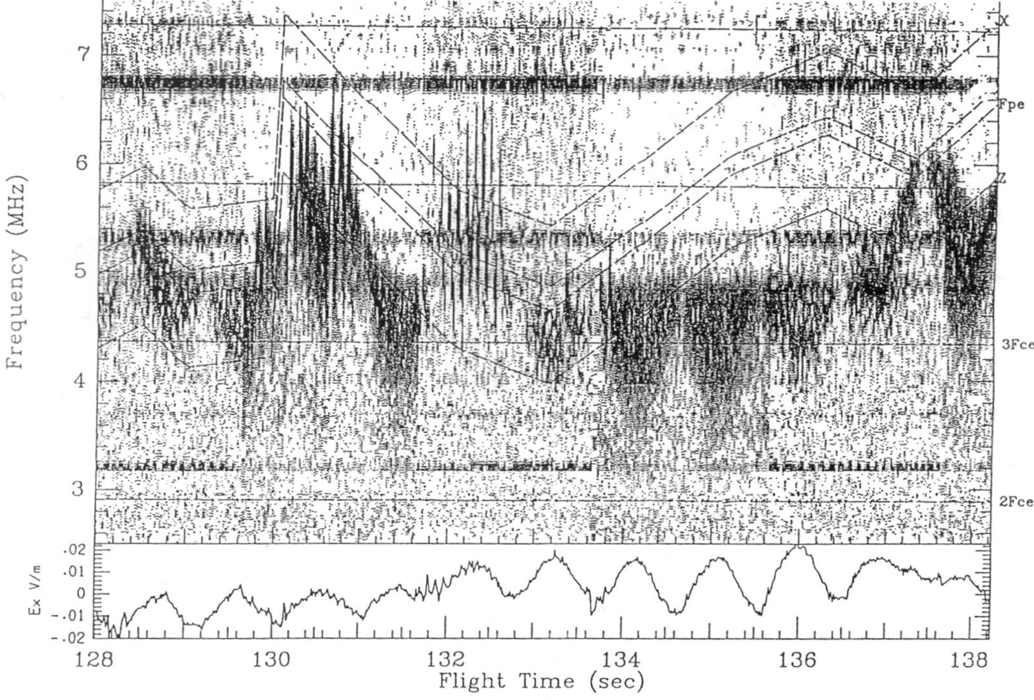

Fig. 3. This is a spectrogram of the forward payload sweeping receiver data from 128 to 138 s. The intensity and frequency limits are the same as those in Figure 2. The "striping" of the emission, observed at ~130.5 and ~132 s corresponds to ~10 Hz oscillations observed in the axial electric field measuresments for the same period.

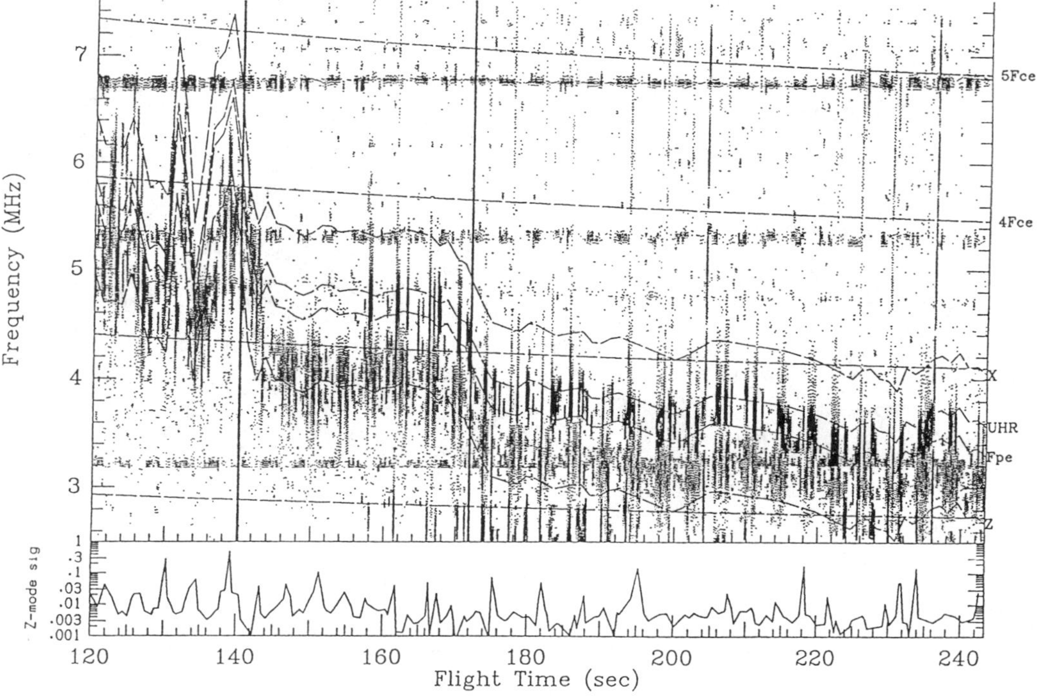

Fig. 4. This spectrogram plot summarizes the entire period (120 to 240 s) of auroral Z-mode emission observations by the forward payload sweeping receiver. In the lower panel the maximum Z-mode signal strength is plotted over the same period. The more intense sporadic emissions are from the "artificial" electron beam which was injected by the aft payload at various currents and energies for short periods of time throughout the "gun active period" 156-450 s.

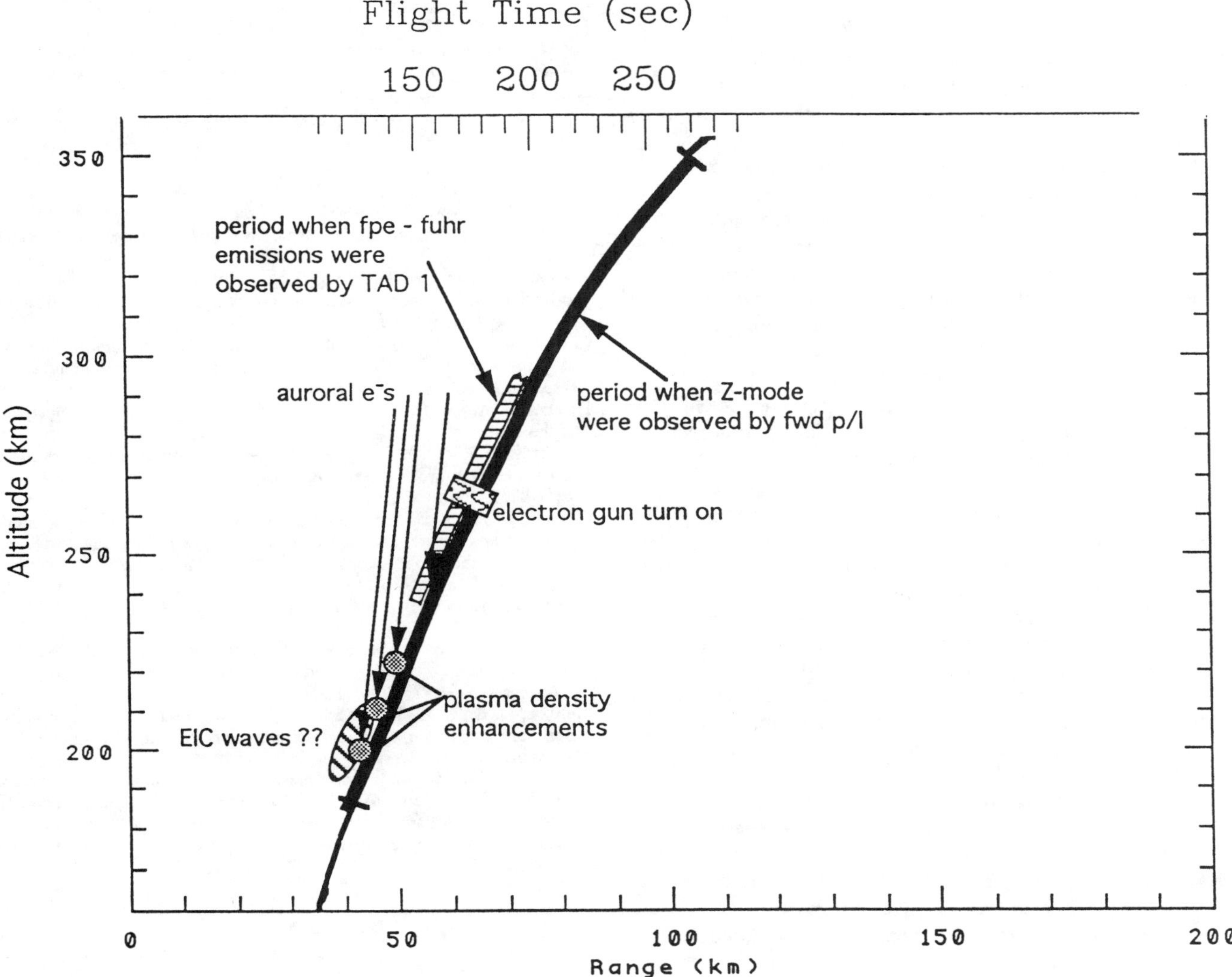

Fig. 5. This "cartoon" illustrates the effects of the auroral activity observed during the SCEX 3 mission.

and the low frequency axial electric field (E_x) which is plotted in the lower panel of Figure 3. The third harmonic of this oscillation was nearly in synch with the sweep frequency of the receiver (39 Hz). It appears from this data that some kind of low frequency (~12 Hz) pulsation in the auroral plasma is modifying the plasma density around the rocket.

Another peculiar feature of the auroral Z-mode emissions is that the saucers are slightly hollow. The signal strength of the Z-mode is attenuated by ~3 dB in the "interior" of the saucers so that a "fingerprint" pattern is very close to forming (see Figure 3 at ~134 s) [Fuselier and Gurnett, 1984]. This implies that the wavelength of these Z-mode waves is on the order of the antenna length (albeit slightly larger than the antenna length). This observation is inconsistent with the apparent frequency of the Z-mode emissions. Near the Z-mode cutoff the wavelength

of the emissions should be on the order of several hundred meters and no fingerprint pattern should be observable. A possible explanation for such an effect is that the waves recently passed through a region of depleted electron density such that the frequency of the waves exceeded the local plasma frequency. It would then be possible for these waves to be refracted in a direction that is near the resonance cone and thereby exhibit much smaller perpendicular wavelengths.

Figure 4 is a spectrogram plot of the whole period over which auroral emissions were observed (120-240 s). The lower panel shows the maximum amplitude of the Z-mode emissions over this period. After 156 s the emissions from the "artificial" electron beam are quite apparent and aid in the identification of the various characteristic frequencies. These induced emissions sometimes fall into the Z-mode frequency range and anomalously increase the Z-mode signal strength plotted in the lower panel. It is still apparent that the amplitude of the Z-mode emissions steadily decreased after a flight time of 145 s and this steady decrease supports the hypothesis that the source of these emissions is the diffuse auroral patch encountered from 120-140 s. The characteristic frequencies are indicated by dashed lines and labeled on the righthand side of the spectrogram. The vertical lines at ~140, ~172, ~204 and ~236 s are calibration signals used to correct the frequency drift of the sweeping receiver. From 120-160 s there is an apparent 0.25 Hz modulation of the maximum signal strength which could be an effect of the gain changing of the receiver for which the data was apparently not properly compensated.

4. Conclusions

The data presented here show evidence for Z-mode emission associated with a diffuse auroral patch. This auroral patch is apparently also responsible for plasma density enhancements observed at ~122, ~132 and ~138 s. There are also low frequency (~10 Hz) plasma density oscillations observed during this period through the striping of the sweeping receiver data as well as by low frequency electric field measurements. These oscillations are tentatively identified as electrostatic ion cyclotron (EIC) candidates. There are also plasma frequency-upper hybrid frequency emissions associated with this aurora which were observed by a nearly field-aligned antenna (TAD 1). The aurora manifested itself in the electron RPA data by two different signatures: an increased ram density and a nearly flat broad diffuse electron flux extending from 10 eV up to the limit of the instrument (600 eV).

A "cartoon" summary of the auroral observations is shown in Figure 5. The payload's altitude is plotted as function of the range (both in kilometers). The associated flight time is indicated by the axis above the plot. The various auroral observation periods are indicated on the plot as well as the electron gun turn-on time. The period of spectrogram striping is roughly indicated by the oval labeled "EIC waves ??", because the source of the oscillation may be electrostatic ion cyclotron waves. However, this identification is extremely tentative. Because of the relatively long duration of the Z-mode emissions in comparison to other auroral observations, and from other evidence presented, we conclude that the observations are of freely propagating electromagnetic waves, and that their source is the auroral patch.

Acknowledgments. This work was supported by NASA under contract NSG-5373 and under the GSRP program by contract NGT-50684. The authors would like to thank the Wallops Island crew, especially Bobbie Flowers and Charles Lankford for their conscientious construction, testing and preparation of the payload. They would also like to thank the crew at Poker Flat Research Range for the support and cooperation they provided.

References

Benson, R. F., Auroral kilometric radiation: wave modes, harmonics and source region electron density structures, *J. Geophys.Res.*, *90*, 2753-2784, 1985.

Benson, R. F. and H. K. Wong, Low-altitude ISIS 1 observations of auroral radio emissions, and their significance to the cyclotron maser instability, *J. Geophys. Res.*, *92*, 1218-1230, 1987.

Cartwright, D. G. and P. J. Kellogg, Observations of radiation from an electron beam artificially injected into the Ionosphere, *J. Geophys.Res.*, *79*, 1439-1457, 1974.

Fuselier, S. A. and D. A. Gurnett, Short wavelength ion waves upstream of the earth's bow shock, *J. Geophys. Res.*, *89*, 91-103, 1984.

Goerke, R. T., P. J. Kellogg and S. J. Monson, An analysis of whistler mode radiation from a 100 mA electron beam, *J. Geophys. Res.*, *95*, 4277-4283, 1990.

Gurnett, D. A., S. D. Shawhan and R. P. Shaw, Auroral hiss, Z mode radiation and auroral kilometric radiation in the polar magnetosphere - DE1 observations, *J. Geophys. Res.*, *88*, 329-340, 1983.

Kellogg, P. J. and S. J. Monson, Rocket borne accelerator results pertaining to the beam plasma discharge, <u>Adv. Space Res.</u>, <u>1</u>, 61-68, 1981.

Kellogg, P. J., S. J. Monson, W. Bernstein and B. A. Whalen, Observations of waves generated by electron beams in the Ionosphere, *J. Geophys. Res.*, *91*, 12065-12077, 1986.

Melrose, D. B., R. G. Hewitt and G. A. Dulk, Electron cyclotron maser emission: Relative growth and damping rates for different modes and harmonics, *J. Geophys. Res.*, *89*, 897-904,1984.

Shawhan, S. D., Magnetospheric plasma waves in *Solar System Plasma Physics*, Vol. III, edited by L. J. Lanzerotti, C. F. Kennel and E. H. Parker, North-Holland, Amsterdam, 1979.

Winglee, R. M., Fundamental and Harmonic electron cyclotron maser emissions, *J. Geophys. Res.*, *90*, 9663-9674, 1985.

R.T. Goerke, International School of Minnesota, 6385 Beach Road, Eden Prairie, MN 55344.

P.J. Kellogg, S.D. Bale and S.J. Monson, School of Physics and Astronomy, University of Minnesota, 116 Church St. S.E., Minneapolis, MN 55455

H.R. Anderson and D.W. Potter, Science Applications International Corporation, 13400B Northrup Way, Suite 36, Bellevue, WA 98005

E.P. Szuszczewicz and G.D. Earle, Plasma Physics Division 157, Science Applications International Corporation, 1710 Goodrich Drive, McLean, VA 22102

Cherenkov Whistler Emissions From Tethers and Magnetic Antennas in Relative Motion to Plasmas

R. L. STENZEL AND J. M. URRUTIA

Department of Physics, University of California, Los Angeles

In a large laboratory plasma the radiation from insulated tethers and magnetic loop antennas excited by short current pulses ($1/\omega_{ci} \gg \Delta t \gg 1/\omega_{ce}$) is measured. The radiation field from moving dc sources is obtained by superimposing the radiation fields from delayed pulses emitted at displaced antenna positions. Whistler wings and whistler shocks from dipoles moving along $\mathbf{B}_0$ are observed.

INTRODUCTION

There is considerable interest in the radiation of large conductors moving through space plasmas [*Drell et al.*, 1965]. Typical examples are electrodynamic tethers [*Penzo and Ammann*, 1989] and Jupiter's moon Io [*Goertz*, 1980]. Although the structures are carrying dc-currents in their moving frame, the stationary plasma supports only transient currents since the conducting body of size D moving with velocity $\mathbf{v}$ ($\perp \mathbf{B}_0$) stays in its flux tube only for a time interval $\Delta t = D/v$. Transient currents are carried by electromagnetic waves which are thought to be Alfvén waves for large objects (e.g., Io) and low-frequency whistlers ($v \simeq 8$ km/sec, $D \approx 1$ m, $f \simeq 8$ kHz $\gg f_{ci}$) for small structures such as man-made tethers. A spectrum of frequencies and wave vectors are excited whose superposition forms a Cherenkov-like structure (Alfvén and whistler wings). The present laboratory experiment [*Stenzel and Urrutia*, 1990] demonstrates the existence of whistler wings by detailed probe measurements at several thousand spatial positions, a task which cannot be done in space plasmas due to the lack of a corresponding number of probing satellites. Furthermore, new physics results are obtained by demonstrating that wings are not only emitted by conducting objects but also by insulated dc-current carrying wires (tethers, magnetic loops). Thus, plasma currents are not only excited by collection or emission of charged particles but also by induction from moving dc magnetic fields (eddy current principle).

Auroral Plasma Dynamics
Geophysical Monograph 80

EXPERIMENTAL ARRANGEMENT

The experiments are performed in a large (1 m diam., 2.5 m length) magnetized afterglow plasma with parameters given in Figure 1. Time-varying magnetic fields are detected with a magnetic probe containing three orthogonal loops movable throughout the central uniform-plasma volume. The field topology $\mathbf{B}(\mathbf{r}, t)$ is assembled from many ($\geq 10^5$) highly repeatable discharges and stored digitally. The current density is calculated from Ampère's law, $\nabla \times \mathbf{B}(\mathbf{r}, t) = \mu_0 \mathbf{J}(\mathbf{r}, t)$.

EXPERIMENTAL RESULTS

Figure 2 displays 'field lines' in the y-z plane orthogonal to a tether wire (Figure 2a) and a magnetic loop antenna (Figure 2b). One can see that the pulsed currents in the insulated wires induce image currents in the plasma which propagate in $\pm z$ direction ($\parallel \mathbf{B}_0$) in the whistler mode. The latter has been confirmed by applying a cw signal to the

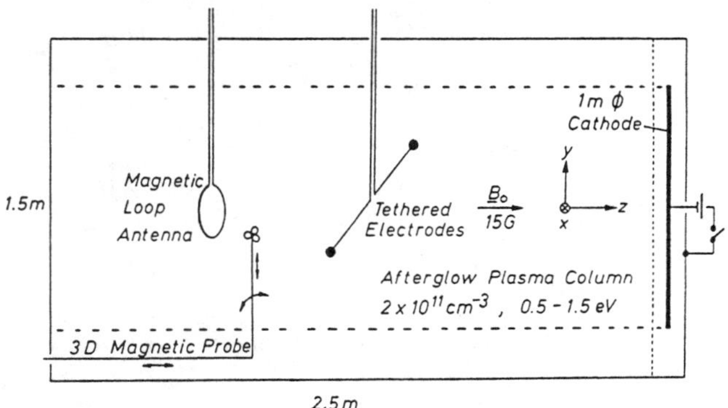

Fig. 1. Experimental arrangement.

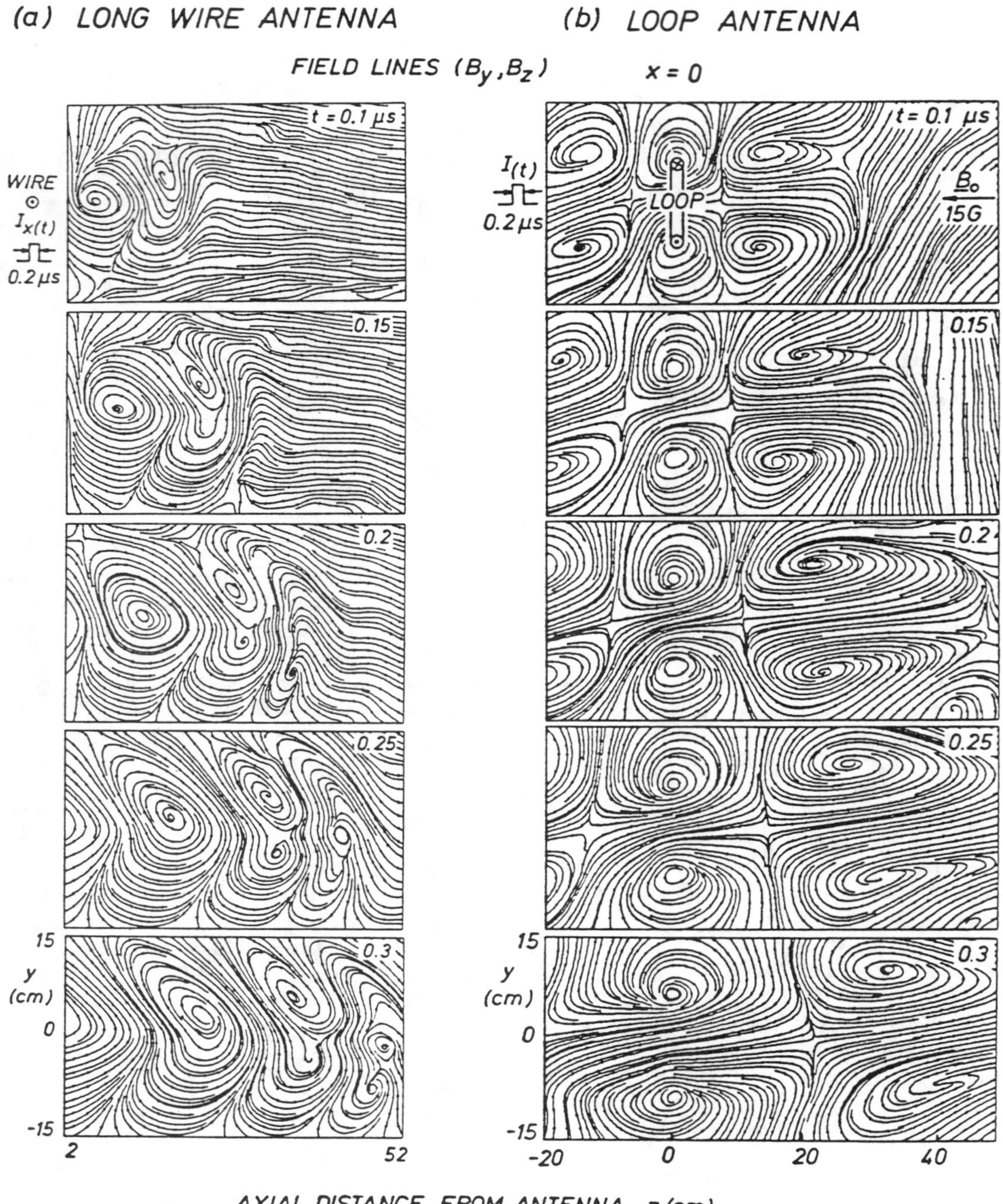

Fig. 2. Magnetic field topology in the y-z plane at different times t after applying a 0.2μs current pulse to (a) a straight insulated wire connecting two tethered electrodes across $\mathbf{B}_0$ (see Figure 1), and (b) an insulated magnetic loop antenna. Image currents/fields are induced which propagate in the whistler mode away from the exciter structure. Note that there is also a B_x component (not shown here) which makes the magnetic field topology three-dimensional.

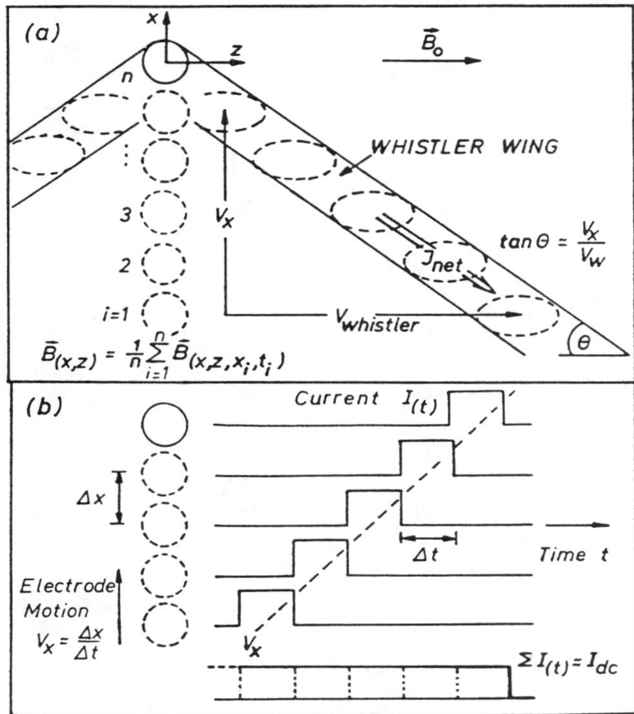

Fig. 3. Schematic diagram for the construction of whistler wings from a dc-current carrying body moving with velocity $v_x \perp \mathbf{B}_0$. (a) At each position $i = 1 \ldots n$ a whistler wavelet is emitted since the current source resides in the flux tube of its diameter D only for a short duration $\Delta t = D/v$. The superposition of wavelets forms a whistler wing which is co-moving with the electrode. (b) The superposition of delayed pulses at the displaced electrode corresponds to a dc-current in the moving frame.

antennas and measuring the wave dispersion interferometrically [*Urrutia and Stenzel*, 1989]. In addition to the B_y and B_z components, a comparably strong B_x component is excited. Thus, the field topology is three-dimensional. However, a two-dimensional cut (Figure 2) has been shown for ease of visualization.

Having measured the pulse response, we proceed to construct the steady-state wave front of a moving source as schematically shown in Figure 3. The continuous source motion across $\mathbf{B}_0$ is approximated by a series of n discrete steps. Analogous to Huygen's principle, the delayed wavelets of each pulse are linearly superimposed so as to form a coherent wave front, termed a whistler wing [*Stenzel and Urrutia*, 1989]. In the moving frame, the adjacent pulses add up to a dc current and the field structure remains stationary.

Figure 4 displays an example of the field and current structure induced by an insulated tether moving across $\mathbf{B}_0$ at the wave speed ($\theta \simeq 45°$). A dispersive whistler wedge is excited with alternating currents J_x flowing across $\mathbf{B}_0$. The currents and field lines close via y and z components not shown here. The Cherenkov-like structure is co-moving

with the tether. The induced currents can be thought of as eddy currents in a magnetoplasma except that the energy is not dissipated by Joule heating but radiated away. The radiation presents a drag on the moving body. The radiating body can be detected over large distances.

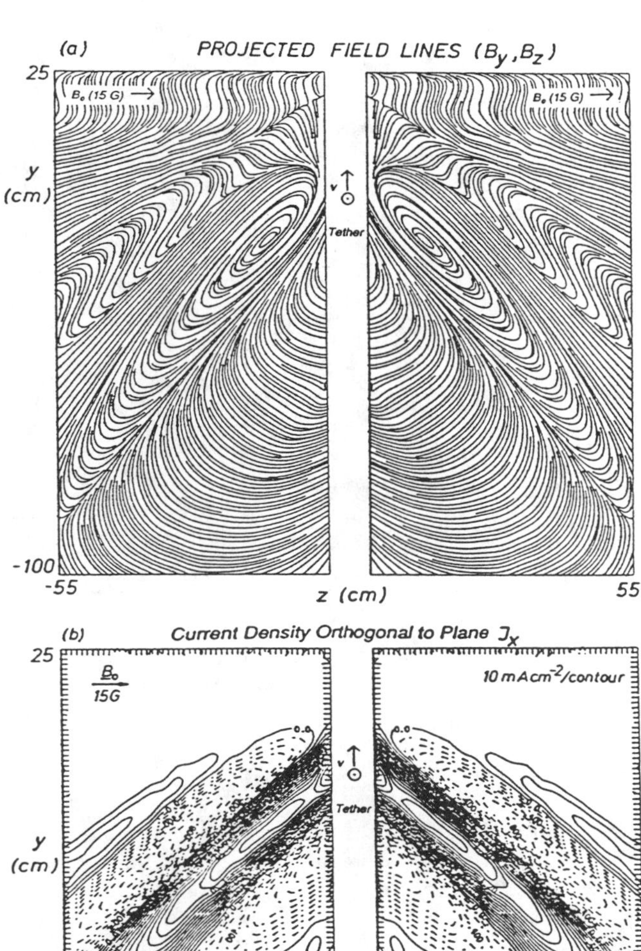

Fig. 4. Cherenkov-like structure of whistler waves generated by an insulated tether with dc-current moving with velocity $\mathbf{v}$ across $\mathbf{B}_0$. (a) Field lines of vector components (B_y, B_z) in the y-z plane obtained from measured pulse responses (Figure 2a) which are linearly superimposed as outlined in Figure 3a. (b) Cross-field current density $J_x = (\nabla \times \mathbf{B})_x/\mu_0 = (\partial B_z/\partial y - \partial B_y/\partial z)/\mu_0$. The induced currents can be thought of as an eddy current system in a collisionless magnetoplasma. Kinetic energy of the moving source is transferred into whistler wave radiation.

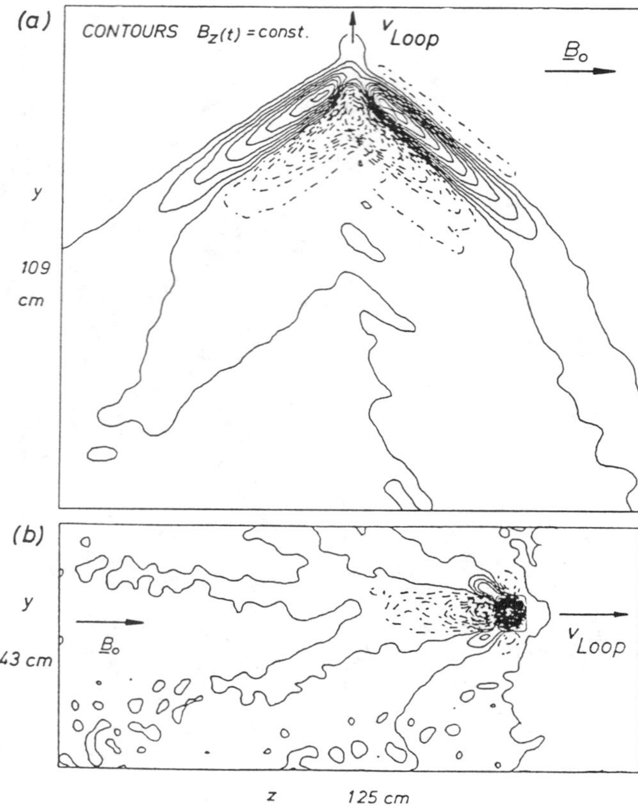

Fig. 5. Radiation patterns of a rapidly moving magnetic loop antenna in a magnetoplasma. The antenna carries a dc-current ($\omega = 0$) and the radiation is due to induced eddy currents as in Fig.4. (a) For motion across $\mathbf{B}_0$, a Cherenkov-like whistler wing is excited. (b) A whistler shock is created upstream and a wake downstream of the dipole for motion along $\mathbf{B}_0$.

Similar to a straight tether wire, a circular loop antenna moving rapidly through a magnetoplasma will emit waves even for dc antenna currents ($\omega = 0$). Figure 5 shows examples of radiation patterns when the loop moves across $\mathbf{B}_0$ (Figure 5a) and along $\mathbf{B}_0$ (Figure 5b). While the former shows a wing structure similar to Figure 4, the latter reveals a field compression upstream of the source and a wake downstream. Although beyond the principle of linear superposition one may suspect the formation of a whistler shock wave ahead of the moving dipole. Note that the broad band whistler mode radiation is only due to the motion of the antenna and *not* due to time-varying antenna currents. When the antenna current oscillates slowly (ELF, ULF), the whistler wings will be modulated and a monochromatic wave at the imposed frequency will also be excited.

CONCLUSIONS

The radiation fields of dc-current carrying conductors moving through collisionless magnetoplasmas have been determined from computer-processed laboratory measurements. The radiation is found to be in the low-frequency whistler wave regime as expected for active experiments (electrodynamic tethers, ELF antennas, space station). The results are also applicable when magnetoplasmas flow rapidly across the conductors, e.g. for spacecraft antennas in the solar wind or in auroral arcs.

Acknowledgements. The authors gratefully acknowledge support for this work by grants from NSF PHY, ATM, and NASA.

REFERENCES

Drell, S. D., H. M. Foley, and M. A. Ruderman, Drag and propulsion of Large satellites in the ionosphere: An Alfvén propulsion engine in space, *J. Geophys. Res.*, **70**, 3131-3146, 1965.

Goertz, C. K., Io's interaction with the plasma torus, *J. Geophys. Res.*, **85**, 2949-2965, 1980.

Penzo, P. A., and P. W. Ammann, editors, *Tethers in Space Handbook, 2nd. edition*, p. 119, NASA, Washington, D.C., 1989.

Stenzel, R. L., and J. M. Urrutia, Whistler wings from moving electrodes in a magnetized laboratory plasma, *Geophys. Res. Lett.*, **16** , 361-364, 1989.

Stenzel, R. L., and J. M. Urrutia, Currents between tethered electrodes in a magnetized laboratory plasma, *J. Geophys. Res.*, **95** , 6209-6226, 1990.

Urrutia, J. M., and R. L. Stenzel, Transport of current by whistler waves, *Phys. Rev. Lett.*, **62**, 272-275, 1989.

R. L. Stenzel and J. M. Urrutia, Department of Physics, University of California, Los Angeles, CA 90024-1547.

Effects of Transverse, Localized, DC Electric Fields
on Current-Driven Ion-Cyclotron Waves

M. E. KOEPKE, W. E. AMATUCCI, J. J. CARROLL III, M. J. ALPORT*, AND T. E. SHERIDAN

Department of Physics, West Virginia University, Morgantown, West Virginia

Two distinct oscillatory modes are identified in a current-carrying, magnetized laboratory plasma column. They are differentiated by the properties of the fluctuation spectrum and fluctuation profile. One mode is excited by a magnetic-field-aligned (i.e., parallel) electron drift. The other mode is destabilized by a combination of parallel electron drift and a localized, transverse electric field. A segmented disk electrode placed at the end of the plasma column is used to simultaneously induce an electron drift with respect to the ions and produce radial electric fields that extend into the plasma.

INTRODUCTION

Wave-particle interactions play an important role in many theoretical models of the auroral ionosphere (Chang 1980, Cornwall 1980, Young 1980, Chang et al. 1990, Temerin et al. 1990). A variety of electromagnetic (Crew and Chang 1990) and electrostatic waves (Lysak 1980) have been connected with mechanisms, such as transverse ion acceleration (Arnoldy et al. 1992), parallel electron acceleration (Collin et al. 1980), and anomalous resistivity (Palmadesso et al. 1974), that have been invoked to explain observations made using rockets and satellites. The physical description of these mechanisms relies on knowledge of the specific properties of the waves. These properties are sometimes established by direct measurements and other times inferred by correlations between the presence of the waves and the characteristics of the background plasma.

Broadband electrostatic waves have been detected together with localized regions of magnetic-field-aligned currents and oblique electric fields in the auroral zone (Basu et al. 1988). The free energy associated with the combination of field-aligned current (FAC) and transverse, localized electric fields (TLE) has been used to explain the presence of these waves (Ganguli et al. 1989). Recent laboratory experiments (Sato et al. 1986, van Niekerk et al. 1991) have demonstrated that waves with frequencies near the ion cyclotron frequency have characteristics that depend on transverse electric fields existing in the plasma. The experimental results reported here demonstrate that

---

*On sabbatical leave from University of Natal, Durban, South Africa.

Auroral Plasma Dynamics
Geophysical Monograph 80
Copyright 1993 by the American Geophysical Union.

broadband waves are excited by the application of TLE in the presence of FAC that is subcritical to narrowband current-driven waves. These two types of waves are observed and are differentiated by their fluctuation profiles and the TLE-dependence of their frequencies.

An attempt is made to draw conclusions from these laboratory results that can be applied to the magnetosphere (e.g., the auroral acceleration region, transition regions such as the interface of the plasma sheet boundary layer with the lobe, and the upwelling of heavy ions from the ionosphere).

ELECTRODE OPERATION

Experiments are performed in a Q-machine (Rynn 1964) plasma column (3-cm radius and 90-cm length) with the following parameters: n ª 10^8 cm^{-3}, $B = 1.5$ kG, $m_i/m_e = 7.15$ ¥ 10^4 (K$^+$ ions), T_i ª T_e ª 0.2 eV. A segmented disk electrode is used to produce FAC and TLE that coexist in space (Koepke and Amatucci 1992). The overall radius of the disk is 1 cm, which is approximately 5 ion gyroradii, and defines the current channel. It is made of two coplanar, concentric circular segments that are heated to prevent surface contamination. The applied voltages on the outer annular segment (0.6 cm $< r <$ 1.0 cm) and on the inner button segment ($r < 0.5$ cm) are independently controlled with precision power supplies. The potential structure $V(r)$ in the plasma is measured with an electrically floating emissive probe. The radial electric field is determined from this structure by $E_r = -\partial V/\partial r$.

Fig. 1a is a sequence of electric field profiles illustrating the capability of the segmented disk electrode to produce TLE of different magnitude and direction. Each profile is associated with a different combination of applied voltages to the outer annular segment V_a and to the inner button segment V_b. The scale length of these electric fields are on the order of 0.4 cm, or 2 ion gyroradii.

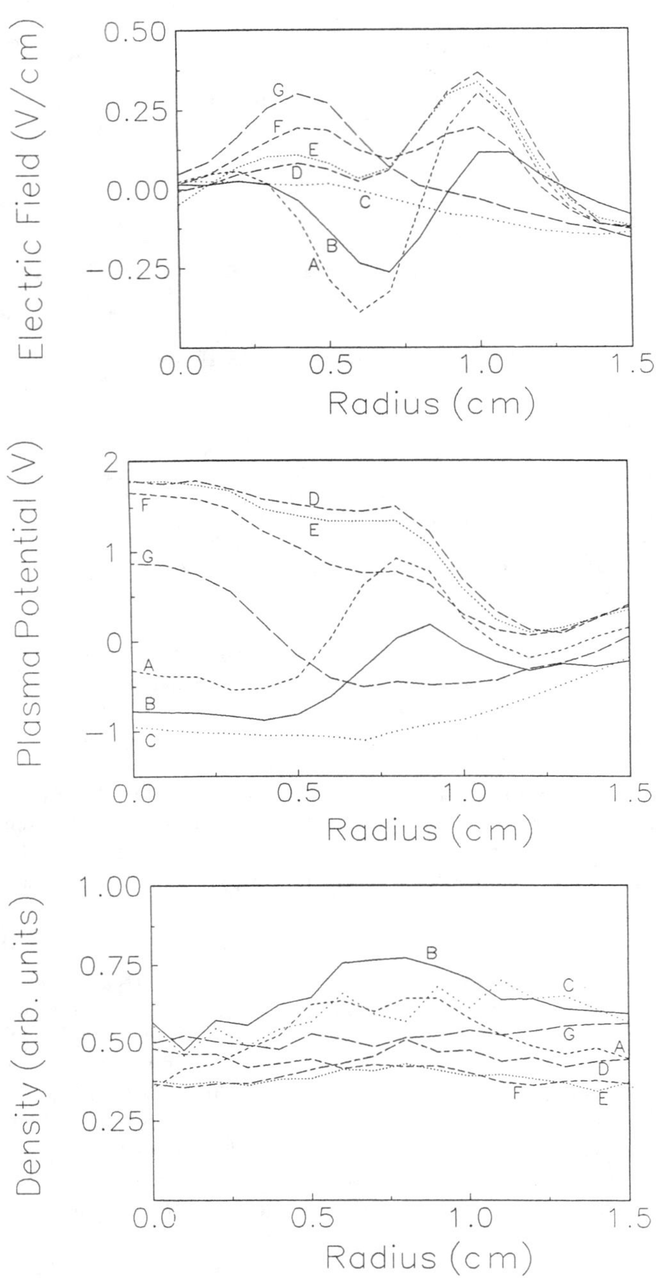

Fig. 1. Radial dependence of plasma properties in the current channel: (a) radial electric field, (b) potential, and (c) density. In volts, (Va, Vb) = A(11, -9), B(1, -9), C(floating, floating), D(21, 31), E(11, 31), F(1, 31), G(-9, 31), for the curves shown.

The measured potential structures for each calculated profile in Fig. 1a are shown in Fig. 1b. Deconvolving these structures to account for the effect of finite probe dimensions would sharpen the potential structures and consequently increase the maximum magnitude of the electric field. Preliminary attempts to deconvolve this data suggest that the effect of finite probe dimensions is slight.

The plasma density is not significantly affected by the inhomogeneous electric fields produced within the FAC channel. Fig. 1c shows the radial profiles of the ion saturation current of a Langmuir probe for the cases shown in Fig. 1a. Density gradients can be ruled out as a primary free-energy source for ion-cyclotron-like waves present at $r < 1$ cm since, by inspection of this data, the density gradients are associated with estimated drift frequencies less than 2 kHz.

Wave Observations

When the annulus and button voltages are approximately equal, the segmented disk electrode behaves similarly to a plain disk electrode (see curves D and E in Fig. 1b). For cases in this category, the two segments can be biased to large enough positive values so that waves can be excited due to sufficiently large electron drift velocity, as previously observed in many experiments (Rasmussen and Schrittwieser 1991). When the annulus and button voltages differ significantly, particularly when the annulus is positively biased and the button voltage is near zero, waves can be excited by the sufficiently large radial electric field. For cases in this category, the electron drift velocity becomes subcritical to the waves mentioned above and a wave with a distinctly different spectrum is destabilized (Koepke et al. 1992).

Figs. 2a and 2b are sequences of spectra from fluctuations in the current collected by the annulus and button segments of the disk electrode, respectively. From the top to the bottom of each sequence, the spectra are associated with segment-voltage combinations from $V_a \approx V_b \gg kT_e/e$ to $V_a \gg V_b \approx 0$, as indicated. The presence of FAC-driven waves is detected in the spectra from both segments for the $V_a \approx V_b$ category of cases (spectra #1-3 in Figs. 2a and 2b). The presence of TLE-driven waves is detected for the $V_a \gg V_b$ category of cases (spectra #8-11 in Figs. 2a and 2b), but only in the spectra from the annulus segment. Spectra #4-14 from the button segment indicate no such waves. Fig. 3 is a plot of the area under the square of the spectra from 40-100 kHz in Fig. 2 and compares the fluctuations measured on the annulus and button for the same sequence of voltage differences between the segments. As expected for current-driven electrostatic ion-cyclotron waves (Motley and D'Angelo 1963), the FAC-driven fluctuations appear on both segments, with larger amplitude detected on the button segment (note that the collecting area of the button segment is 40% that of the annulus segment). This indicates a radial mode structure that is peaked at the center and involves the entire current-channel diameter, which has been well established in the literature (Rasmussen and Schrittwieser 1991). In contrast, the TLE-driven fluctuations appear predominantly on the annulus segment. For the voltage combinations corresponding to the TLE-driven fluctuations, the TLE maximum is shifted slightly outward from the gap (0.5 cm < r < 0.6 cm), i.e., predominantly on the annulus segment (see curves A and B in Fig. 2a). As predicted for inhomogeneous energy-density driven (IEDD)

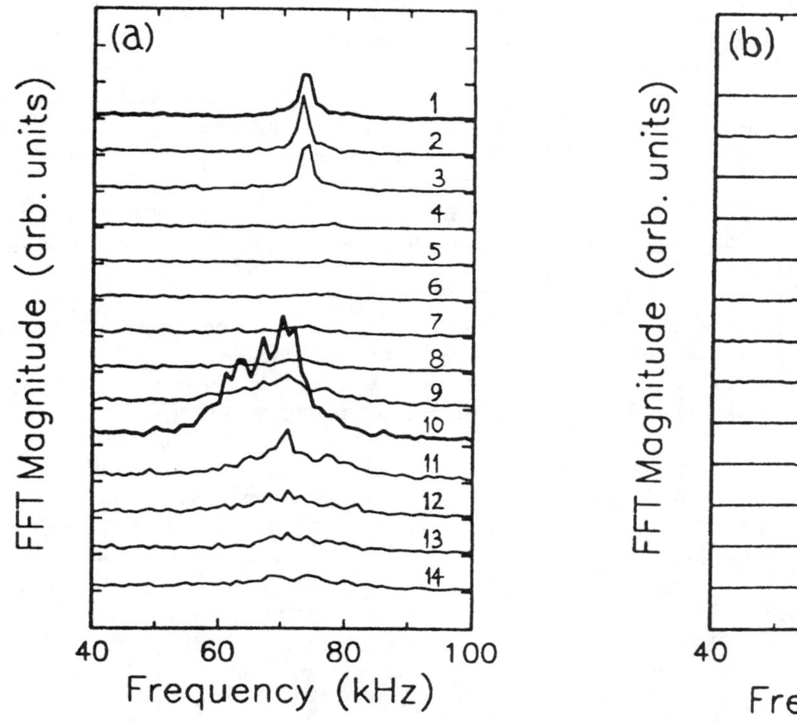

Fig. 2. Vertically displaced spectra of fluctuations in the current collected by (a) annulus and (b) button segments. Annulus voltage is 9.5 V and (Va - Vb) is indicated by a number next to each trace.

waves (Ganguli et al. 1988), the fluctuations appear where the TLE is large. Ongoing experiments are providing better spatial resolution of the radial mode structure of these two modes using small Langmuir probes (Amatucci et al. 1993).

As seen in Fig. 2a, the TLE-driven fluctuations have an amplitude that sensitively depends on the segment-voltage difference. The frequency of these fluctuations is plotted in Fig. 4 for the same sequence of segment-voltage differences as in Fig. 2a. As expected for current-driven electrostatic ion-cyclotron waves, the FAC-driven fluctuations have a frequency approximately independent of the TLE and the FAC. Although the TLE-driven fluctuation frequency is in the vicinity of the ion cyclotron frequency, it differs and shifts linearly with the segment-voltage difference, which is related to the electric field

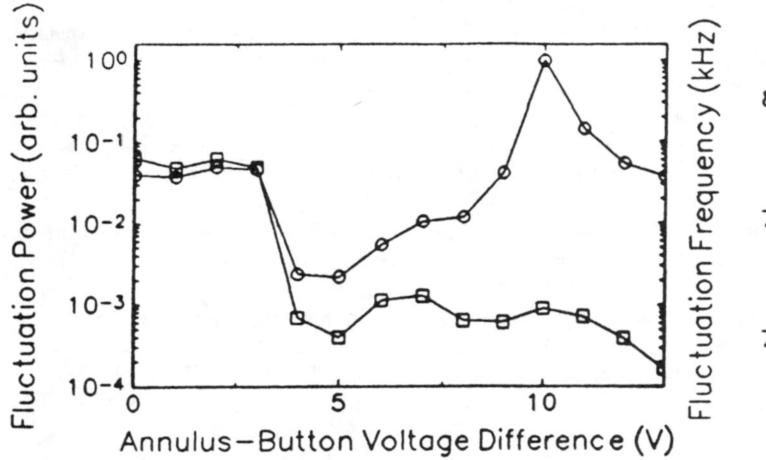

Fig. 3 Behavior of the fluctuation level detected on annulus (○) and button (□) segments from Fig. 2.

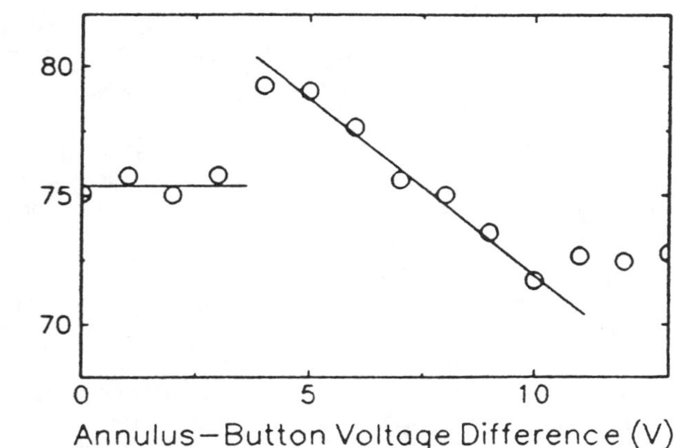

Fig. 4 Behavior of fluctuation frequency detected on annulus segment from Fig 2a.

strength in the plasma.   As predicted for inhomogeneous energy-density driven waves, the Doppler shift of the TLE-driven fluctuation frequency is proportional to the electric field. Additional important information on mode propagation is being determined to verify the identity of the TLE-driven mode.

SUMMARY

A two-segment disk electrode is operated so that a magnetic-field-aligned electron drift velocity and a radial, localized electric field coexist in a Q-machine plasma column.   Two separate modes are identified by distinctly different mode characteristics.   Changes in the mode frequency and in the spatial fluctuation profile of the observed waves as the relative amounts of FAC and TLE are varied are emphasized.   These laboratory results support the idea that broadband ion-cyclotron fluctuations in the high-latitude ionosphere may be explained by waves excited by the presence of magnetic-field-aligned currents and oblique electric fields even if the currents are subcritical to the current-driven electrostatic ion-cyclotron instability.

These laboratory results also suggest that processes involving small-scale density and/or potential structures associated with transition regions in the magnetosphere may be responsible for waves and the anomalous transport these waves can cause (Romero et al. 1990, Ganguli et al. 1991, Ganguli et al. 1993).   These laboratory results also impact the interpretation of the high level of ion heating in the low-altitude ionosphere as indicated by thermal-ion upwelling into the magnetosphere (Tsunoda et al. 1989).   By virtue of the results demonstrating that the FAC threshold is reduced for the TLE-driven waves, it is plausible that IEDD waves are excited in the ionosphere at low altitudes and are partly responsible for ion energization.

Acknowledgments. This research is supported by the Office of Naval Research and the National Science Foundation.

REFERENCES

Amatucci, W. E., M. E. Koepke, T. E. Sheridan, M. J. Alport, and J. J. Carroll III, Self-cleaning Langmuir probe, *Rev. Sci. Instrum.* 64, 1253-1256, 1993.

Arnoldy, R. L., G. P. Garbe, P. M. Kintner, Transverse ion acceleration measured at rocket altitudes, in *Physics of Space Plasmas (1991)*, SPI Conf. Proc. and Reprint Series, No. 11, T. Chang, G. B. Crew, and J. R. Jasperse, eds. (Scientific Publishers, Inc., Cambridge, MA, 1992) p. 3-22.

Basu, S., S. Basu, E. MacKenzie, P. F. Fougere, W. R. Coley, N. C. Maynard, J. D. Winningham, M. Sugiura, W. B. Hanson, and W. R. Hoegy, Simultaneous density and electric field fluctuation spectra associated with velocity shears in the auroral oval, *J. Geophys. Res.* 93, 115-136, 1988.

Chang, T. ed., *Ion Acceleration in the Magnetosphere and Ionosphere*, (AGU, Washington, D.C., 1980).

Chang, T., G. B. Crew, and J. R. Jasperse, eds. *Physics of Space Plasmas (1989)*, SPI Conf. Proc. and Reprint Series, No. 9, (Scientific Publishers, Inc., Cambridge, MA, 1990).

Collin, H. L., E. G. Shelley, A. G. Ghielmetti, and R. D. Sharp, Observation of transverse and parallel acceleration of terrestrial ions at high latitudes, Ref. 1, p. 67-71, 1980.

Cornwall, J. M., Magnetospheric ion acceleration processes, Ref. 1, p. 3-16, 1980.

Crew, G. B., and T. Chang, Particle acceleration by electromagnetic ion cyclotron turbulence, Ref. 4, p. 31-66, 1990.

Ganguli, G., Y. C. Lee and P. J. Palmadesso, Kinetic theory for electrostatic waves due to transverse velocity shears, *Phys. Fluids* 31, 823-838, 1988.

Ganguli, G., Y. C. Lee, P. J. Palmadesso, and S. L. Ossakow, Oscillations in a plasma with parallel currents and transverse velocity shears, in *Physics of Space Plasmas (1988)*, SPI Conf. Proc. and Reprint Series, No. 8, T. Chang, G. B. Crew, and J. R. Jasperse, eds. (Scientific Publishers, Inc., Cambridge, MA, 1989) p. 231-242.

Ganguli, G., Y. C. Lee, and P. J. Palmadesso, Role of small scale processes in global plasma modelling, in *Modeling Magnetospheric Plasma Processes*, AGU Monograph #62 (AGU, Washington, DC, 1991) p. 17-27.

Ganguli, G., H. Romero, and P. Dusenbery, The dynamical plasma sheet boundary layer: A new perspective, in *Micro- and Macro-Scale Phenomena in Space Plasmas*, AGU Monograph (AGU, Washington, DC, submitted).

Koepke, M. E. and W. E. Amatucci, Electrostatic ion-cyclotron wave experiments in the WVU Q Machine, *IEEE Trans. Plasma Sci.* PS-20, 631-635, 1992.

Koepke, M. E., M. J. Alport, T. E. Sheridan, W. E. Amatucci, and J. J. Carroll III, Waves driven by strong transverse potential structures, *Proc. of the 1992 Int. Conf. Plasma Phys.*, W. Freysinger, K. Lackner, R. Schrittwieser, and W. Lindinger, eds. (European Phys. Soc., Innsbruck, 1992) Vol. III, p. 1895-1898.

Lysak, R. L., Ion acceleration by wave-particle interaction, Ref. 1, p. 261-270, 1980.

Motley, R. W. and N. D'Angelo, Excitation of electrostatic plasma oscillations near the ion cyclotron frequency, *Phys. Fluids* 6, 296-299, 1963.

Palmadesso, P. J., T. P. Coffey, S. L. Ossakow, and K. Papadopoulous, Topside ionosphere ion heating due to electrostatic ion cyclotron turbulence, *Geophys. Res. Lett.* 1, 105-108, 1974.

Rasmussen, J. J. and R. W. Schrittwieser, On the current driven electrostatic ion-cyclotron instability, *IEEE Trans. Plasma Sci.* PS-19, 457-501, 1991.

Romero, H., G. Ganguli, P. Palmadesso, and P. B. Dusenbery, Equilibrim structure of the plasma sheet boundary layer-lobe interface, *Geophys. Res. Lett.* 17, 2313-2316, 1990.

Rynn, N., Improved quiescent plasma source, *Rev. Sci. Instrum.* 35, 40-46, 1964.

Sato, N., M. Nakamura, and R. Hatakeyama, Three-dimensional double layers inducing ion-cyclotron oscillations in a collisionless plasma, *Phys. Rev. Lett.* 57, 1227-1230, 1986.

Temerin, M. A., C. W. Carlson, C. A. Cattrell, R. E. Ergun, J. P. McFadden, F. Z. Mozer, D. M. Klumpar, W. K. Peterson, E. G. Shelley, and R. C. Elphic, Wave-particle interactions on the FAST satellite, Ref. 4, p. 343-359, 1990.

Tsunoda, R. T., R. C. Livingston, J. F. Vickrey, R. A. Heelis, W. B. Hansen, F. J. Rich, and P. F. Bythrow, Dayside observations of thermal-ion upwellings at 800-km altitude: An ionospheric signature of the cleft ion fountain, *J. Geophys. Res.* 94, 15277-15290, 1989.

van Niekerk, E. G., P. H. Krumm, and M. J. Alport, Electrostatic ion cyclotron waves driven by a radial electric field, *Plasma Phys. Contr. Fusion* 33, 375-388, 1991.

Young, D. T., Experimental aspects of ion acceleration in the earth's magnetosphere, Ref. 1, p. 17-35, 1980.

M. E. Koepke, W. E. Amatucci, J. J. Carroll III, and T. E. Sheridan, Department of Physics, West Virginia University, Morgantown, WV 26506-6315.

M. J. Alport, Department of Physics, University of Natal, Durban, South Africa 4001.